College Algebra:
Basics to Theory of Equations

2016 Edition

With

Video Reviews & Solutions Manual

@CollegeAlgebraBySchiller.com

by

John J. Schiller & Marie A. Wurster

φ *ImageTec Publishing Systems*

Dedication

This book is dedicated to the inequalities:

$$\frac{a}{b+c} \neq \frac{a}{b} + \frac{a}{c} \quad \text{and} \quad \sqrt{a^2 + b^2} \neq a + b$$

and to:

Gilbert the Frog

Copyright © 2010 - 2016 φ ImageTec Publishing Systems

All rights reserved. No part of this publication may be reproduced, stored in a retrieval system, or transmitted, in any form or means, electronic, mechanical, photocopying, recording, or otherwise, without prior written permission of the copyright owner.

2016 Edition

Schiller, John J.
    College Algebra: Basics to theory of Equations
    John J. Schiller, Marie A. Wurster.

ISBN 978-1-60126-433-6

Printed in the United States of America

Preface

General Goals, Content and Organization

College Algebra: Basics to Theory of Equations is an algebra textbook and accompanying web-based student and teacher resources designed to deal with the diverse mathematical backgrounds of students preparing for calculus, but it is also written for those with technical majors, such as nursing, that do not require calculus.

With this edition, we introduce the concept of "The Inverted Classroom" by providing the student with the opportunity to preview a web-based review of basic theory and exercises for each chapter of the book. The content of these web-based videos is also available in print form at the end of each chapter of the book. With this blended approach we are intentionally moving part of the learning activity outside the traditional classroom environment. Students have the opportunity to become familiar with the content before being introduced to the material in the classroom. We believe this strategy will help to level the playing field for the students regarding preparation and make the material presented in class more comprehensible. Naturally, we expect the instructor to encourage students to review the web-based material before attending class and to allow a more interactive environment within the classroom.

After a thorough review of basic algebra (Chapters R and 1), including properties of real numbers and the arithmetic of algebraic expressions, following chapters then treat equations, functions, graphs, and applications according to a functional hierarchy, from linear (Chapter 2) to quadratic (Chapter 3) to general algebraic (Chapter 4) to exponential and logarithmic (Chapter 5). We feel that this hierarchal arrangement, from linear to transcendental, provides the student with a logical conceptual structure for the subject of algebra. Chapter 6 treats a variety of topics, including the binomial theorem, linear programming, partial fractions, Gaussian elimination, matrix algebra, and the theory of equations.

The Appendices provide an overview of significant digits, common logarithms and their applications. These appendices are designed for students whose majors do not require the more standard pre-calculus treatment of exponential and logarithmic functions.

Use of Textbook

The textbook can be used for various course levels: as a semester course in **College Algebra, Intermediate Algebra**, or even **Basic Algebra**.

As shown in the chart below, chapters 2 through 5, possibly combined with some topics from Chapter 6, and with Chapters R and 1 as reference, provide a robust course in **college algebra**. There are numerous exercises for developing essential skills, as well as thought-provoking exercises for developing critical thinking. Chapters 1 through 3, which cover up through all quadratic topics, and perhaps additional topics from the Appendices, comprise a thorough course in **intermediate algebra**, with Chapter R as a reference. A more thorough treatment of Chapter R, along with 1 and 2, provide the material for a course in **basic algebra**.

Course	Chapter							
	R	1	2	3	4	5	6	A
Basic Algebra	Main course							
Intermediate Algebra	Ref*	Main Course						A/N**
College Algebra		Ref*	Main Course					A/N**

*Ref = Reference   **A/N = As Needed

Use of textbook for various course levels.

We have found that an effective way of dealing with the diverse backgrounds of entering students is to first go over a section of review exercises at the end of Chapter R and back into the corresponding review outlines or text in the chapter as needed; similarly, for Chapter 1. After sufficiently many sessions of "filling in the gaps" in this manner, the class as a whole is ready to start the course proper. Individual students can continue to refer the Chapters R, 1 and the appendices as needed.

Royal Road to Algebra

The first edition of *College Algebra* appeared in 1987, just after the calculus reform movement shifted into high gear with the publication of "Toward a Lean and Lively Calculus" following the Tulane Conference in 1986. Now, 30 years later, despite an intense emphasis on standards and the use of technology in pre-college curricula nationwide, students entering college continue to have a wide range of backgrounds in algebra. Colleges, particularly community colleges, have traditionally met these diverse backgrounds by offering separate courses in basic algebra, intermediate algebra, and college algebra, but a student starting below college algebra will have great difficulty in completing the math requirements for any technical major in four or five years. More recently, the MAA (Mathematical Association of America), alarmed at the reported 50% passing rate in college algebra and utilizing input from partner disciplines, has been supporting revisions of the college algebra curriculum, with more emphasis on modeling or applications than on algebraic manipulations. Corresponding MAA guidelines for college algebra were published in 2007. There are many inviting aspects to the modeling approach, and students can get to work through some very practical and interesting problems, but at the expense of developing essential algebraic skills needed for pre-calculus or calculus. We see the modeling approach as more suitable for a terminal course than as preparation for more advanced mathematics.

The present reform movement in mathematics, starting with calculus, is an attempt to enable more students to experience, along with professional mathematicians, the beauty and utility of mathematics. But no amount of reform can significantly mitigate the hard work needed to master the subject. Our approach is to offer a course in college algebra that emphasizes essential skills needed for more advanced mathematics and also provides students with a ready opportunity to review more basic algebra.

To this point, there is a story told about King Ptolemy (~300 B.C.) in his struggle with geometry: when he asked Euclid if there were an easier way to learn the material, Euclid replied that in the world there are roads for kings and roads for common people, but in geometry there are no royal roads. And there may be no royal road to algebra, but the journey, by whatever road one takes, can be exciting, joyful, and forever rewarding.

Use of Calculators

When the first edition of *College Algebra* appeared, scientific calculators were already in common use in pre-calculus courses, but graphing calculators were not. In the second edition, a generous supply of graphing calculator exercises was added to the review exercises, and these have been retained in further editions and in this text. Today, graphing calculators have been joined by free online graphing calculators such as Desmos, Mathlab, MathPapa, and Symbolab that can run on standard computers, laptops, tablets or even cellular phones. Of course, the exercises in the text can be used with *any* graphing tool. Students taking algebra in college have various backgrounds in the uses of graphing calculators or web-based applications, and we assume that the instructor will supplement or adapt, as needed, the brief instructions given in the exercises regarding the arithmetic, graphing, matrix, and programming capabilities of graphing calculators.

Web-based Digital Supplement

As further help to the student, two web-based digital supplements are provided for viewing on our website @ www.CollegeAlgebraBySchiller.com:

Review Videos: Review of basic theory and related exercises for each chapter of the book. Students are encouraged to view the related video prior to going to class.

Solutions Manual: The manual contains step-by-step solutions to odd numbered exercises. The manual is in the form of an interactive PDF file format. Pages of this manual can be viewed interactively by section and/or by page.

Teacher Resources: Teacher instructional resources are also available on this website for the benefit of the teacher.

Acknowledgements

Overall, we have tried to produce a text that is strong in fundamentals, flexible, and brings out the essential unity of the subject. This would not have been possible without the help of many people. We extend our sincere thanks to all those provided suggestions and assistance throughout the development of the project, especially the following reviewers of *College Algebra*:

James F. Arnold, University of Wisconsin
Nancy J. Bray, San Diego Mesa College
Carl T. Carlson, Moorhead State University
Ben L. Cornelius, Oregon Institute of Technology
Nancy Harbour, Brevard Community College
Nancy G. Hyde, Broward Community College
DeWayne S. Nymann, University of Tennessee
Robert Pruitt, San Jose State University
Ken Seydel, Skyline College
Mahendra Singhal, University of Wisconsin
Shirley C. Sorensen, University of Maryland
Michael D. Taylor, University of Central Florida
Marvel D. Townsend, University of Florida
Thomas J. Woods, Central Connecticut University

In addition, the following mathematicians provided us with comments
and suggestions regarding the topical coverage and organization:
Jasper Adams, Stephen F. Austin State University
Carol A. Edwards, St. Louis Community College
Allen Epstein, West Los Angeles College
E. John Hornsby, Jr., University of New Orleans
Marvin Roof, Mott Community College
John L. Whitcomb, University of North Dakota

We would also like to thank those who helped check the answers at the back of the book: John A. Dersch, Jr., Grand Rapids Junior college; Nancy Harbour, Brevard Community College: Nancy G. Hyde, Broward Community College; and Karen E. Zak, United Sates Naval Academy. Among our colleagues at Temple, we especially wish to acknowledge the work of Elizabeth Van Dusen, who has made significant contributions to an effective use of graphing calculators in pre-calculus.

Contents

Preface   iii
Acknowledgements   vi
Description and Brief History of Algebra   ix
Algebraic Formulas   x

Chapter R   Review of Basic Algebra

R.1   The Real Number System   2
R.2   Basic Properties of Real Numbers   6
R.3   Signed Quantities, Order, and Absolute Value   17
R.4   Fractional Expressions   26
R.5   Powers and Roots   35

Chapter R -- Review Outline and Review Exercises   44

Chapter 1   Algebraic Expressions

1.1   Operations with Polynomials   50
1.2   Factoring Polynomials   62
1.3   Operations with Rational Expressions   70
1.4   Rational Exponents   77
1.5   Operations with Radicals   83

Chapter 1 -- Review Outline and Review Exercises   90

Chapter 2   Algebra and Graphs of Linear Expressions

2.1   Linear Equations   96
2.2   Linear Inequalities   107
2.3   Applications of Linear Equations and Inequalities   116
2.4   The Coordinate Plane   128
2.5   Equations of Straight Lines   140
2.6   Systems of Two Linear Equations in Two Unknowns   150
2.7   Systems of Three Linear Equations in Three Unknowns   163

Chapter 2 -- Review Outline and Review Exercises   170

Chapter 3   Algebra and Graphs of Quadratic Expressions

3.1   Quadratic Equations   176
3.2   Complex Numbers   187
3.3   Equations Reducible to Quadratic Equations   196
3.4   Quadratic Inequalities   200
3.5   Applications of Quadratic Equations and Inequalities   204
3.6   Circles and Parabolas   214

3.7 Translations of Axes and Intersections   226
3.8 Ellipses and Hyperbolas   234

Chapter 3 -- Review Outline and Review Exercises   248

Chapter 4 Functions and their Graphs

4.1 Functions   252
4.2 Polynomial Functions   262
4.3 Rational Functions   270
4.4 Algebraic Functions   280
4.5 One to One and Inverse Functions   288
4.6 Variation   298

Chapter 4 -- Review Outline and Review Exercises   304

Chapter 5 Exponential and Logarithmic Functions

5.1 The Exponential Function and its Graph   310
5.2 The Logarithmic Function and its Graph   318
5.3 Properties of the Logarithm   324
5.4 Applications of Exponential and Logarithmic Equations   331

Chapter 5 -- Review Outline and Review Exercises   343

Chapter 6 Additional Topics

6.1 The Binomial Theorem   348
6.2 Linear Inequalities in the Plane with Applications   356
6.3 Partial Fractions   368
6.4 Linear Systems: Gaussian Elimination   376
6.5 Matrix Algebra   388
6.6 Remainder Theorem, Synthetic Division, and Factor Theorem   402
6.7 Rational Roots, Bounds, and Descartes' Rule of Signs   414

Chapter 6 -- Review Outline and Review Exercises   423

Appendices

A. Significant Digits   429
B. Base-10 Logarithms   432

Answers to Selected Exercises   441

Index   473

Description and Brief History of Algebra

Algebra is a branch of mathematics that builds on arithmetic by incorporating the concepts of variable quantities and equations, thereby providing a powerful tool for solving practical problems and laying the foundation for more advanced mathematics. By associating numbers with points, algebra combines with geometry to enable the use of equations to describe geometric figures. Equations are also used to express quantitative relationships between variables. In classical algebra, arithmetic operations are performed within the real or complex number system. Modern algebra generalizes the notion of arithmetic operation and studies the structure of abstract mathematical systems.

The word "algebra" comes from an Arabic textbook entitled, *Kitâb al Jabr wa'l Muqâbala*, written in 825 AD by the mathematician *Muhammad al-Khwarizmi*. The title represents the two basic operations applied to solving equations: restoration and balancing. The treatise provided symbolic operations for the solution of linear and quadratic equations.

Algebra was used as early as 2000 B.C. in Babylonia and slightly later in Egypt. These ancients solved linear and some quadratic equations that arose in everyday activities. Their equations were written out in words with few symbols, and their number systems contained only positive integers and fractions. By about 200 B.C., mathematicians in China were working with systems of linear equations, and they made limited use of negative numbers.

In Greece starting around 500 B.C., geometric constructions were used to solve quadratic and cubic equations. The Greeks were the first to conceive of irrational magnitudes. Also, around 250 A.D. the Greek mathematician *Diophantus* adopted a simplified algebraic notation consisting of special symbols and abbreviations. He is sometimes called the father of algebra.

Hindu mathematicians in the seventh century solved quadratic equations for both roots, including negative ones, and stated the rules for operations with signed quantities. However, negative numbers were not fully understood or accepted at this time. Our decimal system is based on the Hindu positional base 10 number system, which may have originated in China. The numerals 0, 1, 2, . . , 9 were introduced into Western Europe through Arab mathematicians and are now called the Hindu-Arabic Numerals.

Although the ancient Babylonians solved quadratic equations, it was not until the sixteenth century that algebraic formulas for solving cubic and quartic equations were developed. These equations sometimes contained square roots of negative quantities, thereby forcing mathematicians to study the meaning of such quantities. For example, the equation

$$x^3 = 15x + 4$$

has 4 as a root, but the cubic formula gives this root as

$$\sqrt[3]{2 + 11\sqrt{-1}} + \sqrt[3]{2 - 11\sqrt{-1}}$$

which can be simplified to:

$$2 + \sqrt{-1} + 2 - \sqrt{-1} = 4.$$

In the seventeenth century the theory of equations was developed, and mathematical symbolism, including the use of x to denote the unknown, came close to its present form. Negative roots of equations were called "false roots," and $\sqrt{-1}$ was referred to as "imaginary number."

After solving the cubic and quadratic equations, mathematicians began a search for formulas to solve equations of degree 5 and higher that was to last for 300 years.

Classical algebra culminated in the proof of two major results. In 1799, *Carl Gauss* proved the *Fundamental Theorem of Algebra* which states that in the complex number system, every polynomial equation of degree n has n roots. The second result, proven in stages from about 1824 to 1830, states that general polynomial equations of degree 5 and higher cannot be solved by Algebra. With these two results, algebra ceased to be the science of equations. Also a graphical representation of complex numbers and their operations in the plane led to their general acceptance early in the nineteenth century, and by the middle of the century the defining properties of the real and complex number systems were determined. Other algebraic systems were investigated, and the modern era of abstract algebra began.

Algebraic Formulas

Exponents and Radicals

$x^m \cdot x^n = x^{m+n} \qquad x^0 = 1, \ x \neq 0$

$\dfrac{x^m}{x^n} = x^{m-n} \qquad (xy)^n = x^n y^n$

$y = \sqrt[n]{x} \Leftrightarrow y^n = x$, where $y > 0$ if $x > 0$

$x^{-n} = \dfrac{1}{x^n} \qquad x^{m/n} = \sqrt[n]{x^m} = (\sqrt[n]{x})^m$

$\left(\dfrac{x}{y}\right)^n = \dfrac{x^n}{y^n} \qquad \sqrt[n]{xy} = \sqrt[n]{x}\sqrt[n]{y}$

$(x^m)^n = x^{mn} \qquad \sqrt[n]{\dfrac{x}{y}} = \dfrac{\sqrt[n]{x}}{\sqrt[n]{y}}$

Products

$(a+b)(a-b) = a^2 - b^2$

$(a+b)^2 = a^2 + 2ab + b^2$

$(a+b)^3 = a^3 + 3a^2 b + 3ab^2 + b^3$

Binomial Theorem

$(a+b)^n = a^n + \cdots + {}_nC_r a^{n-r} b^r + \cdots + b^n$

where, ${}_nC_r = \dfrac{n!}{(n-r)! \, r!}$

Factoring

$a^2 x^2 - b^2 = (ax+b)(ax-b)$

$a^2 x^2 + 2abx + b^2 = (ax+b)^2$

$x^3 + a^3 = (x+a)(x^2 - ax + a^2)$

$x^2 + bx + c = (x+m)(x+n)$,

where $m+n = b$ and $mn = c$.

Quadratic Formula

The roots of the quadratic equation $ax^2 + bx + c = 0$, where $a \neq 0$, are:

$$x = \dfrac{-b \pm \sqrt{b^2 - 4ac}}{2a}.$$

Line Segments and Straight Lines

Distance d from $P_1(x_1, y_1)$ to $P_2(x_2, y_2)$:

$$d = \sqrt{(x_2 - x_1)^2 + (y_2 - y_1)^2}$$

Midpoint $M(x,y)$ of segment $P_1 P_2$:

$$x = \tfrac{1}{2}(x_1 + x_2) \text{ and } y = \tfrac{1}{2}(y_1 + y_2)$$

Slope m of line through P_1 and P_2:

$$m = \dfrac{y_2 - y_1}{x_2 - x_1}, \quad x_1 \neq x_2$$

Equations of Lines

Vertical line: $x = c$

Point-slope form: $y - y_1 = m(x - x_1)$

Slope-intercept form: $y = mx + b$

General linear form: $ax + by = c$

Functions

Polynomial function P:

$$P(x) = a_n x^n + a_{n-1} x^{n-1} + \cdots + a_1 x + a_0$$

Rational function f:

$$f(x) = P(x) / Q(x),$$

where P and Q are polynomials.

Exponential function g:

$$g(x) = a^x, \quad \text{where } a > 0 \text{ and } a \neq 1$$

Logarithmic function h:

$$h(x) = \log_a x \Leftrightarrow a^{h(x)} = x$$

Properties of Logarithms

$\log_a(xy) = \log_a x + \log_a y$

$\log_a\left(\dfrac{x}{y}\right) = \log_a x - \log_a y$

$\log_a(x^y) = y \log_a x$

$\log_a 1 = 0, \quad \log_a a = 1$

$\log_b x = \dfrac{\log_a x}{\log_a b}$

R Review of Basic Algebra

R.1 The Real Number System

R.2 Basic Properties of Real Numbers

R.3 Signed Quantities, Order, and Absolute Value

R.4 Fractional Expressions

R.5 Powers and Roots

Hey Students!
You will benefit more from the classroom lectures and have a better understanding of the material if you take the time to view the Gilbert Review Videos @
www.CollegeAlgebraBySchiller.com
before attending the class lecture for this chapter.

Chapter R. Review of Basic Algebra

This chapter is a review of the most basic notions in algebra: addition, subtraction, multiplication, and division of real numbers. Since many readers are already familiar with most of this material, we will emphasize the "why" of algebra rather than the "how." Fortifying our understanding of the fundamentals at this stage will help to improve our skills in the chapters that follow.

R.1 The Real Number System

"God created the integers; the rest is the work of man."

–Leopold Kronecker

The Set of Real Numbers
Order and the Real Number Line
Algebraic Operations

The Set of Real Numbers. Children are introduced to real numbers when they learn to count objects, and the counting numbers 1, 2, 3, ... are called **positive integers**. The absence of objects to count suggests the number 0, an integer that is not positive. Parts of objects suggest fractional numbers such as 1/2, 2/3, or 3/4. These are ratios of integers, and any such number is called a **rational number**. The positive integers and 0 are special cases of rational numbers. For example, $5 = 5/1$ and $0 = 0/1$.

Geometric measurements lead to numbers other than rational ones. For example, in a square whose side measures 1 unit, the diagonal measures $\sqrt{2}$ units, and it can be shown that $\sqrt{2}$ is not a rational number. Also, a circle whose diameter is 1 unit has a circumference of π units, and π is not a rational number.[1] These are just two examples from an infinite collection of nonrational numbers called **irrational numbers**.

Both rational and irrational numbers can be expressed as **decimals**. The decimal expansion of a rational number either will terminate, as in $3/4 = .75$, or will eventually have a pattern that repeats indefinitely, as in $25/99 = .25252525\ldots$. The decimal expansion of an irrational number neither terminates nor becomes a repeating pattern. For example, although π is approximately 3.14159, its true decimal expansion is infinitely long and has no repeating pattern.[2]

[1] *The Babylonians of 2000 B.C. set 3 1/8 as the value of π. In 1761 the Swiss mathematician Johann Lambert proved that π is irrational.*

[2] *In 1984 a team of mathematicians at the University of Tokyo, using a supercomputer, determined π to 16 million decimal places. At 5000 decimals per page, a printout of this expansion would require 3200 pages.*

Example 1 A standard 8-place calculator will give 1.4142136 as the decimal value of $\sqrt{2}$. This is accurate enough for most purposes, but in fact

$$(1.4142136)^2 = 2.0000010642496.$$

Hence, the rational number 1.4142136 is slightly larger than the irrational number $\sqrt{2}$. ∎

Example 2 Express the repeating decimal .49999999... as the ratio of two integers.

Solution: Let $x = .49999999\ldots$. Then $10x = 4.9999999\ldots$ and $100x = 49.9999999\ldots$. Subtract $10x$ from $100x$ to obtain $90x = 45.0$. Hence, $x = 45/90 = 1/2$. We note that $1/2$ also has the *terminating* decimal expansion .5 (see Exercises 13-16 at the end of this section). ■

Example 3 The infinite decimal expansion 1.01001000100001... has a nonrepeating pattern because the number of 0's between successive pairs of 1's is always increasing by one. Hence, the decimal represents an irrational number. ■

The **negative integers** $-1, -2, -3\ldots$ are suggested by debts and budget deficits, and every positive rational and irrational number has its negative counterpart: for example, $3/4$ and $-3/4$, and π and $-\pi$. The total collection of rational and irrational numbers forms the **set of real numbers**. These are the numbers with which we work in algebra.

Comment Starting with the positive integers, we arrived at the entire set of real numbers in less than two pages of text. The actual development of the real numbers was a gradual process spread out over many centuries. A classic account of this development is contained in the book *Number, The Language of Science*, by Tobias Dantzig.

Order and the Real Number Line. Geometrically, the real numbers can be identified with the points on a directed line marked off with a number scale (Figure 1). Each point on the line corresponds to a decimal expansion of a real number, and conversely, a decimal expansion of a real number occupies one point on the line. Positive numbers are to the right of 0 on the line and negative numbers to the left. The number 0 is neither positive nor negative. The rational and irrational numbers are densely packed along the real number line. In fact, between any two rational numbers there is an irrational number, and between any two irrational numbers there is a rational number.

The real numbers are arranged in increasing order from left to right along the number line. Since the number 8 is to the right of 4 on the number line, we say 8 is **greater than** 4 and write $8 > 4$. Equivalently, we say 4 is **less than** 8 and write $4 < 8$. Any positive number is greater than any negative number, regardless of magnitude. Hence, $7 > -15$ and $1 > -100$. Also, $0 > -3$ and $-5 > -20$ because 0 is to the right of -3 and -5 is to the right of -20. Since any two unequal real numbers can be compared in this manner, the set of real numbers is called an **ordered set**.

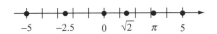

Figure 1 The Real Number Line

The magnitude of a real number, without regard to sign, is called the **absolute value** of the number and is denoted by vertical bars | |. For example, $|5| = 5$ and $|-5| = 5$. The absolute value of a number is its distance form 0 on the number line.

Example 4   Arrange the numbers $2, -11, 0, -5, -1, 6, \pi$ in increasing order.

Solution: $-11 < -5 < -1 < 0 < 2 < \pi < 6$ ■

Example 5   Arrange the numbers $-3, \sqrt{2}, -7, 4.3, 0, |-4|, -2$ in decreasing order.

Solution: $4.3 > |-4| > \sqrt{2} > 0 > -2 > -3 > -7$ ■

Algebraic Operations.   The real numbers form a **closed system** with respect to the operations of addition, subtraction, and multiplication. That is, when any one of these operations is performed on two real numbers, the result always *remains* within the set of real numbers. Also, division of any real number by any real number *except* 0 always will result in a real number. The ordered set of real numbers, together with the four operations of addition, subtraction, multiplication, and division, make up the **real number system.**

Question 1   *Do the integers by themselves form a closed system under addition, subtraction, and multiplication?*

Answer: Yes, because the sum, difference, and product of any two integers is still an integer.   ■

Question 2   *Do the odd integers form a closed system under the operation of addition?*

Answer: No, they do not form a closed system because the sum of two odd integers is not an odd integer.   ■

The usual symbols $+$ and $-$ will be used for addition and subtraction, respectively. The product of two numbers a and b will be denoted by $a \cdot b$ or ab, and the quotient by $a \div b$, or a/b, provided $b \neq 0$. The properties of addition, subtraction, multiplication, and division within the real number system constitute our subject matter for the study of algebra. In the next section we set down nine *basic properties* of algebra. In succeeding sections and chapters we discover further properties that follow from the basic ones. Your ability to "do" algebra will be directly related to your understanding of these properties.

Exercises R.1

Fill in the blanks to make each statement true.

1. Every real number is either a _____ number or an _____ number.
2. A calculator can generate only numbers that are terminating decimals. Hence, all numbers generated on a calculator are _____ numbers.
3. An irrational number has a decimal expansion that _____.
4. The only real number that is neither positive nor negative is _____.
5. The real numbers form a closed system with respect to the operations of _____, _____, and _____.

Write true *or* false *for each statement.*

6. Every integer is a rational number.
7. The decimal .123456789101112131415..., which continues through all the positive integers, is a rational number.
8. The decimals .4999999... and .5 occupy different points on the number line.
9. Every positive number is the absolute value of two real numbers.
10. There is no largest or smallest real number.

The Set of Real Numbers

Supply the missing numbers in Exercises 11 and 12.

11. A student has just finished the seventh problem of a test, and there are three more to go. The part of the test that has been finished corresponds to the rational number _____, and the part yet to be finished corresponds to _____.
12. A pizza is cut into eight equal slices. Each slice corresponds to the rational number _____. If Terry and Pam finish the entire pizza, but Terry eats three times as many slices as Pam, then this suggests the rational number _____ for Terry's portion and _____ for Pam's.

Every terminating decimal can also be written as a repeating decimal (each represents the same point on the number line). For example, $1.28 = 1.27999999...$ and $.241 = .240999999...$. Express each number in Exercises 13-16 as a repeating decimal.

13. .35
14. .724
15. .667
16. 1.0

17. Express the rational number 1/3 as a decimal.
18. Express the rational number 2/7 as a decimal.

Express each decimal in Exercises 19-22 as the ratio of two integers.

19. .123
20. 1.25
21. .555555...
22. .2121212121...

23. Give the decimal expansion of a rational number between the rational numbers .5 and .5001.
24. Give the decimal expansion of a rational number between the irrational numbers .101001000100001... and .102003000400005....

Order and the Real Number Line

25. Arrange the following numbers in increasing order.
    $\frac{22}{7}, 3.14, \pi, 1.41, \sqrt{2}, -3, -3.5, -\sqrt{9.1}$
26. Arrange the following numbers in decreasing order.
    $-\frac{1}{2}, -.501, -.499, 0, .249, \frac{1}{4}, \frac{1}{3}, |-.333|$
27. Locate the following numbers on a real number line.
    $3.5, 2.75, -5, -4.25, 0, 1.5, -2, .75$
28. Locate the following numbers on a real number line.
    $\frac{3}{5}, -\frac{1}{2}, -2\frac{2}{3}, 5\frac{3}{4}, \frac{7}{8}, \frac{10}{3}, -\frac{7}{2}$

*The middle value of an odd number of distinct values is called the **median** of the collection. For example, the median of the numbers 1, 3, 6, 8, and 15 is 6. In Exercises 29-32, arrange the numbers in increasing order and find the median.*

29. weights: 110 lb, 140 lb, 135 lb, 150 lb, 100 lb
30. heights: 5'4", 6', 4'11", 5'7", 6'3", 5'10", 5'2"
31. ages: 17 yr, 18 yr, 22 yr, 19 yr, 20 yr, 21 yr, 16 yr
32. grades: 60, 98, 75, 45, 85, 70, 53, 67, 72, 88, 77

*The median of an even number of distinct values is the **average** of the two middle values of the collection. For example, the **median** of the numbers 5, 7, 9, 14, 17, 21, 32, 38 is $(14 + 17)/2 = 15.5$. In Exercises 33-36, arrange each list of numbers in increasing order and find the median.*

33. wages: $325, $475, $250, $625, $125, $750, $350, $500
34. salaries: $20,000, $15,000, $30,000, $10,000, $25,000, $35,000

35. interest rates: 8.5%, 5.5%, 5.25%, 9.75%, 10%, 10.5%
36. stock prices:
$16\frac{3}{4}, 17\frac{1}{8}, 16\frac{7}{8}, 17\frac{3}{4}, 17\frac{1}{4}, 16\frac{1}{2}, 16\frac{1}{8}, 17\frac{1}{2}$

Algebraic Operations

In Exercises 37-40, state the operation(s) $(+, -, \cdot, \div)$ under which the given system is closed.
37. the real numbers with 0 deleted
38. the rational numbers
39. the rational numbers that have terminating decimals
40. the real numbers between 0 and 1
41. Find a subset of the real numbers that is closed under multiplication but not under addition.
42. Find a subset of the real numbers that is closed under addition but not under multiplication.
43. When Andrew received his savings account statement for the month of June, he observed that his account was closed with respect to addition. How much money was in Andrew's account?

R.2 Basic Properties of Real Numbers

"It must be realized that the essence of algebra is in its generality."

–Philip E. B. Jourdain

Commutative, Associative, and Distributive Properties

Identities and Inverses

Grouping Symbols and Priority Conventions

Extended Commutative, Associative, and Distributive Properties

Commutative, Associative, and Distributive Properties. The only difference between algebra and arithmetic is that in algebra we often use letters in place of numbers. The arithmetic rules that govern the use of real numbers are the same rules that apply to letters representing real numbers. For example, we know that $7 + 5 = 5 + 7$, and if we replace 7 or 5 by other numbers, the equality still holds. In letters, we say

$$a + b = b + a, \quad \text{Commutative Property of Addition} \quad (1)$$

where a and b stand for *any two real numbers*. This basic property illustrates the power and importance of algebra. The particular equation $7 + 5 = 5 + 7$ is a statement about the specific real numbers 7 and 5, whereas the general equation $a + b = b + a$ is a statement about the entire set of real numbers. The general equation contains infinitely more information than the particular one.

In the following basic properties, $a, b,$ and c represent any real numbers.

$$(a + b) + c = a + (b + c) \quad \text{Associative Property of Addition} \quad (2)$$

For example, $(8 + 4) + 16 = 8 + (4 + 16)$.

$$a \cdot b = b \cdot a \quad \text{Commutative Property of Multiplication} \quad (3)$$

For example, $5 \cdot 12 = 12 \cdot 5$.

$$(a \cdot b) \cdot c = a \cdot (b \cdot c) \qquad \text{Associative Property of Multiplication} \qquad (4)$$

For example, $(3 \cdot 4) \cdot 10 = 3 \cdot (4 \cdot 10)$.

$$a \cdot (b + c) = (a \cdot b) + (a \cdot c) \qquad \text{Left Distributive Property} \qquad (5)$$

For example, $6 \cdot (5 + 7) = (6 \cdot 5) + (6 \cdot 7)$.

Question 1 *Is either subtraction or division a commutative or an associative operation?*

Answer: No. For example, $8 - 2 = 6$ but $2 - 8 = -6$, and $8/2 = 4$ but $2/8 = 1/4$. Also, $(12 - 6) - 2 = 4$ but $12 - (6 - 2) = 8$, and $(12 \div 6) \div 2 = 1$ but $12 \div (6 \div 2) = 4$. ∎

Question 2 *Does the left distributive property hold if $+$ and $\cdot$ are interchanged? In other words, is it true that $a + (b \cdot c) = (a + b) \cdot (a + c)$ for all values of $a, b,$ and c?*

Answer: No. For example, $3 + (4 \cdot 5) = 23$ but $(3 + 4) \cdot (3 + 5) = 56$. ∎

We can combine basic properties (3) and (5) to derive the following property.

$$(a + b) \cdot c = (a \cdot c) + (b \cdot c) \qquad \text{Right Distributive Property}$$

For example, $(6 + 8) \cdot 5 = (6 \cdot 5) + (8 \cdot 5)$.

Identities and Inverses. The number 0 is called the additive identity or the neutral element for addition because for any real number a,

$$a + 0 = a = 0 + a. \qquad \text{Additive Identity Property} \qquad (6)$$

Similarly, the number 1 is called the multiplicative identity or the neutral element for multiplication because for any real number a,

$$a \cdot 1 = a = 1 \cdot a. \qquad \text{Multiplicative Identity Property} \qquad (7)$$

It can be shown that both 0 and 1 are unique in that they are the only identity elements for addition and multiplication, respectively.

If -7 is added to 7, the result is the additive identity 0. For this reason -7 is called the additive inverse of 7, and conversely, 7 is the additive inverse of -7. Every real number a has a *unique* additive inverse, which is denoted by $-a$ and which satisfies

$$a + (-a) = 0 = -a + a. \qquad \text{Additive Inverse Property} \qquad (8)$$

It is important to note that $-a$ is not necessarily a negative number. For example, if $a = -7$, then $-a = 7$ because $-7 + 7 = 0$. Hence, $-(-7) = 7$.

In general, a and $-a$ are the same distance from 0 on the number line but on opposite sides of 0. In other words, the effect of placing a minus sign before a number is to reverse its direction on the number line. If a is positive, then $-a$ is negative, but if a is negative, then $-a$ is positive. Always,

$$-(-a) = a. \qquad \text{Double Inverse Property}$$

That is, the additive inverse of the additive inverse of a is a itself, or a double reversal of a on the number line leads back to a.

If 5 is multiplied by its reciprocal 1/5, the result is the multiplicative identity 1. For this reason 1/5 is called the multiplicative inverse of 5, and conversely, 5 is the multiplicative inverse of 1/5. The reciprocal 1/5 is also denoted by 5^{-1}. In general, every nonzero number a has a *unique* multiplicative inverse, which is denoted by $1/a$ or a^{-1}, and which satisfies

$$a \cdot a^{-1} = 1 = a^{-1} \cdot a. \qquad \text{Multiplicative Inverse Property} \qquad (9)$$

We can use inverses to define subtraction and division in terms of addition and multiplication, respectively, that is, $a - b = a + (-b)$ and $\frac{a}{b} = a \cdot b^{-1}$ ($b \neq 0$).

The number 0 does not have a multiplicative inverse. If it did, then 0^{-1} would have to be a real number that satisfies the equation $0 \cdot 0^{-1} = 1$. However, 0 times any real number is equal to 0, as is shown in the following example.

Example 1 Show that, for any real number a,

$$0 \cdot a = 0. \qquad \text{Zero multiplier Property}$$

Solution: An actual proof of this apparently simple result is by no means simple. We give one here in order to illustrate the use of the basic properties. At this stage in your study of algebra you would not be expected to think up such a proof, but you should be able to follow it step by step.

By the additive identity property,

$$0 + 0 = 0.$$

Multiply both sides of the above equation by a:

$$(0+0) \cdot a = 0 \cdot a.$$

Apply the right distributive property to the left side:

$$(0 \cdot a) + (0 \cdot a) = 0 \cdot a.$$

Let x stand for $0 \cdot a$. The above equation says that

$$x + x = x.$$

Add $-x$ to both sides:

$$(x + x) + (-x) = x + (-x).$$

Apply the associative property for addition to the left side and the additive inverse property to the right:

$$x + [x + (-x)] = 0.$$

Apply the additive inverse property to the quantity in brackets on the left side:

$$x + 0 = 0.$$

Apply the additive identity property to the left side:

$$x = 0.$$

Hence, $0 \cdot a = 0$, which completes the proof. ∎

$\dfrac{0}{a} = 0$ *for any* $a \neq 0$, *but* $\dfrac{a}{0}$ *is not defined for any a whatsoever.*

Another way of saying that 0 has no multiplicative inverse is to say that 1/0 is a meaningless expression or that division by 0 is not possible. Remember that the real numbers form a closed system under addition, subtraction, and multiplication, but division by 0 is not defined.

In the proof in Example 1 above, we assumed the following algebraic and logical properties of equality of real numbers.

Algebraic Properties of Equality

If $a = b$, then $a \pm c = b \pm c$.	Any number may be added or subtracted on both sides of an equation.
If $a = b$, then $a \cdot c = b \cdot c$.	An equation may be multiplied on both sides by any real number.
If $a = b$, and $c \neq 0$, then $\dfrac{a}{c} = \dfrac{b}{c}$.	An equation may be divided on both sides by a nonzero real number.

Logical Properties of Equality

$a = a$ for any real number a.	Reflexive Property
If $a = b$, then $b = a$.	Symmetric Property
If $a = b$ and $b = c$, then $a = c$.	Transitive Property

These properties also justify the familiar process of **substitution**, that is, if $a = b$, we may substitute b for a in any algebraic equation.

By the zero-multiplier property and the commutative property for multiplication we know that if either $a = 0$ or $b = 0$, then $ab = 0$. Conversely, if $ab = 0$, then either $a = 0$ or $b = 0$ (see Exercise 32). Therefore, the statement "$ab = 0$" is equivalent to the statement "$a = 0$ or $b = 0$."

We will use the double arrow $\Leftrightarrow$ (read "if and only if") to indicate that two statements are equivalent. Hence,

$$\mathbf{ab = 0 \Leftrightarrow a = 0 \text{ or } b = 0.} \qquad \text{Zero-product Property}$$

For example,

$$(x-1)(x+2) = 0 \Leftrightarrow x - 1 = 0 \quad \text{or} \quad x + 2 = 0$$
$$\Leftrightarrow \quad x = 1 \quad \text{or} \quad x = -2.$$

The zero-product property extends to more than two factors. For instance,

$$x(x-3)(x+4) = 0 \Leftrightarrow x = 0 \quad \text{or} \quad x - 3 = 0 \quad \text{or} \quad x + 4 = 0$$
$$\Leftrightarrow x = 0 \quad \text{or} \quad x = 3 \quad \text{or} \quad x = -4.$$

Analogous to the equation $-(-a) = a$ for additive inverses, we have, for multiplicative inverses, the equation

$$\mathbf{\left(a^{-1}\right)^{-1} = a} \quad (\mathbf{a \neq 0}). \qquad \text{Double Inverse Property}$$

The double inverse property for multiplication follows because the expression $(a^{-1})^{-1}$ means the unique number that when multiplied by a^{-1} results in 1. Since we know that $a^{-1} \cdot a = 1$, it follows that $(a^{-1})^{-1} = a$. Hence, $(5^{-1})^{-1} = 5$, which is another way of saying that $1/(1/5) = 5$. In general, $1/(1/a) = a$ for any real number $a \neq 0$.

If $a \neq 0$, then the numbers a and $1/a$ are on the same side of 0 on the number line. That is, if a is positive, so is $1/a$, and if a is negative, so is $1/a$. However, a positive number and its reciprocal are on opposite sides of 1 on the number line (*why?*), whereas a negative number and its reciprocal are on opposite sides of -1. The reciprocal $1/a$ of a number a can be located on the number line by means of the geometric construction indicated in Figure 2 (see Exercise 50). The circle in Figure 2 has its center at 0, and its radius is 1. Line L is tangent to the circle.

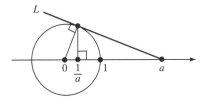

Figure 2 Construction of $1/a$

Basic properties (1) through (9) are the basis for all other algebraic properties of the real number system, including those properties related to order (see Section R.3). In Example 1, we saw how the basic properties are used to derive further ones. The following example is another illustration of this process. See if you can supply the reason for each step. Each reason is one of the nine basic properties or one of the properties already derived from them.

Example 2   Show that, for any real number a,

$$-a = (-1)a. \qquad \text{Negative Multiplier Property}$$

That is, the additive inverse of a is equal to a multiplied by -1.

Solution: By definition, $-a$ is the unique number that when added to a results in 0. Hence, we would like to show that $a + (-1)a = 0$. We proceed as follows.

$$\begin{aligned}
a + (-1)a &= (a \cdot 1) + (-1)a && \text{Basic Property} \\
&= (a \cdot 1) + a(-1) && \text{Basic Property} \\
&= a \cdot [1 + (-1)] && \text{Basic Property} \\
&= a \cdot 0 && \text{Basic Property} \\
&= 0 && \text{Derived Property}
\end{aligned}$$

We will not labor this process of deriving new properties in the following sections and chapters. Our main goal is to use algebra as a tool rather than to forge this tool. However, it is important to keep in mind that every algebraic property we learn can be traced to the nine basic ones.

Question 3   *Why do the basic properties involve only addition and multiplication, not subtraction or division?*

Answer: The operations of subtraction and division can be defined in terms of addition and multiplication, respectively (see Section R.3).

When we say that $a - b = c$, we mean that c is the unique number that when *added* to b results in a, that is $a - b = c$ means $c + b = a$. Similarly, $a \div b = c$ means that $c \cdot b = a$. Hence, it is not necessary to state basic properties of subtraction and division. The properties of these operations can be worked out by using the established properties of addition and multiplication. ■

Grouping Symbols and Priority Conventions. When several operations are applied to numbers or letters, grouping symbols such as parentheses ( ), brackets [ ], or braces { } are used to indicate the order in which the operations are to be performed. For example, $(6 \cdot 3) + (8 \div 2) = 18 + 4 = 22$. The operations inside the parentheses are performed first. If grouping symbols appear within other grouping symbols, the operations are performed inside out; that is, the operations within the innermost grouping symbols are performed first, then the operations within the remaining innermost grouping symbols, and so on. For example,

$$\begin{aligned} 24 \div \{32 - [8 + (6 \cdot 3)]\} &= 24 \div \{32 - [8 + 18]\} \\ &= 24 \div \{32 - 26\} \\ &= 24 \div 6 \\ &= 4. \end{aligned}$$

We can reduce the number of grouping symbols required by establishing the following **order priority**[1] for the operations of addition, subtraction, multiplication, and division.

Unless grouping symbols indicate otherwise,

1. all multiplications and divisions will be performed before any additions or subtractions; for example, $6 + 5 \cdot 3 = 6 + 15 = 21$;
2. multiplications and divisions will be performed in the order in which they appear from left to right in an algebraic expression; for example, $8 \div 2 \cdot 3 \div 6 = 4 \cdot 3 \div 6 = 12 \div 6 = 2$;
3. additions and subtractions will be performed in the order in which they appear from left to right in an algebraic expression; for example, $18 - 4 + 16 - 7 = 14 + 16 - 7 = 30 - 7 = 23$.

For visual clarity, we will sometimes use grouping symbols even though they are not absolutely necessary. For instance, we might write
$$3 \cdot 6 - 20 \div 5 + 2 \cdot 8 - 28$$
as
$$(3 \cdot 6) - (20 \div 5) + (2 \cdot 8) - 28.$$

Question 4 *On a calculator that has the above priorities, which of the following three*

---

[1] *This is the priority that is programmed into most scientific calculators. Also, special functions, such as the square root, exponential, logarithmic, and trigonometric functions, precede multiplication and division in priority.*

procedures give the correct answer for $\dfrac{256}{16 \cdot 8}$?

(a) Enter 256   (b) Enter 256   (c) Enter 256
    Enter ÷16   Enter ÷   Enter ÷ 16
    Enter * 8   Enter (16*8)   Enter ÷ 8

Answer: Procedures (b) and (c) give the correct answer, 2. Procedure (a) gives 128, which is equal to $(256/16) \cdot 8$. ∎

Extended Commutative, Associative, and Distributive Properties. The commutative and associative properties for addition and multiplication can be extended to three or more terms. That is, in a sum or product of three or more terms, the terms can be arranged or grouped in any manner whatsoever without changing the results. For example, $b+y+a+x = a+b+x+y$, and $byax = abxy$. Also, $a+b+x+y = (a+b)+(x+y) = a+(b+x)+y = (a+b+x)+y$, and similarly for multiplication. The left and right distributive properties can be extended to three or more terms. For example, $a(x+y+z) = ax+ay+az$, and $(a+b+c)x = ax+bx+cx$. The following chart illustrates how the priority conventions and extended properties simplify algebraic notation.

Unsimplified Notation	Simplified Notation
$2(xy)$ or $(2x)y$	$2xy$
$2+(x+y)$ or $(2+x)+y$	$2+x+y$
$[(2a)+(3b)]+[(4x)+(5y)]$	$2a+3b+4x+5y$
$a[(x+y)+z]$	$a(x+y+z) = ax+ay+az$
$\quad = [a(x+y)]+(az)$	
$\quad = [(ax)+(ay)]+(az)$	
$(2x)(3y) = 2[x(3y)]$	$2x \cdot 3y = 6xy$
$\quad = 2[(3y)x]$	
$\quad = 2[3(yx)]$	
$\quad = (2 \cdot 3)(yx)$	
$\quad = 6(xy)$	

We will, of course, use the simplified notation throughout the remainder of the text except when emphasis or visual clarity suggests otherwise.

Comment As children, we learned to speak words before we were taught the alphabet or basic elements of words. Similarly, mathematicians worked with algebra long before the basic properties of real numbers were discovered. It was not until the nineteenth century that mathematicians finally took time out from *doing* algebra to ask the question, *"What is algebra?"*. They went from the *"how"* of algebra to the "why." The nine basic properties described in this section became known as **the field axioms** of the real number system. Other mathematical systems satisfying different basic properties also were investigated, which led to the present day axiomatic development of abstract algebra. Hence, almost 4000 years after the Babylonians were solving equations, the guiding principles of algebra were finally determined. We might say that it took mathematicians 4000 years to put the horse before the cart!

Exercises R.2

Fill in the blanks to make each statement true.

1. The property $a + b = b + a$ is called the _____ of addition, and $(a + b) + c = a + (b + c)$ is called the _____ of addition.
2. The left distributive property is _____, and the right distributive property is _____.
3. The additive inverse of -3 is _____, and the multiplicative inverse of $1/3$ is _____.
4. The additive inverse of 0 is_____; that is, $-0 =$ _____; the multiplicative inverse of 1 is _____; that is, $1^{-1} =$ _____.
5. If $1/0$ were a real number, it would have to satisfy $0 \cdot 1/0 =$ _____. But $0 \cdot a =$ _____ for any real number a. Therefore, _____.

Write true or false for each statement.

6. Subtraction is a commutative operation.
7. Division is not an associative operation.
8. The additive inverse of a number is always a negative quantity.
9. Every real number has an additive inverse.
10. Every real number has a multiplicative inverse.

Basic Properties

11. Name the basic property illustrated by each of the following equations.

    (a) $3 \cdot 5 = 5 \cdot 3$
    (b) $(4 + 6) + 3 = 4 + (6 + 3)$
    (c) $2(5 + 7) = 2 \cdot 5 + 2 \cdot 7$

12. Given $a = 4$, $b = 3$, and $c = 8$, name the basic property illustrated by each of the following.

    (a) $7 + 8 = 4 + 11$ (b) $12 \cdot 8 = 4 \cdot 24$
    (c) $4 \cdot 11 = 12 + 32$

State the basic property or properties that justify each statement in Exercises 13-16. The values of numerical sums may be assumed.

13. $(5x + 4) + 3 = 5x + 7$
14. $(y + 6) + (x + 1) = (x + 1) + (y + 6)$
15. $2x + 3x = 5x$
16. $x + x = 2x$
17. Which of the following statements are equivalent to the statement "0 has no multiplicative inverse"?

    (a) $0 \cdot a = 0$ for any real number a.
    (b) $1/0$ has no meaning as a real number.
    (c) Division by 0 is an invalid operation.

18. Which of the following statements are equivalent to the statement "a is negative"?

    (a) a is to the left of 0 on the number line.
    (b) $-a$ is positive.
    (c) a is less than 0.
    (d) $-a > a$

In each of the following, find all values of x that satisfy the equation.

19. $(x - 2)(x + 2) = 0$
20. $(x + 3)(x - 1) = 0$
21. $(|x| - 1)(|x| - 2) = 0$
22. $(|x| - 3)(|x| + 4) = 0$

Here is a simple application of the distributive property. If there is a 6% sales tax on an item whose price is $15, then the total cost of the item in dollars is

$$15 + 15(.06) = 15 \cdot 1 + 15(.06) = 15(1.06) = 15.90$$

That is, instead of computing the sales tax separately and adding it to the price, just multiply the price by 1.06 to obtain the total cost. Use this method in Exercises 23-26.

23. The price of a shirt is $25 and the sales tax is 6%. Find the total cost.
24. The price of a rare coin is $170 and the sales tax is 7%. Find the total cost.
25. The total cost of a used car is $2662.50, which includes a sales tax of 6.5%. Find the price of the car.
26. The total cost of a new car is $9675, which includes a 7.5% sales tax. Find the price of the car.
27. In attempting to solve an algebraic equation, Adam arrives at the conclusion that $5 = 3$. What is wrong with Adam's "proof"?

$$\text{Given}: 5x + 10 = 3x + 6.$$
$$\text{Then } 5(x+2) = 3(x+2)$$
$$\text{(Left Distributive Property)}$$
$$[5(x+2)](x+2)^{-1} = [3(x+2)](x+2)^{-1}$$
$$\text{(Property of Equality)}$$
$$5[(x+2)(x+2)^{-1}] = 3[(x+2)(x+2)^{-1}]$$
$$\text{(Associative Property)}$$
$$5 \cdot 1 = 3 \cdot 1$$
$$\text{(Multiplicative Inverse Property)}$$
$$5 = 3. \text{ (Multiplicative Identity Property)}$$

Use the basic properties (and properties of equality) to prove the statements in Exercises 28-31.

28. If $x + 5 = y + 5$, then $x = y$.
29. If $a + b = a + c$, then $b = c$.
30. If $4x = 4y$, then $x = y$.
31. If $ac = bc$ and $c \neq 0$, then $a = b$.

32. Example 1 and the commutative property for multiplication say that if either a or $b = 0$, then $ab = 0$. The converse is also true. That is,

   if $ab = 0$, then either $a = 0$ or $b = 0$.

   Fill in the blanks to complete the following proof. Each reason is a basic property, a property of equality, or a derived property.
   Given: $ab = 0$.
   If $b \neq 0$, then b has a multiplicative inverse b^{-1} and

$$(ab)b^{-1} = 0 \cdot b^{-1} \quad \text{_____}$$
$$a(bb^{-1}) = 0 \quad \text{_____}$$
$$a \cdot 1 = 0 \quad \text{_____}$$
$$a = 0. \quad \text{_____}$$

   Hence, if $b \neq 0$, then $a = 0$. Therefore, if $ab = 0$, then either $a = 0$ or $b = 0$.

33. Because $0 + 0 = 0$, it follows that $-0 = 0$. Here we show that 0 is the *only* real number with this property. *Fill in the reasons for each step.*
   Given: $-x = x$.
   Then $x + x = 0$ by the _____ (Basic Property).
   Therefore, $2x = 0$ by Exercise _____.
   Therefore, $x = 0$ by Exercise _____.

Grouping Symbols and Priority Conventions

Evaluate each expression in Exercises 34-35.

34. (a) $(48 \div 3)[(6 + 12) \div (4 - 2)]$
    (b) $48 \div \{3[6 + 12 \div (4 - 2)]\}$
    (c) $48 \div 3[6 + 12 \div 4 - 2]$
    (d) $48 \div 3 \cdot 6 + 12 \div 4 - 2$
35. (a) $8 \cdot [(18 + 9) \div (3 \cdot 3)]$
    (b) $8 \cdot [18 + 9 \div (3 \cdot 3)]$
    (c) $8 \cdot [18 + 9 \div 3 \cdot 3]$
    (d) $8 \cdot 18 + 9 \div 3 \cdot 3$

Use a calculator to evaluate each expression in Exercises 36-37.

36. (a) $\dfrac{152.24 + 5.018}{12.687 - 8.972}$

    (b) $\dfrac{152 + 5.018}{12.687} - 8.972$

36. (c) $152.24 + 5.018 \div (12.687 - 8.972)$
    (d) $152.24 + 5.018 \div 12.687 - 8.972$
37. (a) $(12.52 + 6.8) \cdot (5.75 \div 3.21)$
    (b) $(12.52 + 6.8 \cdot 5.75) \div 3.21$
    (c) $(12.52 + 6.8) \cdot 5.75 \div 3.21$
    (d) $12.52 + 6.8 \cdot 5.75 \div 3.21$

Extended Properties

State the extended property or properties that justify each of the following. The values of numerical sums and products may be assumed.

38. $2y \cdot 5x = 10xy$
39. $(x + 1) + (3 + y) = x + 4 + y$
40. $4x + 5x + 7x = 16x$
41. $2x + 7 + 3x + 4 = 5x + 11$

Miscellaneous

Place all the even integers into one category called E and all the odd integers into another category called O. Addition is defined in this system by: $E + E = E$, $E + O = O$, $O + E = O$, $O + O = E$. *Multiplication is defined by:* $E \cdot E = E$, $E \cdot O = E$, $O \cdot E = E$, $O \cdot O = O$.

42. Verify that the above system satisfies basic properties (1) – (5).
43. Verify that E is the identity element for addition in the above system. What is the additive inverse of E? What is the additive inverse of O?
44. Verify that O is the identity element for multiplication in the above system. Does E have a multiplicative inverse? If so, what is it? Does O have a multiplicative inverse? If so, what is it?

*In Exercises 45–49, let a A b denote the **average** of the two numbers a and b, that is, $a\,A\,b = (a+b)/2$. Also, let a M b denote the **maximum** of a and b. That is, $a\,M\,b = a$ if $a > b$ or $a = b$, and $a\,M\,b = b$ if $b > a$ or $b = a$.*

45. In the system above, compute each of the following.

    (a) 6 A 2
    (b) (6 A 2) A 12
    (c) 6 A (2 A 12)
    (d) 6 M 2
    (e) (6 M 2) M 12
    (f) 6 M (2 M 12)

46. In the system above, find b in each of the following cases, if possible.

    (a) 2 A b = 2
    (b) 5 M b = 0
    (c) 2 M b = 2
    (d) 7 A b = 1
    (e) 5 A b = 0
    (f) 7 M b = 1

47. Answer each question for the system above.

    (a) Is A a commutative operation?
    (b) Is A an associative operation?
    (c) Is M a commutative operation?
    (d) Is M an associative operation?

48. It can be shown that the left distributive property holds when · is replaced by A and + is replaced by M. Verify that 6 A (2 M 12) = (6 A 2) M (6 A 12).

49. It can be shown that the left distributive property does *not* always hold when · is replaced by M and + is replaced by A. Verify that
    6 M (2 A 12) ≠ (6 M 2) A (6 M 12).

50. The location of $1/a$ as indicated by x in the figure can be demonstrated by elementary geometry as follows: right triangles $0xP$ and $0Pa$ are similar because they have acute angle $P0x$ in common. Therefore, the ratio of the length of base x to that of hypotenuse $0P$ in triangle $0xP$ is equal to the ratio of the length of base $0P$ to that of hypotenuse a in triangle $0Pa$. That is,

$$\frac{x}{\text{length of } 0P} = \frac{\text{length of } 0P}{a}.$$

Since the length of $0P = 1$, it follows that $x = 1/a$, as stated.

Now $x = 1/a$ implies that $a = 1/x$. What happens to point a as x approaches 0? If x becomes 0, what happens to the triangle $0Pa$? This shows that the algebraic property "0 has no multiplicative inverse" is related to the geometric property "parallel lines do not intersect."

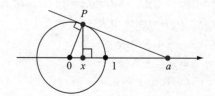

R.3 Signed Quantities, Order and Absolute Value

"The algebraic rules of operation with negative numbers are generally admitted by everyone and acknowledged as exact, whatever idea we may have about these quantities."

—Jean Le Rond d'Alembert

Addition and Subtraction
Multiplication and Division
Algebraic Properties of Order
Absolute Value

Addition and Subtraction. Every nonzero real number has both a magnitude and a sign. However, because $+a = a$ for every real number a, the leading plus sign is rarely used. The main use of the symbol $+$ is for the operation of addition. Hence, the equation $+6 + (+5) = +11$ is normally written as $6 + 5 = 11$. The minus sign $-$ is used for negative numbers or, more generally, for additive inverses, and it is also used for the operation of subtraction. Now $-6 + (-5) = (-1)6 + (-1)5 = (-1)(6+5) = (-1)11 = -11$. These special cases illustrate the general rule for adding quantities with the same sign: to compute $a + b$ when a and b have the *same sign*, add their absolute values and precede this result by their common sign. For example,

$$-7 + (-8) = -\underbrace{(7+8)}_{\text{Sum of the absolute values}} = -15.$$

Common sign

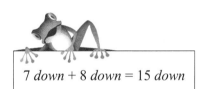

7 *down* + 8 *down* = 15 *down*

The special cases $-5 + 9 = -5 + 5 + 4 = 0 + 4 = 4$ and $-5 + 3 = -2 + (-3) + 3 = -2 + 0 = -2$ illustrate the general rule for adding quantities with opposite signs: to compute $a + b$ when a and b have *opposite signs*, subtract the smaller absolute value from the larger, and precede this result by the sign of the term with the larger absolute value. If both terms have the same absolute value, their sum is 0. For example,

$$7 + (-8) = -\underbrace{(8-7)}_{\text{Difference of the absolute values}} = -1.$$

Sign of the term with the larger absolute value (sign of -8)

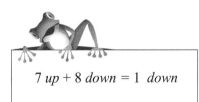

7 *up* + 8 *down* = 1 *down*

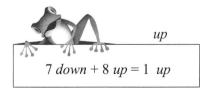

7 *down* + 8 *up* = 1 *up*

Similarly, $-7 + 8 = +(8 - 7) = 1$.

As stated in Section R.2, subtraction is defined in terms of addition. By definition,

$$a - b = a + (-b). \quad \text{Definition of Subtraction}$$

Hence, to subtract b from a, we *add* the additive inverse of b to a. For example,

$$3 - 14 = 3 + (-14) = -11,$$

and
$$-4 - 12 = -4 + (-12) = -16.$$

The result $-(7 + 8) = -7 + (-8)$ can be generalized as follows.

Example 1 Show that

$$-(a + b) = -a + (-b). \quad \text{Inverse of a Sum}$$

That is, show that the additive inverse of the sum of two numbers is equal to the sum of their additive inverses.

Solution:

$$\begin{aligned}
-(a + b) &= (-1)(a + b) && \text{Negative Multiplier Property} \\
&= (-1)a + (-1)b && \text{Left Distributive Property} \\
&= -a + (-b) && \text{Negative Multiplier Property}
\end{aligned}$$ ∎

Some applications of Example 1 are listed below.

1. $-(x + 6) = -x + (-6) = -x - 6$
2. $-(-3 + y) = -(-3) + (-y) = 3 - y$
3. $-(10 - y) = -[10 + (-y)] = -10 + -(-y) = -10 + y$
4. $-(-x - 2) = -[-x + (-2)] = -(-x) + -(-2) = x + 2$

The above four special cases illustrate the general rule that *when there is a minus sign before parentheses, the parentheses can be removed by changing the sign of each term within the parentheses*. This rule can be extended to more than two terms. For instance,

$$-(2x - 3y + 4) = -2x + 3y - 4$$
and $\quad -(-x + y - a - b) = x - y + a + b.$

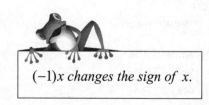

$(-1)x$ *changes the sign of* x.

Multiplication and Division. In the statement

$$(-1)(-1) = -(-1) = 1,$$

the first equality follows from the negative multiplier property, and the second from the double inverse property $-(-a) = a$. In a similar way, we can show that for any two real numbers a and b,

$$(-a)(-b) = ab \quad \text{and} \quad (-a)b = a(-b) = -(ab).$$
Sign Rules for Multiplication

In particular, these sign rules tell us that *the product of two factors having the same sign is positive, and the product of two factors having opposite signs is negative.* For example,

$$(-6)(-8) = 6 \cdot 8 = 48 \quad \text{and} \quad (-6)8 = 6(-8) = -48.$$

Also, $\quad (-5)(-x) = 5x \quad \text{and} \quad (-5)x = 5(-x) = -5x.$

As previously stated, division is defined in terms of multiplication. By definition, if $b \neq 0$, then

$$\frac{a}{b} = a \cdot \frac{1}{b} = a \cdot b^{-1}. \qquad \text{Definition of Division}$$

Hence, to divide a by b, we *multiply* a by the reciprocal or multiplicative inverse of b. From the definition of division and the sign rules for multiplication, we can show that if $b \neq 0$, then

$$\frac{-a}{-b} = \frac{a}{b} \quad \text{and} \quad \frac{-a}{b} = \frac{a}{-b} = -\frac{a}{b}. \qquad \text{Sign Rules for Division}$$

As in products, *the quotient of two terms with the same sign is positive, and the quotient of two terms with opposite signs in negative.*

For example, $\quad \dfrac{-28}{-14} = \dfrac{28}{14} = 2 \quad \text{and} \quad \dfrac{-28}{14} = \dfrac{28}{-14} = -2.$

Also, $\quad \dfrac{-1}{-x} = \dfrac{1}{x} \quad \text{and} \quad \dfrac{-1}{x} = \dfrac{1}{-x} = -\dfrac{1}{x} \text{ if } x \neq 0.$

In Example 1 it was shown that the additive inverse of the sum of two numbers is equal to the sum of their additive inverses. In a similar way, we could show that the multiplicative inverse of a product of two numbers is the product of their multiplicative inverses. That is, if $a \neq 0$ and $b \neq 0$, then

$$(ab)^{-1} = a^{-1}b^{-1} \quad \text{or} \quad \frac{1}{ab} = \frac{1}{a} \cdot \frac{1}{b}. \qquad \text{Inverse of a Product}$$

We can also apply the above inverse property to quotients. For instance,

$$\left(\frac{3}{4}\right)^{-1} = (3 \cdot 4^{-1})^{-1} \qquad \text{Definition of Division}$$
$$= 3^{-1}(4^{-1})^{-1} \qquad \text{Inverse of a Product}$$
$$= 3^{-1} \cdot 4 \qquad \text{Double Inverse}$$

$$= 4 \cdot 3^{-1} \qquad \text{Commutative Property}$$

$$= \frac{4}{3}. \qquad \text{Definition of Division}$$

In general, we can conclude that if $a \neq 0$ and $b \neq 0$, then

$$\left(\frac{a}{b}\right)^{-1} = \frac{b}{a}. \qquad \text{Inverse of a Quotient}$$

For example,

$$\left(\frac{x+2}{x-2}\right)^{-1} = \left(\frac{x-2}{x+2}\right) \text{ if } x \neq 2 \text{ or } -2,$$

and

$$\left(-\frac{x}{7}\right)^{-1} = \left(\frac{-x}{7}\right)^{-1} = \frac{7}{-x} = -\frac{7}{x} \text{ if } x \neq 0.$$

We can also apply the zero-product property from Section R.2 to quotients. The quotient a/b is equal to the product $a \cdot b^{-1}$, and therefore a/b is zero if and only if $a \cdot b^{-1}$ is zero; but neither b nor b^{-1} can equal to zero, or a/b is not defined. We arrive at the following result.

$$\frac{a}{b} = 0 \Leftrightarrow a = 0 \text{ and } b \neq 0 \qquad \text{Zero-quotient Property}$$

For example,

$$\frac{2x-1}{2x+1} = 0 \Leftrightarrow 2x - 1 = 0 \text{ and } 2x + 1 \neq 0$$

$$\Leftrightarrow x = \frac{1}{2} \text{ and } x \neq -\frac{1}{2}$$

$$\Leftrightarrow x = \frac{1}{2}.$$

Example 2 Show that $(x-2)/(x^2-4)$ cannot equal 0 for any value of x.

Solution: Now

$$\frac{x-2}{x^2-4} = 0 \Leftrightarrow x - 2 = 0 \text{ and } x^2 - 4 \neq 0.$$

But it is impossible to have $x - 2 = 0$ and $x^2 - 4 \neq 0$ (*why?*). Therefore, $(x-2)/(x^2-4)$ can never equal 0. ∎

Algebraic Properties of Order. As stated in Section R.1, the real number a is greater than the real number b, denoted $a > b$ or $b < a$, if a is located to the right of b on the real number line. For example, $8 > 3$ and $-4 > -11$ (Figure 3). Note that $8 = 3 + 5$ and $-4 = -11 + 7$, where 5 and 7 are *positive* quantities. That is, adding a positive quantity increases the value of a number. From this we can deduce that our positional definition of greater than in Section R.1 is equivalent to the following algebraic definition.

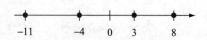

Figure 3 $8 > 3$ and $-4 > -11$

$$a > b \Leftrightarrow a = b + p, \text{ where p is positive.} \quad \text{Algebraic Definition of "Greater Than"}$$

Example 3 Show that $x + 5 > x + 2$ for any value of x.

Solution: We can write

$$\underbrace{x+5}_{a} = \underbrace{x+2}_{b} + \underbrace{3}_{p},$$

and since 3 is positive, it follows that $x + 5 > x + 2$ for any value of x. ■

The above definition can be used to prove the relationship "greater than" is preserved under addition (see Exercise 49). That is,

if c is any real number, then
$$a > b \Leftrightarrow a + c > b + c. \quad \text{Preservation Property for Addition}$$

Example 4 Show that $x - 6 > 4$ is equivalent to $x > 10$.

Solution:

$$x - 6 > 4 \Leftrightarrow x - 6 + 6 > 4 + 6 \quad \textit{Add 6 to both sides}$$
$$\Leftrightarrow x > 10 \quad ■$$

Question 1 *Is "greater than" preserved under subtraction? That is, is $a > b$ equivalent to $a - c > b - c$?*

Answer: Yes, since subtracting c is the same as adding $-c$. ■

If we multiply both sides of $8 > 5$ by the positive quantity 10, we obtain $10 \cdot 8 > 10 \cdot 5$ or $80 > 50$. On the other hand, if we multiply both sides by the *negative* quantity -10, we obtain $-10 \cdot 8 < -10 \cdot 5$ or $-80 < -50$. In general (see Exercise 50), we have the following results.

If c is any positive real number, then
$$a > b \Leftrightarrow a \cdot c > b \cdot c. \quad \text{Preservation Property for Multiplication}$$
If c is any negative real number, then
$$a > b \Leftrightarrow a \cdot c < b \cdot c. \quad \text{Reversal Property for Multiplication}$$

Example 5 Show that (a) $2x > 5$ is equivalent to $x > 5/2$ and (b) $-2x > 6$ is equivalent to $x < -3$.

Solution:

(a) $2x > 5 \Leftrightarrow 2x \cdot \dfrac{1}{2} > 5 \cdot \dfrac{1}{2}$ *Multiply both sides by the positive quantity 1/2.*

$\Leftrightarrow x > \dfrac{5}{2}$

(b) $-2x > 6 \Leftrightarrow -2x \left(\dfrac{1}{2}\right) < 6 \left(\dfrac{1}{2}\right)$ *Multiply both sides by the negative quantity $-1/2$ and change $>$ to $<$.*

$\Leftrightarrow x < -3$ ∎

Question 2 *Does the above multiplication property hold for division by c?*

Answer: Yes, since $1/c$ has the same sign as c and division by c is the same as multiplication by $1/c$. ∎

In addition to the above algebraic properties, the relationship "greater than" also has the following logical property (see Exercise 51).

If ***a > b*** and ***b > c,*** then ***a > c.***   Transitive Property

For example, if $x + 2 > y$ and $y > 7$, then $x + 2 > 7$. Compare the above algebraic and logical properties of "greater than" with those of "equality" in Section R.2.

We will use the notation ***a ≥ b*** to mean that a is greater than or equal to b. Similarly, ***b ≤ a*** means that b is less than or equal to a. The above properties hold if $>$ and $<$ are replaced by $\geq$ and $\leq$, respectively.

Comment The real number system has a nonalgebraic property (one that cannot be derived from the nine basic properties) related to "greater than" which, along with its algebraic properties, makes the real number system different from any other number system. This property, called the **completion axiom**, states that if some real number is greater than or equal to every number in a given nonempty subset of real numbers, then there is a smallest number that is greater than or equal to all the numbers in the given subset. For example, let the subset consist of all numbers $a_1, a_2, a_3, a_4, \ldots,$ where

$$a_1 = \frac{1}{1 \cdot 2},$$
$$a_2 = \frac{1}{1 \cdot 2} + \frac{1}{3 \cdot 4},$$
$$a_3 = \frac{1}{1 \cdot 2} + \frac{1}{3 \cdot 4} + \frac{1}{5 \cdot 6}$$
$$a_4 = \frac{1}{1 \cdot 2} + \frac{1}{3 \cdot 4} + \frac{1}{5 \cdot 6} + \frac{1}{7 \cdot 8},$$
$$\vdots$$

It can be shown that 1 is greater than or equal to every number in this subset. Therefore, there is a smallest real number that is greater than or equal to every number in the subset. The completion axiom is important for calculus, but it will not be needed in our study of algebra.

Absolute Value. As stated in Section R.1, the magnitude of a real number x without regard to its sign is called the absolute value of x and is denoted by $|x|$. If x is positive or 0, then $|x| = x$. However, if x is negative, then multiplying x by -1 will produce a positive quantity of the same magnitude. Since $(-1)x = -x$, we arrive at the following algebraic definition.

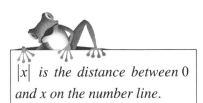

$|x|$ is the distance between 0 and x on the number line.

$$|x| = \begin{cases} x & \text{if } x \geq 0 \\ -x & \text{if } x < 0 \end{cases} \qquad \text{Definition of Absolute Value}$$

For example, $|-10| = -(-10) = 10$, and $|-2/3| = -(-2/3) = 2/3$.

The rules for multiplication and division with absolute values, which can be derived from the sign rules for multiplication and division, are as follows:

$$|ab| = |a||b| \qquad \text{and} \qquad \left|\frac{a}{b}\right| = \frac{|a|}{|b|}.$$

That is, the absolute value of a product equals the product of the absolute values, and the absolute value of a quotient equals the quotient of the absolute values. For example,

$$|-2x| = |-2||x| = 2|x| \qquad \text{and} \qquad \left|\frac{x}{-5}\right| = \frac{|x|}{|-5|} = \frac{|x|}{5}.$$

We note that

$$|b - a| = |(-1)(a - b)| = |-1||a - b| = |a - b|.$$

Geometrically, both $|b - a|$ and $|a - b|$ represent the **distance** between a and b on the real number line (Figure 4).

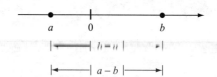

Figure 4 $|b - a| = |a - b|$

The rules for addition and subtraction with absolute values, which can be deduced from the sign rules for addition and subtraction, are as follows.

$$|a+b| \leq |a|+|b| \quad \text{and} \quad |a-b| \geq ||a|-|b|| \quad \text{Triangle Inequality}$$

The absolute value of a sum is less than or equal to the sum of the absolute values, but the absolute value of a difference is greater than or equal to the difference of the absolute values.

Equality of $|a + b|$ and $|a| + |b|$ occurs when either a or $b = 0$ or a and b have the same sign. Inequality occurs when they have opposite signs. The same holds true for $|a - b|$ and $||a| - |b||$. For example, equality occurs in both cases for $a = -10$ and $b = -5$, while inequality occurs if $a = -10$ and $b = 5$.

Question 3 *What is the meaning of the quotation at the beginning of this section?*

Answer: Although today we work freely with negative numbers, their acceptance by mathematicians was a gradual process. The ancient Greeks could not conceive of negative numbers; to Diophantus (ca. 275 A.D.) the equation $4x + 20 = 4$ had no solution. The Hindus realized that negative numbers could be used to signify debts or losses, but they had misgivings. Bhaskara (ca. 1150 A.D.) rejected -5 as a solution to an equation, stating that "people do not approve of negative numbers." Even by the seventeenth century negative numbers were not completely accepted. The French philosopher and mathematician René Descartes (1596-1650) referred to negative solutions of equations as "false" solutions. The statement by d'Alembert appeared in the French *Encyclopedia* in 1751 and shows that even at this late date mathematicians had reservations about negative numbers. ■

Exercises R.3

Fill in the blanks to make each statement true.

1. To compute $a + b$ when a and b are unequal and have opposite signs, we _____.
2. When there is a minus sign before parentheses, the parentheses can be removed by _____ of each term inside the parentheses.
3. The subtraction $a - b$ is equal to the addition _____.
4. The division a/b is equal to the multiplication _____.
5. The product of two nonzero terms having the same sign is always _____, and if the terms have opposite signs, their product is _____.

Write true *or* false *for each statement.*

6. If a and b have opposite signs but the same absolute value, then $a + b = 0$.
7. Since the minus sign is used to denote negative numbers, it follows that $-a$ is always a negative number.
8. If a is to the right of b on the number line, then $a - b$ is positive.
9. If a and b have opposite signs, then $|a + b| < |a| + |b|$.
10. If $|a| > |b|$, then $a > b$.

Addition, Subtraction, Multiplication, and Division

Evaluate each of the following.

11. $-7 + (-2)$
12. $-4 + (-4)$
13. $5 - (-8)$
14. $3 + (-8)$
15. $(-3)(-5)$
16. $\dfrac{24}{-6}$
17. $4(-2)^{-1}$
18. $(-6)3^{-1}$
19. $(-4)(-4)(-4)$
20. $(-3)^{-1}[-(-3)]$
21. $8 - (4 - 2)$
22. $-3 - (8 - 16)$
23. $(-2)[17 + (-3)]$
24. $(-3)(8 - 12)$
25. $\dfrac{-2(3 - 7)}{3(-4) + 4(-3)}$
26. $\dfrac{-3(4 - 12)}{3(12 - 4)}$

Express each of the following without parentheses or brackets.

27. $-(-x - 7)$
28. $-(-x - 3)$
29. $-(6 - y)$
30. $-[(-x) + (-4)]$
31. $-[(-x) - (-y)]$
32. $(-5)(-y)$
33. $(-4)b$
34. $(-x)[-(-3)]$
35. $[-(-5)](-y)$
36. $-(-1)[-(-y)]$

In each of the following, find all possible values of x that satisfy the equation.

37. $\dfrac{x - 1}{x - 2} = 0$
38. $\dfrac{x + 3}{x - 4} = 0$
39. $\dfrac{x + 5}{|x| + 5} = 0$
40. $\dfrac{x + 1}{|x| - 1} = 0$
41. $\dfrac{(|x| - 2)(|x| - 3)}{x - 3} = 0$
42. $\dfrac{(|x| - 4)(|x| - 5)}{(x + 4)(x + 5)} = 0$

Properties of Order

In Exercises 43-48, use the algebraic properties of order to complete each statement.

43. $x + 8 > x + 3$ because $x + 8 = (x + 3) + 5$, and 5 is _____.
44. $x + 5 > 2$ is equivalent to $x >$ _____.
45. $3x > 18$ is equivalent to $x >$ _____.
46. $-3x > 18$ is equivalent to x _____ -6.
47. $-2x > -6$ is equivalent to x _____.
48. $x > x - 1$ because _____.

49. Fill in the reason for each step in the following proof that "greater than" is preserved under addition.

$a > b$, and c is any real number.	Given
$a = b + p$, where p is positive.	_____
$a + c = (b + p) + c$	_____
$a + c = (b + c) + p$	_____
$a + c > b + c$	_____

50. The first proof below shows that "greater than" is preserved under multiplication by a positive quantity. The second proof shows that a negative multiplier changes "greater than" to "less than." Fill in the reason for each step.

    Given: $a > b$, and c is a *positive* real number. Then, by definition, $a = b + p$, where p is positive.

    (i) $ac = (b + p)c$ _____
    $ac = bc + pc$ _____
    pc is positive. _____
    $ac > bc$ _____

(ii) $a(-c) = (b+p)(-c)$ ———

$a(-c) = b(-c) + p(-c)$ ———

$a(-c) + pc = b(-c)$ ———

pc is positive. ———

$a(-c) < b(-c)$ ———

51. The transitive property of "greater than" says that if $a > b$ and $b > c$, then $a > c$. Fill in a reason for each step in the following proof of the transitive property.

$a > b$ and $b > c$	Given
$a = b + p$, where p is positive.	———
$b = c + q$, where q is positive.	———
$a = (c + q) + p$	———
$a = c + (q + p)$,	———
and $q + p$ is positive.	———
$a > c$	———

52. Use the algebraic definition of "greater than" to conclude that $0 > a \Leftrightarrow a$ is negative. [*Hint*: start with the equation $0 = a + (-a)$.]

Absolute Value

53. Show that $|-x| = |x|$ for any real number x.
54. Show that $a \leq |a|$ for any real number a. (*Hint*: consider separately the cases $a > 0$, $a = 0$, and $a < 0$.)
55. Show that $|a^{-1} + b^{-1}| \leq \dfrac{|a| + |b|}{|a||b|}$.
56. Show that $|a^{-1} + b^{-1}| \geq \dfrac{||a| - |b||}{|a||b|}$.
57. Show that $|a + b + c| \leq |a| + |b| + |c|$. [*Hint*: write $a + b + c = (a + b) + c$.]
58. Show that $|a - b| \geq |a| - |b|$ by writing $|a| = |a - b + b|$ and using the triangle inequality.

Miscellaneous

59. *Subtraction Is Not Commutative*. Show that if $a - b = b - a$, then $a = b$. (*Hint*: let $x = a - b$ and show that $-x = b - a$.)
60. *Subtraction Is Not Associative*. Show that if $(a - b) - c = a - (b - c)$, then $c = 0$.
61. *Definition of Subtraction*. Use the fact that $a - b = c$ means $c + b = a$ to arrive at the definition $a - b = a + (-b)$. That is, starting with the equation $c + b = a$, use the basic properties to conclude that $c = a + (-b)$.
62. *Definition of Division*. Use the fact that $a/b = c$ means $c \cdot b = a$ to arrive at definition $a/b = a \cdot b^{-1}$. That is, starting with the equation $c \cdot b = a$, use the basic properties to conclude that $c = a \cdot b^{-1}$.
63. Fill in a reason for each step in the proof of the following algebraic property of "greater than": if $a > b$ and $c > d$, then $a + c > b + d$.

$a > b$	and	$c > d$	Given
$a + c > b + c$	and	$b + c > b + d$	———
$a + c > b + d$			———

64. Fill in a reason for each step in the proof of the following algebraic property of "greater than": if $a > b > 0$ and $c > d > 0$, then $ac > bd$. Also, show by examples that the result does not hold in general if the condition $b > 0$ is removed or if the condition $d > 0$ is removed.

$a > b > 0$	and	$c > d > 0$	Given
$ac > bc$	and	$bc > bd$	———
$ac > bd$			———

R.4 Fractional Expressions

"After making the denominator of the divisor its numerator, the operation to be performed is as in multiplication."

—Rule for dividing fractions in a ninth-century work by the mathematician Mahāvira

Equality of Fractions
Addition and Subtraction
Multiplication and Division
Simplifying Complex Fractions

Equality of Fractions. We know from arithmetic that different ratios can have the same value. For example,

$$\frac{3}{4} = \frac{6}{8} = \frac{9}{12} = \cdots = \frac{3n}{4n}$$

for any integer n except 0. Similarly, if $b \neq 0$, then

$$\frac{a}{b} = \frac{ac}{bc}, \qquad \text{common factor property}$$

where c can be any real number except 0. This property follows from

$$\frac{ac}{bc} = ac(bc)^{-1} = acb^{-1}c^{-1} = ab^{-1}cc^{-1} = ab^{-1} = \frac{a}{b}.$$

In arithmetic, the common factor property is used to reduce rational numbers to lowest terms. For instance,

$$\frac{45}{60} = \frac{3 \cdot \cancel{15}}{4 \cdot \cancel{15}} = \frac{3}{4}$$

and

$$\frac{165}{462} = \frac{\cancel{3} \cdot 5 \cdot \cancel{11}}{2 \cdot \cancel{3} \cdot 7 \cdot \cancel{11}} = \frac{5}{14}.$$

Similarly, the property can be used to cancel common factors of the numerator and denominator of an algebraic fraction.

Example 1 (a) $\dfrac{3ax}{12xy} = \dfrac{\cancel{3}a\cancel{x}}{\cancel{3} \cdot 4\cancel{x}y} = \dfrac{a}{4y} \quad (x \neq 0, y \neq 0)$

(b) $\dfrac{ac}{ac+bc} = \dfrac{a\cancel{c}}{(a+b)\cancel{c}} = \dfrac{a}{a+b} \quad (c \neq 0, a+b \neq 0)$ ∎

Example 2 (a) $\dfrac{ax+3ay}{4bx+12by} = \dfrac{a\cancel{(x+3y)}}{4b\cancel{(x+3y)}} = \dfrac{a}{4b} \quad (b \neq 0, x+3y \neq 0)$

(b) $\dfrac{a-b}{b-a} = \dfrac{-1\cancel{(b-a)}}{1\cancel{(b-a)}} = \dfrac{-1}{1} = -1 \quad (b-a \neq 0)$ ∎

Caution It is very important to note that only common *factors* of the numerator and denominator can be canceled. For example, in the expression $a/(a+b)$, a is not a factor of the denominator and therefore cannot be canceled. That is,

$$\frac{a}{a+b} \neq \frac{1}{1+b};$$

for instance,
$$\frac{3}{3+7} \neq \frac{1}{1+7}.$$

For a to be canceled, a would have to be a factor of both terms in the denominator. That is, if $a \neq 0$, then
$$\frac{a}{a+ab} = \frac{\not{a}}{\not{a}(1+b)} = \frac{1}{1+b} \quad (1+b \neq 0);$$

for instance,
$$\frac{3}{3+3\cdot 7} = \frac{\not{3}}{\not{3}(1+7)} = \frac{1}{1+7}.$$

Similarly,
$$\frac{a+b}{b} \neq a+1,$$

whereas
$$\frac{ab+b}{b} = \frac{(a+1)\not{b}}{\not{b}}$$
$$= \frac{a+1}{1} = a+1 \quad (b \neq 0),$$

and
$$\frac{a+b}{a+c} \neq \frac{1+b}{1+c},$$

whereas $\quad \dfrac{a+ab}{a+ac} = \dfrac{\not{a}(1+b)}{\not{a}(1+c)} = \dfrac{1+b}{1+c} \quad (a \neq 0, 1+c \neq 0).$

In short, *if a quantity cannot be factored out, then it cannot be canceled.*

Another criterion for equality of fractions is

$$\frac{a}{b} = \frac{c}{d} \Leftrightarrow ad = bc \quad (b \neq 0, d \neq 0) \qquad \text{Equality of Fractions}$$

which follows from

$$\frac{a}{b} = \frac{c}{d} \Leftrightarrow ab^{-1} = cd^{-1} \Leftrightarrow ab^{-1}bd = cd^{-1}bd \Leftrightarrow ad = bc.$$

For example,

$$\frac{2x+1}{7} = \frac{3}{5} \Leftrightarrow (2x+1)\cdot 5 = 7\cdot 3$$
$$\Leftrightarrow 10x+5 = 21.$$

Addition and Subtraction. For two fractions with the same denominator, the respective procedures for addition and subtraction are

$$\frac{a}{b} + \frac{c}{b} = \frac{a+c}{b} \quad \text{and} \quad \frac{a}{b} - \frac{c}{b} = \frac{a-c}{b}, \quad \text{Addition and Subtraction, Common Denominator}$$

where $b \neq 0$. For example,

$$\frac{3}{5} + \frac{7}{5} = \frac{10}{5}$$

and

$$\frac{x}{x-1} - \frac{2}{x-1} = \frac{x-2}{x-1} \quad (x - 1 \neq 0).$$

The rule for addition follows from

$$\frac{a}{b} + \frac{c}{b} = ab^{-1} + cb^{-1} = (a+c)b^{-1} = \frac{a+c}{b},$$

and the subtraction rule is similarly derived.

Caution There are no corresponding rules for adding or subtracting fractions with the same numerators. In general,

$$\frac{a}{b} + \frac{a}{c} \neq \frac{a}{b+c} \quad \text{and} \quad \frac{a}{b} - \frac{a}{c} \neq \frac{a}{b-c}.$$

(See Section 1.1, Exercise 101.) For example,

$$\frac{5}{3} + \frac{5}{7} \neq \frac{5}{10}$$

and

$$\frac{|x|+1}{x} - \frac{|x|+1}{2} \neq \frac{|x|+1}{x-2}.$$

Two of the most common mistakes in algebra are to equate $a/(b+c)$ with $a/b + a/c$, and $a/(b-c)$ with $a/b - a/c$. Be sure to avoid making these errors.

The general procedures for adding and subtracting fractions whose denominators are not necessarily equal are

$$\frac{a}{b} + \frac{c}{d} = \frac{ad+bc}{bd} \quad \text{and} \quad \frac{a}{b} - \frac{c}{d} = \frac{ad-bc}{bd}, \quad \text{Addition and Subtraction, General Rule}$$

where $b \neq 0$ and $d \neq 0$. The addition rule follows from

$$\frac{a}{b} + \frac{c}{d} = \frac{ad}{bd} + \frac{bc}{bd} = \frac{ad+bc}{bd},$$

and the subtraction rule is derived in a similar way.

Example 3 (a) $\dfrac{3}{2} + \dfrac{5}{x} = \dfrac{3x+10}{2x}$ $\qquad (x \neq 0)$

(b) $\dfrac{2}{x} + \dfrac{4}{x+1} = \dfrac{2(x+1) + 4x}{x(x+1)} = \dfrac{6x+2}{x(x+1)}$ $\quad (x \neq 0, x+1 \neq 0)$

Example 4 (a) $\dfrac{x}{2} - \dfrac{x}{3} = \dfrac{3x - 2x}{6} = \dfrac{x}{6}$ ∎

(b) $\dfrac{3}{x+1} - \dfrac{2}{x+2} = \dfrac{3(x+2) - 2(x+1)}{(x+1)(x+2)}$

$\qquad\qquad\qquad = \dfrac{x+4}{(x+1)(x+2)}$ $\quad (x+1 \neq 0, x+2 \neq 0)$ ∎

Multiplication and Division. The procedure used for multiplying fractions is

$$\frac{a}{b} \cdot \frac{c}{d} = \frac{ac}{bd}, \qquad \text{Multiplication}$$

where $b \neq 0$ and $d \neq 0$. That is, the product of two fractions is equal to the product of their numerators divided by the product of their denominators. This follows from

$$\frac{a}{b} \cdot \frac{c}{d} = ab^{-1} \cdot cd^{-1} = ac \cdot b^{-1}d^{-1} = ac \cdot (bd)^{-1} = \frac{ac}{bd}.$$

For instance,

$$\frac{3}{2} \cdot \frac{x}{x+1} = \frac{3x}{2(x+1)}$$

Multiply both the numerator fraction and the denominator fraction by $\dfrac{d}{c}$ to get the desired result.

and $\quad \dfrac{4}{x+1} \cdot \dfrac{5}{x-2} = \dfrac{20}{(x+1)(x-2)} \quad (x+1 \neq 0, x-2 \neq 0).$

To divide one fraction by another, the procedure is

$$\frac{\frac{a}{b}}{\frac{c}{d}} = \frac{a}{b} \cdot \frac{d}{c}, \qquad \text{Division}$$

where $b \neq 0$, $c \neq 0$, and $d \neq 0$. This is so because

$$\frac{\frac{a}{b}}{\frac{c}{d}} = \frac{a}{b} \cdot \left(\frac{c}{d}\right)^{-1} = \frac{a}{b} \cdot \frac{d}{c}.$$

Hence, to divide two fractions, *invert the fraction in the denominator and proceed as in multiplication.*

Example 5 (a) $\dfrac{\frac{1}{x}}{\frac{1}{y}} = \dfrac{1}{x} \cdot \dfrac{y}{1} = \dfrac{y}{x}$ $(x \neq 0, y \neq 0)$; for instance, $\dfrac{\frac{1}{2}}{\frac{1}{3}} = \dfrac{1}{2} \cdot \dfrac{3}{1} = \dfrac{3}{2}$.

(b) $\dfrac{\frac{x+1}{x}}{\frac{x-3}{x+5}} = \dfrac{x+1}{x} \cdot \dfrac{x+5}{x-3}$

$= \dfrac{(x+1)(x+5)}{x(x-3)}$ $(x \neq 0, x - 3 \neq 0, x + 5 \neq 0)$ ∎

Just as any integer n can be written as a rational number $n/1$, so can any algebraic expression x be written as a fractional expression $x/1$. We use this property in divisions that involve fractional and non-fractional expressions, as is illustrated in the following example, where $b \neq 0$ and $c \neq 0$.

Example 6 (a) $\dfrac{\frac{a}{b}}{c} = \dfrac{\frac{a}{b}}{\frac{c}{1}} = \dfrac{a}{b} \cdot \dfrac{1}{c} = \dfrac{a}{bc}$; for instance, $\dfrac{\frac{2}{3}}{5} = \dfrac{2}{15}$.

(b) $\dfrac{a}{\frac{b}{c}} = \dfrac{\frac{a}{1}}{\frac{b}{c}} = \dfrac{a}{1} \cdot \dfrac{c}{b} = \dfrac{ac}{b}$; for instance, $\dfrac{2}{\frac{3}{5}} = \dfrac{10}{3}$. ∎

Simplifying Complex Fractions. Ratios of algebraic combinations of fractions, such as

$$\dfrac{\frac{a}{b} + \frac{c}{d}}{\frac{a}{b} - \frac{c}{d}} \quad \text{and} \quad \dfrac{a^{-1} + b^{-1}}{a^{-1} - b^{-1}}$$

are called **complex fractions**. These can be simplified by applying the previous rules for addition, subtraction, multiplication, division, and cancellation. For example,

$$\dfrac{\frac{a}{b} + \frac{c}{d}}{\frac{a}{b} - \frac{c}{d}} = \dfrac{\frac{ad+bc}{bd}}{\frac{ad-bc}{bd}} = \dfrac{ad+bc}{bd} \cdot \dfrac{bd}{ad-bc} = \dfrac{ad+bc}{ad-bc}.$$

The above result can also be obtained by multiplying the numerator and denominator of the original complex fraction by bd. Also,

$$\frac{a^{-1}+b^{-1}}{a^{-1}-b^{-1}} = \frac{\dfrac{1}{a}+\dfrac{1}{b}}{\dfrac{1}{a}-\dfrac{1}{b}}$$

$$= \frac{\dfrac{b+a}{ab}}{\dfrac{b-a}{ab}} = \frac{b+a}{ab} \cdot \frac{ab}{b-a} = \frac{b+a}{b-a}.$$

Example 7
$$\frac{a+\dfrac{b}{c}}{\dfrac{a}{c}+b} = \frac{\dfrac{a}{1}+\dfrac{b}{c}}{\dfrac{a}{c}+\dfrac{b}{1}}$$

$$= \frac{\dfrac{ac+b}{c}}{\dfrac{a+bc}{c}} = \frac{ac+b}{c} \cdot \frac{c}{a+bc} = \frac{ac+b}{a+bc} \blacksquare$$

Example 8
$$\frac{a+b^{-1}}{a^{-1}+b} = \frac{\dfrac{a}{1}+\dfrac{1}{b}}{\dfrac{1}{a}+\dfrac{b}{1}}$$

$$= \frac{\dfrac{ab+1}{b}}{\dfrac{1+ab}{a}} = \frac{ab+1}{b} \cdot \frac{a}{1+ab} = \frac{a}{b} \blacksquare$$

Example 9
$$\frac{\dfrac{1}{x+1}+\dfrac{1}{x+2}}{\dfrac{1}{x+1}-\dfrac{1}{x+2}} = \frac{\dfrac{x+2+x+1}{(x+1)(x+2)}}{\dfrac{x+2-(x+1)}{(x+1)(x+2)}}$$

$$= \frac{\dfrac{2x+3}{(x+1)(x+2)}}{\dfrac{1}{(x+1)(x+2)}}$$

$$= \frac{2x+3}{(x+1)(x+2)} \cdot \frac{(x+1)(x+2)}{1} = 2x+3 \blacksquare$$

Comment Equalities obtained by simplifying algebraic expressions are **conditional equalities**. For instance, in Example 9, the simplified expression $2x+3$ is meaningful for all values of x, but $1/(x+1)$ and $1/(x+2)$ are meaningful only if $x \neq -1$ or -2, respectively. Hence, the equality is true for all x except -1 and -2. In general, *when we say that two algebraic expressions in x are equal, we mean that equality holds for all values of x for which both expressions are meaningful*. Also, whether stated explicitly or not, it is understood that the denominator of a fractional expression is restricted to nonzero values. See if you can determine the corresponding restrictions placed on a, b, and c in each of the above complex fractions.

Exercises R.4

Fill in the blanks to make each statement true.

1. If $a/b = 0$, then $a =$ _____, and $b \neq$ _____.
2. If $a/b = 1$, then $a =$ _____, and $b \neq$ _____.
3. The sum $a/b + c/b$ is equal to _____, whereas $a/b + a/c =$ _____.
4. The product of two fractions is equal to the product of their _____ divided by the product of their _____.
5. To divide one fraction by another, _____ the fraction in the denominator and proceed as in _____.

Write true *or* false *for each statement.*

6. $\dfrac{ac}{b+c} = \dfrac{a}{b+1}$, provided $b + c \neq 0$ and $b + 1 \neq 0$.
7. $\dfrac{1}{a+b} = \dfrac{1}{a} + \dfrac{1}{b}$, provided $a \neq 0$, $b \neq 0$, and $a + b \neq 0$.
8. $\dfrac{1}{a^{-1} + b^{-1}} = a + b$ if $a \neq 0$ and $b \neq 0$.
9. $\dfrac{a}{b} \cdot \dfrac{c}{d} = ab^{-1}cd^{-1}$ if $b \neq 0$ and $d \neq 0$.
10. $\dfrac{\frac{a}{b}}{\frac{c}{d}} = ab^{-1}c^{-1}d$, provided $b \neq 0$, $c \neq 0$, and $d \neq 0$.

Equality of Fractions

In each of the following, find all possible values of x for which the equality holds.

11. $\dfrac{3}{2} = \dfrac{x}{2}$.
12. $\dfrac{x}{15} = \dfrac{1}{5}$
13. $\dfrac{x-7}{2} = 1$
14. $\dfrac{1}{x+4} = 1$
15. $\dfrac{(x-2)(x-3)}{x-3} = 0$
16. $\dfrac{x-1}{x+1} = 1$

Simplify each of the following by canceling all factors common to the numerator and denominator. State any restrictions on a, b, c, x, y, or z.

17. $\dfrac{2ab}{6a}$
18. $\dfrac{-6axy}{8y}$
19. $\dfrac{3a + 6ab}{6ab}$
20. $\dfrac{x-2}{2-x}$
21. $\dfrac{ab - ac}{ab + ac}$
22. $\dfrac{2x+3}{4x+6}$
23. $\dfrac{2x - 3y + 4z}{4ax - 6ay + 8az}$
24. $\dfrac{2ax - 2x}{2ax + 2x}$
25. $\dfrac{ax - ay + bx - by}{x - y}$

Operations with Fractions

Perform each of the following additions and subtractions. State any restrictions on x.

26. $\dfrac{4}{x} + \dfrac{2}{x}$
27. $\dfrac{3}{x+1} + \dfrac{1}{x+1}$
28. $\dfrac{3}{x-3} + \dfrac{2}{3-x}$
29. $\dfrac{x}{2x-1} - \dfrac{x}{1-2x}$
30. $\dfrac{x+3}{x+2} - \dfrac{1}{x+2}$
31. $\dfrac{3x}{3x-1} - \dfrac{1}{3x-1}$
32. $\dfrac{x-2}{(2x+1)(x+3)} + \dfrac{x+2}{(2x+1)(x+3)}$
33. $\dfrac{x+5}{3x+1} - \dfrac{x-2}{3x+1} + \dfrac{2x-1}{3x+1}$
34. $\dfrac{x}{x-1} + \dfrac{3}{x-1} + \dfrac{2}{1-x}$
35. $\dfrac{3}{x(x+1)} + \dfrac{4x+2}{x(x+1)} - \dfrac{x+2}{x(x+1)}$

Perform each of the following additions and subtractions. State any restrictions on x, a, or b.

36. $\dfrac{x}{3} + \dfrac{1}{5}$
37. $\dfrac{x}{3} + \dfrac{x}{5}$
38. $\dfrac{3}{x} - \dfrac{2}{x+1}$
39. $\dfrac{x+1}{2} - \dfrac{x}{3}$
40. $\dfrac{1}{x+1} + \dfrac{1}{x-1}$
41. $1 + \dfrac{1}{x}$

42. $5 - \dfrac{2}{x+1}$

43. $\dfrac{4}{x+2} + \dfrac{3}{x-2}$

44. $\dfrac{1}{2x+3} - \dfrac{1}{3x+2}$

45. $x + \dfrac{2x-3}{5}$

46. $\dfrac{1}{x+a} + \dfrac{1}{x-a}$

47. $\dfrac{1}{x-a} - \dfrac{1}{x-a}$

48. $\dfrac{1}{ax-1} - \dfrac{1}{ax-2}$

49. $\dfrac{a}{2x+a} + \dfrac{b}{2x-b}$

50. $\dfrac{a}{ax+2} - \dfrac{b}{bx+2}$

Perform each of the following multiplications. Simplify answers. State any restrictions on a, b, c, or x.

51. $\dfrac{a}{8} \cdot \dfrac{2}{a}$

52. $\dfrac{3b}{4a} \cdot \dfrac{2c}{9x}$

53. $2a \cdot \dfrac{3}{4}$

54. $a \cdot \dfrac{2}{a}$

55. $\dfrac{2x+3}{6} \cdot \dfrac{3}{4x+6}$

56. $\dfrac{2}{x} \cdot \dfrac{x+3}{6}$

57. $\dfrac{4}{2x-1} \cdot \dfrac{10x-5}{16}$

58. $\dfrac{x-1}{x+2} \cdot \dfrac{x+2}{x-2}$

59. $\dfrac{2x-1}{2x-6} \cdot \dfrac{x-3}{4x-2}$

60. $\dfrac{10x}{5x+10} \cdot \dfrac{2x+1}{2x}$

Perform each of the following divisions. Simplify answers. State any restrictions on a, b, c, d, or x.

61. $\dfrac{\frac{a}{8}}{\frac{a}{2}}$

62. $\dfrac{\frac{a}{2b}}{\frac{a}{4}}$

63. $\dfrac{\frac{3a}{4b}}{\frac{2c}{9d}}$

64. $\dfrac{\frac{9a}{4b}}{\frac{3a}{10b}}$

65. $\dfrac{\frac{5a}{a}}{5}$

66. $\dfrac{\frac{10}{2x+3}}{\frac{4}{10x+5}}$

67. $\dfrac{\frac{10}{2x+3}}{5}$

68. $\dfrac{\frac{10}{2x+3}}{\frac{5}{4}}$

69. $\dfrac{\frac{x+1}{8}}{\frac{4}{x+1}}$

70. $\dfrac{\frac{4}{x+1}}{\frac{x+1}{8}}$

71. $\dfrac{\frac{x-1}{x-2}}{\frac{x+2}{x-2}}$

72. $\dfrac{\frac{3}{x-2}}{\frac{6}{x-2}}$

73. $\dfrac{\frac{2x}{x+1}}{\frac{3x}{4x+4}}$

74. $\dfrac{\frac{3x-5}{6x-10}}{\frac{x}{5x}}$

75. $\dfrac{\frac{4x+2}{3x-6}}{\frac{2x+1}{4x-8}}$

Complex Fractions

Simplify each expression. State any restrictions on a, b, c, or x.

76. $\dfrac{\dfrac{a}{b+c} + \dfrac{b}{b+c}}{\dfrac{c}{b+c}}$

77. $\dfrac{\dfrac{a-b}{a+b}}{\dfrac{a}{a+b} - \dfrac{b}{a+b}}$

78. $\dfrac{\dfrac{1}{a} - \dfrac{1}{b}}{\dfrac{1}{b} - \dfrac{1}{a}}$

79. $\dfrac{a + \dfrac{1}{b}}{b + \dfrac{1}{a}}$

80. $\dfrac{\dfrac{a}{b} - 1}{\dfrac{a}{b} + 1}$

81. $\dfrac{\dfrac{1}{x+3} + \dfrac{1}{x+4}}{\dfrac{1}{x+3} - \dfrac{1}{x+4}}$

82. $\dfrac{\dfrac{1}{x+3} - \dfrac{1}{x+4}}{\dfrac{1}{x+4} - \dfrac{1}{x+3}}$

83. $\dfrac{1 + \dfrac{1}{x+2}}{2 + \dfrac{1}{x+1}}$

84. $\dfrac{1 - \dfrac{1}{x+1}}{2 - \dfrac{1}{x+2}}$

85. $\dfrac{\dfrac{x}{x-1} - \dfrac{1}{1-x}}{\dfrac{x+1}{x-1}}$

Express each of the following without the superscript −1 and simplify. State any restrictions on a, b, c, or x.

86. $\dfrac{a}{b^{-1} + c^{-1}}$

87. $\dfrac{a^{-1}}{b^{-1} + c^{-1}}$

88. $\dfrac{a^{-1} + b^{-1}}{c}$

89. $\dfrac{a + b^{-1}}{c^{-1}}$

90. $\dfrac{a^{-1} + b^{-1}}{a^{-1} - b^{-1}}$

91. $\dfrac{\dfrac{1}{a^{-1}} + \dfrac{1}{b^{-1}}}{\dfrac{1}{a^{-1}} - \dfrac{1}{b^{-1}}}$

92. $\dfrac{a + \dfrac{b^{-1}}{c}}{a - \dfrac{b^{-1}}{c}}$

93. $\dfrac{a + \dfrac{b^{-1}}{c}}{c + \dfrac{b^{-1}}{a}}$

94. $\dfrac{(x+1)^{-1}}{(x+2)^{-1} + (x+3)^{-1}}$

95. $\dfrac{(x+1)^{-1} + (x+2)^{-1}}{(x+3)^{-1}}$

96. $\dfrac{1 + (x-1)^{-1}}{2 + (x-1)^{-1}}$

97. $\dfrac{(x-1)^{-1} + (x-2)^{-1}}{(x-1)^{-1} - (x-2)^{-1}}$

Miscellaneous

98. What values of a, b, and c are excluded in the equation of Example 7 of the text? What values of a and b are excluded in Example 8?

99. Prove the subtraction rule: $\dfrac{a}{b} - \dfrac{c}{d} = \dfrac{ad - bc}{bd}$ ($b \neq 0$, $d \neq 0$).

100. Prove each of the following.

(a) $\dfrac{a}{a+b} = \dfrac{1}{1+b} \Leftrightarrow a = 1$ or $b = 0$ (Assume $a \neq -b$ and $b \neq -1$.)

(b) $\dfrac{a+b}{a+c} = \dfrac{1+b}{1+c} \Leftrightarrow a = 1$ or $b = c$ (Assume $a \neq -c$ and $c \neq -1$.)

R.5 Powers and Roots

"Though people do a great deal of talking, the total output since the beginning of gabble to the present day, including all baby talk, love songs, and Congressional debates, totals about 10^{16}."

—*from* Mathematics and the Imagination *by Edward Kasner and James R. Newman (1956)*

Positive Exponents
Zero and Negative Exponents
Roots
Algebraic Expressions

Positive Exponents. A sum such as $2+2+2+2+2$ can be written more compactly as $5 \cdot 2$. Similarly, we write the product $2 \cdot 2 \cdot 2 \cdot 2 \cdot 2$ in more compact form as 2^5, where the **exponent** 5 indicates the number of times that the **base** 2 occurs as a factor. Thus, $3^4 = 3 \cdot 3 \cdot 3 \cdot 3 = 81$, and $(-6)^3 = (-6)(-6)(-6) = -216$. In general, if x stands for a real number and n is a positive integer, then we define the nth power of x as follows.

$$x^n = \underbrace{x \cdot x \cdot x \cdot \ldots \cdot x}_{n \text{ factors}} \qquad n\text{th power of } x$$

We note that $3^2 \cdot 3^5 = (3 \cdot 3)(3 \cdot 3 \cdot 3 \cdot 3 \cdot 3) = 3^7$, whereas $(3^2)^5 = (3 \cdot 3)(3 \cdot 3)(3 \cdot 3)(3 \cdot 3)(3 \cdot 3) = 3^{10}$. In general, if m and n are positive integers, then

$$x^m x^n = x^{m+n} \qquad \text{Same-base Rule (multiplication)}$$

and

$$(x^m)^n = x^{mn}. \qquad \text{Power-of-a-Power Rule}$$

Hence, to multiply powers of the same base, you *add* the exponents, and to raise a power of x to another power, you *multiply* the exponents. These rules can be extended to more than two exponents. For instance,

$$x^2 \cdot x^3 \cdot x^4 = x^{2+3+4} = x^9 \quad \text{and} \quad [(x^2)^3]^4 = x^{2 \cdot 3 \cdot 4} = x^{24}.$$

Example 1 (a) $3^5 \cdot 2^4 \cdot 3^2 \cdot 2^6 = 3^5 \cdot 3^2 \cdot 2^4 \cdot 2^6 = 3^7 \cdot 2^{10}$
(b) $(-3)^{2n} = [(-3)^2]^n = 9^n$
(c) $(x^2)^5 (x^3)^4 = x^{10} \cdot x^{12} = x^{22}$

We note that

$$(3 \cdot 2)^4 = (3 \cdot 2)(3 \cdot 2)(3 \cdot 2)(3 \cdot 2)$$
$$= (3 \cdot 3 \cdot 3 \cdot 3)(2 \cdot 2 \cdot 2 \cdot 2)$$
$$= 3^4 \cdot 2^4. \quad \blacksquare$$

In general, if n is any positive integer, then

$$(xy)^n = x^n y^n. \quad \text{Same-exponent Rule (multiplication)}$$

That is, the nth power of a product is equal to the product of the nth powers of the factors. The rule extends to more than two base factors, as in

$$(xyz)^4 = x^4 y^4 z^4.$$

Example 2 (a) $(2y)^3 = 2^3 y^3 = 8y^3$
(b) $(-x)^n = [(-1)x]^n = (-1)^n x^n = \begin{cases} x^n & \text{if } n \text{ is even} \\ -x^n & \text{if } n \text{ is odd} \end{cases}$ $\blacksquare$

Example 3 Express $(2^5 4^3)^2$ as a power of 2.

Solution:

$$(2^5 4^3)^2 = (2^5)^2 (4^3)^2 = 2^{10} 4^6 = 2^{10}(2^2)^6 = 2^{10} 2^{12} = 2^{22} \quad \blacksquare$$

The product of fractions is equal to the product of the numerators divided by the product of the denominators. For instance,

$$\left(\frac{3}{2}\right)^4 = \frac{3}{2} \cdot \frac{3}{2} \cdot \frac{3}{2} \cdot \frac{3}{2} = \frac{3 \cdot 3 \cdot 3 \cdot 3}{2 \cdot 2 \cdot 2 \cdot 2} = \frac{3^4}{2^4}.$$

In general, if n is any positive integer and $y \neq 0$, then

$$\left(\frac{x}{y}\right)^n = \frac{x^n}{y^n}. \quad \text{Same-exponent Rule (division)}$$

That is, the nth power of a quotient equals the quotient of the nth powers.

Example 4 (a) $\left(\dfrac{x}{2}\right)^6 = \dfrac{x^6}{2^6} = \dfrac{x^6}{64}$

(b) $\left(\dfrac{2x+1}{3x+4}\right)^2 = \dfrac{(2x+1)^2}{(3x+4)^2}$, provided $3x+4 \neq 0$ ∎

Common factors of the numerator and denominator of a fraction can be canceled without changing the value of the fraction. For instance,

$$\dfrac{3^6}{3^2} = \dfrac{3 \cdot 3 \cdot 3 \cdot 3 \cdot \cancel{3} \cdot \cancel{3}}{\cancel{3} \cdot \cancel{3}} = 3^4 = 3^{6-2}.$$

More generally, if $x \neq 0$ and $m > n$, then

$$\dfrac{x^m}{x^n} = x^{m-n}. \quad \text{Same-base Rule (division)}$$

Hence, to divide powers of the same base, we *subtract* the exponents. Furthermore, the restriction $m > n$ can be removed with the introduction of zero and negative exponents, as we shall soon see.

Example 5 (a) $\dfrac{(4x-5)^5}{(4x-5)^3} = (4x-5)^{5-3} = (4x-5)^2$, provided $4x-5 \neq 0$

(b) $\dfrac{(x^2+1)^4}{x^2+1} = \dfrac{(x^2+1)^4}{(x^2+1)^1} = (x^2+1)^{4-1} = (x^2+1)^3$ ∎

Zero and Negative Exponents. If we formally apply the same-base rule for quotients to the case $m = n$, we obtain

(a) $\dfrac{x^n}{x^n} = x^{n-n} = x^0 \ (x \neq 0)$.

But (b) $\dfrac{x^n}{x^n} = 1 \ (x \neq 0)$.

Therefore, in order to make (a) consistent with (b), we are led to the following definition. If $x \neq 0$, then

$$x^0 = 1.$$

That is, any nonzero quantity raised to the zero power is 1 by definition. The expression 0^0 is undefined.

Example 6 (a) $7^0 = 1, (-5)^0 = 1$, and $\left(\dfrac{3}{4}\right)^0 = 1$

(b) $(x+12)^0 = 1$, provided $x+12 \neq 0$ ∎

If we formally apply the same-base rule to the expression x^2/x^6, we obtain

(a) $\dfrac{x^2}{x^6} = x^{2-6} = x^{-4}$ $(x \neq 0)$.

But (b) $\dfrac{x^2}{x^6} = \dfrac{x \cdot x}{x \cdot x \cdot x \cdot x \cdot x \cdot x} = \dfrac{1}{x^4}$ $(x \neq 0)$.

In order to make (a) consistent with (b), we are led to the following definition. If n is a positive integer and $x \neq 0$, then by definition

$$x^{-n} = \dfrac{1}{x^n}.$$

That is, x^{-n} is the reciprocal of x^n. The expression 0^{-n} is undefined.

Example 7 (a) $2^{-3} = \dfrac{1}{2^3} = \dfrac{1}{8}$

(b) $(3x+4)^{-2} = \dfrac{1}{(3x+4)^2}$, provided $3x+4 \neq 0$ ∎

Comment In Section R.2, we defined x^{-1} as the multiplicative inverse of x. This agrees with our new definition of x^{-n} when $n = 1$.

Three equivalent interpretations of x^{-n}, which can be derived from our definitions, are

$$x^{-n} = \begin{cases} 1/x^n \\ (x^n)^{-1} \\ (x^{-1})^n \end{cases}.$$

Example 8 (a) $4^{-3} = \dfrac{1}{4^3} = \dfrac{1}{64}$

(b) $4^{-3} = (4^3)^{-1} = 64^{-1} = \dfrac{1}{64}$

(c) $4^{-3} = (4^{-1})^3 = \left(\dfrac{1}{4}\right)^3 = \dfrac{1}{64}$ ∎

It can be shown that *all of the previous exponent rules for products, quotients, and powers hold for all integers m and n, positive, negative, or zero, provided the base is not* 0 *when the exponent is* 0 *or negative*. In particular, we can rewrite the same-base rule for division as

$$\dfrac{x^m}{x^n} = x^{m-n} = \dfrac{1}{x^{n-m}} \ (x \neq 0),$$

with no restrictions placed on m and n. The following examples illustrate how the exponent rules are used to simplify algebraic expressions.

Example 9 (Same-Base Rule)
If $x \neq 0$, then

(a) $x^5 x^{-2} = x^{5+(-2)} = x^3$;

(b) $\dfrac{x^{-5}}{x^{-2}} = x^{-5-(-2)} = x^{-3} = \dfrac{1}{x^3}$ or $\dfrac{x^{-5}}{x^{-2}} = \dfrac{1}{x^{-2-(-5)}} = \dfrac{1}{x^3}$;

(c) $\dfrac{1}{x^{-2}} = \dfrac{x^0}{x^{-2}} = x^{0-(-2)} = x^2$ [see Example 10(c)]. ∎

Example 10 (Power-of-a-Power Rule)

If $a \neq 0$, then

(a) $(a^{-3})^4 = a^{(-3)4} = a^{-12} = \dfrac{1}{a^{12}}$;

(b) $(a^{-3})^{-4} = a^{(-3)(-4)} = a^{12}$;

(c) $\dfrac{1}{a^{-2}} = (a^{-2})^{-1} = a^{(-2)(-1)} = a^2$. ∎

Example 11 (Same-Exponent Rule)

If $a \neq 0$, then

(a) $(2a^{-1})^3 = 2^3 (a^{-1})^3 = 2^3 a^{-3} = \dfrac{8}{a^3}$;

(b) $\left(\dfrac{1}{a^{-3}}\right)^{-2} = \dfrac{1^{-2}}{(a^{-3})^{-2}} = \dfrac{1}{a^6}$. ∎

Example 12 (Miscellaneous)

If $a \neq 0$, $x \neq 0$, and $y \neq 0$, then

(a) $(3^2 a^3 x)(3^{-2} a^{-1} x^{-4}) = 3^2 3^{-2} \cdot a^3 a^{-1} \cdot x x^{-4}$
$= 3^{2+(-2)} \cdot a^{3+(-1)} \cdot x^{1+(-4)}$
$= 3^0 \cdot a^2 \cdot x^{-3}$
$= \dfrac{a^2}{x^3}$;

(b) $\left(\dfrac{a^{-1}}{y}\right)^2 \cdot \dfrac{x^3 y^{-3}}{a^{-4} x^{-2}} = \dfrac{a^{-2}}{y^2} \cdot \dfrac{x^3 y^{-3}}{a^{-4} x^{-2}}$
$= \dfrac{a^{-2}}{a^{-4}} \cdot \dfrac{x^3}{x^{-2}} \cdot \dfrac{y^{-3}}{y^2}$
$= a^{-2-(-4)} \cdot x^{3-(-2)} \cdot y^{-3-2}$
$= a^2 \cdot x^5 \cdot y^{-5}$
$= \dfrac{a^2 x^5}{y^5}$. ∎

Example 13 (Complex Fractions)

If $x \neq 0$, $y \neq 0$, $a \neq 0$, and $b \neq 0$, then

(a) $\dfrac{x^{-2} + y^{-3}}{y^{-2}} = \dfrac{\dfrac{1}{x^2} + \dfrac{1}{y^3}}{\dfrac{1}{y^2}} = \dfrac{\dfrac{y^3 + x^2}{x^2 y^3}}{\dfrac{1}{y^2}}$

$= \dfrac{y^3 + x^2}{x^2 y^3} \cdot \dfrac{y^2}{1} = \dfrac{x^2 + y^3}{x^2 y}$;

(b) $\dfrac{a^2 + b^{-3}}{a^{-2} + b^3} = \dfrac{a^2 + \dfrac{1}{b^3}}{\dfrac{1}{a^2} + b^3} = \dfrac{\dfrac{a^2 b^3 + 1}{b^3}}{\dfrac{1 + a^2 b^3}{a^2}}$

$= \dfrac{a^2 b^3 + 1}{b^3} \cdot \dfrac{a^2}{1 + a^2 b^3} = \dfrac{a^2}{b^3}.$ ∎

Roots. If $y^2 = x$, then y is called a **square root** of x, and if $y^3 = x$, then y is called a **cube root** of x. In general, a number y is called an **nth root** of x if $y^n = x$ ($n = 2, 3, 3, \ldots$). Hence, 3 and -3 are fourth roots of 81 since $3^4 = 81$ and $(-3)^4 = 81$. Similarly, -2 is a fifth root of -32 since $(-2)^5 = -32$.

When n is odd, every real number has exactly one real nth root, which is denoted by $\sqrt[n]{x}$. For example, $\sqrt[5]{-32} = -2$. When n is even, it is "double or nothing"; that is, every *positive* real number will have *two* real nth roots, one the negative of the other, but a *negative* number has *no* even roots in the real number system. For example, 81 has the fourth roots 3 and -3, but -81 has no real fourth roots. The nth root of 0 is 0 itself for both even and odd n.

The symbol $\sqrt[n]{x}$ is also used to denote the *positive* nth root when n is even and x is positive. For example, $\sqrt[4]{81} = 3$. In general, the expression $\sqrt[n]{x}$ is called a **radical**, $\sqrt{}$ is the **radical sign**, n is the **index**, and x is the **radicand**. For square roots, the radical sign is written without an index.

Example 14 (a) $\sqrt[4]{16} = 2$, but $\sqrt[4]{-16}$ does not exist as a real number.

(b) $\sqrt[5]{32} = 2$ and $\sqrt[5]{-32} = -2$ ∎

Example 15 (a) $\sqrt{3^2} = \sqrt{9} = 3$ and $\sqrt{(-3)^2} = \sqrt{9} = 3$

(b) $\sqrt[3]{3^3} = \sqrt[3]{64} = 4$ and $\sqrt[3]{(-4)^3} = \sqrt[3]{-64} = -4$ ∎

Example 15 illustrates an important difference between even and odd roots. When n is even, a^n is positive whether a is positive or negative, and $\sqrt[n]{a^n}$ means the positive nth root of a^n. Therefore

$$\sqrt[n]{a^n} = |a| \quad \text{when } n \text{ is even}.$$

On the other hand, when n is odd, the nth root of a^n is a itself, whether a is positive or negative. That is,

$$\sqrt[n]{a^n} = a \text{ if } n \text{ is odd}.$$

We note that

$$\sqrt{25 \cdot 4} = 5 \cdot 2 = \sqrt{25}\sqrt{4} \quad \text{and} \quad \sqrt{\frac{25}{4}} = \frac{5}{2} = \frac{\sqrt{25}}{\sqrt{4}}.$$

In general,

$$\sqrt[n]{ab} = \sqrt[n]{a}\sqrt[n]{b} \quad \text{and} \quad \sqrt[n]{\frac{a}{b}} = \frac{\sqrt[n]{a}}{\sqrt[n]{b}},$$

where $a > 0$ and $b > 0$ if n is even. That is, the nth root of a product equals the product of the nth roots of the factors, and the nth root of a quotient of two terms equals the quotient of their nth roots. These rules can be used to simplify expressions that are not exact nth roots, as is illustrated in the following examples.

Example 16 (a) $\sqrt{20} = \sqrt{4 \cdot 5} = \sqrt{4}\sqrt{5} = 2\sqrt{5}$

(b) $\sqrt[3]{32} = \sqrt[3]{8 \cdot 4} = \sqrt[3]{8}\sqrt[3]{4} = 2\sqrt[3]{4}$ ∎

Example 17 (a) $\sqrt[3]{a^7 b^5} = \sqrt[3]{a^3 a^3 a \cdot b^3 b^2} = \sqrt[3]{a^3}\sqrt[3]{a^3}\sqrt[3]{a} \cdot \sqrt[3]{b^3}\sqrt[3]{b^2}$
$= a \cdot a \sqrt[3]{a} \cdot b \sqrt[3]{b^2} = a^2 b \sqrt[3]{ab^2}.$

(b) $\sqrt{\dfrac{a^3}{b^2}} = \dfrac{\sqrt{a^3}}{\sqrt{b^2}} = \dfrac{\sqrt{a^2 a}}{\sqrt{b^2}} = \dfrac{|a|\sqrt{a}}{|b|} = \dfrac{a\sqrt{a}}{|b|} \quad (b \neq 0)$

(Why is $|a|$ equal to a?) ∎

Algebraic Expressions. The extraction of roots and the basic operations of addition, subtraction, multiplication, and division are called **algebraic operations**. Any expression constructed out of numbers and letters representing numbers by these operations is called an **algebraic expression**. Any process that utilizes these operations, for example, solving equations, is called an **algebraic process**.

In a given algebraic expression, some letters may represent real numbers that are unspecified but fixed in value. These letters are called **constants**. Also, any real number itself is called a constant. Letters that represent real numbers that may assume different values in a given context are called **variables**. We usually let letters toward the beginning of the alphabet, such as a, b, and c, denote constants, and letters toward the end of the alphabet, such as x, y, and z, denote variables. For example, in the algebraic expression

$$\sqrt{ax + b},$$

a and b are constants and x is a variable. Also, whether stated explicitly or not, it is understood that, in the case of an even root, x can only assume values that make the radicand non-negative.

Exercises R.5

1. Write each of the following equations in compact form.

   (a) $(4 \cdot 4 \cdot 4)(4 \cdot 4 \cdot 4 \cdot 4 \cdot 4) = 4 \cdot 4 \cdot 4 \cdot 4 \cdot 4 \cdot 4 \cdot 4 \cdot 4$
   (b) $(3 \cdot 3)(3 \cdot 3)(3 \cdot 3)(3 \cdot 3) = 3 \cdot 3 \cdot 3 \cdot 3 \cdot 3 \cdot 3 \cdot 3 \cdot 3$
   (a) $(5 \cdot y)(5 \cdot y)(5 \cdot y)(5 \cdot y)(5 \cdot y) = 5 \cdot 5 \cdot 5 \cdot 5 \cdot 5 \cdot y \cdot y \cdot y \cdot y \cdot y$

Fill in the blanks to make each statement true.

2. x^{-n} is the multiplicative inverse of _____, provided $x \neq 0$.
3. The number 2 is a sixth root of 64 because _____. In general, y is an nth root of x if _____.
4. $\sqrt[n]{x^n} =$ _____ if n is odd; $\sqrt[n]{x^n} =$ _____ if n is even.
5. If n is an even positive integer, then every positive real number has _____ nth root(s) and every negative real number has _____ nth root(s); if n is odd, then every positive real number has _____ nth root(s) and every negative number has _____ nth root(s).

Write true or false for each statement.

6. To raise one power of x to another power, multiply the two exponents.
7. If x is positive, then x^{-7} is negative.
8. To divide two powers of x, divide the exponents.
9. $\sqrt[5]{ab} = \sqrt[5]{a}\sqrt[5]{b}$ for all real numbers a and b, but $\sqrt[4]{ab} = \sqrt[4]{a}\sqrt[4]{b}$ only if a and b are both ≥ 0.
10. If y is a square root of x, then y is also a fourth root of x^2.

Positive Exponents

11. Express each of the following as a power of -2.

    (a) -8
    (b) 16
    (c) -32
    (d) 64
    (e) $(-32)(64)$

12. Express each of the following as a power of -3.

    (a) 9
    (b) -27
    (c) 81
    (d) -243
    (e) $(-27)(-243)$

13. Express each of the following as a power of 2.

    (a) $2^3 \cdot 4$
    (b) $2^2 \cdot 4^3$
    (c) $2^2 \cdot 4^4 \cdot 16^2$

14. Express each of the following as a power of 3.

    (a) $27 \cdot 3^2$
    (b) $3^2 \cdot 9^3 \cdot 81$
    (c) $(3 \cdot 9)^2 (3 \cdot 27)^3$

15. Express each of the following as a product of a power of 2 by a power of 3.

    (a) $2^3 \cdot 3^2 \cdot 6$
    (b) $4 \cdot 6 \cdot 12 \cdot 16$
    (c) $18^2 \cdot 24^3$

Simplify each of the following.

16. $(2^2)^4$
17. $(x^3)^5$
18. $7^2 \cdot 7^3 \cdot 7^5$
19. $(-3)^7(-3)^{10}$
20. $(-5)(-5)^2(-5)^3$
21. $a \cdot a \cdot a^2 \cdot a^2 \cdot a^3 \cdot a^3$
22. $x^2 \cdot x \cdot x^3 \cdot x \cdot x^4 \cdot x$
23. $[(-2)^3]^5$
24. $[(4^2)^3]^4$
25. $[(x^2)^n]^5$
26. $2^8 \cdot 5^8$
27. $(-3)^9(-4)^9$
28. $2^6 a^6 b^6$
29. $(a^2 b)^4 (ab^2)^4$
30. $(a^2 b)^5 (ac)^3 (b^2 c^3)^4 (ab)^2$

Zero and Negative Exponents

Simplify each of the following. Express answers in terms of positive exponents.

31. $3^5 \cdot 2^7 \cdot 3^{-2} \cdot 2^{-4}$
32. $(2^3 a^0 b^6)(2^{-1} a^{-3} b^{-4})$
33. $(x^2 y^{-1})^3 (x^{-1} y^4)^2$
34. $(3^{-2})^5 (3^{-2})^{-3} (3^2)^0$
35. $3^8 \cdot 9^{-2}$
36. $16 \cdot 2^{-3} \cdot 4^3 \cdot 32^{-1}$
37. $\dfrac{2^0}{2^{-3}}$
38. $\dfrac{3}{3^2}$
39. $\dfrac{8^{-7}}{8^{-9}}$
40. $\dfrac{a^{-1}}{(a^{-1})^2}$
41. $\dfrac{a^{-1}}{(a^{-1})^2}$
42. $\dfrac{(b^{-2})^3}{(b^{-4})^2}$
43. $\dfrac{x^{-1} y}{x^{-2} y^{-3}}$
44. $\left(\dfrac{b^{-2}}{b^{-3}}\right)^{-4}$

45. $\left(\dfrac{a^2b^4}{ab^{-4}}\right)^2$ 47. $\dfrac{(a^{-1}b^2)^{-3}}{(ab^{-2})^{-2}}$

46. $\left(\dfrac{x^{-1}y^2}{x^3y^4}\right)^{-3}$ 48. $\dfrac{2^{-1}a^2b^6}{(2^{-2}a^{-1}b)^4}$

Roots

Evaluate, if possible, or simplify each of the following.

49. $\sqrt{25 \cdot 36 \cdot 49 \cdot 64}$
50. $\sqrt{16a^2b^4c^5}$
51. $\sqrt[3]{8 \cdot 27 \cdot (-a)^3 b^6}$
52. $\sqrt{\dfrac{16 \cdot 36}{25 \cdot 49}}$
53. $\sqrt[3]{\dfrac{8a^6}{125b^3}}$
54. $\sqrt[4]{\dfrac{16a^4}{(-3)^4 b^8}}$
55. $\sqrt[3]{(-15)^3}$
56. $\sqrt[4]{(-15)^4}$
57. $\left(\sqrt[3]{-15}\right)^3$
58. $\left(\sqrt[4]{-15}\right)^4$
59. $\sqrt{128}$
60. $\sqrt[3]{96}$
61. $\sqrt[3]{16a^4b^5c^6}$
62. $\sqrt{18a^3b^4}$
63. $\sqrt{a^2b^3c^4d^5}$
64. $\sqrt{\dfrac{a^3b^3}{c^3}}$
65. $\sqrt[3]{\dfrac{a^7}{b^5}}$
66. $\sqrt{\dfrac{a^3b^4}{c^2d^6}}$

Miscellaneous

*Science deals with some very large and some very small numbers. In order to avoid many zeros, we use **scientific notation**. In **scientific notation**, a number is written as a product of a decimal between 1 and 10 and a power of 10 (the symbol × is used for multiplication). For numbers greater than 1, positive exponents and used. For example, 520 is written as 5.2×10^2 and $5,200,000 = 5.2 \times 10^6$. For numbers between 0 and 1, negative exponents are used. For example, $.0125 = 1.25 \times 10^{-2}$ and $.00000125 = 1.25 \times 10^{-6}$. Write each of the following numbers in scientific notation.*

67. 5,280
68. 52,800
69. 176
70. .025
71. .00025
72. .76
73. 176,000
74. 4,125,271
75. 25,000,000
76. .000076
77. .0000001
78. .00000000001

Compute each of the following. Express answers in scientific notation.

79. $(1.2 \times 10^5)(1.1 \times 10^7)$
80. $(1.5 \times 10^4)(8 \times 10^6)$
81. $(2.5 \times 10^8)(2.5 \times 10^{-2})$
82. $(6.2 \times 10^{-4})(4 \times 10^5)$
83. $(2.5 \times 10^{-3})(4 \times 10^{-5})$
84. $\dfrac{3.3 \times 10^8}{6.6 \times 10^2}$
85. $\dfrac{4.5 \times 10^3}{2 \times 10^7}$
86. $\dfrac{5.1 \times 10^6}{3 \times 10^{-4}}$
87. $\dfrac{7.5 \times 10^{-5}}{2.5 \times 10^6}$
88. $\dfrac{2.13 \times 10^{-10}}{8.52 \times 10^{-5}}$

89. The speed of light is 29,977,600,000 cm/sec. Express this value in scientific notation.
90. **Avogadro's number** (the number of atoms of an element in a mass equal to its gram atomic weight) is 602,300,000,000,000,000,000,000. Write this number in scientific notation.

Each of the following numbers represents the corresponding planet's average distance from the sun. Express each value in standard decimal notation.

91. Mercury: 3.6×10^7 mi
92. Venus: 6.72×10^7 mi
93. Earth: 9.2956×10^7 mi
94. Mars: 1.416×10^8 mi
95. Jupiter: 4.834×10^8 mi
96. Saturn: 8.86×10^8 mi
97. Uranus: 1.782×10^9 mi
98. Neptune: 1.792×10^9 mi

99. Pluto: 3.664×10^9 mi

100. The electric charge of an electron is .00000000000000000016 coulombs. Express this number in scientific notation.

101. **Planck's constant** (the increase in maximum energy of a photoelectron per unit increase in the frequency of absorbed light) is .00000000000000000000000006624. Write Planck's constant in scientific notation.

Write the given mass of each of the following in decimal notation.

102. hydrogen: 1.673×10^{-24} g
103. oxygen: 1.328×10^{-23} g
104. silver: 1.791×10^{-22} g
105. gold: 3.27×10^{-22} g

Use a calculator to do Exercises 106–7.

106. The formula for **compound interest** is

$$A = P\left(1 + \frac{R}{N}\right)^{Nt},$$

where

$$\begin{cases} P = \text{amount originally invested}, \\ A = \text{amount accumulated after } t \text{ years}, \\ R = \text{annual interest rate (annual percentage rate/100)}, \\ N = \text{number of compound periods per year}. \end{cases}$$

For example, if $250 is invested at 9.6% annual rate, compounded monthly, then the amount of the investment after ten years is

$$250\left(1 + \frac{.096}{12}\right)^{12 \cdot 10} = 250(1.008)^{120} \approx 250(2.6)$$
$$= 650 \text{ dollars}.$$

Compute A in each of the following cases.

(a) $P = \$500$, $R = 16\%$, $N = 1$, $t = 10$
(b) $P = \$500$, $R = 15\%$, $N = 12$, $t = 10$
(c) $P = \$1{,}000$, $R = 12\%$, $N = 1$, $t = 5$
(d) $P = \$1{,}000$, $R = 12\%$, $N = 360$, $t = 5$

107. The **monthly payment formula** for repaying a loan is

$$P = \frac{rL(1+r)^n}{(1+r)^n - 1},$$

where

$$\begin{cases} P = \text{monthly payment}, \\ L = \text{amount of loan}, \\ r = \frac{1}{12} \cdot R, R = \text{annual interest rate}, \\ n = \text{the total number of payments}. \end{cases}$$

Compute P in each of the following cases.

(a) auto: $L = \$8{,}000$, $R = 10.75\%$, $n = 36$
(b) auto: $L = \$8{,}000$, $R = 10.75\%$, $n = 48$
(c) house: $L = \$60{,}000$, $R = 12\%$, $n = 360$
(d) house: $L = \$60{,}000$, $R = 9\%$, $n = 360$

Chapter R Review Outline

R.1 The Real Number System

The set of real numbers is made up of the rational numbers (terminating or repeating decimals) and the irrational numbers. It corresponds to the set of points on a directed line. The real numbers are closed under addition, subtraction, and multiplication. Division by any real number except 0 is possible.

The real numbers form an ordered set. That is, either any two real numbers are equal, or one is greater than the other. $a > b$ (a is greater than b) if a is to the right of b on the number line.

$|a|$ (the absolute value of a) is the distance between 0 and a on the number line.

R.2 Basic Properties of Real Numbers

Basic Properties

Commutative Properties:
$$a + b = b + a$$
$$a \cdot b = b \cdot a$$

Associative Properties:
$$(a + b) + c = a + (b + c)$$
$$(a \cdot b) \cdot c = a \cdot (b \cdot c)$$

Left Distributive Property:
$$a \cdot (b + c) = (a \cdot b) + (a \cdot c)$$

Identities:
$$a + 0 = a = 0 + a$$
$$a \cdot 1 = a = 1 \cdot a$$

Inverses:
$$a + (-a) = 0 = -a + a$$
$$a \cdot a^{-1} = 1 = a^{-1} \cdot a \qquad (a \neq 0)$$

Derived Properties

Right Distributive Property:
$$(a + b) \cdot c = (a \cdot c) + (b \cdot c)$$

Zero-Multiplier Property:
$$0 \cdot a = 0$$

Negative Multiplier Property:
$$(-1)a = -a$$

Double Inverse Properties:
$$-(-a) = a, \text{ and } (a^{-1})^{-1} = a \quad (a \neq 0)$$

Zero-Product Property:
$$ab = 0 \Leftrightarrow a = 0 \text{ or } b = 0$$

R.3 Signed Quantities, Order, and Absolute Value

Definitions

Subtraction: $a - b = a + (-b)$

Division: $\dfrac{a}{b} = a \cdot \dfrac{1}{b} = ab^{-1} \quad (b \neq 0)$

Greater Than: $a > b \Leftrightarrow a = b + p$, where p is a positive real number.

Absolute Value:
$$|x| = \begin{cases} x & \text{if } x \geq 0 \\ -x & \text{if } x < 0 \end{cases}$$

Derived Properties

Inverses of Sums, Products, and Quotients:
$$-(a + b) = -a + (-b)$$
$$(ab)^{-1} = a^{-1}b^{-1} \quad (a \neq 0, b \neq 0)$$
$$\left(\frac{a}{b}\right)^{-1} = \frac{a^{-1}}{b^{-1}} = \frac{b}{a} \quad (a \neq 0, b \neq 0)$$

Sign Rules:
$$(-a)(-b) = ab$$
$$(-a)b = a(-b) = -(ab)$$
$$\frac{-a}{-b} = \frac{a}{b} \quad (b \neq 0)$$
$$\frac{-a}{b} = \frac{a}{-b} = -\frac{a}{b} \quad (b \neq 0)$$

Zero-Quotient Property:
$$\frac{a}{b} = 0 \Leftrightarrow a = 0 \text{ and } b \neq 0$$

Properties of Greater Than:

$$a > b \Leftrightarrow a + c > b + c$$

If $c > 0$, then $a > b \Leftrightarrow ac > bc$.

If $c < 0$, then $a > b \Leftrightarrow ac < bc$.

If $a > b$ and $b > c$, then $a > c$.

Properties of Absolute Value:

$|ab| = |a||b|$

$\left|\dfrac{a}{b}\right| = \dfrac{|a|}{|b|}$ $(b \neq 0)$

$|a + b| \leq |a| + |b|$

$|a - b| \geq ||a| - |b||$

R.4 Fractional Expressions

Equality

$\dfrac{ac}{bc} = \dfrac{a}{b}$ $(b \neq 0, c \neq 0)$

$\dfrac{a}{b} = \dfrac{c}{d} \Leftrightarrow ad = bc$ $(b \neq 0, d \neq 0)$

Addition and Subtraction:

$\dfrac{a}{b} \pm \dfrac{c}{b} = \dfrac{a \pm c}{b}$ $(b \neq 0)$

$\dfrac{a}{b} \pm \dfrac{c}{d} = \dfrac{ad \pm bc}{bd}$ $(b \neq 0, d \neq 0)$

Multiplication and Division:

$\dfrac{a}{b} \cdot \dfrac{c}{d} = \dfrac{ac}{bd}$ $(b \neq 0, d \neq 0)$

$\dfrac{\frac{a}{b}}{\frac{c}{d}} = \dfrac{a}{b} \cdot \dfrac{d}{c}$ $(b \neq 0, c \neq 0, d \neq 0)$

R.5 Powers and Roots

Definitions (n a positive integer)

nth Power: $x^n = xx\ldots x$ (n factors)

Zero Exponent: $x^0 = 1$ $(x \neq 0)$

Negative Exponent:

$x^{-n} = \dfrac{1}{x^n}$ $(x \neq 0)$

nth Root: $y = \sqrt[n]{x} \Leftrightarrow y^n = x$ (x, y positive if n is even)

Properties of nth Roots:

$\sqrt[n]{ab} = \sqrt[n]{a}\sqrt[n]{b}$ (a and b positive if n is even)

$\sqrt[n]{\dfrac{a}{b}} = \dfrac{\sqrt[n]{a}}{\sqrt[n]{b}}$ $(b \neq 0;$ a and b positive if n is even)

Derived Rules for Powers

In each of the following, m and n are any integers (positive, negative, or 0) for which both sides of the equation are defined.

Same-Base Rules:

$x^m x^n = x^{m+n}$

$\dfrac{x^m}{x^n} = x^{m-n}$

Same-Exponent Rules:

$(xy)^n = x^n y^n$

$\left(\dfrac{x}{y}\right)^n = \dfrac{x^n}{y^n}$

Power-of-a-Power Rule:

$(x^m)^n = x^{mn}$

Chapter R Review Exercises

1. Which of the following numbers are real numbers? Classify those that are real numbers as either rational or irrational.

   $\dfrac{2}{3}, 3.14159, \sqrt{5}, \sqrt{-8}, \sqrt[3]{-8}, \dfrac{0}{4}, \dfrac{4}{0}, |-3|$

2. Express the repeating decimal $.757575\ldots$ as the ratio of two integers.

3. Express the rational number 4/7 as a decimal.

4. Show that the repeating decimal $.9999\ldots$ is equal to 1.

5. Without actually dividing, show that $\dfrac{3}{8} = \dfrac{33}{88} = \dfrac{333}{888}$.

6. Name the basic property of algebra illustrated by each of the following.

   (a) $3 + 7 = 7 + 3$ (c) $3 \cdot (4+5) = 3 \cdot 4 + 3 \cdot 5$

   (b) $(2 \cdot 3) \cdot 4 = 2 \cdot (3 \cdot 4)$ (d) $5 + 0 = 5$

7. Use the basic properties of algebra and equality to verify each of the following.

   (a) If $7 + a = 7 + b$, then $a = b$.

   (b) If $5 \cdot a = 5 \cdot b$, then $a = b$.

8. Use the algebraic definition of "greater than" to show that

   (a) $\dfrac{1}{2} > \dfrac{1}{3}$ (b) $-\dfrac{1}{3} > -\dfrac{1}{2}$

9. Show that the inequality $2x - 18 > 5x - 3$ is equivalent to $x < -5$.

10. Answer True or False for each of the following. Explain your answer.

    (a) If $a + c = b + c$, then $a = b$.
    (b) If $a \cdot c = b \cdot c$, then $a = b$.
    (c) $a - b = a + (-b)$
    (d) $-a = (-1)a$
    (e) If $a \neq 0$, then $a^{-1} = 1/a$.
    (f) If $a < 0$, then $-a > 0$.
    (g) $\sqrt{a^2} = a$
    (h) If $a > 0$ and $b > 0$, then $\sqrt{a+b} = \sqrt{a} + \sqrt{b}$.
    (i) If $a < 0$ and $b < 0$, then $|a + b| = -(a + b)$
    (j) If $a > 0$ and $b < 0$, then $|a - b| = a - b$

11. Evaluate each of the following.

    (a) $\dfrac{3}{4} + \dfrac{2}{5}$ (c) $72 \div (12 \div 6)$

    (b) $\dfrac{3+2}{4+5}$ (d) $72 \div 12 \div 6$

    (e) $\dfrac{-3(5-9)}{3 \cdot 5 + 5(-2)}$

    (f) $[52 + 24 \div (8 - 2)] \div 8 - 2$

12. Express each of the following without grouping symbols.

    (a) $2[5(a+b) + 6c]$
    (b) $3\{-2 + 4[(a+1) + (b+5)]\}$

13. Show that if $a \neq b$, then $\dfrac{a-b}{b-a} = -1$.

14. Explain why $\dfrac{x+3}{x^2-9} = 0$ cannot be true for any value of x.

15. Write each of the following as a single fractional expression.

    (a) $\dfrac{x-7}{x-3} + \dfrac{4}{x-3}$ (b) $\dfrac{x-1}{2} - \dfrac{x}{3}$

    (c) $\dfrac{1}{2} + \dfrac{2}{x}$

16. Write the expression $2^3 \cdot 6^2 \cdot 12^3 \cdot 18^2$ as a product of powers of 2 and 3.

17. Simplify each of the following. Express answers in terms of positive exponents. Assume $x \neq 0$, $a \neq 0$, $b \neq 0$.

    (a) $2x^2 x^{-3}$ (d) $\dfrac{ab^{-1}}{a^{-1}b}$

    (b) $\dfrac{5}{x^{-2}}$

    (c) $(2x)^2(x^2)^3$ (e) $\dfrac{a+b^{-1}}{a^{-1}+b}$

18. Evaluate

    (a) $\sqrt{(-2)^2}$ (b) $\sqrt[3]{(-2)^3}$ (c) $\sqrt[4]{(-2)^0}$

19. Simplify each of the following.

    (a) $\sqrt{50a^2b^3}$ (c) $\sqrt{(-x)^2}$

    (b) $\sqrt[3]{16a^5b^3}$ (d) $\sqrt[3]{(-x)^3}$

20. Simplify each of the following.

    (a) $\dfrac{\sqrt{50}}{\sqrt{8}}$ (b) $\sqrt{8} \cdot \sqrt{50}$

    (c) $\sqrt{8} + \sqrt{50}$

Calculator Exercises

21. Approximate $\sqrt{2}$ to nine decimal places. Can you determine the digit in the 10th decimal place? Hint: Consider $\sqrt{2} - 1$.

22. Suppose x is one of the positive integers 2, 3, 4, ..., 9, and n is any positive integer that is not a multiple of x. If x has the property that the decimal representation of n/x is always a repeating decimal, then $x = $ _____. Hint: You need only consider $n = 1, 2, \ldots, x-1$. Why? If the length of the repeating pattern is one digit, then $x = $ _____. If $x = 7$, then the length of the repeating pattern is _____.

23. Compute each of the following.

   (a) 24^5
   (b) 32^{-1}
   (c) 21^{-2}
   (d) $\sqrt{10}$
   (e) $\sqrt[3]{25}$

24. *Compute each of the following and explain your answer.*

   (a) $\dfrac{7 \cdot 5}{(3 \cdot 3)(4 \cdot 5)}$
   (b) $(7 \cdot 5)/(3 \cdot 3)/(4 \cdot 5)$
   (c) $(7 \cdot 5)/(3 \cdot 3)(4 \cdot 5)$
   (d) $(7 \cdot 5)/(3 \cdot 3/4 \cdot 5)$
   (e) $(7 \cdot 5/4 \cdot 5)/(3 \cdot 3)$

25. Compute each of the following.

   (a) $\dfrac{17 \cdot 5}{1 \cdot 4 + 2 \cdot 5^{-1}}$
   (b) $17 \cdot 5/1 \cdot 4 + 2 \cdot 5^{-1}$
   (c) $17 \cdot 5/(1 \cdot 4 + 2 \cdot 5)^{-1}$

Use the information in the Miscellaneous Exercises of Section R5 to do the following exercises.

26. Enter the following numbers in standard notation. How are the numbers denoted on the calculator?

   (a) the speed of light in cm/sec
   (b) Avogadro's number
   (c) the charge of an electron in coulombs
   (d) Planck's constant

27. Enter the following numbers in scientific notation. How are the numbers denoted on the calculator?

   (a) the distance between Earth and the sun in miles
   (b) the distance between Mars and the sun in miles
   (c) the mass of a hydrogen atom in grams
   (d) the mass of an oxygen atom in grams

28. Compute each of the following.

   (a) (the speed of light)(Avagadro's number)
   (b) (the charge of an electron)/(Planck's constant)
   (c) (the distance between the Earth and the sun)(the distance between Mars and the sun)
   (d) (the mass of a hydrogen atom)/(the mass of an oxygen atom)

29. Suppose $10,000 is invested at 5% annual interest, compounded monthly. What is the amount accumulated in

   (a) 5 years   (b) 10 years   (c) 20 years?

30. Suppose $10,000 is borrowed at 8% annual interest. Find the monthly payments if the loan is to be repaid in

   (a) 3 years   (b) 4 years   (c) 5 years.

1 Algebraic Expressions

1.1	Operations with Polynomials
1.2	Factoring Polynomials
1.3	Operations with Rational Expressions
1.4	Rational Exponents
1.5	Operations with Radicals

Hey Students!
You will benefit more from the classroom lectures and have a better understanding of the material if you take the time to view the Gilbert Review Videos @
www.CollegeAlgebraBySchiller.com
before attending the class lecture for this chapter.

Chapter 1. Algebraic Expressions

In the preceding chapter we showed how the basic properties of addition, subtraction, multiplication, and division of real numbers are used to perform operations with elementary algebraic expressions. We now proceed to more general algebraic expressions. However, no matter how complex an expression may appear, both the expression and its components represent real numbers and therefore follow the same basic rules. With practice, manipulating algebraic expressions becomes as routine as working with integers.

1.1 Operations with Polynomials

Polynomials are the integers of algebra.

Polynomials
Addition and Subtraction
Multiplication
Division
Polynomial Functions and Zeros of Polynomials

Polynomials. Polynomials are algebraic expressions that arise when the operations of addition, subtraction, and multiplication are applied to numbers and letters representing numbers. For example, $x + 2$, $x^3 - 2xy + 4y^2$, and $ax^2 + bx + c$ are polynomials. Generalizing gives the following definition.

A polynomial of degree n in x is an algebraic expression of the form

$$a_n x^n + a_{n-1} x^{n-1} + a_{n-2} x^{n-2} + \ldots + a_1 x + a_0,$$

where n is a nonnegative integer and $a_n \neq 0$.

In this definition, the real numbers a_k ($k = 0, 1, 2, \ldots, n$) are called the **coefficients** of the polynomials, and the $a_k x^k$ ($k = 0, 1, 2, \ldots, n$) are called the **terms** of the polynomial. Polynomials of degree 1, such as $2x + 3$, are called **linear polynomials**, and those of degree 2, such as $x^2 + 4x + 7$, are called **quadratic polynomials**. Since $x^0 = 1$, polynomials of degree 0 are simply nonzero real numbers. Polynomials can be added, subtracted, and multiplied, and the result is always a polynomial. Hence, the collection of all polynomials forms a closed system under these operations.[1] Division of polynomials is also possible, but we shall see that the result is not always a polynomial. That is, the collection of all polynomials is not a closed system under division.

For example, the fractions

$$\frac{x+1}{x-1} \quad \text{and} \quad \frac{1}{x^2+2}$$

---

[1] *In order to be correct when saying that polynomials are closed under addition and subtraction, we must admit the constant 0 as a polynomial. However, we will not attach degree to the polynomial 0. (Some authors assign the degree $-\infty$ to the polynomial 0 for reasons that do not concern us here.)*

are *not* polynomials, nor are the radicals

$$\sqrt{x^2+4} \quad \text{and} \quad \sqrt{x+3}.$$

Addition and Subtraction. Polynomials are added and subtracted by means of the commutative and associative properties for addition and the distributive properties.

Example 1 $(3x^2 + 2x + 1) + (5x^2 - 7x + 4)$
$= 3x^2 + 5x^2 + 2x - 7x + 1 + 4$ Extnded Commutative and
  Associative Properties
  for Addition

$= (3+5)x^2 + (2-7)x + 5$ Right Distributive Property
$= 8x^2 - 5x + 5$ ∎

Example 2 $(4x^2 - 3x + 1) - (x^2 - 7x + 4)$
$= 4x^2 - 3x + 1 - x^2 + 7x - 4$ *When removing parentheses following a minus sign, change the sign of each term within the parentheses.*

$= 4x^2 - x^2 + (-3x + 7x) + 1 - 4$ Extended Commutative
  and Associative Properties
  for Addition

$= (4-1)x^2 + (-3+7)x - 3$ Right Distributive Property
$= 3x^2 + 4x - 3$ ∎

Example 3 $(ax^2y^2 + bxy + cx + dy + e) + (2x^2y^2 + 3xy + 4x + 5y + 6)$
$= ax^2y^2 + 2x^2y^2 + bxy + 3xy + cx + 4x + dy + 5y + e + 6$
$= (a+2)x^2y^2 + (b+3)xy + (c+4)x + (d+5)y + e + 6$ ∎

Similar (like) terms have the same variables raised to the same powers but can have different coefficients.

In the above examples, the terms of the polynomials were first rearranged and regrouped in order to bring together terms corresponding to similar powers, and then the respective similar terms were combined by means of the right distributive property. This same technique of rearranging, regrouping, and combining applies when adding or subtracting any number of polynomials. In short, to add or subtract polynomials, simply add or subtract similar terms.

Example 4 $(4x+7) + (3x-12) - (2x-7) - (-3x+4)$
$= 4x + 7 + 3x - 12 - 2x + 7 + 3x - 4$
$= 4x + 3x - 2x + 3x + 7 - 12 + 7 - 4$
$= 8x - 2$ ∎

Multiplication. To multiply two polynomials, we first apply the distributive properties and then proceed as in addition.

Example 5
$$(3x + 4)(x + 6) = 3x(x + 6) + 4(x + 6) \quad \text{Right Distributive Property}$$
$$= 3x^2 + 18x + 4x + 24 \quad \text{Left Distributive Property}$$
$$= 3x^2 + 22x + 24 \quad \text{Addition}$$

Example 6
$$(x^2 + 2x - 1)(3x^2 - 5x + 4)$$
$$= x^2(3x^2 - 5x + 4) + 2x(3x^2 - 5x + 4) \quad \text{Right distributive property}$$
$$-1(3x^2 - 5x + 4)$$

$$= 3x^4 - 5x^3 + 4x^2 + 6x^3 - 10x^2 + 8x \quad \text{Left distributive property}$$
$$-3x^2 + 5x - 4$$

$$= 3x^4 + x^3 - 9x^2 + 13x - 4 \quad \text{Addition} \quad \blacksquare$$

In both of the above examples, the effect of applying the distributive properties is that each term of the polynomial factor on the left multiplies each term of the polynomial factor on the right. Thus, in Example 5,

$$(3x + 4)(x + 6) = 3x^2 + 18x + 4x + 24,$$
↑ ↑ ↑ ↑   ↑   ↑   ↑   ↑
1 2 a b   1a 1b 2a 2b

and in Example 6,

$$(x^2 + 2x - 1)(3x^2 - 5x + 4)$$
↑ ↑ ↑ ↑ ↑ ↑
1 2 3 a b c
$$= 3x^4 - 5x^3 + 4x^2 + 6x^3 - 10x^2 + 8x - 3x^2 + 5x - 4.$$
↑   ↑   ↑   ↑   ↑   ↑   ↑   ↑   ↑
1a   1b   1c   2a   2b   2c 3a 3b 3c

This principle applies when multiplying any two polynomials P and Q, regardless of their respective number of terms.

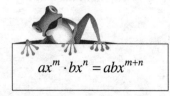

$ax^m \cdot bx^n = abx^{m+n}$

> To form the **product** PQ of polynomials P and Q, first multiply each term of P by each term of Q and then proceed as in addition.

Polynomials with two terms are called **binomials**, and binomials occur frequently in mathematics. If we apply the above product rule to the binomials $a + b$ and $a - b$, we obtain the following important results.

1. $(a+b)(a+b) = a^2 + ab + ba + b^2 = a^2 + ab + ab + b^2$; therefore,

$$(a+b)^2 = a^2 + 2ab + b^2. \qquad \text{Square of a Sum}$$

2. $(a-b)(a-b) = a^2 - ab - ba + (-b)^2 = a^2 - ab - ab + b^2$; therefore,

$$(a-b)^2 = a^2 - 2ab + b^2. \qquad \text{Square of a Difference}$$

3. $(a-b)(a+b) = a^2 + ab - ba - b^2 = a^2 + ab - ab - b^2$; therefore,

$$(a-b)(a+b) = a^2 - b^2. \qquad \text{Difference of Squares}$$

Example 7
(a) $(3x+4)^2 = 9x^2 + 24x + 16$
(b) $(3x-4)^2 = 9x^2 - 24x + 16$
(c) $(3x-4)(3x+4) = 9x^2 - 16$ ■

Example 8
(a) $(2x^2 + 3y)^2 = 4x^4 + 12x^2y + 9y^2$
(b) $(2x^2 - 3y)^2 = 4x^4 - 12x^2y + 9y^2$
(c) $(2x^2 - 3y)(2x^2 + 3y) = 4x^4 - 9y^2$ ■

As illustrated in Example 8, a binomial need not be a linear polynomial in x. Any polynomial with exactly two terms is a binomial, regardless of its degree or the number of variables in each term.

Caution From the formulas for the square of a sum and the square of a difference, we see that in general,

$$(a+b)^2 \neq a^2 + b^2 \quad \text{and} \quad (a-b)^2 \neq a^2 - b^2.$$

Similarly,

$$\sqrt{a^2 + b^2} \neq a + b \quad \text{and} \quad \sqrt{a^2 - b^2} \neq a - b.$$

Treating these inequalities as equalities is a very common mistake in algebra. Be sure to avoid doing this. For example,

$$(7+3)^2 = 100, \quad \text{but} \quad 7^2 + 3^2 = 58,$$

and

$$(10-4)^2 = 36, \quad \text{but} \quad 10^2 - 4^2 = 84.$$

Also,

$$\sqrt{3^2 + 4^2} = 5 \neq 3 + 4 \quad \text{and} \quad \sqrt{13^2 - 5^2} = 12 \neq 13 - 5.$$

Products of three or more polynomials can be computed by taking the factors two at a time.

Example 9
$$(x+1)(x+2)(x+3) = (x^2 + 3x + 2)(x+3)$$
$$= x^3 + 3x^2 + 3x^2 + 9x + 2x + 6$$
$$= x^3 + 6x^2 + 11x + 6 \quad \blacksquare$$

Example 10 Verify the formula

$$(a+b)^3 = a^3 + 3a^2b + 3ab^2 + b^3. \quad \text{Cube of a Sum}$$

Solution:
$$(a+b)^3 = (a+b)^2(a+b)$$
$$= (a^2 + 2ab + b^2)(a+b)$$
$$= a^3 + a^2b + 2a^2b + 2ab^2 + b^2a + b^3$$
$$= a^3 + 3a^2b + 3ab^2 + b^3 \quad \blacksquare$$

Example 11 Use the result of Example 10 to expand the following:

(a) $(2x+5)^3$ and
(b) $(x-2)^3$.

Solution:

(a) If we substitute $a = 2x$ and $b = 5$ in the formula in Example 10, then

$$(2x+5)^3 = (2x)^3 + 3(2x)^2 \cdot 5 + 3(2x) \cdot 5^2 + 5^3$$
$$= 8x^3 + 60x^2 + 150x + 125.$$

(b) Here we substitute $a = x$ and $b = -2$ to obtain

$$(x-2)^3 = [x + (-2)]^3$$
$$= x^3 + 3x^2(-2) + 3x(-2)^2 + (-2)^3$$
$$= x^3 - 6x^2 + 12x - 8. \quad \blacksquare$$

1.1. Operations with Polynomials

Division. We know that if one integer is divided by another, the result need not be an integer. For instance,

$$\frac{27}{4} = 6 + \frac{3}{4},$$

and we say that when 27 is divided by 4, the quotient or integer part is 6 and the remainder is 3. The above equation is equivalent to

$$27 = 6 \cdot 4 + 3,$$

and in general, if n is any integer and m is a positive integer, then there are unique integers q and r for which

$$n = qm + r \quad (0 \leq r < m).$$

The result stated in this equation is called the **division algorithm for integers** and can be seen geometrically by marking off a number line in multiples of m as shown in Figure 1. The integer n is located between qm and $(q+1)m$, and r is the difference between n and qm.

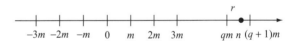

Figure 1 Division Algorithm for Integers

There is also a **division algorithm for polynomials**. It says that if P is any polynomial and T is a polynomial of degree greater than 0, then there are unique polynomials Q and R for which

$$P = Q \cdot T + R \quad (R = 0 \text{ or degree } R < \text{ degree } T).$$

For example, if $P = 6x^4 + 7x^3 + 16x - 11$ and $T = x + 2$, then,

$$6x^4 + 7x^3 + 16x - 11 = (6x^3 - 5x^2 + 10x - 4)(x + 2) - 3,$$

which is equivalent to

$$\frac{6x^4 + 7x^3 + 16x - 11}{x + 2} = 6x^3 - 5x^2 + 10x - 4 + \frac{-3}{x + 2}.$$

Here $Q = 6x^3 - 5x^2 + 10x - 4$ and $R = -3$. We will soon see how Q and R are determined, but first some terminology. The numerator $6x^4 + 7x^3 + 16x - 11$ of the ratio

$$\frac{6x^4 + 7x^3 + 16x - 11}{x + 2}$$

is called the **dividend**, and the denominator $x + 2$ is called the **divisor**. The polynomial part

$$6x^3 - 5x^2 + 10x - 4$$

of the result is called the **quotient**, and the numerator -3 of the fractional part of the result is called the **remainder**. That is,

> dividend = quotient · divisor + remainder
>
> or
>
> $$\frac{\text{dividend}}{\text{divisor}} = \text{quotient} + \frac{\text{remainder}}{\text{divisor}}.$$

Degree of quotient = degree of dividend – degree of divisor.

Remainder = 0 or has degree less than degree of divisor.

Hence, the ratio of dividend to divisor will itself be a polynomial precisely when the remainder is equal to 0. We now describe a step-by-step process for finding Q and R.

1. Express both the dividend and the divisor in terms of decreasing powers of x, and use 0 for the coefficient of any missing power of x in the dividend. Then position the divisor and dividend as follows:

$$x + 2 \overline{\smash{\big)}\, 6x^4 + 7x^3 + 0x^2 + 16x - 11}.$$

2. Divide the lead term x of the divisor into the lead term $6x^4$ of the dividend, and place the result $6x^3$ above the division line:

$$\begin{array}{r} 6x^3 \\ x + 2 \overline{\smash{\big)}\, 6x^4 + 7x^3 + 0x^2 + 16x - 11}. \end{array}$$

3. Multiply the divisor $x + 2$ by the result $6x^3$ of step 2, and subtract this product $6x^4 + 12x^3$ from the dividend to obtain the reduced dividend $-5x^3 + 0x^2 + 16x - 11$:

$$\begin{array}{r} 6x^3 \\ x + 2 \overline{\smash{\big)}\, 6x^4 + 7x^3 + 0x^2 + 16x - 11} \\ \underline{6x^4 + 12x^3 } \\ -5x^3 + 0x^2 + 16x - 11. \quad \text{Reduced Dividend} \end{array}$$

4. Repeat steps 2 and 3 with the reduced dividend taking the place of the dividend until the degree of the reduced dividend becomes less than the degree of the divisor. The last reduced dividend is the reminder.

$$
\begin{array}{r}
6x^3 - 5x^2 + 10x - 4 \\
x+2 \overline{\smash{\big)}\, 6x^4 + 7x^3 + 0x^2 + 16x - 11} \\
\underline{6x^4 + 12x^3 } \\
-5x^3 + 0x^2 + 16x - 11 \\
\underline{-5x^3 - 10x^2 } \\
10x^2 + 16x - 11 \\
\underline{10x^2 + 20x } \\
-4x - 11 \\
\underline{-4x - 8} \\
-3
\end{array}
$$

with labels: Reduced Dividend, Reduced Dividend, Reduced Dividend, Remainder.

Example 12   Perform the division $\dfrac{5x^5 + 6x^4 - 2x^2 + 4}{x^2 - 3}$.

Solution:

$$
\begin{array}{r}
5x^3 + 6x^2 + 15x + 16 \\
x^2 - 3 \overline{\smash{\big)}\, 5x^5 + 6x^4 + 0x^3 - 2x^2 + 0x + 4} \\
\underline{5x^5 - 15x^3 } \\
6x^4 + 15x^3 - 2x^2 + 0x + 4 \\
\underline{6x^4 - 18x^2 } \\
15x^3 + 16x^2 + 0x + 4 \\
\underline{15x^3 - 45x } \\
16x^2 + 45x + 4 \\
\underline{16x^2 - 48} \\
45x + 52
\end{array}
$$

with labels: Reduced Dividend, Reduced Dividend, Reduced Dividend, Remainder.

Hence, the quotient is $5x^3+6x^2+15x+16$ and the remainder is $45x+52$.

That is, $\dfrac{5x^5 + 6x^4 - 2x^2 + 4}{x^2 - 3} = 5x^3 + 6x^2 + 15x + 16 + \dfrac{45x+52}{x^2 - 3}$

or $\quad 5x^5 + 6x^4 - 2x^2 + 4 = (5x^3 + 6x^2 + 15x + 16)(x^2 - 3) + 45x + 52.$ ∎

Note that in Example 12 the divisor $x^2 - 3$ has degree 2 and the remainder $45x+52$ has degree 1. In general, as stated above, the remainder is either 0 or a polynomial of degree less than the degree of the divisor. Also, the degree of the quotient *plus* the degree of the divisor is equal to the degree of the dividend.

Example 13   Perform the division $\dfrac{3x^3 - 3x^2 + 5x + 7}{2x + 1}$.

Solution:

$$
\begin{array}{r}
1.5x^2 - 2.25x + 3.625 \\
2x+1 \overline{\smash{\big)}\, 3x^3 - 3x^2 + 5x + 7} \\
\underline{3x^3 + 1.5x^2 } \\
-4.5x^2 + 5x + 7 \\
\underline{-4.5x^2 - 2.25x } \\
7.25x + 7 \\
\underline{7.25x + 3.625} \\
3.375
\end{array}
$$

Here the quotient is $1.5x^2 - 2.25x + 3.625$, and the remainder is 3.375, which illustrates that the quotient and remainder need not have integer coefficients, even though the coefficients of the dividend and divisor are integers. ∎

Polynomial Functions and Zeros of Polynomials.   Every integer is a polynomial in the base 10. For example, $3258 = 3 \cdot 10^3 + 2 \cdot 10^2 + 5 \cdot 10 + 8$. That is, the integer 3258 corresponds to the polynomial $P = 3x^3 + 2x^2 + 5x + 8$ evaluated at $x = 10$. In this sense we can think of polynomials as generalized integers. For $x = -1$, the value of P is $3(-1)^3 + 2(-1)^2 + 5(-1) + 8 = 2$, and for $x = \sqrt{2}$, the value of P is $3(\sqrt{2})^3 + 2(\sqrt{2})^2 + 5\sqrt{2} + 8 = 11\sqrt{2} + 12$. **The notation $P(x)$, read "P of x" not "P times x," is used to denote the numerical dependence of P on x.** Hence, $P(10) = 3258$, $P(-1) = 2$, and $P(\sqrt{2}) = 11\sqrt{2} + 12$.

Example 14   Let $P(x) = 4x^2 - 6x - 15$. Find (a) $P(-2)$, (b) $P(0)$, (c) $P(1/2)$, and (d) $P(1 + \sqrt{3})$.

Solution:

(a) $P(-2) = 4(-2)^2 - 6(-2) - 15 = 16 + 12 - 15 = 13$

(b) $P(0) = 4 \cdot 0^2 - 6 \cdot 0 - 15 = -15$

(c) $P\left(\dfrac{1}{2}\right) = 4\left(\dfrac{1}{2}\right)^2 - 6\left(\dfrac{1}{2}\right) - 15 = 1 - 3 - 15 = -17$

(d) $P(1 + \sqrt{3}) = 4(1 + \sqrt{3})^2 - 6(1 + \sqrt{3}) - 15$
$= 4(1 + 2\sqrt{3} + 3) - 6 - 6\sqrt{3} - 15$
$= 4 + 8\sqrt{3} + 12 - 6 - 6\sqrt{3} - 15$
$= -5 + 2\sqrt{3}$ ∎

Hence, for a given polynomial P, each value of x produces a corresponding value of $P(x)$. In this sense, we can think of a polynomial as a *rule that assigns to a real number x a corresponding real number* $P(x)$. In mathematics, such rules are called **functions**, and the concept of a function is fundamental to all branches of mathematics. We will study functions more formally in Chapter 4. In this chapter we are more concerned with polynomials as **algebraic expressions**, that is, with how polynomials are added, subtracted, multiplied, and divided.

A value of x that makes a given polynomial $P(x)$ equal to zero is called a **root of the equation $P(x) = 0$** or a **zero of $P(x)$**. For example, -3 is a root of the equation $2x + 6 = 0$ since $2(-3) + 6 = 0$. Similarly, $\sqrt{2}$ and $-\sqrt{2}$ are zeros of the polynomials $x^2 - 2$. The problem of finding roots of polynomial equations is the oldest problem in algebra. Indeed, the development of algebra from ancient to modern times is the story of mathematicians' efforts to solve polynomials for their zeros. We will trace this development throughout the rest of the text.

Note that the proper terminology is ROOT of an equation and ZERO of a polynomial.

Example 15 Verify that $1 + \sqrt{2}$ and $1 - \sqrt{2}$ are zeros of $P(x) = x^2 - 2x - 1$.

Solution: $P(1 + \sqrt{2}) = (1 + \sqrt{2})^2 - 2(1 + \sqrt{2}) - 1$
$= 1 + 2\sqrt{2} + 2 - 2 - 2\sqrt{2} - 1 = 0$
and $P(1 - \sqrt{2}) = (1 - \sqrt{2})^2 - 2(1 - \sqrt{2}) - 1$
$= 1 - 2\sqrt{2} + 2 - 2 + 2\sqrt{2} - 1 = 0$ ∎

Question *Why does $P(x) = x^2 + 1$ not have any zeros in the real number system?*

Answer: For $x^2 + 1$ to be 0, x^2 would have to equal -1. But x^2 cannot be negative for any real number x. ∎

As illustrated in the question and answer above, some polynomials do not have any zeros in the real numbers system. This situation led mathematicians to consider the possibility of a larger number system in which all polynomials do have zeros. Efforts to develop such a system resulted in the **complex number system**, which will be introduced in Chapter 3.

$\sqrt{-1}$ makes sense in the complex number system.

Exercises 1.1

Fill in the blanks to make each statement true.

1. Polynomials are generated from real numbers and letters by the operations of _____, _____, and _____.
2. If $P = 3x^2 - 5x - 2$, then $-P =$ _____.
3. The Product PQ of two polynomials P and Q is obtained by _____.
4. dividend = divisor · _____ + _____
5. If $P(r) = 0$, then r is called a _____ of the equation $P(x) = 0$ or a _____ of $P(x)$.

Write true or false for each statement.

6. If P is a polynomial, then $P(x)$ does not mean $P \cdot x$.
7. If P and Q are polynomials, then $P - Q = P + (-Q)$.
8. $(x+3)^2 = x^2 + 9$
9. $\sqrt{4x^2 + 25} = |2x + 5|$
10. $\dfrac{ax^2 + b}{x + b} = ax + 1$

Addition and Subtraction

Add and subtract as indicated in Exercises 11-20.

11. $(x^2 + 4xy + y) + (x^2 + xy + 3y)$
12. $(x^2 - 2y^2) + (x^2 + y^2) - (x^2 + 4y^2)$
13. $(x^2 - 2x - 3) - (4x^2 + 5x - 1) + (3x^2 + 7)$
14. $(2x^2 + 3x - 7) - (x^2 + 4) + (3x^2 + 2x)$
15. $(3x - 2) - (2x^2 + 2x + 5) + (3x^3 - 1)$
16. $(3x + 5) + (x^4 - 1) + (x^3 - 2x) + (5x^2 - 2)$
17. $(ax^2 + bx + c) + (2ax^2 + 3bx + 4c)$
18. $(ax^2 + bx + c) + (ax^2 + bx) + (ax^2 + c)$
19. $(x^3 + x^2 + x + 1) + (ax^3 + bx^2 + cx + d)$
20. $(ax + 1) + (ax^2 + 2x + 3) - (x^2 + x + a)$

Multiplication

Perform each of the following multiplications.

21. $2x(3x + 4)$
22. $a(2b + 3)$
23. $(x^2 + 1)2xy$
24. $(a^2 - 1)b^2$
25. $(x + 1)(x + 2)$
26. $(2a + 3)(4a + 5)$
27. $(x - 3)(x + 2)$
28. $(2a - 1)(3a + 5)$
29. $(xy^2 + y)(x^2y + x)$
30. $(a^2b^2 + 2ab)(2a^3 + 3b^2)$
31. $(5x + 4)^2$
32. $(5x - 4y)^2$
33. $(2x^2 + 3)^2$
34. $(3x^3 - 2y^2)^2$
35. $(2x^2 - 5)(2x^2 + 5)$
36. $(x^3 - y^3)(x^3 + y^3)$
37. $(x + 2)^3$
38. $(x - y)^3$
39. $(x + 2)^4$
40. $(x - y)^4$
41. $6x(x + 3)(x - 5)$
42. $(x - 1)(x - 2)(x - 3)$
43. $5x^2(x^2 + 1)(3x + 4)$
44. $2x(3x - 5)(4x^2 + 2x + 7)$
45. $(2x - 3)(3x - 4)(4x - 5)$
46. $x(x - 2)(x + 1)(x^2 - 3)$
47. $(x - 1)(x + 2)(x + 1)(x - 2)$
48. $3x(2x + 1)(3x - 2)(x^2 - x + 1)$
49. $x(x + 1)(x + 2)(x + 3)(x + 4)$
50. $(3x - 1)(2x^2 + 3x + 1)(x + 2)(x^2 - 2x - 1)$

Division

Perform each of the following divisions.

51. $x + 3 \,\overline{\smash{)}\, x^3 + 4x^2 + 2x - 3}$
52. $x - 2 \,\overline{\smash{)}\, 3x^3 - 2x^2 + x - 1}$
53. $x - 1 \,\overline{\smash{)}\, x^4 + x^3 + x^2 + x + 1}$
54. $x + 5 \,\overline{\smash{)}\, 2x^4 + 3x^2 + 2x - 2}$
55. $x - 4 \,\overline{\smash{)}\, x^5 - 5x^3 + 2x^2 + 4}$
56. $x + 1 \,\overline{\smash{)}\, x + 2}$
57. $2x - 3 \,\overline{\smash{)}\, 4x + 6}$
58. $2x + 3 \,\overline{\smash{)}\, 5x^2 + 8x + 11}$
59. $2x - 1 \,\overline{\smash{)}\, 6x^3 + 5x^2 - 8x + 2}$
60. $2x - 1 \,\overline{\smash{)}\, x^2 + 3x - 5}$
61. $x^2 + 2x - 1 \,\overline{\smash{)}\, 3x^4 + 2x^3 - 10x + 5}$
62. $x^2 + 3x + 1 \,\overline{\smash{)}\, x^4 + 3x^3 - 5x^2 + 8x + 2}$
63. $x^2 + x + 1 \,\overline{\smash{)}\, x^5 + x^3 + x + 1}$
64. $x^2 - 1 \,\overline{\smash{)}\, 4x^4 + 3x^3 + 2x^2 - x + 15}$
65. $x^2 + 1 \,\overline{\smash{)}\, x^8 - 4x^6 - 2x^4 - 3x^2 - 1}$
66. $x^2 + 1 \,\overline{\smash{)}\, 4x^2 + 3x + 2}$
67. $x^2 - 2x - 1 \,\overline{\smash{)}\, 5x^2 + 7}$
68. $x^2 \,\overline{\smash{)}\, x^3 - 3x^2 + 2x + 11}$

69. $x^3 + x^2 + x + 1 \overline{\smash{\big)}\, x^4}$.

70. $x^2 - 1 \overline{\smash{\big)}\, x^8 - 1}$.

In each of the following, fill in the blanks with the quotient and remainder.

71. $\dfrac{5x^2 + 3x + 2}{x + 1} = \underline{} + \dfrac{}{x + 1}$

72. $\dfrac{x^3 - x^2 + 1}{x + 3} = \underline{} + \dfrac{}{x + 3}$

73. $\dfrac{x^7 - 1}{x - 1} = \underline{} + \dfrac{}{x - 1}$

74. $\dfrac{x^5 + 3x^4 + 2x^3 - x^2 + 7x - 5}{x^2 + 3x + 4}$
$= \underline{} + \dfrac{}{x^2 + 3x + 4}$

75. $\dfrac{2x^2 + 7x + 5}{x^2 + 1} = \underline{} + \dfrac{}{x^2 + 1}$

76. $4x^2 + 3x + 1 = (x + 2)(\underline{}) + \underline{}$

77. $x^3 + 2x^2 - x + 3 = (x^2 + x + 1)(\underline{}) + \underline{}$

78. $x^4 - 5x^2 + 2 = (x^2 - 3)(\underline{}) + \underline{}$

79. $x^5 + 1 = (x + 1)(\underline{}) + \underline{}$

80. $x^5 + 1 = (x - 1)(\underline{}) + \underline{}$

Polynomial Functions and Zeros of Polynomials

In Exercises 81-85, find each value of P for the given polynomial.

81. $P(x) = 2x^2 - 7x + 4$

   (a) $P(1)$ (b) $P(3)$ (c) $P(0)$

82. $P(x) = x^3 + 3x^2 + 2x + 1$

   (a) $P(-2)$ (b) $P(4)$ (c) $P(10)$

83. $P(x) = x^2 - 10$

   (a) $P(5)$ (b) $P(-7)$ (c) $P(\sqrt{10})$

84. $P(x) = 3x^4 + 2x^2 + 5$

   (a) $P(\sqrt{2})$ (b) $P(-\sqrt{3})$ (c) $P(2a)$

85. $P(x) = x^2 - 4x - 1$

   (a) $P(1 + \sqrt{2})$ (b) $P(2 + \sqrt{3})$ (c) $P(2 - \sqrt{3})$

Find all possible real zeros of each of the following polynomials.

86. $P(x) = (2x - 3)(4x + 5)$

87. $P(x) = (3x + 2)(3x - 5)(x + 2)(x - 7)$

88. $P(x) = x(x^2 - 4)(x^2 - 10)$

89. $P(x) = (x^2 - 5)(x^2 + 5)$

90. $P(x) = x^2 + 10x + 25$

Miscellaneous

91. If $(ax + b)^2 = 4x^2 + 12x + 9$, then either $a = \underline{}$ and $b = \underline{}$, or $a = \underline{}$ and $b = \underline{}$.

92. Use the rule $(a + b)^2 = a^2 + 2ab + b^2$ to compute each of the following.

   (a) 16^2 (b) 17^2 (c) 21^2 (d) 26^2

93. Use the rule $(a - b)^2 = a^2 - 2ab + b^2$ to compute each of the following.

   (a) 19^2 (b) 18^2 (c) 29^2 (d) 28^2

94. Use the rule $a^2 - b^2 = (a - b)(a + b)$ to compute each of the following.

   (a) $18 \cdot 22$ (b) $27 \cdot 33$ (c) $36 \cdot 44$ (d) $45 \cdot 55$

95. Use the formula for the difference of squares to show that

$$x^2 = y^2 \Leftrightarrow x = y \quad \text{or} \quad x = -y.$$

96. From Section R.2, Exercise 33, it follows that $x = -x \Leftrightarrow x = 0$. Now show that

$$x = x^{-1} \Leftrightarrow x = 1 \quad \text{or} \quad x = -1.$$

97. *Division Is Not Commutative.* Show that if $a/b = b/a$, then either $a = b$ or $a = -b$.

98. *Division Is Not Associative.* Show that if $(a/b)/c = a/(b/c)$, then either $a = 0$, or $c = 1$ or -1.

99. Show that + and · are not interchangeable in the left distributive property. That is, show that if $a + bc = (a+b)(a+c)$, then either $a = 0$ or $a + b + c = 1$.

100. A square with side b is placed next to a square with side a. If two rectangles, each with length a and width b, are placed as shown in the figure, a square with side $a+b$ is obtained. The geometric statement "the area of the large square is equal to the sum of the areas of its parts" becomes the algebraic equation

$$(a+b)^2 = a^2 + 2ab + b^2.$$

Draw a figure to illustrate the equation $(a-b)^2 = a^2 - 2ab + b^2$.

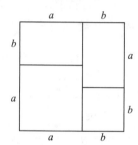

101. $a/b + a/c \neq a/(b+c)$ unless $a = 0$. The following argument shows that the assumption $a/b + a/c = a/(b+c)$ leads to a contradiction. Fill in the reasons for each step. Suppose $a \neq 0$. Then

$$\frac{a}{b} + \frac{a}{c} = \frac{a}{b+c} \Leftrightarrow \frac{1}{b} + \frac{1}{c} = \frac{1}{b+c} \quad \underline{}$$

$$\Leftrightarrow \frac{c+b}{bc} = \frac{1}{b+c} \quad \underline{}$$

$$\Leftrightarrow (b+c)^2 = bc. \quad \underline{}$$

Now, assuming that $b \neq 0$ and $c \neq 0$, show that $(b+c)^2 > bc$. (*Hint:* consider the two cases $bc < 0$ and $bc > 0$ separately.)

1.2 Factoring Polynomials

$$\begin{array}{r} a-b \\ a+c \\ \hline \end{array} \quad \begin{array}{l} = aa - ba \\ + ca - bc \end{array}$$

The factoring of $a^2 - ab + ac - bc$ into $(a-b)(a+c)$ was depicted this way in Artis Analyticae Praxis, by Thomas Harriott (1631).

Perfect Squares and the Difference of Squares
Common Factors and Grouping
Quadratics $x^2 + Bx + C$
Quadratics $Ax^2 + Bx + C$
Miscellaneous Factoring

Up to now we have multiplied factors to obtain products. Now we consider the problem in reverse: given a product, we are to find its factors. We will consider polynomials whose coefficients are integers and look first for factors $ax + b$, where a and b are also integers. Later we consider cases where a and b are arbitrary real numbers.

Perfect Squares and the Difference of Squares. From the rules for multiplying binomials, we already have three important factorizations.

1.2. Factoring Polynomials

$$a^2x^2 + 2abx + b^2 = (ax+b)(ax+b)$$
$$a^2x^2 - 2abx + b^2 = (ax-b)(ax-b)$$
Perfect Squares

$$a^2x^2 - b^2 = (ax-b)(ax+b)$$ Difference of Squares

Example 1
(a) $4x^2 + 12x + 9 = (2x+3)(2x+3)$
(b) $4x^2 - 12x + 9 = (2x-3)(2x-3)$
(c) $4x^2 - 9 = (2x-3)(2x+3)$ ∎

Caution It is important to remember that in the square of a difference, $a^2x^2 - 2abx + b^2$, only the *middle term* $-2abx$ has the minus sign. For instance $x^2 - 10x - 25$ is *not* the square of a difference, nor is $4x^2 - 12x - 9$. Also, we can factor the difference of squares, but there is no factorization of the sum of squares into linear factors in the real number system. For instance, the expressions $x^2 + 25$ and $4x^2 + 9$ *cannot* be factored into linear factors in the real number system.

Common Factors and Grouping. The terms of a polynomial may have a factor in common. For example, each term of $6x^2 - 2x + 8$ has 2 as a factor. Therefore, $6x^2 - 2x + 8 = 2(3x^2 - x + 4)$. After removing a common factor from the terms of a polynomial, the resulting expression may still be factorable, for example, $5x^2 + 20x + 20 = 5(x^2 + 4x + 4) = 5(x+2)(x+2)$.

Example 2
(a) $3x^2 - 12x + 12 = 3(x^2 - 4x + 4) = 3(x-2)(x-2)$
(b) $7x^2 - 7 = 7(x^2 - 1) = 7(x-1)(x+1)$ ∎

If a polynomial has more than three terms, it may be possible to group the terms in such a way that each grouping has a common polynomial factor. For example,

$$2x^3 + 8x^2 + 3x + 12 = 2x^2(x+4) + 3(x+4)$$
$$= (2x^2 + 3)(x+4)$$

Here, after removing the common factor $2x^2$ from the first two terms and 3 from the last two, we see that the binomial $x + 4$ is a common factor for each pair in the grouping. Similarly,

$$4x^2 - 3 + 3x - 4x = 4x^2 - 4x + 3x - 3$$
$$= 4x(x-1) + 3(x-1)$$
$$= (4x+3)(x-1).$$

Here, the first and last terms have $4x$ as a common factor, and the middle two terms have 3 as a common factor. The grouped terms each have $x-1$ as a factor.

Example 3
$$\begin{aligned} 5x^3 + x^2 - 20x - 4 &= x^2(5x+1) - 4(5x+1) \\ &= (x^2 - 4)(5x+1) \\ &= (x-2)(x+2)(5x+1) \quad \blacksquare \end{aligned}$$

Example 4
$$\begin{aligned} x^3 + 10x^2 + 25x + 2x^2 + 20x + 50 &= x(x^2 + 10x + 25) \\ &\quad + 2(x^2 + 10x + 25) \\ &= (x+2)(x^2 + 10x + 25) \\ &= (x+2)(x+5)(x+5) \quad \blacksquare \end{aligned}$$

Quadratics $x^2 + Bx + C$. We now turn our attention to quadratics other than perfect squares or the difference of squares. For example, $x^2 + 4x + 3$ is not a perfect square but factors into $(x+3)(x+1)$. If we start with a quadratic $x^2 + Bx + C$, where B and C are integers, we want to determine whether there are integers m and n for which $x^2 + Bx + C = (x+m)(x+n)$. Now $(x+m)(x+n) = x^2 + (m+n)x + mn$, which will equal $x^2 + Bx + C$ precisely when $m + n = B$ and $mn = C$. Hence, the quadratic $x^2 + Bx + C$ is factorable by integers if there are two integers m and n whose sum is B and whose product is C; that is,

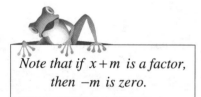

Note that if $x+m$ is a factor, then $-m$ is zero.

$$x^2 + Bx + C = (x+m)(x+n). \qquad (1)$$
$$\uparrow\uparrow$$
$$m+n\ \ mn$$

Example 5 Factor each of the following by integers:

(a) $x^2 - 5x + 4$
(b) $x^2 + x - 12$.

Solution:

(a) We want two integers whose sum is -5 and whose product is 4. The integers -4 and -1 have these properties. Hence, $x^2 - 5x + 4 = (x-4)(x-1)$.
(b) The integers 4 and -3 have their sum equal to 1 and their product equal to -12. Therefore, $x^2 + x - 12 = (x+4)(x-3)$. $\blacksquare$

Example 6 If possible, factor each of the following by integers:

(a) $3x^2 - 6x - 45$
(b) $x^2 + 2x + 3$.

1.2. Factoring Polynomials

Solution:

(a) First, we can factor 3 from each term to obtain $3(x^2 - 2x - 15)$. Next, two integers whose sum is -2 and whose product is -15 are -5 and 3. Hence, $3x^2 - 6x - 45 = 3(x-5)(x+3)$.

(b) There are no two integers whose sum is 2 and whose product is 3. Therefore, $x^2 + 2x + 3$ is not factorable by integers. ∎

Question 1 *Is it possible to have two different factorizations of $x^2 + Bx + C$ by integers?*

Answer: No. If $x + m$ and $x + n$ are factors of $x^2 + Bx + C$, then $-m$ and $-n$ are *roots* of $x^2 + Bx + C = 0$. We will see in Chapter 3 that a quadratic equation cannot have more than two roots. Therefore, except for the order in which we write the factors, the factorization is *unique*. ∎

Quadratics $Ax^2 + Bx + C$. In the previous section we factored quadratics in which the coefficient of x^2 is 1. We now consider arbitrary quadratics $Ax^2 + Bx + C$ in which A, B and C are integers. For example, $10x^2 + 9x + 2$ factors into $(5x+2)(2x+1)$, and $6x^2 + x - 2$ factors into $(2x-1)(3x+2)$. In general, if $Ax^2 + Bx + C$ factors into $(px+m)(qx+n)$, where p, q, m, and n are integers, then

$$Ax^2 + Bx + C = (px+m)(qx+n)$$
$$= pqx^2 + (pn+qm)x + mn. \quad (2)$$
$$\quad\uparrow\qquad\qquad\uparrow\qquad\quad\uparrow$$
$$\quad A\qquad\qquad B\qquad\quad C$$

Hence, p, q, m and n must satisfy the equations

$$A = pq, \quad B = pn + qm, \quad C = mn.$$

The integers p, q, m, and n are often found by inspection, as in the previous case $x^2 + Bx + C$. Now, finding *four* numbers satisfying the conditions in equation (2) is much more difficult than finding *two* numbers satisfying equation (1). However, if we note in (2) that the product $(pn)(qm)$ of the terms whose sum is B is equal to the product $AC = (pn)(qm)$, then we are led to the following procedure.

To factor $Ax^2 + Bx + C$, first find two numbers M and N whose sum is B and whose product is AC, then write

$$Ax^2 + Bx + C = Ax^2 + Mx + Nx + C$$

and proceed by the method of grouping.

Example 7
$$2x^2 + 11x + 12 = 2x^2 + 8x + 3x + 12 \quad M = 8, N = 3; 8 + 3 = 11$$
$$= 2x(x + 4) + 3(x + 4) \quad = B \text{ and } 8 \cdot 3 = 2 \cdot 12 = AC$$
$$= (2x + 3)(x + 4) \quad \blacksquare$$

Example 8
$$6x^2 + x - 2 = 6x^2 + 4x - 3x - 2 \quad M = 4, N = -3$$
$$= 2x(3x + 2) - 1(3x + 2) \quad \text{Note that}$$
$$\quad\quad\quad\quad\quad\quad\quad\quad\quad\quad\quad -3x - 2 = -1(3x + 2)$$
$$= (2x - 1)(3x + 2) \quad \blacksquare$$

Example 9
$$10x^2 - 29x + 10 = 10x^2 - 25x - 4x + 10 \quad M = -25, N = -4$$
$$= 5x(2x - 5) - 2(2x - 5)$$
$$= (5x - 2)(2x - 5) \quad \blacksquare$$

Question 2 *How many pairs of numbers M and N can be found whose sum is B and whose product is AC?*

Answer: If $M + N = B$ and $MN = AC$, then $x^2 - Bx + AC$ factors into $(x - M)(x - N)$. Hence, M and N are roots of the quadratic equation $x^2 - Bx + AC = 0$. As noted in the answer to Question 1, a quadratic equation cannot have more than two roots. Therefore, there can be only one pair of numbers whose sum is B and whose product is AC. $\blacksquare$

In general,
  ax + b is a factor
if and only if −b/a is zero.

Miscellaneous Factoring. The previous rules for factoring quadratics in x can be applied to a wide variety of polynomials. For example, $x^6 - 7x^3 + 12$ can be written as $(x^3)^2 - 7x^3 + 12$, which is a quadratic in x^3 that factors into $(x^3 - 4)(x^3 - 3)$. Similarly, $x^4 - 1$ factors into $(x^2 - 1)(x^2 + 1)$, which can be further factored into $(x - 1)(x + 1)(x^2 + 1)$. The polynomial $x^2 + 5xy + 4y^2$ in x and y factors into $(x + 4y)(x + y)$, and $x^2y^2 - 2xy - 3$ factors into $(xy - 3)(xy + 1)$. By letting $x = a$ in the Thomas Harriott example cited in the quotation at the beginning of this section, we obtain $x^2 - bx + cx - bc = x(x - b) + c(x - b) = (x + c)(x - b)$.

Example 10
$$16x^4 - 81y^4 = (4x^2 - 9y^2)(4x^2 + 9y^2)$$
$$= (2x - 3y)(2x + 3y)(4x^2 + 9y^2) \quad \blacksquare$$

Example 11 $\quad x^4 - 13x^2 + 36 = (x^2 - 9)(x^2 - 4) = (x - 3)(x + 3)(x - 2)(x + 2) \quad \blacksquare$

1.2. Factoring Polynomials

Example 12
$$ax^2 + by^2 - bx^2 - ay^2 = (a-b)x^2 + (b-a)y^2$$
$$= (a-b)x^2 - (a-b)y^2 \quad \text{Note that}$$
$$= (a-b)(x^2 - y^2) \qquad b-a = -(a-b).$$
$$= (a-b)(x-y)(x+y) \quad \blacksquare$$

Example 13
$$6x^2 + 7xy + y^2 = 6x^2 + 6xy + xy + y^2$$
$$= 6x(x+y) + y(x+y)$$
$$= (6x+y)(x+y) \quad \blacksquare$$

Example 14
$$12ax^3 - 2ax^2 - 2ax = 2ax(6x^2 - x - 1)$$
$$= 2ax(6x^2 - 3x + 2x - 1)$$
$$= 2ax[3x(2x-1) + 1(2x-1)]$$
$$= 2ax(3x+1)(2x-1) \quad \blacksquare$$

As illustrated in the following example, our factoring procedures can also be applied to polynomials whose coefficients are not necessarily integers. We will consider the problem of factoring polynomials whose coefficients are arbitrary real numbers in the chapter on roots of polynomial equations.

Example 15
(a) $x^2 - \dfrac{4}{9} = \left(x - \dfrac{2}{3}\right)\left(x + \dfrac{2}{3}\right)$
(b) $x^2 - 5 = (x - \sqrt{5})(x + \sqrt{5})$
(c) $x^2 + .5x - .5 = (x+1)(x-.5)$ $\quad \blacksquare$

If $x^3 - a^3$ is divided by $x - a$ according to the method in Section 1.1, the quotient is $x^2 + ax + a^2$ and the remainder is zero (see Exercise 103). Therefore, we obtain the factorization

$$\boldsymbol{x^3 - a^3 = (x - a)(x^2 + ax + a^2)}. \qquad \text{Difference of Cubes}$$

Similarly, if we divide $x^3 + a^3$ by $x + a$, then the quotient is $x^2 - ax + a^2$ and the remainder is zero (Exercise 104). That is,

$$\boldsymbol{x^3 + a^3 = (x + a)(x^2 - ax + a^2)}. \qquad \text{Sum of Cubes}$$

The factorization for the sum of cubes could also be obtained by replacing a by $-a$ in the factorization of the difference of two cubes *(why?)*

Example 16
(a) $x^3 - 27 = x^3 - 3^3$
$= (x-3)(x^2 + 3x + 9)$
(b) $x^3 + 27 = x^3 + 3^3$
$= (x+3)(x^2 - 3x + 9)$

(c) $a^3 - 8b^3 = a^3 - (2b)^3$
$= (a - 2b)(a^2 + 2ab + 4b^2)$
(d) $a^3 + 8b^3 = a^3 + (2b)^3$
$= (a + 2b)(a^2 - 2ab + 4b^2)$ ∎

Note that both $x^3 - a^3$ and $x^2 - a^2$ have $x - a$ as a factor, and $x^3 + a^3$ has $x + a$ as a factor but $x^2 + a^2$ does not. In general, if n is *any* positive integer, then $x^n - a^n$ has $x - a$ as a factor (see Exercise 105). Also, if n is an *odd* positive integer, then $x^n + a^n$ has $x + a$ as a factor (Exercise 106). However, if n is an *even* positive integer and $a \neq 0$, then $x + a$ is not a factor of $x^n + a^n$ (Exercise 107).

Exercises 1.2

Fill in the blanks to make each statement true.

1. If $m + n = 4$ and $mn = -5$, then $(x + m)(x + n) =$ _____.
2. $x^2 + (6 - a)x - 6a$ factors into _____.
3. $a(x - y) + b(y - x) = a(x - y)$ _____ $b(x - y) = ($_____$)$ $($_____$)$.
4. If $m + n = B$ and $mn = C$, then $x^{2k} + Bx^k + C =$ $($_____$)($_____$)$.
5. To factor $Ax^2 + Bx + C$, first find two integers M and N, where $M + N =$ _____ and $MN =$ _____, then write $Ax^2 + Bx + C = Ax^2 + Mx + Nx + C$ and proceed by the method of _____.

Write true or false for each statement.

6. The expression $x^2 + 2x + 2$ is not factorable by integers.
7. $(a - b)(x - y) + (b - a)(y - x) = 2(a - b)(x - y)$
8. If $Ax^2 + Bx + C = (px + m)(qx + n)$, then $m + n = B$ and $mn = C$.
9. If $A = 1$, then the procedure for factoring $Ax^2 + Bx + C$ will give an answer different from the one given by the procedure for factoring $x^2 + Bx + C$.
10. The quadratic $Ax^2 + (A + C)x + C$ has the same factors as $x^2 + (A + C)x + AC$.

Perfect Squares and the Difference of Squares

Some of the following are perfect squares and some are not.

Write the ones that are perfect squares in factored form.

11. $x^2 + 4x + 4$
12. $x^2 + 4x - 4$
13. $x^2 + 4$
14. $x^2 - 2x + 1$
15. $x^2 - 2x - 1$
16. $x^2 + x + 1$
17. $4x^2 - 20xy + 25y^2$
18. $4x^2 + 20xy - 25y^2$
19. $x^4 + 6x^2 + 9$
20. $49x^8 - 14x^4y^2 + y^4$

Factor by integers each of the following.

21. $x^2 - 64$
22. $4x^2 - 25$
23. $x^4 - y^4$
24. $4x^4 - 625$
25. $(a + x)^2 - (a - x)^2$
26. $(a + x)^2 - (a + y)^2$

Common Factors and Grouping

Factor by integers each of the following.

27. $3x^2 - 12$
28. $125 - 20y^2$
29. $2x^2 + 8x + 8$
30. $3x^2 - 12x + 12$
31. $x^2 + x + 4x + 4$
32. $x^2 - 4x - x + 4$
33. $3x^2 + 9x - 12x - 36$
34. $4x^2 + 16x + 12x + 48$
35. $x^3 + x^2 - 9x^2 - 9x$
36. $x^3 - 9x^2 - x^2 + 9x$

Factoring Quadratics $Ax^2 + Bx + C$

Some of the following are factorable by integers and some are not. Write each factorable polynomial in factored form.

37. $x^2 - 3x + 2$
38. $x^2 + 3x - 2$
39. $x^2 - x - 2$
40. $x^2 - x + 2$
41. $x^2 - 2$
42. $2x^2 - 50$
43. $3x^2 - 3x - 6$
44. $3x^2 - 3x + 6$
45. $2x^2 - 5x + 3$
46. $2x^2 + 5x + 3$

Factor by integers each of the following.

47. $x^2 - 18x + 80$
48. $x^2 + 18x + 80$
49. $x^2 + 29x + 100$
50. $x^2 - 48x - 100$
51. $x^2 + 23x + 132$
52. $x^2 + x - 132$
53. $3x^2 + 15x + 12$
54. $4x^2 - 20x + 16$
55. $2x^2 - 16x - 18$
56. $5x^2 - 50x + 45$
57. $2x^2 + 9x + 4$
58. $3x^2 + 17x + 10$
59. $5x^2 + 21x + 4$
60. $3x^2 + 25x + 8$
61. $5x^2 + 12x + 7$
62. $6x^2 + 5x + 1$
63. $4x^2 - 12x + 5$
64. $8x^2 - 15x - 2$
65. $10x^2 - 21x - 10$
66. $12x^2 + 7x - 10$

Miscellaneous Factoring

Factor by integers each of the following.

67. $x^3 + 8$
68. $x^3 - 8$
69. $8x^3 - 27y^3$
70. $64x^3 + 125y^3$
71. $16a^6 - 54b^3$
72. $(x+a)^3 - 8a^3$
73. $x^2 - 5bx + 4b^2$
74. $x^2 + 2ax - 3a^2$
75. $x^2 - (a+b)x + ab$
76. $x^2 - (2a + 3b)x + 6ab$
77. $a^2x^2 + 5ax + 6$
78. $a^2x^2 - 9ax + 20$
79. $(a-b)x^2 - 3(a-b)x + 2(a-b)$
80. $(a+b)x^2 - 7(a+b)x + 10(a+b)$
81. $(x+a)^2 - a^2$
82. $(a^2 + 2a + 1)x^2 - 25$
83. $(x^2 + 6x + 9) - (a^2 + 2a + 1)$
84. $(a-b)x^4 + (b-a)y^4$
85. $ax^2 - ay^2 + 2x^2 - 2y^2$
86. $ax^2 - ay^2 - 2x^2 + 2y^2$
87. $(a+b)x^2 - (a+b)y^2 + (a-b)x^2 - (a-b)y^2$
88. $(a+b)x^2 - (a+b)y^2 - (a-b)x^2 + (a-b)y^2$
89. $6ax^2 + 13ax + 6a$
90. $10ax^2 + 3ax - 4a$
91. $5x^3 + 18x^2 - 8x$
92. $9x^3 - 24x^2 + 12x$
93. $3x^4 + 8x^2 + 5$
94. $4x^4 - 17x^2 + 4$
95. $2x^4 + (2a^2 + 1)x^2 + a^2$
96. $4x^4 + (12 - b^2)x^2 - 3b^2$

Factor the expressions in Exercises 97-102 by real numbers.

97. $x^2 - 2$
98. $9x^2 - \dfrac{1}{4}$
99. $4x^2 - 12$
100. $x^2 + \dfrac{6}{5}x + \dfrac{1}{5}$
101. $x^2 + 2x + \dfrac{3}{4}$
102. $x^2 - 1.5x - 1$

103. Verify that
$$\dfrac{x^3 - a^3}{x - a} = x^2 + ax + a^2$$
by performing the division
$$x - a \overline{)x^3 + 0x^2 + 0x - a^3}.$$
Hence, conclude that
$x^3 - a^3 = (x - a)(x^2 + ax + a^2).$

104. Verify that
$$\dfrac{x^3 + a^3}{x + a} = x^2 - ax + a^2$$
by performing the division
$$x + a \overline{)x^3 + 0x^2 + 0x + a^3}.$$
Hence, conclude that
$x^3 + a^3 = (x + a)(x^2 - ax + a^2).$

105. Let n be *any* positive integer. Show that
$$\dfrac{x^n - a^n}{x - a} = x^{n-1} + ax^{n-2} + a^2x^{n-3} + \ldots$$
$$+ a^{n-2}x + a^{n-1}$$
by performing the division
$$x - a \overline{)x^n + 0x^{n-1} + 0x^{n-2} + \ldots + 0x - a^n}.$$
Hence, $x - a$ is a factor of $x^n - a^n$.

70   Chapter 1. Algebraic Expressions

106. Let n be an *odd* positive integer. Show that

$$\frac{x^n + a^n}{x + a} = x^{n-1} - ax^{n-2} + a^2x^{n-3} - \ldots$$
$$- a^{n-2}x + a^{n-1}$$

by performing the division

$$x + a \overline{\smash{\big)}\, x^n + 0x^{n-1} + 0x^{n-2} + \ldots + 0x + a^n}.$$

Hence, $x + a$ is a factor of $x^n + a^n$ when n is odd.

107. Let n be an *even* positive integer. Show that

$$\frac{x^n + a^n}{x + a} = x^{n-1} - ax^{n-2} + a^2x^{n-3} - \ldots$$
$$+ a^{n-2}x - a^{n-1} + \frac{2a^n}{x + a}$$

by performing the division

$$x + a \overline{\smash{\big)}\, x^n + 0x^{n-1} + 0x^{n-2} + \ldots + 0x + a^n}.$$

Hence, if $a \neq 0$, then $x + a$ is *not* a factor of $x^n + a^n$ when n is even.

1.3 Operations with Rational Expressions

Rational Expressions
Least Common Multiple
Addition of Rational Expressions
Multiplication and Division of Rational Expressions

Polynomials are the integers of algebra, and rational expressions are the fractions of algebra.

To add rational expressions with the least amount of work, use the least common denominator.

Rational Expressions. As defined in the preceding chapter, a rational number is the ratio of two integers. Similarly, the ratio of two polynomials is a **rational expression.** For instance, 3/4 and 5/6 are rational numbers, whereas

$$\frac{x + 2}{x^2 - 1} \quad \text{and} \quad \frac{1}{x^2 + 3x + 2}$$

are rational expressions. In general, any algebraic expression generated by the operations of addition, subtraction, multiplication, and division is called a rational expression. Of course, it is understood that a denominator in a rational expression cannot be equal to zero.

Example 1   Which of the following algebraic expressions are rational expressions?

(a) $\dfrac{x^2 + x + 1}{x + 1}$

(b) $\sqrt{x^2 + 3}$

(c) $(x^3 - x + 2)^2$

(d) $(x + 4)^{-1}$

(e) $\dfrac{\sqrt{x}}{x}$

Solution: (a), (c), and (d) are rational expressions. ∎

Least Common Multiple.   Following the rules for addition of fractions in Section R.4, we have

$$\frac{5}{6} + \frac{3}{4} = \frac{5 \cdot 4 + 6 \cdot 3}{6 \cdot 4} = \frac{38}{24} = \frac{19}{12},$$

and

$$\frac{x}{x^2 + 3x + 2} + \frac{1}{x^2 - 1} = \frac{x(x^2 - 1) + x^2 + 3x + 2}{(x^2 + 3x + 2)(x^2 - 1)}$$

$$= \frac{x^3 + x^2 + 2x + 2}{(x^2 + 3x + 2)(x^2 - 1)}$$

$$= \frac{x^2(x + 1) + 2(x + 1)}{(x + 2)(x + 1)(x - 1)(x + 1)}$$

$$= \frac{(x^2 + 2)\cancel{(x+1)}}{(x + 2)(x + 1)(x - 1)\cancel{(x+1)}}$$

$$= \frac{x^2 + 2}{(x + 2)(x + 1)(x - 1)}.$$

In both cases, the addition can be simplified by using the common factor property (Section R.4) to introduce a **common denominator** as follows:

$$\frac{5}{6} + \frac{3}{4} = \frac{5 \cdot 2}{6 \cdot 2} + \frac{3 \cdot 3}{4 \cdot 3} = \frac{10}{12} + \frac{9}{12} = \frac{19}{12},$$

and

$$\frac{x}{x^2 + 3x + 2} + \frac{1}{x^2 - 1}$$

$$= \frac{x}{(x + 2)(x + 1)} + \frac{1}{(x - 1)(x + 1)} \qquad \textit{Factor each denominator.}$$

$$= \frac{x \cdot (x - 1)}{(x + 2)(x + 1) \cdot (x - 1)} + \frac{1 \cdot (x + 2)}{(x - 1)(x + 1) \cdot (x + 2)} \qquad \begin{array}{l}\textit{Common}\\\textit{Factor}\\\textit{Property}\end{array}$$

$$= \frac{x(x - 1) + x + 2}{(x + 2)(x + 1)(x - 1)} \qquad \textit{Addition with Common Denominator}$$

$$= \frac{x^2 + 2}{(x + 2)(x + 1)(x - 1)}.$$

In the first case, the common denominator 12 is the *smallest* positive integer that has for its factors the factors of 6 and 4. Hence, 12 is called the **Least Common Multiple (LCM)** of 6 and 4. In the second case, the common denominator $(x+2)(x+1)(x-1)$ is the polynomial of *smallest degree* that has for its only factors the factors of $x^2 + 3x + 2$ and $x^2 - 1$. Hence, $(x+2)(x+1)(x-1)$ is called the least common multiple (LCM) of $x^2 + 3x + 2$ and $x^2 - 1$.

Comment In the case of integers, the LCM is defined for *positive* integers only. Similarly, for polynomials of degree n, we assume that the coefficient of x^n is a *positive* integer.

Example 2 Show that the LCM of 56 and 98 is 392.

Solution: First factor each term.

$$56 = 2^3 \cdot 7 \text{ and } 98 = 2 \cdot 7^2$$

Therefore, $2^3 \cdot 7^2 = 392$ is the smallest positive integer that has 2^3, 7, 2, and 7^2 as factors. ∎

Example 3 Find the LCM of $2x - 4$, $x^2 - 4x + 4$, and $x^3 - 8$.

Solution: First factor each polynomial.

$$2x - 4 = 2(x - 2)$$
$$x^2 - 4x + 4 = (x - 2)^2$$
$$x^3 - 8 = (x - 2)(x^2 + 2x + 4)$$

Therefore, $2(x-2)^2(x^2 + 2x + 4)$ is the polynomial of smallest degree that has 2, $x - 2$, $(x-2)^2$, $x - 2$, and $x^2 + 2x + 4$ as its only factors. ∎

As these examples illustrate, the LCM of a set of integers or polynomials is computed according to the following procedure.

---

[1] *A prime factor is a factor other than 1 whose only factors are itself and 1.*

---

1. Factor each member of the set into powers of prime factors.[1]
2. The **least common multiple (LCM)** is the product of all the different prime factors, each raised to the highest power to which it appears among all the members of the set.

Example 4 Find the LCM of 54, $24x + 12$, and $2x^2 - x - 1$.

1.3. Operations with Rational Expressions

Solution:

$$54 = 2 \cdot 3^3$$
$$24x + 12 = 12(2x + 1)$$
$$= 2^2 \cdot 3(2x + 1)$$
$$2x^2 - x - 1 = (2x + 1)(x - 1)$$

Therefore, the LCM $= 2^2 \cdot 3^3(2x + 1)(x - 1)$. ∎

Example 5 Find the LCM of $(x - 1)^3(2x + 3)(x + 4)$, $(x - 1)(2x + 3)^3(x + 4)^2$, and $(x - 1)^2(2x + 3)^2(x + 4)^3$.

Solution: The terms are already in factored form. Therefore,

$$\text{LCM} = (x - 1)^3(2x + 3)^3(x + 4)^3. \quad \blacksquare$$

Addition of Rational Expressions. As observed at the beginning of this section, the easiest way to add (or subtract) rational expressions is to use as a common denominator the LCM of the given denominators. In this context, the LCM is called the **least common denominator** (LCD). Hence, to add or subtract rational expressions, we first factor each denominator and compute the LCD. Then, by multiplying each numerator and denominator by the necessary factor, we replace each rational expression by an equivalent one whose denominator is the LCD. We then follow the procedure for adding or subtracting fractional expressions having a common denominator. That is, we combine the numerators and place this result over the LCD. Of course, answers should be simplified as much as possible.

Example 6
$$\frac{x + 1}{x^2 + x - 2} + \frac{x - 2}{x^2 - 2x + 1} = \frac{x + 1}{(x + 2)(x - 1)} + \frac{x - 2}{(x - 1)^2}$$
$$\text{LCD} = (x+2)(x-1)^2 \qquad \textit{Factor each denominator}$$
$$= \frac{(x + 1) \cdot (x - 1)}{(x + 2)(x - 1) \cdot (x - 1)}$$
$$+ \frac{(x - 2) \cdot (x + 2)}{(x - 1)^2 \cdot (x + 2)}$$
$$\qquad\qquad\qquad\qquad \textit{Common Factor Property}$$
$$= \frac{x^2 - 1 + x^2 - 4}{(x + 2)(x - 1)^2}$$
$$\qquad\qquad\qquad\qquad \textit{Addition with LCD}$$
$$= \frac{2x^2 - 5}{(x + 2)(x - 1)^2} \quad \blacksquare$$

Example 7

$$\frac{5}{2x-4} + \frac{4}{3x+6} - \frac{x}{x^2-4} = \frac{5}{2(x-2)} + \frac{4}{3(x+2)} - \frac{x}{(x-2)(x+2)}$$

LCD = $6(x-2)(x+2)$

$$= \frac{5 \cdot 3(x+2)}{2(x-2) \cdot 3(x+2)} + \frac{4 \cdot 2(x-2)}{3(x+2) \cdot 2(x-2)}$$

$$- \frac{x \cdot 6}{(x-2)(x+2) \cdot 6}$$

$$= \frac{15x + 30 + 8x - 16 - 6x}{6(x-2)(x+2)}$$

$$= \frac{17x + 14}{6(x-2)(x+2)} \quad \blacksquare$$

Example 8 Add $\dfrac{2x+1}{x+3} + \dfrac{1}{9-x^2}$

Solution: Here we first multiply the numerator and denominator of $\dfrac{1}{9-x^2}$ by -1 to obtain $\dfrac{-1}{x^2-9}$ (see previous Comment). We then continue as in the preceding examples.

$$\frac{2x+1}{x+3} + \frac{1}{9-x^2} = \frac{2x+1}{x+3} + \frac{-1}{x^2-9}$$

LCD = $(x-3)(x+3)$

$$= \frac{2x+1}{x+3} + \frac{-1}{(x-3)(x+3)}$$

$$= \frac{(2x+1) \cdot (x-3)}{(x+3) \cdot (x-3)} + \frac{-1}{(x-3)(x+3)}$$

$$= \frac{2x^2 - 5x - 3 + (-1)}{(x-3)(x+3)}$$

$$= \frac{2x^2 - 5x - 4}{(x-3)(x+3)} \quad \blacksquare$$

For fractions,

$$\frac{a}{b} \cdot \frac{c}{d} = \frac{ac}{bd}$$

and for rational expressions,

$$\frac{P}{Q} \cdot \frac{R}{S} = \frac{PR}{QS}.$$

Multiplication and Division of Rational Expressions. Unlike addition and subtraction, multiplication of rational expressions does not require a common denominator. As discussed in Section R.4, we simply multiply the numerators and multiply the denominators. Also, to divide rational expressions, we first invert the divisor and proceed as in multiplication. Of course, we simplify the results as far as possible by canceling any common factors, *which we assume are never zero*.

Example 9 Perform the multiplication $\dfrac{x^2 - 2xy}{3x + 3y} \cdot \dfrac{x^2 - y^2}{xy - 2y^2}$.

Solution:

$$\frac{x^2 - 2xy}{3x + 3y} \cdot \frac{x^2 - y^2}{xy - 2y^2} = \frac{(x^2 - 2xy)(x^2 - y^2)}{(3x + 3y)(xy - 2y^2)}$$

$$= \frac{x\cancel{(x - 2y)}\cancel{(x + y)}(x - y)}{3\cancel{(x + y)}y\cancel{(x - 2y)}}$$

$$= \frac{x(x - y)}{3y} \quad \blacksquare$$

In Example 9, we were able to simplify the result by canceling common factors *after* the multiplication was performed. As the following example illustrates, we can sometimes simplify the computations by canceling in each rational expression *before* the operations are performed.

Example 10 Perform the division $\dfrac{x^2 - 16}{x^2 + 5x + 4} \div \dfrac{x^2 - 3x - 4}{x^3 + 1}$.

Solution:

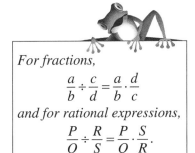

For fractions,

$$\frac{a}{b} \div \frac{c}{d} = \frac{a}{b} \cdot \frac{d}{c}$$

and for rational expressions,

$$\frac{P}{Q} \div \frac{R}{S} = \frac{P}{Q} \cdot \frac{S}{R}.$$

$$\frac{x^2 - 16}{x^2 + 5x + 4} \div \frac{x^2 - 3x - 4}{x^3 + 1}$$

$$= \frac{\cancel{(x + 4)}(x - 4)}{\cancel{(x + 4)}(x + 1)} \div \frac{(x - 4)\cancel{(x + 1)}}{\cancel{(x + 1)}(x^2 - x + 1)}$$

$$= \frac{x - 4}{x + 1} \div \frac{x - 4}{x^2 - x + 1} \quad \text{\textit{Simplify each expression before performing the division.}}$$

$$= \frac{x - 4}{x + 1} \cdot \frac{x^2 - x + 1}{x - 4} \quad \text{\textit{Invert the divisor.}}$$

$$= \frac{\cancel{(x - 4)}(x^2 - x + 1)}{(x + 1)\cancel{(x - 4)}} \quad \text{\textit{Cancel after the division.}}$$

$$= \frac{x^2 - x + 1}{x + 1} \quad \blacksquare$$

Exercises 1.3

Fill in the blanks to make each statement true.

1. The LCM of two positive integers is the _____ positive integer that has both integers as a _____.
2. To add rational numbers, we use the LCD, which is the _____ of the given denominators.
3. The LCD of two denominators must be a _____ of both.
4. The rational expression $(x^2 - 36)(2x + 12)(x^2 + 12x + 36)$ is a multiple of $x^2 - 36$, $2x + 12$, and $x^2 + 12x + 36$, but the LCM is _____.
5. Operations with rational expressions follow the same rules as operations with _____.

Chapter 1. Algebraic Expressions

Write true or false for each statement.

6. If m and n integers, then $m \cdot n$ is a multiple of m and n.
7. If m and n are positive integers, then $m \cdot n$ is always their LCM.
8. In adding two rational expressions with polynomial denominators $P(x)$ and $Q(x)$, the common denominator $P(x) \cdot Q(x)$ can be used.
9. In adding two rational expressions with polynomial denominators $P(x)$ and $Q(x)$, the simplest denominator to use is the LCM of $P(x)$ and $Q(x)$.
10. When dividing rational expressions, we divide the LCM of the numerators by the LCM of the denominators.

Least Common Multiple

Find the LCM of the given terms in each of the following.

11. $14, 42, 9$
12. $3^4 \cdot 2,\; 3 \cdot 2^3 \cdot 5,\; 3^2 \cdot 5^2$
13. $x^2 - 9,\; x^2 - 2x - 3,\; x^2 + 2x - 3$
14. $2x^3 + 16,\; 6x^2 + 9x - 6,\; 4x^2 - 8x + 16$

Find the LCD to be used when adding fractions with the given denominators.

15. $3x^2,\; x^3 y,\; xy^2$
16. $3x - 3y,\; x^2 - y^2$
17. $x^2 - x - 2,\; x^2 - 4,\; 2x + 2$
18. $x^3 - 8,\; x^2 - 4x + 4,\; x^2 + x - 2$

Additional of Rational Expressions

Combine each of the following by using the LCD, and simplify.

19. $\dfrac{1}{14} + \dfrac{5}{42} - \dfrac{2}{9}$

20. $\dfrac{3x}{20} - \dfrac{3x}{15}$

21. $\dfrac{1}{x+h} - \dfrac{1}{x}$

22. $\dfrac{1}{(x+h)^2} - \dfrac{1}{x^2}$

23. $\dfrac{x+1}{x^3} - \dfrac{3}{x^2}$

24. $\dfrac{4}{x^3} + \dfrac{1}{x^4}$

25. $\dfrac{w+2}{2x-2} + \dfrac{3w-1}{x^2 - 4x + 3}$

26. $\dfrac{x}{(x^2 - 1)^2} + \dfrac{3}{(x+1)^2}$

27. $\dfrac{4x - 3}{4x^2 - 1} + \dfrac{3x}{2 - 4x}$

28. $\dfrac{2}{2x - x^2} - \dfrac{3}{x^2 - 4}$

29. $\dfrac{1}{x^2 + 3x + 2} + \dfrac{x+2}{x^2 + 5x + 4}$

30. $\dfrac{x-2}{x^2 + x - 2} + \dfrac{x-3}{x^2 + 2x - 3}$

31. $\dfrac{x+1}{2x^2 - 5x - 3} - \dfrac{x-2}{x^2 - 3x}$

32. $\dfrac{x-4}{3x^2 + 7x + 2} + \dfrac{x+4}{x^2 + 2x}$

Perform the indicated addition and subtractions.

33. $\dfrac{1}{x^2 - 4} + \dfrac{x}{4 - x^2} + \dfrac{1}{x - 2}$

34. $\dfrac{x}{x^2 - 9} - \dfrac{x}{3 - x} + \dfrac{x}{x + 3}$

35. $\dfrac{1}{x - 2} + \dfrac{x + 4}{x^2 - 5x + 6} - \dfrac{1}{x^2 - 4x + 4}$

36. $\dfrac{1}{x^2 - 1} - \dfrac{x}{x^2 + 2x + 1} - x$

37. $\dfrac{3}{x} - \dfrac{3}{x(x-2)} + \dfrac{x}{(x-2)^2}$

38. $\dfrac{1}{x(x+1)} + \dfrac{x+1}{x^2} - \dfrac{x}{(x+1)^2}$

39. $\dfrac{x}{x+3} - \dfrac{2}{x(x+3)} + \dfrac{1}{x^2} - \dfrac{5}{(x+3)^2}$

40. $\dfrac{x+3}{x(x-1)} + \dfrac{2}{x} + \dfrac{x+1}{(x-1)^2} + \dfrac{1}{x^2(x-1)}$

41. $\dfrac{1}{x^2+3x+2} - \dfrac{1}{x+2} + \dfrac{2}{x^2+4x+3} - \dfrac{2}{x+1}$

42. $\dfrac{x}{x^2-1} - \dfrac{1}{x^2+x-2} - \dfrac{x}{x+1} + \dfrac{1}{x^2-x-2}$

43. $\dfrac{x}{x^2+(a+b)x+ab} - \dfrac{1}{x+a} + \dfrac{1}{x+b}$

44. $\dfrac{ax}{x^2-a^2} + \dfrac{1}{x-a} - \dfrac{1}{x+a}$

45. $\dfrac{x^2}{x^3-a^3} - \dfrac{x}{x^2+ax+a^2} + \dfrac{1}{x-a}$

46. $\dfrac{x^2}{x^3+a^3} - \dfrac{x}{x^2-ax+a^2} + \dfrac{1}{x+a}$

Multiplication and Division of Rational Expressions

Perform the indicated operations in each of the following, and simplify.

47. $\dfrac{3xy^2}{5ab} \cdot \dfrac{20a^3}{9x^2}$

48. $\dfrac{2x+y}{x^2-4y^2} \cdot \dfrac{x+2y}{4x^2+4xy+y^2}$

49. $\dfrac{10x^2-29x+10}{9x^2-4} \cdot \dfrac{6x^2+x-2}{4x^2-20x+25}$

50. $\dfrac{2x^2+11x+12}{2x^2-3x-20} \cdot \dfrac{4x^2+16x+15}{x^2-16}$

51. $\dfrac{3ab^2c^3}{7xy} \div \dfrac{12abc}{35y^2z}$

52. $\dfrac{6x^2+7xy+y^2}{6x+6y} \div \dfrac{(x+y)^2}{6x+y}$

53. $\dfrac{x^4-13x^2+36}{x^2-2x} \div \dfrac{x^2-x-6}{5x}$

54. $\left(\dfrac{1}{x}-x\right) \div \left(\dfrac{2}{x+1}-1\right)$

55. $\dfrac{\dfrac{x+1}{x^2+x-2} - \dfrac{x-2}{x^2-2x+1}}{\dfrac{1}{x-1} - \dfrac{1}{x^2+x-2}}$

Miscellaneous

In each of the following, simplify and perform the indicated operations

56. $\dfrac{1}{1+\dfrac{2}{x+1}} + \dfrac{1}{x+1-\dfrac{4}{x+1}}$

57. $\dfrac{1}{1-\dfrac{1}{x^2}} - \dfrac{1}{1+\dfrac{1}{x}}$

58. $\dfrac{x^{-1}}{1-x^{-1}} + \dfrac{1}{1+x^{-1}} + \dfrac{x^{-1}}{2}$

59. $\left(1+\dfrac{1}{1+\dfrac{1}{1+\dfrac{1}{x}}}\right) \cdot \left(3-\dfrac{2}{1-\dfrac{1}{1-\dfrac{1}{x}}}\right)$

60. $\dfrac{1+\dfrac{1}{1+\dfrac{1}{x}}}{1+\dfrac{1}{1-\dfrac{1}{x}}} \div \dfrac{1-\dfrac{1}{1+\dfrac{1}{x}}}{1-\dfrac{1}{1-\dfrac{1}{x}}}$

1.4 Rational Exponents

Definition of $x^{1/n}$
Definition of $x^{m/n}$
Power-Root Rule
Rules for Rational Exponents

With rational exponents, roots behave like powers.

The meaning of integral exponents, as in

$$3^4 = 81, \qquad 10^{-2} = \dfrac{1}{100}, \qquad \text{and} \qquad 2^0 = 1,$$

and the meaning of roots, such as

$$\sqrt{4} = 2, \quad \sqrt[3]{-8} = -2, \quad \text{and} \quad \sqrt[4]{(-2)^4} = 2,$$

were explained in Section R.5. We now show how exponents and roots are related.

Definition of $x^{1/n}$. We know that $(\sqrt{5})^2 = 5$, and if we formally apply the exponent rule $(x^m)^n = x^{mn}$ to the expression $(5^{1/2})^2$, the result is

$$(5^{1/2})^2 = 5^{(1/2)2} = 5^1 = 5.$$

Therefore, since the squares of both $5^{1/2}$ and $\sqrt{5}$ must equal 5, consistency requires that we set $5^{1/2} = \sqrt{5}$. This example suggests the following definition.

> If n is an integer greater than or equal to 2, then by definition
>
> $$x^{1/n} = \sqrt[n]{x}.$$

The same restriction must be applied to $x^{1/n}$ as to $\sqrt[n]{x}$. That is, if n is *even*, x cannot be negative, and for n even and x positive, both $x^{1/n}$ and $\sqrt[n]{x}$ mean the *positive* nth root of x.

Example 1
(a) $36^{1/2} = \sqrt{36} = 6$, whereas $(-36)^{1/2}$ has no meaning in the real number system.
(b) $8^{1/3} = \sqrt[3]{8} = 2$ and $(-8)^{1/3} = \sqrt[3]{-8} = -2$
(c) $10^{1/4} = \sqrt[4]{10} = 1.778$ (to three decimal places) ∎

Definition of $x^{m/n}$. Now let us figure out how to give meaning to $x^{m/n}$ for any rational number m/n. For example, the expression $9^{3/2}$ should equal $(9^{1/2})^3$ since $3/2 = (1/2)3$, and, similarly, we want $(-8)^{2/3}$ to mean $[(-8)^{1/3}]^2$. But we note a necessary restriction. For instance, since $2/3 = 4/6$, we would expect $(-8)^{2/3}$ and $(-8)^{4/6}$ to have the same meaning. But

$$(-8)^{2/3} = [(-8)^{1/3}]^2 = (-2)^2 = 4,$$

whereas $(-8)^{4/6} = [(-8)^{1/6}]^4$ can't be a real number (*why?*). To avoid this problem, we assume that the rational exponent m/n is in *lowest terms*, that is, all common factors (except 1) have been canceled from the numerator and denominator, and of course $n \neq 0$. These considerations lead us to the following definition.

1.4. Rational Exponents

> If m and n are positive integers with no common factors (except 1), then by definition
> $$x^{m/n} = (x^{1/n})^m,$$
> where x cannot be negative when n is even. If, in addition, $x \neq 0$, then
> $$x^{-m/n} = \frac{1}{x^{m/n}}.$$

Note that the last part of this definition agrees with the corresponding definition in Section R.5 when m/n is an integer.

Example 2 (a) $4^{5/2} = (4^{1/2})^5 = 2^5$, but $(-4)^{5/2}$ has no meaning in the real number system.

(b) $32^{-3/5} = \dfrac{1}{32^{3/5}} = \dfrac{1}{(32^{1/5})^3} = \dfrac{1}{2^3} = \dfrac{1}{8}$ and

$(-32)^{-3/5} = \dfrac{1}{(-32)^{3/5}} = \dfrac{1}{[(-32)^{1/5}]^3} = \dfrac{1}{(-2)^3} = \dfrac{1}{(-8)} = -\dfrac{1}{8}$ ∎

Example 3 $x^{3/2} = (\sqrt{x})^3$ for $x \geq 0$, and $x^{2/3} = (\sqrt[3]{x})^2$ for *every* real number x. ∎

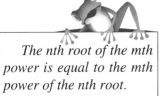

The nth root of the mth power is equal to the mth power of the nth root.

Power-Root Rule. If we note that $(4^3)^{1/2} = 64^{1/2} = 8$, and also that $(4^{1/2})^3 = 2^3 = 8$, we have an example of the following general rule (see Exercise 40).

$$(x^m)^{1/n} = (x^{1/n})^m \qquad \text{Power-root Rule}$$

Hence, we can interpret $x^{m/n}$ as either $(x^{1/n})^m$ or $(x^m)^{1/n}$, provided the restrictions in the definition of $x^{m/n}$ are observed. The power-root rule in terms of radicals is

$$\sqrt[n]{x^m} = (\sqrt[n]{x})^m.$$

Example 4 By definition, $8^{4/3} = (8^{1/3})^4 = 16$. Also, by the power-root rule, $8^{4/3} = (8^4)^{1/3} = (4096)^{1/3} = 16$. ∎

Example 5 Show that $8^{-4/3}$, $(8^{1/3})^{-4}$, and $(8^{-4})^{1/3}$ are equal to $1/16$.

Solution:

$$8^{-4/3} = \frac{1}{8^{4/3}} = \frac{1}{2^4} = \frac{1}{16}$$

$$(8^{1/3})^{-4} = 2^{-4} = \frac{1}{2^4} = \frac{1}{16}$$

$$(8^{-4})^{1/3} = \sqrt[3]{\frac{1}{8^4}} = \sqrt[3]{\frac{1}{4096}} = \frac{1}{16} \quad \blacksquare$$

Example 6 $[(-9)^2]^{1/4} = \sqrt[4]{81} = 3$, but $[(-9)^{1/4}]^2$ is not a real number. This example does not contradict the power-root rule because $2/4$ is not yet in lowest terms. $\blacksquare$

Rules for Rational Exponents. All the basic rules given for integral exponents in Section R.5 continue to hold for rational exponents, provided we avoid even roots of negative numbers. These rules are as follows, where it is understood that p, q and r are positive or negative rational numbers in lowest terms.

For products with the same base, we *add* the exponents and for quotients with the same base, we *subtract* the exponents:

$$x^p \cdot x^q = x^{p+q} \quad \text{and} \quad \frac{x^p}{x^q} = x^{p-q} \quad \text{Same-base Rules}$$

Example 7 (a) $8^{2/3} \cdot 8^{1/3} = 8^{2/3+1/3} = 8^1 = 8$

(b) $\dfrac{16^{-3/4}}{16^{1/2}} = 16^{-3/4-1/2} = 16^{-5/4} = \dfrac{1}{16^{5/4}} = \dfrac{1}{32}$ $\blacksquare$

Also, the power of a product equals the *product* of the powers, and the power of a quotient equals the *quotient* of the powers:

$$(x \cdot y)^r = x^r \cdot y^r \quad \text{and} \quad \left(\frac{x}{y}\right)^r = \frac{x^r}{y^r}. \quad \text{Same-exponent Rules}$$

1.4. Rational Exponents

Example 8 (a) $(16 \cdot 9)^{3/2} = 16^{3/2} \cdot 9^{3/2} = 64 \cdot 27 = 1728$
(b) $\left(\dfrac{4}{25}\right)^{3/2} = \dfrac{4^{3/2}}{25^{3/2}} = \dfrac{8}{125}$ ∎

Finally, if $x > 0$, then

$$(x^p)^q = x^{pq}. \qquad \text{Power-of-a-Power Rule}$$

Example 9 (a) $(1024^{5/2})^{1/5} = 1024^{(5/2)(1/5)} = 1024^{1/2} = 16$
(b) $[(-64)^{2/3}]^{3/4} = 16^{3/4} = 8$ but $(-64)^{(2/3)(3/4)} = (-64)^{1/2}$ has no meaning in the real number system. Therefore, the power-of-a-power rule may not apply when the base is negative. ∎

Examples 10-14 are included to show how the definitions and rules for rational exponents can be used to simplify algebraic expressions.

Example 10
$\sqrt{x} \cdot \sqrt[3]{x} = x^{1/2} \cdot x^{1/3}$ Definition $(x > 0)$
$\phantom{\sqrt{x} \cdot \sqrt[3]{x}} = x^{1/2+1/3}$ Same-base Rule
$\phantom{\sqrt{x} \cdot \sqrt[3]{x}} = x^{5/6}$
$\phantom{\sqrt{x} \cdot \sqrt[3]{x}} = (\sqrt[6]{x})^5$ Definition ∎

Example 11
$\sqrt{x}/\sqrt[3]{x} = x^{1/2}/x^{1/3}$ Definition $(x > 0)$
$\phantom{\sqrt{x}/\sqrt[3]{x}} = x^{1/2-1/3}$ Same-base Rule
$\phantom{\sqrt{x}/\sqrt[3]{x}} = x^{1/6}$
$\phantom{\sqrt{x}/\sqrt[3]{x}} = \sqrt[6]{x}$ Definition ∎

Example 12
$\sqrt{4a^2} = (4a^2)^{1/2}$ Definition
$\phantom{\sqrt{4a^2}} = 4^{1/2}(a^2)^{1/2}$ Same-exponent Rule
$\phantom{\sqrt{4a^2}} = 2\sqrt{a^2}$ Definition
$\phantom{\sqrt{4a^2}} = 2|a|$ *Section R.5* ∎

Example 13
$\sqrt[3]{\sqrt{2}} = (\sqrt{2})^{1/3}$ Definition
$\phantom{\sqrt[3]{\sqrt{2}}} = (2^{1/2})^{1/3}$ Definition
$\phantom{\sqrt[3]{\sqrt{2}}} = 2^{(1/2)(1/3)}$ Power-of-a Power Rule
$\phantom{\sqrt[3]{\sqrt{2}}} = 2^{1/6}$
$\phantom{\sqrt[3]{\sqrt{2}}} = \sqrt[6]{2}$ Definition ∎

Example 14
$x^{2/3} + x^{5/3} = x^{2/3} + x^{2/3} \cdot x^{3/3}$ Same-base Rule
$\phantom{x^{2/3} + x^{5/3}} = x^{2/3}(1 + x^{3/3})$
$\phantom{x^{2/3} + x^{5/3}} = x^{2/3}(1 + x)$ ∎

Exercises 1.4

Fill in the blanks to make each statement true.

1. $x^{1/n}$ means _____ with the restriction that x cannot be negative when n is _____.
2. If n is an even positive integer and x is positive, then $x^{1/n}$ means the _____ root of x
3. By definition, $x^{m/n}$ means _____ when m and n are positive integers with no common factor; also, $x^{-m/n}$ means _____, provided x is not _____.
4. If m/n is a positive or negative rational number in lowest terms, then $x^{m/n}$ may be interpreted as either _____ or _____.
5. The rule $(x^{m/n})^{p/q} = x^{(m/n)(p/q)}$ is valid for all rational exponents if we restrict the base x to be _____.

Write true or false for each statement.

6. For every real number x, $x^{2/3}$ means $(\sqrt[3]{x})^2$.
7. For every real number x, $x^{3/2}$ means $(\sqrt{|x|})^3$.
8. For every real number x, $x^{-2/3} = (1/\sqrt[3]{x})^2$.
9. For every real number x, $x^{3/5} = |x|^{3/5}$.
10. For every real number $x^{2/3} = |x|^{2/3}$.

Definition of $x^{m/n}$

Evaluate, if possible, each of the following.

11. (a) $64^{1/6}$ (b) $(-64)^{1/6}$ (c) $(-64)^{1/3}$
12. (a) $32^{1/5}$ (b) $(-32)^{1/5}$ (c) $(-32)^{-1/5}$
13. (a) $81^{3/2}$ (b) $81^{-3/2}$ (c) $(-81)^{3/2}$
14. (a) $125^{2/3}$ (b) $125^{-2/3}$ (c) $(-125)^{-2/3}$

Power-Root Rule

Verify that $(x^m)^{1/n} = (x^{1/n})^m$ in each of the following. Use a calculator for Exercises 18-20.

15. $x = 9, n = 2, m = 3$
16. $x = -8, n = 3, m = 2$
17. $x = -32, n = 5, m = 2$
18. $x = 100, n = 4, m = 3$
19. $x = 1250, n = 5, m = 2$
20. $x = 707, n = 8, m = 4$

Rules for Rational Exponents

Use the rules for rational exponents to simplify each of the following.

21. (a) $64^{1/2} \cdot 64^{-1/3}$ (c) $(-1)^{1/3}(-1)^{2/9}$
    (b) $\sqrt[3]{64} \cdot \sqrt[6]{64}$
22. (a) $\dfrac{8^{2/3}}{8^{1/3}}$ (b) $\dfrac{16^{1/2}}{16^{1/4}}$ (c) $\dfrac{27^{-1/3}}{27^{2/3}}$
23. (a) $(8 \cdot 64)^{4/3}$ (b) $(9 \cdot 81)^{-3/2}$ (c) $(\sqrt[4]{16})^3$
24. (a) $\left(-\dfrac{125}{8}\right)^{2/3}$ (b) $\left(\dfrac{9}{225}\right)^{-1/2}$ (c) $(\sqrt[3]{16})^4$
25. $2^{1/2} \cdot 3^{1/4} \cdot 2^{-1/3} \cdot 3^{1/2} \cdot 2^{1/6}$
26. $(5^{-6/7})^{-7/2}(5^{-2/3})^3$

Assuming x and y are positive numbers, write each expression in Exercises 27-31 with only positive exponents.

27. (a) $(x^{2/3})^6$ (b) $(x^{3/2})^6$ (c) $(\sqrt{x})^{-1/2}$
28. $\dfrac{(x^{2/3})^6 (x^{-3/2})^4}{(x^{-4/3})^{3/2}}$
29. $\left(\dfrac{8x^{1/3}x^{-1/2}}{x^{-1/6}}\right)^{-1/3}$
30. $(9x^{1/3}y^{-1/3})^{1/2}(y^{-2}x^6)^{-1}$
31. $\sqrt[3]{\dfrac{(7^{-2}y^{-6})^6}{5^{-2}x^4}}$

32. Explain why $[(-8)^2]^{1/6} \neq (-8)^{2 \cdot 1/6}$.

33. Find a real number x for which $(x^2)^{1/2} \neq x$.

Simplify each of the following. Use factoring where possible.

34. $x^{3/4} + x^{7/4}$ $(x > 0)$

35. $(xy)^{2/3} + y^{5/3}$

36. $(x^{3/2}y)^{1/3} + x^{1/2}$ $(x > 0)$

37. $\dfrac{\left(\dfrac{x^2}{y}\right)^3 + \left(\dfrac{x}{y}\right)^6}{\left(\dfrac{x}{y}\right)^4}$ $(x \neq 0, y \neq 0)$

38. $\dfrac{(x^{1/2}y^{1/3})^3 + (x^3y^2)^{1/2}}{x^{3/2}}$ $(x > 0, y > 0)$

39. $\dfrac{(x^{1/2}y^{1/3})^3 + (x^3y^2)^{1/2}}{x^{3/2}}$ $(x > 0, y > 0)$

40. *The Power-Root Rule.* $(x^m)^{1/n} = (x^{1/n})^m$ can be proven for $x > 0$, $m = 25$, and any positive n as follows.

Let $y = (x^{25})^{1/n}$. Then, by the definition of $\sqrt[n]{\ }$, y is the positive number that satisfies $y^n = x^{25}$. Therefore, to prove that $y = (x^{1/n})^{25}$, we must show that $[(x^{1/n})^{25}]^n = x^{25}$. But

$$[(x^{1/n})^{25}]^n = (x^{1/n})^{25n}$$
$$= [(x^{1/n})^n]^{25} \quad \begin{cases} \text{Rule } (a^m)^n = a^{mn} \text{ applied} \\ \text{to base } (x)^{1/n} \text{ and } integral \\ \text{exponents 25 and } n. \end{cases}$$
$$= [(\sqrt[n]{x})^n]^{25} \quad \text{Definition of } x^{1/n}$$
$$= x^{25}. \quad \text{Definition of } \sqrt[n]{x}$$

Use this same method to prove that $(x^m)^{1/n} = (x^{1/n})^m$ for *any* positive integral exponents m and n, given that x is greater than 0.

41. *The Same-Exponent Rule.* $(xy)^{m/n} = x^{m/n} \cdot y^{m/n}$ can be proven for the case $m = 1$ and $n = 6$ as follows (since 6 is even, we are assuming x and y are both positive).

Let $z = (xy)^{1/6}$. Then, by the definition of $\sqrt[6]{\ }$, z is the positive number satisfying $z^6 = xy$. Hence, to prove that $z = x^{1/6} \cdot y^{1/6}$, we must show that $(x^{1/6} \cdot y^{1/6})^6 = xy$. But

$$\left(x^{1/6} \cdot y^{1/6}\right)^6$$
$$= (x^{1/6})^6 \cdot (y^{1/6})^6 \quad \begin{cases} \text{Rule } (ab)^n = a^n b^n \text{ applied} \\ \text{to base } (x)^{1/6} \text{ and } (y)^{1/6} \\ \text{and } integral \text{ exponent 6.} \end{cases}$$
$$= \left(\sqrt[6]{x}\right)^6 \cdot \left(\sqrt[6]{x}\right)^6 \quad \text{Definition of } x^{1/6} \text{ and } y^{1/6}$$
$$= xy. \quad \text{Definition of sixth roots}$$

Use this method to prove that $(xy)^{1/n} = x^{1/n} \cdot y^{1/n}$ for any positive integer n, given that x and y are both positive if n is even.

1.5 Operations with Radicals

Radicals and Rational Exponents
Rationalizing Denominators

Working with radicals is the nitty-gritty of algebra

Radicals were introduced in Section R.5. Here we make use of rational exponents in order to perform algebraic operations on radicals. For all radicals in this section, we assume that the base is positive whenever the root is even or the power-of-a-power rule is applied.

Radicals and Rational Exponents. To add, subtract, multiply, or divide algebraic expressions containing radicals, we can convert to rational exponents and apply our previous rules. Answers can then be converted back to radical form.

Example 1
$$\sqrt{x-1}\sqrt[4]{(x-1)^3} = (x-1)^{1/2}(x-1)^{3/4} \quad \text{Convert to rational exponents.}$$
$$= (x-1)^{1/2+3/4} \quad \text{Same-base Rule}$$
$$= (x-1)^{5/4} \quad \text{Answer in exponent form.}$$
$$= (x-1)^{1+1/4}$$
$$= (x-1)(x-1)^{1/4} \quad \text{Same-base Rule}$$
$$= (x-1)\sqrt[4]{x-1} \quad \text{Answer in simplified radical form.} \blacksquare$$

Example 2
$$\frac{\sqrt[4]{(x-1)^3}}{\sqrt{x-1}} = \frac{(x-1)^{3/4}}{(x-1)^{1/2}} \quad \text{Convert to rational exponents.}$$
$$= (x-1)^{3/4-1/2} \quad \text{Same-base Rule}$$
$$= (x-1)^{1/4} \quad \text{Answer in exponent form.}$$
$$= \sqrt[4]{x-1} \quad \text{Answer in simplified radical form.} \blacksquare$$

Example 3
$$\sqrt{\frac{a+b}{a-b}}\left(\sqrt{a+b}\right)^3\sqrt{(a-b)^3} = \frac{(a+b)^{1/2}}{(a-b)^{1/2}}(a+b)^{3/2}(a-b)^{3/2}$$
$$= (a+b)^{1/2+3/2}(a-b)^{3/2-1/2}$$
$$= (a+b)^2(a-b) \quad \blacksquare$$

Example 4 $x^2\sqrt{x-1} - 2x\sqrt{x-1} + 4\sqrt{x-1} = (x^2 - 2x + 4)\sqrt{x-1}$

Here there was no need to covert to rational exponents. We simply factored out $\sqrt{x-1}$ from each term. $\blacksquare$

Example 5
$$2\sqrt{x-1} + 3\sqrt{(x-1)^3}$$
$$= 2(x-1)^{1/2} + 3(x-1)^{3/2}$$
$$= 2(x-1)^{1/2} + 3(x-1)(x-1)^{1/2}$$
$$= (x-1)^{1/2} \cdot [2 + 3(x-1)] \quad \text{Factor out } (x-1)^{1/2}$$
$$= (x-1)^{1/2}(3x-1)$$
$$= (3x-1)\sqrt{x-1} \quad \blacksquare$$

Rationalizing Denominators. If terms to be added have radicals in their denominators, as in

$$\frac{1}{\sqrt{x}} + \frac{1}{\sqrt{x^3}} + \frac{1}{\sqrt[4]{x^3}},$$

we can first replace each term by an equivalent one whose denominator contains only *integral powers*. That is, we can first **rationalize** each denominator as follows.

$$\frac{1}{\sqrt{x}} = \frac{1}{\sqrt{x}} \cdot \frac{\sqrt{x}}{\sqrt{x}} = \frac{\sqrt{x}}{x}$$

$$\frac{1}{\sqrt{x^3}} = \frac{1}{x^{3/2}} = \frac{1}{x^{3/2}} \cdot \frac{x^{1/2}}{x^{1/2}} = \frac{x^{1/2}}{x^2} = \frac{\sqrt{x}}{x^2}$$

$$\frac{1}{\sqrt[4]{x^3}} = \frac{1}{x^{3/4}} = \frac{1}{x^{3/4}} \cdot \frac{x^{1/4}}{x^{1/4}} = \frac{x^{1/4}}{x} = \frac{\sqrt[4]{x}}{x}$$

Therefore,

$$\frac{1}{\sqrt{x}} + \frac{1}{\sqrt{x^3}} + \frac{1}{\sqrt[4]{x^3}} = \frac{\sqrt{x}}{x} + \frac{\sqrt{x}}{x^2} + \frac{\sqrt[4]{x}}{x}. \tag{1}$$

We can now proceed with the addition by means of a common denominator. In this case the LCD of the terms on the right side of equation (1) is x^2. Therefore,

$$\frac{1}{\sqrt{x}} + \frac{1}{\sqrt{x^3}} + \frac{1}{\sqrt[4]{x^3}} = \frac{\sqrt{x}}{x} \cdot \frac{x}{x} + \frac{\sqrt{x}}{x^2} + \frac{\sqrt[4]{x}}{x} \cdot \frac{x}{x}$$

$$= \frac{x\sqrt{x} + \sqrt{x} + x\sqrt[4]{x}}{x^2} \qquad \textit{Addition with LCD.}$$

$$= \frac{(x+1)\sqrt{x} + x\sqrt[4]{x}}{x^2} \qquad \textit{Answer with rationalized denominator.}$$

Example 6 Perform the addition $\dfrac{1}{x^2\sqrt{x-1}} + \dfrac{1}{x\sqrt{(x-1)^3}}$

Solution: First we rationalize the denominators.

$$\frac{1}{x^2\sqrt{x-1}} = \frac{1}{x^2\sqrt{x-1}} \cdot \frac{\sqrt{x-1}}{\sqrt{x-1}} = \frac{\sqrt{x-1}}{x^2(x-1)}$$

$$\frac{1}{x\sqrt{(x-1)^3}} = \frac{1}{x(x-1)^{3/2}}$$

$$= \frac{1}{x(x-1)^{3/2}} \cdot \frac{(x-1)^{1/2}}{(x-1)^{1/2}} = \frac{\sqrt{x-1}}{x(x-1)^2}$$

Therefore,

$$\frac{1}{x^2\sqrt{x-1}} + \frac{1}{x\sqrt{(x-1)^3}} = \frac{\sqrt{x-1}}{x^2(x-1)} + \frac{\sqrt{x-1}}{x(x-1)^2}$$

$$\underset{LCD=x^2(x-1)^2}{=} \frac{\sqrt{x-1}}{x^2(x-1)} \cdot \frac{x-1}{x-1} + \frac{\sqrt{x-1}}{x(x-1)^2} \cdot \frac{x}{x}$$

$$= \frac{(x-1)\sqrt{x-1} + x\sqrt{x-1}}{x^2(x-1)^2}$$

$$= \frac{(2x-1)\sqrt{x-1}}{x^2(x-1)^2} \quad \text{Answer with rationalized denominator.} \blacksquare$$

Example 7 Perform the subtraction

$$\frac{1}{\sqrt{a}-\sqrt{b}} - \frac{\sqrt{a}}{a-b}.$$

Solution: First we rationalize the denominator of the term on the left.

$$\frac{1}{\sqrt{a}-\sqrt{b}} = \frac{1}{\sqrt{a}-\sqrt{b}} \cdot \frac{\sqrt{a}+\sqrt{b}}{\sqrt{a}+\sqrt{b}} = \frac{\sqrt{a}+\sqrt{b}}{a-b}$$

Here we substituted $\sqrt{a} = x$, $\sqrt{b} = y$ in the rule $(x-y)(x+y) = x^2 - y^2$ to obtain $(\sqrt{a}-\sqrt{b})(\sqrt{a}+\sqrt{b}) = a-b$. The expressions $\sqrt{a}-\sqrt{b}$ and $\sqrt{a}+\sqrt{b}$ are called **conjugates** of each other. Therefore,

$$\frac{1}{\sqrt{a}-\sqrt{b}} - \frac{\sqrt{a}}{a-b} = \frac{\sqrt{a}+\sqrt{b}}{a-b} - \frac{\sqrt{a}}{a-b} \quad \text{Terms with rationalized denominators.}$$

$$= \frac{\sqrt{a}+\sqrt{b}-\sqrt{a}}{a-b} \quad \text{Subtraction with common denominator.}$$

$$= \frac{\sqrt{b}}{a-b} \quad \text{Answer with rationalized denominator.} \blacksquare$$

Question *Should the denominator of an algebraic expression containing radicals always be rationalized?*

Answer: No. When adding algebraic expression with radical denominators, it is sometimes easier to obtain a common denominator by first rationalizing the denominators. However, this is not always the case, as is illustrated in Examples 8 and 9 below. Also, if rationalizing a denominator results in an expression that is considerably more complicated than the original one, then the process should be avoided. For instance, when

$$\frac{1}{\sqrt{a}+\sqrt{b}+\sqrt{c}}$$

is rationalized, the result is

$$\frac{(a-b-c)\sqrt{a} + (b-a-c)\sqrt{b} + (c-a-b)\sqrt{c} + 2\sqrt{abc}}{a^2 + b^2 + c^2 - 2ab - 2ac - 2bc}$$

Here the unrationalized form is obviously preferable. $\blacksquare$

1.5. Operations with Radicals

In Examples 8 and 9 below we further illustrate the various approaches to performing with radicals.

Example 8   Perform the addition

$$\frac{x}{\sqrt{2x+1}} + \sqrt{2x+1}$$

in two ways: (a) by first rationalizing the denominator, and (b) without rationalizing the denominator.

Solution:

(a) First rationalize the denominator, then add:

$$\frac{x}{\sqrt{2x+1}} + \sqrt{2x+1} = \frac{x\sqrt{2x+1}}{2x+1} + \frac{\sqrt{2x+1}}{1} \quad \text{Write each term as a fraction.}$$

$$= \frac{x\sqrt{2x+1} + (2x+1)\sqrt{2x+1}}{2x+1} \quad \text{Addition of Fractions}$$

$$= \frac{(3x+1)\sqrt{2x+1}}{2x+1}. \quad \text{Answer with rationalized denominator.}$$

(b) Add without rationalizing the denominator:

$$\frac{x}{\sqrt{2x+1}} + \sqrt{2x+1} = \frac{x}{\sqrt{2x+1}} + \frac{\sqrt{2x+1}}{1} \quad \text{Write each term as a fraction.}$$

$$= \frac{x + \sqrt{2x+1}\sqrt{2x+1}}{\sqrt{2x+1}} \quad \text{Addition of Fractions.}$$

$$= \frac{x + 2x + 1}{\sqrt{2x+1}}$$

$$= \frac{3x+1}{\sqrt{2x+1}}. \quad \text{Answer in non rationalized form.}$$

Example 9   Simplify $\dfrac{\sqrt{x^2+1} - \dfrac{x}{\sqrt{x^2+1}}}{x^2+1}$ in three ways:

(a) by first rationalizing the denominator of $\dfrac{x}{\sqrt{x^2+1}}$,

(b) without rationalizing, and

(c) by using rational exponents.

Chapter 1. Algebraic Expressions

Solution:

(a) Simplify by first rationalizing the denominator of $\dfrac{x}{\sqrt{x^2+1}}$:

$$\dfrac{\sqrt{x^2+1} - \dfrac{x}{\sqrt{x^2+1}}}{x^2+1}$$

$$= \dfrac{\dfrac{\sqrt{x^2+1}}{1} - \dfrac{x\sqrt{x^2+1}}{x^2+1}}{\dfrac{x^2+1}{1}} \qquad \text{Write each term as a fraction.}$$

$$= \dfrac{\dfrac{(x^2+1)\sqrt{x^2+1} - x\sqrt{x^2+1}}{x^2+1}}{\dfrac{x^2+1}{1}}$$

$$= \dfrac{(x^2+1)\sqrt{x^2+1} - x\sqrt{x^2+1}}{x^2+1} \cdot \dfrac{1}{x^2+1}$$

$$= \dfrac{(x^2 - x + 1)\sqrt{x^2+1}}{(x^2+1)^2} \qquad \text{Answer in rationalized form.}$$

(b) Simplify without rationalizing:

$$\dfrac{\sqrt{x^2+1} - \dfrac{x}{\sqrt{x^2+1}}}{x^2+1}$$

$$= \dfrac{\dfrac{\sqrt{x^2+1}}{1} - \dfrac{x}{\sqrt{x^2+1}}}{\dfrac{x^2+1}{1}} \qquad \text{Write each term as a fraction.}$$

$$= \dfrac{\sqrt{x^2+1}\sqrt{x^2+1} - x}{\sqrt{x^2+1}} \cdot \dfrac{1}{x^2+1}$$

$$= \dfrac{x^2+1-x}{\sqrt{x^2+1}(x^2+1)}$$

$$= \dfrac{x^2-x+1}{(x^2+1)\sqrt{x^2+1}}. \qquad \text{Answer in nonrationalized from.}$$

(c) Simplify by using rational exponents:

$$\dfrac{\sqrt{x^2+1} - \dfrac{x}{\sqrt{x^2+1}}}{x^2+1}$$

$$= \dfrac{\dfrac{(x^2+1)^{1/2}}{1} - \dfrac{x}{(x^2+1)^{1/2}}}{\dfrac{x^2+1}{1}} \qquad \text{Write each term as a fraction.}$$

$$= \frac{(x^2+1)^{1/2}(x^2+1)^{1/2} - x}{(x^2+1)^{1/2}} \cdot \frac{1}{x^2+1}$$

$$= \frac{x^2 + 1 - x}{(x^2+1)^{1/2}(x^2+1)}$$

$$= \frac{x^2 - x + 1}{(x^2+1)^{3/2}}. \qquad \textit{Answer with rational exponents.} \qquad \blacksquare$$

Exercises 1.5

All radicands in these exercises are assumed to be positive.

Fill in the blanks to make each statement true.

1. The rationalized form of $1/\sqrt{x}$ is _____.
2. The rationalized form of $1/\sqrt[3]{x}$ is _____.
3. The rationalized form of $1/x^{5/4}$ is _____.
4. The conjugate of $\sqrt{a} - \sqrt{b}$ is _____.
5. The rationalized form of $(\sqrt{x} + \sqrt{a})/(\sqrt{x} - \sqrt{a})$ is _____.

Write true or false for each statement.

6. $(x+2)^{-1/2} = \dfrac{\sqrt{x+2}}{x+2}$
7. $\dfrac{1}{(x-1)^{2/3}} = \dfrac{\sqrt[3]{x-1}}{x-1}$
8. $x - \dfrac{1}{\sqrt{x}} = \dfrac{(x^2-1)\sqrt{x}}{x}$
9. When working with radical denominators, the first step is always to rationalize the denominator.
10. When working with radicals, it is usually convenient to switch to rational exponents.

Radicals and Rational Exponents

Express each of the following in terms of rational exponents.

11. $\sqrt{x^2+1}$
12. $\sqrt[3]{(x-2)^2}$
13. $(\sqrt[4]{x^2+3})^3$
14. $(\sqrt{(x+1)^3})^5$
15. $\sqrt{x + \sqrt{x}}$
16. $\sqrt[5]{\sqrt{x-1}}$
17. $\sqrt{(\sqrt{x+2})^3}$
18. $\sqrt{\sqrt{\sqrt{x+1}}}$

Perform the indicated operations in each of the following. Simplify and express answers in radical form.

19. $\sqrt{x^2+1}\sqrt[3]{x^2+1}$
20. $\sqrt{(x-1)^3}\sqrt{x-1}$
21. $\sqrt{x+1}\sqrt[3]{(x+1)^2}$
22. $\sqrt[4]{(x-2)^3}(\sqrt[4]{x-2})^3$
23. $\sqrt{(x^2-1)^3}$
24. $\dfrac{\sqrt{x^2+1}}{\sqrt[3]{x^2+1}}$
25. $\dfrac{\sqrt{(x-2)^3}}{\sqrt{x-2}}$
26. $\dfrac{\sqrt{x+1}}{\sqrt[3]{(x+1)^2}}$
27. $\dfrac{\sqrt[4]{(x-2)^3}}{(\sqrt[4]{x-2})^3}$
28. $\dfrac{\sqrt{(x^2-1)^3}}{\sqrt[3]{(x^2-1)^2}}$
29. $5\sqrt[3]{x^2-1} + \sqrt[3]{x^2-1}$
30. $3\sqrt[4]{x+2} - 5\sqrt[4]{x+2} + 4\sqrt[4]{x+2}$
31. $2a\sqrt{x^2+x+1} + 3b\sqrt{x^2+x+1}$
32. $a\sqrt{x^2+5} - 2b\sqrt{x^2+5} + (3a-b)\sqrt{x^2+5}$
33. $x\sqrt[3]{x-4} + 2\sqrt[3]{x-4}$
34. $(x+1)\sqrt{x+1} - (2x+3)\sqrt{x+4} + (x+4)\sqrt{x+4}$

Rationalizing Denominators

Rationalize the denominator in each of the following.

35. $\dfrac{1}{\sqrt{x-3}}$
36. $\dfrac{2}{\sqrt{2x}}$
37. $\dfrac{x^2}{\sqrt{x}}$
38. $\dfrac{x-1}{\sqrt{x-1}}$

39. $\dfrac{1}{(\sqrt[3]{x+1})^2}$

40. $\dfrac{x+1}{\sqrt[3]{x^2+1}}$

41. $\dfrac{1}{\sqrt{x+1}+\sqrt{x-1}}$

42. $\dfrac{1}{\sqrt{a}+\sqrt{b}+\sqrt{c}}$

Perform the indicated operations in each of the following. Express answers in radical form with rationalized denominators.

43. $\dfrac{1}{\sqrt{x}} + \dfrac{x}{\sqrt{x}}$

44. $\dfrac{1}{\sqrt{a}} + \dfrac{\sqrt{a}}{a}$

45. $\dfrac{\sqrt{a}}{b} - \sqrt{\dfrac{b}{a}}$

46. $\dfrac{1}{\sqrt{x}} + \dfrac{x^2-1}{x\sqrt{x}}$

47. $\dfrac{1}{a} + \dfrac{1}{\sqrt{a}} + \dfrac{1}{a\sqrt{a}}$

48. $\dfrac{1}{\sqrt[3]{x^2}} - \dfrac{1}{\sqrt{x^3}}$

49. $\dfrac{x}{\sqrt{x}} + \dfrac{x^2}{\sqrt[4]{x}} + \dfrac{x^4}{\sqrt[6]{x}}$

50. $\dfrac{1}{\sqrt{x-1}} - \dfrac{1}{x^2-1} + \dfrac{1}{\sqrt{x+1}}$

51. $\dfrac{1}{\sqrt{a^2-b^2}} + \sqrt{\dfrac{a+b}{a-b}} + \sqrt{\dfrac{a-b}{a+b}}$

52. $\dfrac{1}{x\sqrt{x+2}} - \dfrac{1}{(x+2)\sqrt{x}}$

53. $\dfrac{1}{\sqrt{x-1}} - \dfrac{1}{\sqrt{x+1}}$

54. $\dfrac{1}{\sqrt{a}-\sqrt{b}} + \dfrac{1}{\sqrt{a}+\sqrt{b}}$

Miscellaneous

The following expressions appear in calculus. Simplify each expression and express answers in radical form.

55. $\sqrt{x^2+1} + \dfrac{x^2}{\sqrt{x^2+1}}$

56. $\sqrt{x+1} + \dfrac{x}{2\sqrt{x+1}}$

57. $\dfrac{\dfrac{x}{2\sqrt{x+5}} - \sqrt{x+5}}{x^2}$

58. $\dfrac{\dfrac{x^2}{\sqrt{x^2+5}} - \sqrt{x^2+5}}{x^2}$

59. $(x^2-1)^{1/3} + \dfrac{2}{3}x^2(x^2-1)^{-2/3}$

60. $(x+3)^{2/3} + \dfrac{2}{3}(x+2)(x+3)^{-1/3}$

61. $\dfrac{x^2(x^2+2)^{-1/2} - (x^2+2)^{1/2}}{x^2}$

62. $\dfrac{x\sqrt{x+1}(x^2+2)^{-1/2} - \sqrt{x^2+2}(x+1)^{-1/2}}{x+1}$

Chapter 1 Review Outline

1.1 Operations with Polynomials

Definitions

A polynomial $P(x)$ of degree n has the form

$$a_n x^n + a_{n-1} x^{n-1} + \ldots + a_1 x + a_0,$$

where n is a nonnegative integer and $a_n \neq 0$. A zero of $P(x)$ is a root of the equation $P(x) = 0$.

Operations

To add or subtract polynomials, combine similar terms. To multiply polynomials P and Q, multiply each term of P by each term of Q and proceed as in addition.

dividend = quotient · divisor + remainder

$$\dfrac{\text{dividend}}{\text{divisor}} = \text{quotient} + \dfrac{\text{remainder}}{\text{divisor}}$$

Square of a Binomial:

$$(a+b)^2 = a^2 + 2ab + b^2$$
$$(a-b)^2 = a^2 - 2ab + b^2$$

Difference of Squares:

$$(a-b)(a+b) = a^2 - b^2$$

Caution : $(a \pm b)^2 \neq a^2 \pm b^2$

$$\sqrt{a^2 \pm b^2} \neq a \pm b$$

Cube of a Sum:

$$(a+b)^3 = a^3 + 3a^2b + 3ab^2 + b^3$$

1.2 Factoring Polynomials

Perfect Squares:

$$a^2x^2 + 2abx + b^2 = (ax+b)(ax+b)$$
$$a^2x^2 - 2abx + b^2 = (ax-b)(ax-b)$$

Difference of squares:

$$a^2x^2 - b^2 = (ax-b)(ax+b)$$

Factoring a Quadratic:

$$x^2 + Bx + C = (x+m)(x+n)$$
$$\uparrow \quad \uparrow$$
$$m+n \quad mn$$

$$Ax^2 + Bx + C = Ax^2 + Mx + Nx + C,$$
$$M + N = B, MN = AC$$

(finish by grouping)

Sum and Difference of Cubes:

$$x^3 + a^3 = (x+a)(x^2 - ax + a^2)$$
$$x^3 - a^3 = (x-a)(x^2 + ax + a^2)$$

1.3 Operation with Rational Expressions

Definitions

A rational expression is the ratio of two polynomials. The least common multiple (LCM) of a set of polynomials is the polynomial of smallest degree that has for its only factors the factors of each of the polynomials in the set. The least common denominator (LCD) of a set of rational expressions is the LCM of their denominators.

Operations

To add or subtract rational expressions, first factor each denominator and determine the LCD. Second, replace each rational expressions by an equivalent one with the LCD. Third, proceed as in addition and subtraction of simple fractions.

Multiplication and division of rational expressions follow the same rules as in the case of simple fractions.

1.4 Rational Exponents

Definitions for Positive Integers n, m

$$x^{1/n} = \sqrt[n]{x} \quad (n = 2, 3, 4 \ldots)$$

(If n is an even integer, then x must be greater than or equal to zero and $x^{1/n}$ is the nonnegative nth root.)

$$x^{m/n} = (x^{1/n})^m$$

(m and n in lowest terms)

$$x^{-m/n} = 1/x^{m/n} \quad (x \neq 0)$$

Rules for Rational Exponents

In the following m and n are positive integers; p, q, and r are rational numbers

Power-Root Rule:

$$(x^m)^{1/n} = (x^{1/n})^m$$

Same-Base Rules:

$$x^p \cdot x^q = x^{p+q}$$
$$\frac{x^p}{x^q} = x^{p-q}$$

Same-Exponent Rules:

$$(x \cdot y)^r = x^r \cdot y^r$$
$$\left(\frac{x}{y}\right)^r = \frac{x^r}{y^r}$$

Power of-a-Power Rule:

$$(x^p)^q = x^{pq}$$

(The usual restrictions against dividing by 0 and taking even roots of negative numbers apply to all of the above).

1.5 Operations with Radicals

To add, subtract, multiply, or divide radicals, we can convert to rational exponents, perform the operations, and then convert back to radicals.

When adding radicals, it is sometimes helpful to rationalize the denominators.

Chapter 1 Review Exercises

1. Perform the operations
$(3ab+2a-1)-(4a+6b^2+8)+(5ab-2a+6b^2+10)$.
2. Multiply: $3x(x^2-2)(x^2+2)(x^4+4)$.
3. Divide $x^3 - 3x^2 + 5$ by $x+3$ and express the result in two equivalent ways.
4. Find the quotient and remainder when $x^5 + 5x^3 + 10x - 3$ is divided by $x^2 + 2$.
5. Expand $(2x+1)^2$ and $(2x-1)^3$.
6. If $P(x) = x^3 - 2x^2 + 5$, find $P(-2)$ and $P(\sqrt{2})$.
7. Verify that $\sqrt{2} - 2$ is a zero of $x^2 + 4x + 2$.
8. Explain why the polynomial $P(x) = x^6 + 3x^4 + 4x^2 + 1$ has no zeros in the real number system.

In Exercises 9-14, factor by integers as much as possible.

9. $16x^2 - 49y^2$
10. $5x^2 - 30x + 45$
11. $6x^2 + 11x - 10$
12. $2x^3 - 16y^3$
13. $5x^2 - 30x + 2x - 12$
14. $a^4 + 2a^2 + 1 - b^4$
15. Show that $x - 2$ is a factor of $x^5 - 32$.

In Exercises 16 and 17, factor by real numbers.

16. $2x^2 - 6$
17. $x^2 + 2.1x + .2$

18. Find the LCM of $6x + 6$, $x^2 + 2x + 1$, and $8x^2 - 8$.
19. Subtract by using the LCD and then simplify:

$$\frac{x+1}{x^2+x-2} - \frac{x}{x^2+2x-3}.$$

20. Combine by using the LCD and then simplify:

$$\frac{x}{x^2-4} + \frac{3x}{x-2} - \frac{x-2}{x+2}.$$

21. Perform the indicated operations and simplify:

$$\frac{2x^2-18}{x^2+2x-15} \cdot \frac{3x+15}{4x+12} \div \frac{6x+3}{4x^2-1}.$$

22. Simplify $\left\{\dfrac{2}{x} + \dfrac{2}{1-x} + \dfrac{1}{(x-1)^2}\right\} \div \left\{\dfrac{3}{x-1} - \dfrac{3}{x}\right\}$.
23. Evaluate each expression as a real number if possible.

 (a) $16^{3/2}$
 (b) $16^{-3/2}$
 (c) $(-16)^{3/2}$
 (d) $(-16)^0$
 (e) 0^{-16}

24. Simplify and then evaluate:

$$2^{1/2} \cdot 5^{3/4} \cdot 2^{-1/3} \cdot 10^{5/6} \div 5^{7/12}.$$

25. Evaluate $64^{-2/3}$ in three different ways.
26. Now $[(-5)^2]^{1/2} = 25^{1/2} = 5$, whereas $(-5)^{2/2} = (-5)^1 = -5$, and $[(-5)^{1/2}]^2$ is undefined. Explain why these results do not violate the rule $(x^m)^{1/n} = (x^{m/n}) = (x^{1/n})^m$.

27. Simplify $(x^{3/2}y^3)^{1/3} + (9xy^2)^{1/2}$.
28. Simplify $\dfrac{x^{2/3} + x^{5/3}}{1+x}$.
29. Simplify $5\sqrt{9a^4} - 3\sqrt[4]{16a^8}$.
30. Simplify $\sqrt{\dfrac{x+y}{x-y}} \cdot \dfrac{(\sqrt{x-y})^5}{(\sqrt{x+y})^3}$.
31. Rationalize the denominator of $\dfrac{\sqrt{x+2}}{\sqrt{x+2} - \sqrt{2}}$.
32. Rationalize the denominator and combine:

$$\frac{1}{\sqrt{x}-1} - \frac{1}{\sqrt{x}+1} + \frac{1}{x-1}.$$

Calculator Exercises

In each of the following, use a calculator to verify your answer to the indicated review exercises of this sections.

33. Exercises 6. *Hint* : Your Calculator will give a numerical answer for $P(\sqrt{2})$. Check that it is the calculator value of $1 + 2\sqrt{2}$.
34. Exercise 7
35. Exercise 23. For (b), see the hint in Exercise 33.
36. Exercise 24

2 Algebra and Graphs of Linear Expressions

2.1	Linear Equations
2.2	Linear Inequalities
2.3	Applications of Linear Equations and Inequalities
2.4	The Coordinate Plane
2.5	Equations of Straight Lines
2.6	Systems of Two Linear Equations in Two Unknowns
2.7	Systems of Three Linear Equations in Three Unknowns

Hey Students!
You will benefit more from the classroom lectures and have a better understanding of the material if you take the time to view the Gilbert Review Videos @
www.CollegeAlgebraBySchiller.com
before attending the class lecture for this chapter.

Chapter 2. Algebra and Graphs of Linear Expressions

Chapters R and 1 were devoted to manipulating various types of algebraic expression by means of the basic rules. In this chapter we concentrate on **linear expressions,** which are polynomials of degree 1 in one or more variables. Specifically, we consider

and
$$\textbf{(1)} \ \ ax + b \qquad (a \neq 0)$$
$$\textbf{(2)} \ \ ax + by + c \qquad (a \neq 0 \text{ or } b \neq 0),$$

where the coefficients a, b, and c are constant real numbers, and x and y are the variables.

From Chapters R and 1 we know how to manipulate polynomials algebraically. Our main concern here is with the *zeros* of these expressions. We will see that expression (1) has a single point on the real line for its zero, whereas the zeros of expression (2) form a straight line in the plane. Hence, from a geometric of view, this chapter is devoted to points and lines.

2.1 Linear Equations

Solving equations is the most fundamental process in algebra.

Equations
The Linear Form
Equations with Fractional Expressions
Expressions with Radicals
Equations with Absolute Values

Equations. We have been working with equations since Section R.1, but without the aid of a formal definition. We now define an **equation** as a statement that one algebraic expression is equal to another; for example,

$$5x - 1 = 3x + 6.$$

Here x is called the **unknown quantity,** and to **solve** the equation means to find all values of x for which the statement is true. We proceed as follows:

$$
\begin{aligned}
5x - 1 + 1 &= 3x + 6 + 1 &&\text{Add } 1 \text{ to both sides.} \\
5x &= 3x + 7 \\
5x - 3x &= 3x + 7 - 3x &&\text{Subtract } 3x \text{ from both sides.} \\
2x &= 7 \\
\frac{2x}{2} &= \frac{7}{2} &&\text{Divide both sides by } 2 \\
x &= \frac{7}{2}. &&\text{Solution}
\end{aligned}
$$

Check by substituting 7/2 for x in the original equation:

$$5\left(\frac{7}{2}\right) - 1 = \frac{35}{2} - 1 = \frac{33}{2} \quad \text{and} \quad 3\left(\frac{7}{2}\right) + 6 = \frac{21}{2} + 6 = \frac{33}{2}.$$

The method for solving the previous equation is based on the following algebraic properties of equality, which were introduced in Section R.2. Here M and N are any two algebraic expressions.

Algebraic Properties of Equality

$M = N \Leftrightarrow M \pm c = N \pm c$ for any value of c	Adding or subtracting a quantity on both sides of an equation results in an equivalent equation.
if $c \neq 0$, then $M = N \Leftrightarrow \begin{cases} M \cdot c = N \cdot c \\ \text{and} \\ \dfrac{M}{c} = \dfrac{N}{c} \end{cases}$	Multiplying or dividing on both sides by a nonzero quantity results in an equivalent equation.

Addition and subtraction are reversible steps, as are multiplication and division by a nonzero quantity.

Squaring both sides is not a reversible step and may lead to extraneous (false) roots.

For a given equation in x, the object is to utilize the above properties in order to reduce the given equation to one in which all the terms involving the unknown x are on one side and those not involving x are on the other. Since each step in this process is reversible, the reduced form of the equation and the original one are by definition **equivalent**. Equivalent equations have the same solution.

Example 1 Solve $x + 8 = 4(x - 1)$.

Solution:

$$\begin{aligned} x + 8 &= 4x - 4 & &\text{\textit{Apply distributive property on right.}} \\ x + 8 - 8 &= 4x - 4 - 8 & &\text{\textit{Subtract 8 from both sides.}} \\ x &= 4x - 12 \\ x - 4x &= 4x - 12 - 4x & &\text{\textit{Subtract $4x$ from both sides.}} \\ -3x &= -12 \\ \frac{-3x}{-3} &= \frac{-12}{-3} & &\text{\textit{Divide both sides by -3.}} \\ x &= 4 & &\text{\textit{Solution}} \end{aligned}$$

Check: $4 + 8 = 12$ and $4(4 - 1) = 12$. ∎

Comment Technically, the solution in Example 1 is 4, not $x = 4$. The statement $x = 4$ is an *equation* that is equivalent to the given equation

$x + 8 = 4(x - 1)$. When we say "the solution is $x = 4$," we mean that 4 is the value of the unknown x that satisfies the equation.

Example 2 Solve $3x + \dfrac{1}{2} = \dfrac{x}{2} + \dfrac{1}{3}$.

Solution:

$$6\left(3x + \frac{1}{2}\right) = 6\left(\frac{x}{2} + \frac{1}{3}\right) \qquad \text{Multiply both sides by 6 to clear the fractions.}$$

$$18x + 3 = 3x + 2$$

$$18x + 3 - 3 = 3x + 2 - 3 \qquad \text{Subtract 3 from both sides.}$$

$$18x = 3x - 1$$

$$18x - 3x = 3x - 1 - 3x \qquad \text{Subtract } 3x \text{ from both sides.}$$

$$15x = -1$$

$$\frac{15x}{15} = \frac{-1}{15} \qquad \text{Divide both sides by 15.}$$

$$x = -\frac{1}{15} \qquad \text{Solution}$$

Check: $3\left(\dfrac{-1}{15}\right) + \dfrac{1}{2} = -\dfrac{1}{5} + \dfrac{1}{2} = \dfrac{3}{10}$

and $\dfrac{\frac{-1}{15}}{2} + \dfrac{1}{3} = -\dfrac{1}{30} + \dfrac{1}{3} = \dfrac{3}{10}$. ∎

The Linear Form. The equations in the previous examples are called **linear equations in x** because each is equivalent to an equation of the form

$$\boldsymbol{ax + b = 0 \ (a \neq 0)}.$$

For instance, Example 1 is equivalent to $-3x + 12 = 0$, and Example 2 is equivalent to $(5/2)x + 1/6 = 0$. We will see later why the word "linear" is used to describe these equations. For a given equation $ax + b = 0$, there are three possibilities:

(a) If the equation is linear ($a \neq 0$), then the solution is $-b/a$.
(b) If $a = 0$ and $b \neq 0$, then the equation has *no* solution (*why?*).
(c) If $a = 0$ and $b = 0$, then *every real number* is a solution (*why?*).

Because of the extreme nature of the "solution" in cases (b) and (c), equations of the form $0 \cdot x + b = 0$ are not called linear equations.

2.1. Linear Equations

Example 3 Solve $c(x-1) = x+1$ for x in terms of c.

Solution:

$$cx - c = x + 1 \qquad \textit{Apply distributive property on left.}$$

$$cx - c + c = x + 1 + c \qquad \textit{Add c to both sides.}$$

$$cx = x + c + 1$$

$$cx - x = x + c + 1 - x \qquad \textit{Subtract x from both sides.}$$

$$(c-1)x = c + 1 \qquad \textit{Factor out x on left side.}$$

If $c \neq 1$, then the solution is $(c+1)/(c-1)$, but if $c = 1$, then there is no solution. ∎

We now consider several types of equations that lead to linear equations.

Equations with Fractional Expressions. If an equation contains fractional expressions in x, we can first clear the fractions by multiplying both sides by the LCM of the denominators. Then, if the resulting equation is linear, we can proceed as before.

Caution Any value of x obtained as a solution after clearing fractions must be tested in the original equation. If it makes a denominator equal to zero, the value is a false solution called an **extraneous root**, and it must be discarded.

Example 4 Solve $\dfrac{5}{x+1} - \dfrac{3}{x} = \dfrac{1}{x(x+1)}$.

Solution:

$$x(x+1)\left(\frac{5}{x+1} - \frac{3}{x}\right) = \frac{x(x+1)}{x(x+1)} \qquad \textit{Multiply both sides by } x(x+1).$$

$$\frac{5x(x+1)}{x+1} - \frac{3x(x+1)}{x} = 1$$

$$5x - 3(x+1) = 1$$

The linear equation $5x - 3(x+1) = 1$ and the original fractional equation are equivalent provided $x \neq 0$ or -1 *(why?)*. We can now solve the linear equation as before.

$$5x - 3x - 3 = 1$$

$$2x = 4$$

$$x = 2$$

Since $2 \neq 0$ or -1, 2 is the solution of the original equation. ∎

Example 5 Solve $\dfrac{2x}{x-1} + \dfrac{x+1}{x-2} = \dfrac{3x-1}{x-1}$.

Solution: Multiply both sides by $(x-2)(x-1)$ and then simplify.

$$(x-2)(x-1)\left(\dfrac{2x}{x-1} + \dfrac{x+1}{x-2}\right) = (x-2)(x-1)\left(\dfrac{3x-1}{x-1}\right)$$

$$\dfrac{(x-2)\cancel{(x-1)}2x}{\cancel{x-1}} + \dfrac{\cancel{(x-2)}(x-1)(x+1)}{\cancel{x-2}} = \dfrac{(x-2)\cancel{(x-1)}(3x-1)}{\cancel{x-1}}$$

$$2x^2 - 4x + x^2 - 1 = 3x^2 - 7x + 2$$

$$3x^2 - 4x - 1 = 3x^2 - 7x + 2$$

$$-4x - 1 = -7x + 2$$

$$3x = 3$$

$$x = 1$$

The solution 1 of the linear equation $-4x-1 = -7x+2$ is an extraneous root of the original equation. Therefore, the original equation has no solution. ∎

Example 6 Solve $\dfrac{x}{x-1} = \dfrac{1}{x-1} + 1$.

Solution:

$$(x-1)\left(\dfrac{x}{x-1}\right) = (x-1)\left(\dfrac{1}{x-1} + 1\right) \quad \textit{Multiply both sides by } x-1.$$

$$\dfrac{\cancel{(x-1)}x}{\cancel{x-1}} = \dfrac{\cancel{x-1}}{\cancel{x-1}} + x - 1$$

$$x = 1 + x - 1$$

$$x = x$$

The equation $x = x$ is true for all real numbers x. However, the original equation is meaningful only if $x \neq 1$. Therefore, the solution of the original equation consists of all real numbers except 1. ∎

Equations with Radicals. The equation $\sqrt{2x+3} = 5$ can be transformed into a linear equation by squaring both sides.

$$\left(\sqrt{2x+3}\right)^2 = 5^2$$

$$2x + 3 = 25$$

$$x = 11$$

Check: $\sqrt{2 \cdot 11 + 3} = \sqrt{25} = 5$.

Caution Squaring both sides of an equation is not a reversible step and may lead to false solutions (extraneous roots). For example, $\sqrt{2x+3} = -5$ also becomes $2x + 3 = 25$ when both sides are

2.1. Linear Equations

squared, but in this case 11 is not a solution. Because of the possibility of introducing extraneous roots, all values obtained by squaring must be tested for acceptability by substitution in the *original* equation.

Example 7 Solve $\sqrt{x^2 + 4} + x = 1$.

Solution:

$$\sqrt{x^2 + 4} = 1 - x \qquad \text{\textit{Isolate the radical by subtracting } x \text{ \textit{from both sides.}}}$$

$$x^2 + 4 = 1 - 2x + x^2 \qquad \text{\textit{Square both sides.}}$$

$$2x = -3$$

$$x = -\frac{3}{2}$$

Check: We have $\sqrt{(-3/2)^2 + 4} + (-3/2) = \sqrt{9/4 + 4} - 3/2 = \sqrt{25/4} - 3/2 = 1$, so $-3/2$ is a solution. ■

Example 8 Solve $\sqrt{x^2 - 8} + x = 2$.

Solution:

$$\sqrt{x^2 - 8} = 2 - x \qquad \text{\textit{Isolate the radical by subtracting } x \text{ \textit{from both sides.}}}$$

$$x^2 - 8 = 4 - 4x + x^2 \qquad \text{\textit{Square both sides.}}$$

$$4x = 12$$

$$x = 3$$

Check: $\sqrt{3^2 - 8} + 3 = 4 \neq 2$; therefore, 3 is an extraneous root, and the original equation has no solution. ■

Example 9 Solve $\sqrt{x + 1} + \sqrt{x - 1} = 2$.

Solution:

$$\sqrt{x + 1} = 2 - \sqrt{x - 1} \qquad \text{\textit{Isolate } } \sqrt{x+1} \text{ \textit{on left.}}$$

$$x + 1 = 4 - 4\sqrt{x - 1} + x - 1 \quad \text{\textit{Square both sides.}}$$

$$4\sqrt{x - 1} = 2 \qquad \text{\textit{Isolate } } 4\sqrt{x-1} \text{ \textit{on left.}}$$

$$\sqrt{x - 1} = \frac{1}{2}$$

$$x - 1 = \frac{1}{4} \qquad \text{\textit{Square again.}}$$

$$x = \frac{5}{4}$$

Check: $\sqrt{5/4 + 1} + \sqrt{5/4 - 1} = \sqrt{9/4} + \sqrt{1/4} = 3/2 + 1/2 = 2$; therefore, $5/4$ is a solution. ■

102   Chapter 2. Algebra and Graphs of Linear Expressions

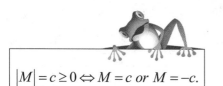

$|M| = c \geq 0 \Leftrightarrow M = c \text{ or } M = -c.$

Two squarings were necessary in Example 9. The general procedure for solving equations with radicals is to eliminate one radical at a time by isolating it on side of the equation and then squaring both sides.

Equations with Absolute Values. The solution to $|x| = 3$ is $x = 3$ or $x = -3$. In general, if M is an algebraic expression in x and $c \geq 0$, then the solution to $|M| = c$ is the set of all real numbers x for which either $M = c$ or $M = -c$.

Example 10 Solve $|2x + 3| = 4$.

Solution:

$$2x + 3 = 4 \quad \text{or} \quad 2x + 3 = -4 \qquad \textit{Each equation}$$
$$2x = 1 \qquad\qquad 2x = -7 \qquad \textit{is solved separately.}$$
$$x = \frac{1}{2} \quad \textbf{(1)} \qquad x = -\frac{7}{2} \quad \textbf{(2)}$$

Check:

(1) $|2(1/2) + 3| = |4| = 4$, so $1/2$ is a solution;
(2) $|2(-7/2) + 3| = |-4| = 4$, so $7/2$ is also a solution. ∎

$|M| = |N| \Leftrightarrow M = N \text{ or } M = -N.$

Similarly, if M and N are algebraic expressions in x, then the solution to $|M| = |N|$ consists of all real numbers x for which either $M = N$ or $M = -N$.

Example 11 Solve $|3x - 4| = |x - 6|$.

Solution:

$$3x - 4 = x - 6 \quad \text{or} \quad 3x - 4 = -(x - 6)$$
$$2x = -2 \qquad\qquad 3x - 4 = -x + 6$$
$$x = -1 \quad \textbf{(1)} \qquad\qquad 4x = 10$$
$$x = \frac{5}{2} \quad \textbf{(2)}$$

Check:

(1) $|3(-1) - 4| = |-7| = 7$ and $|-1 - 6| = |-7| = 7$;
(2) $|3 \cdot 5/2 - 4| = |7/2| = 7/2$ and $|5/2 - 6| = |-7/2| = 7/2$.

Therefore, both -1 and $5/2$ are solutions. ∎

The method of solution in each of the previous examples is based on the following property of **absolute value** that was introduced in Section R.3.

2.1. Linear Equations

> If M an algebraic expression in x, then
>
> $$|M| = \begin{cases} M \text{ for all } x \text{ in which } M \geq 0, \\ -M \text{ for all } x \text{ in which } M < 0. \end{cases}$$

In each example, the above property was used to replace one equation with absolute values by two separate equations not involving absolute values. For instance, the single equation $|M| = c$ becomes $M = c$ or $-M = c$; equivalently, $M = c$ or $M = -c$.

The general procedure for solving an equation with absolute values is to eliminate one absolute value $|M|$ at a time by replacing the equation with two auxiliary equations, one with M in place of $|M|$ and the other with $-M$ in place of $|M|$. If there is another absolute value $|N|$ in the original equation, then four auxiliary equations result, corresponding to M, N; $M, -N$; $-M, N$; and $-M, -N$, respectively. Three absolute value terms $|M|, |N|$ and $|P|$ would generate eight auxiliary equations (*why?*), and so on, until all absolute value terms have been eliminated. *The final auxiliary equations are solved separately, and all solutions must be tested in the original equation to eliminate extraneous roots.*

Example 12   Solve $|x + 1| + |2x - 5| = 4$.

Solution: This equation generates the four auxiliary equations (a), (b), (c), and (d) that are listed below. We solve each of these and check our answers in the original equation.

$$
\begin{aligned}
\text{(a)} \quad x + 1 + 2x - 5 &= 4 \\
3x - 4 &= 4 \\
3x &= 8 \\
x &= \frac{8}{3}
\end{aligned}
$$

Check: $|8/3 + 1| + |2(8/3) - 5| = |11/3| + |1/3| = 12/3 = 4$. Hence, 8/3 *is* a solution.

$$
\begin{aligned}
\text{(b)} \quad x + 1 - (2x - 5) &= 4 \\
-x + 6 &= 4 \\
-x &= -2 \\
x &= 2
\end{aligned}
$$

Check: $|2 + 1| + |2 \cdot 2 - 5| = |3| + |-1| = 4$. Hence, 2 *is* a solution.

(c) $\quad -(x+1) + 2x - 5 = 4$

$$x - 6 = 4$$
$$x = 10$$

Check: $|10 + 1| + |2 \cdot 10 - 5| = |11| + |15| = 26 \neq 4$. Therefore, 10 is *not* a solution.

(d) $\quad -(x+1) - (2x - 5) = 4$

$$-3x + 4 = 4$$
$$-3x = 0$$
$$x = 0$$

Check: $|0 + 1| + |2 \cdot 0 - 5| = |1| + |-5| = 6 \neq 4$. Therefore, 0 is *not* a solution.

Our checks tell us to accept 8/3 and 2 and to reject 10 and 0. Therefore, 8/3 and 2 are the only solutions to the original equation. ∎

Exercises 2.1

Fill in the blanks to make each statement true.

1. As statement that one algebraic expression is equal to another is called an _____.
2. If one equation can be transformed into another by a series of reversible algebraic steps, then the two equations are said to be _____.
3. An equation $ax + b = 0$, where $a \neq 0$, is called a _____ equation.
4. The solution to $ax + b = 0$, with $a \neq 0$, is _____.
5. If $a = 0$, then the solution to $ax + b = 0$ is _____ if $b = 0$, and _____ if $b \neq 0$.

Write true or false for each statement.

6. The equation $M = N$ is equivalent to the equation $M + c = N + c$, for any real number c.
7. The equation $M = N$ is equivalent to the equation $Mc = Nc$, for any real number c.
8. The equation $M = N$ is equivalent to the equation $M - N = 0$.
9. The equation $M = N$ is equivalent to the equation $M/N = 1$.
10. The equations $M = N$ and $M^2 = N^2$ are equivalent.

Linear Equations

Solve each of the following equations.

11. $3x + 5 = 11$
12. $2x - 6 = 7$
13. $2x - 5 - (x + 5) = 0$
14. $3x + 11 - (6x - 11) = 0$
15. $\frac{2}{3}x - 7 = x + \frac{5}{2}$
16. $\frac{3}{4}x + \frac{1}{2} = 5x - \frac{1}{4}$
17. $2.4(x - 1) - 4.2 = 1.8 - 3.6(x - 2)$
18. $5.2(2x + 1) - 3.2 = -2.2(3x + 4) + 3.8$
19. $\frac{2}{5}(x - 1) + \frac{5}{2}(x + 2) = \frac{3}{4}(x - 2)$
20. $\frac{5}{6}(2x + 3) + \frac{2}{3}(2x - 1) = \frac{1}{2}(x + 2)$
21. $3x + 4 - (x + 1) = 2x + 3$
22. $5(x - 2) + 2(3x - 1) = 4(x - 3) + 7x$
23. $2x + 4 = 4(x - 1) - (2x - 3)$
24. $5x + 2(x - 1) = 7(x - 1)$

25. $\sqrt{3}x + 2(x-4) = 5x + \sqrt{3}$
26. $\sqrt{2}(3x-1) + 4x = \sqrt{2}(2x-3) + 1$
27. $\dfrac{2}{3}(x+\sqrt{5}) + \dfrac{1}{3}(x-\sqrt{5}) = 1$
28. $\dfrac{1}{2}(x+\sqrt{2}) + \dfrac{1}{4}(\sqrt{3}x - 1) = 0$
29. $x + 4[x - 2(x+7)] = 5[1 + 2(x-1)]$
30. $3x - 2[5 - 2(x-1)] = 2 - 3(x+1)$

Solve for x in terms of the remaining constants and variables.

31. $cx + 5 = 2x - 7$ $\quad (c \neq 2)$
32. $3x + c = cx - 4$ $\quad (c \neq 3)$
33. $c^2 x + 5 = 25x + c$
34. $c^2 x - 3 = 9x - c$
35. $ax + by + c = 0$ $\quad (a \neq 0)$
36. $\dfrac{x}{a} + \dfrac{y}{b} + \dfrac{1}{c} = 0$ $\quad (a \neq 0,\, b \neq 0,\, c \neq 0)$
37. $a(x+b) + b(x+a) = 0$ $\quad (a \neq -b)$
38. $a(x-b) + b(x-a) = 1$ $\quad (a \neq -b)$
39. $\dfrac{x}{a} + \dfrac{y}{b} = \dfrac{x+y}{a+b}$ $\quad (a \neq 0,\, b \neq 0,\, a \neq -b)$
40. $\dfrac{x}{a} - \dfrac{y}{b} = \dfrac{x-y}{ab}$ $\quad (a \neq 0,\, b \neq 0)$

Fractional Equations

Solve each of the following equation.

41. $\dfrac{1}{2x+3} = 1$
42. $\dfrac{7}{3x-5} = 2$
43. $\dfrac{3}{2x-4} - 1 = 0$
44. $\dfrac{1}{x+1} - 2 = 0$
45. $5 - \dfrac{2}{3x+1} = 0$
46. $4 - \dfrac{6}{x-1} = 0$
47. $\dfrac{1}{x+3} = \dfrac{2}{3x+1}$
48. $\dfrac{3}{2x-5} = \dfrac{1}{2x+5}$
49. $x + \dfrac{3}{x-1} = x - \dfrac{2}{x-1}$
50. $x + \dfrac{1}{x} = x - \dfrac{1}{x}$
51. $\dfrac{1}{x} + \dfrac{1}{x-2} = \dfrac{3}{x} - \dfrac{2}{x-2}$
52. $\dfrac{4}{4x-1} - \dfrac{5}{4x+1} = \dfrac{2}{4x-1} - \dfrac{1}{4x+1}$
53. $\dfrac{1}{x-3} + \dfrac{1}{x+3} = \dfrac{2}{x^2 - 9}$
54. $\dfrac{2}{x-5} - \dfrac{1}{x+5} = \dfrac{1}{x^2 - 25}$

55. $\dfrac{4}{x^2 - 4} = \dfrac{1}{x-2} - \dfrac{1}{x+2}$
56. $\dfrac{3x+5}{x^2 + 3x + 2} = \dfrac{1}{x+2} + \dfrac{2}{x+1}$
57. $\dfrac{1}{x+3} - \dfrac{2}{x-2} = \dfrac{x+1}{x^2 + x - 6}$
58. $\dfrac{x}{x-5} - \dfrac{x+1}{x-4} = \dfrac{2x+3}{x^2 - 9x + 20}$
59. $\dfrac{1}{x^2 - 5x + 6} = \dfrac{x-1}{x-3} - \dfrac{x}{x-2}$
60. $\dfrac{3x+2}{x^2 + 6x + 8} = \dfrac{x+1}{x+2} - \dfrac{x}{x+4}$
61. $\dfrac{4}{2x^2 + 7x + 3} - \dfrac{3}{2x+1} + \dfrac{2}{x+3} = 0$
62. $\dfrac{10}{5x^2 + x - 4} + \dfrac{x}{x+1} - \dfrac{5x}{5x-4} = 0$
63. $\dfrac{1}{x - \dfrac{1}{x}} = \dfrac{4}{1 + \dfrac{1}{x}}$
64. $\dfrac{2x}{x+1 - \dfrac{1}{x+1}} = \dfrac{2}{1 + \dfrac{2}{x}}$
65. $\dfrac{\dfrac{1}{x+1} - \dfrac{1}{x+4}}{x^2 + 5x + 4} = \dfrac{1}{x+2}$
66. $\dfrac{\dfrac{1}{(x+2)(x+3)}}{\dfrac{1}{x+2} + \dfrac{1}{x+3}} = \dfrac{1}{5x+2}$
67. $\dfrac{1 + (x+2)^{-1}}{1 - (x+2)^{-1}} = \dfrac{x+3}{x+1}$
68. $\dfrac{(x+1)^{-1} + (x+2)^{-1}}{(x+2)^{-1} + (x+3)^{-1}} = \dfrac{x+2}{x+3}$
69. $\dfrac{x + 1 - \dfrac{2}{x+2}}{\dfrac{4}{x+3}} = 1 + \dfrac{4}{x} + \dfrac{4}{x^2}$
70. $\dfrac{(x+2)^{-1} + (x+3)^{-1}}{\dfrac{1}{2x+5}}^{-1} = x + 1 + \dfrac{2}{x+4}$

Solve for x in terms of the remaining constants and variables.

71. $y = \dfrac{x}{x+1}$
72. $y = \dfrac{x-1}{x}$

73. $\dfrac{1}{x} + \dfrac{1}{y} = 1$

74. $\dfrac{a}{x} + \dfrac{b}{y} = 1 \quad (a \neq 0,\ b \neq 0)$

75. $\dfrac{a}{x} - \dfrac{b}{x} = c \quad (a \neq b,\ c \neq 0)$

76. $\dfrac{x}{x+a} + \dfrac{y}{x+a} = c \quad (c \neq 1)$

77. $\dfrac{1}{x+a} + \dfrac{1}{x-a} = \dfrac{a}{x^2-a^2} \quad (a \neq 0)$

78. $\dfrac{a}{x+y} - \dfrac{b}{x-y} = \dfrac{1}{x^2-y^2} \quad (a \neq b)$

79. $\dfrac{x+y}{x-y} - \dfrac{x-y}{x+y} = \dfrac{1}{x-y} - \dfrac{1}{x+y}$

80. $\dfrac{x+y}{x-y} - \dfrac{x-y}{x+y} = \dfrac{2y}{x-y} + \dfrac{2y}{x+y}$

Equations with Radicals

Solve each equation.

81. $\sqrt{2x+1} = 4$

82. $\sqrt{3x-2} = 5$

83. $\sqrt{5x-1} + 2 = 0$

84. $\sqrt{2x-3} + 1 = 0$

85. $\sqrt{5-2x} - \sqrt{9-x} = 0$

86. $\sqrt{4x+11} - \sqrt{1-x} = 0$

87. $\sqrt{3x+2} - \sqrt{3x-1} = 1$

88. $\sqrt{5x+8} - \sqrt{5x+3} = 1$

89. $\sqrt{2x+6} + \sqrt{2x+1} = 5$

90. $\sqrt{2x+6} - \sqrt{2x+1} = 5$

91. $5 + \sqrt{x^2-4} = 3 - x$

92. $\sqrt{x^2+2x+2} = 3 - x$

93. $x + \sqrt{4x^2+4} = 3x - 2$

94. $3 + \sqrt{x^2+2x+2} = 1 + x$

95. $3x + \sqrt{4x^2+4} = x + 2$

96. $3 + \sqrt{x^2-4} = 5 + x$

97. $\dfrac{\sqrt{x^2+1}}{x+1} - \dfrac{1}{x+1} = 1$

98. $\dfrac{\sqrt{x^2+1}}{x+1} - \dfrac{1}{x+1} = 1$

99. $\sqrt{x+1} - \dfrac{x}{\sqrt{x+1}} = 2$

100. $\sqrt{x+1} + 2 = \dfrac{x}{\sqrt{x+1}}$

101. $\dfrac{x}{\sqrt{x^2+1}} + \dfrac{1}{x\sqrt{x^2+1}} = \dfrac{\sqrt{x^2+1}}{x}$

102. $\dfrac{1}{\sqrt{x^2+x+1}} + \dfrac{1}{x} = \dfrac{2}{x\sqrt{x^2+x+1}}$

Equations with Absolute Values

Solve each equation.

103. $|x-2| = 3$

104. $|x+1| = 5$

105. $|3x-7| = 5$

106. $|4x+3| = 8$

107. $|x+2| + 3 = 0$

108. $|2x-1| + 4 = 0$

109. $\left|\dfrac{2}{3}x + 7\right| = \dfrac{5}{6}$

110. $\left|\dfrac{4}{5}x - \dfrac{1}{5}\right| = \dfrac{3}{4}$

111. $\left|\dfrac{2x+1}{3x-4}\right| = 5$

112. $\left|\dfrac{4x-3}{x+7}\right| = 12$

113. $2 + \dfrac{1}{|x-1|} = 4$

114. $3 - \dfrac{1}{|2x+1|} = 1$

115. $|4x-3| = |3x+2|$

116. $|5x-7| = |7x-5|$

117. $\dfrac{1}{|x+2|} = \dfrac{2}{|3x-1|}$

118. $\dfrac{5}{|2x-3|} = \dfrac{4}{|3x+2|}$

119. $|2x+1| + |5x-7| = 10$

120. $|3x-2| + |x+4| = 3$

121. $|5x-3| - |2x+1| = 4$

122. $|6x+7| - |3x-4| = 5$

2.2 Linear Inequalities

Not all algebraic expressions are created equal.

Inequalities
Linear Inequalities and Infinite Intervals
Finite Intervals
Inequalities with Fractional Expressions
Inequalities with Absolute Values

Inequalities. In the previous section, we applied the relationship of equality to algebraic expressions, and now we do the same for the relationship of inequality. An **inequality** is a statement that one algebraic expression is *not* equal to another. Four basic types of inequalities are listed below.

	Type	Definition	Example
(1)	$M > N$	M is greater than N.	$3x - 7 > x + 5$
(2)	$M \geq N$	M is greater than or equal to N.	$5x \geq 3(2x + 1)$
(3)	$M < N$	M is less than N.	$-3x + 7 < 2(4x + 3)$
(4)	$M \leq N$	M is less than or equal to N.	$x - \frac{1}{2} \leq -\frac{x}{2} + 2$

Even though types (2) and (4) combine inequality with equality, we still call them inequalities. By contrast, types (1) and (3) are called **strict inequalities.** The method for solving an inequality is similar to that for solving an equation and is based on the following properties of "greater than" that were introduced in Section R.3

Adding any quantity (positive or negative) to both sides of an inequality or multiplying both sides by a positive quantity preserves the direction of the inequality.

If c is any real number, then
$$M > N \Leftrightarrow M + c > N + c.$$
Preservation Property for Addition

If $c > 0,$ then
$$M > N \Leftrightarrow M \cdot c > N \cdot c.$$
Preservation Property for Multiplication

The preservation property for addition also holds for subtraction and that for multiplication also holds for division (*why?*). These properties also apply to inequality types (2), (3), and (4). That is, the type of an inequality is not changed by addition and subtraction, or by multiplication and division by a positive quantity.

108   Chapter 2. Algebra and Graphs of Linear Expressions

Example 1   Solve $3x - 7 > x + 5$.

Solution:

$$3x > x + 12 \quad \text{Add 7 to both sides.}$$
$$2x > 12 \quad \text{Subtract } x \text{ from both sides.}$$
$$x > 6 \quad \text{Divide both sides by 2.}$$

Hence, the solution consists of all real numbers to the right of 6 on the number line, as indicated in Figure 1. The parenthesis symbol (is used to indicate that the endpoint 6 is *not* included in the solution. ∎

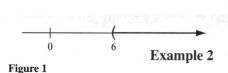

Figure 1

Example 2   Solve $x - 1/2 \leq -x/2 + 2$.

Solution:

$$2x - 1 \leq -x + 4 \quad \text{Multiply both sides by 2.}$$
$$2x \leq -x + 5 \quad \text{Add 1 to both sides.}$$
$$3x \leq 5 \quad \text{Add } x \text{ to both sides.}$$
$$x \leq \frac{5}{3} \quad \text{Divide both sides by 3.}$$

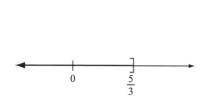

Figure 2

The solution consists of 5/3 and all real numbers to the left of 5/3 on the number line, as indicated in Figure 2. Here the bracket symbol] is used to indicate that the endpoint 5/3 *is* included in the solution. ∎

Example 3   Solve $5x + 4 \geq 5(x + 1)$.

Solution:

$$5x + 4 \geq 5x + 5 \quad \text{Apply distributive property on right.}$$
$$5x \geq 5x + 1 \quad \text{Subtract 4 from both sides.}$$
$$0 \geq 1 \quad \text{Subtract } 5x \text{ from both sides.}$$

Since 0 is not greater than or equal to 1, regardless of the value of x, we must conclude that there is *no* solution, or we can say that the solution is the empty set. ∎

Example 4   Solve $x + 2(x - 1) < 3x + 5$.

Solution:

$$x + 2x - 2 < 3x + 5 \quad \text{Apply distributive property on left.}$$
$$3x - 2 < 3x + 5$$
$$3x < 3x + 7 \quad \text{Add 2 to both sides.}$$
$$0 < 7 \quad \text{Subtract } 3x \text{ from both sides.}$$

2.2. Linear Inequalities   109

Since 0 is less than 7, regardless of x, the solution consists of the set of *all* real numbers. ∎

Up to now we have multiplied and divided inequalities by positive quantities only. Now suppose $c < 0$. If $M > N$, then $M - N$ is positive and therefore the product $(M - N)c$ is negative. Hence, $Mc - Nc$ is negative, which means that $Mc < Nc$. We can reverse the argument and thereby obtain the following result.

If $c < 0$, then $M > N \iff M \cdot c < N \cdot c$.   Reversal Property for Multiplication

Multiplying both sides of an inequality by a negative quantity reverses *the direction of the inequality.*

The reversal property also holds for division by a negative quantity (why?). Also, if $c < 0$, then $M \geq N$ becomes $Mc \leq Nc$, $M < N$ becomes $Mc > Nc$, and $M \leq N$ becomes $Mc \geq Nc$. Similar results apply to division. Hence, *the effect of multiplying or dividing any inequality by a negative quantity is to reverse the direction of the inequality.*

Example 5   Solve $5x - 3(2x + 1) \geq 0$.

Solution:

$5x - 6x - 3 \geq 0$ *Apply distributive property on left.*
$-x - 3 \geq 0$
$-x \geq 3$ *Add 3 to both sides.*
$x \leq -3$ *Multiply both sides by -1, and reverse the direction of the inequality.*

The solution is graphed in Figure 3. ∎

Figure 3

Example 6   Solve $-3x + 7 \leq 2(4x + 3)$.

Solution:

$-3x + 7 \leq 8x + 6$ *Apply distributive property on right.*
$-3x \leq 8x - 1$ *Subtract 7 from both sides.*
$-11x \leq -1$ *Subtract $8x$ from both sides.*
$x \geq \dfrac{1}{11}$ *Divide both sides by -11, and reverse the direction of the inequality.*

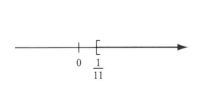

Figure 4

The solution is graphed in Figure 4. ∎

Linear Inequalities and Infinite Intervals.   Any inequality that can be reduced to one of the four basic inequalities in which $M = ax + b$

($a \neq 0$) and $N = 0$ is called a **linear inequality**. All of our previous examples except Examples 3 and 4 are linear inequalities. The solution of a linear inequality is always an **infinite interval** of points along the real line. The following chart indicates the solution for each of the four basic types of inequalities when $a > 0$. (If the coefficient of x is negative in a linear inequality, we can obtain a positive coefficient $a > 0$ by multiplying both sides by -1 and reversing the direction of the inequality. For example, $-2x + 1 > 0$ is equivalent to $2x - 1 < 0$.)

Basic Type ($a > 0$)	Solution	Graph	Interval
(1) $ax + b > 0$	$x > -\dfrac{b}{a}$		$\left(-\dfrac{b}{a}, \infty\right)$
(2) $ax + b \geq 0$	$x \geq -\dfrac{b}{a}$		$\left[-\dfrac{b}{a}, \infty\right)$
(3) $ax + b < 0$	$x < -\dfrac{b}{a}$		$\left(-\infty, -\dfrac{b}{a}\right)$
(4) $ax + b \leq 0$	$x \leq -\dfrac{b}{a}$		$\left(-\infty, -\dfrac{b}{a}\right]$

The symbol ∞ (read "infinity") in the chart indicates that the solution extends to all real numbers *greater than* $-b/a$. Similarly, $-\infty$ indicates that the solution extends to all real numbers *less than* $-b/a$. We note that ∞ and $-\infty$ are not real numbers. The parentheses in the graphs and the intervals for (1) and (3) indicate that the endpoint $-b/a$ is *not* included in the solution, whereas the brackets in (2) and (4) mean that $-b/a$ is included in the solution.

We note that two extreme cases of "intervals" were met in Examples 3 and 4. In Example 4, the solution is the *all* real numbers, which can be denoted in interval notation as $(-\infty, \infty)$. Example 3 has for its solution the **empty set**, usually denoted by $\emptyset$.

Finite Intervals. We have seen that the solution to a linear inequality is a real line interval extending infinitely in one direction. We will soon see that *two* infinite intervals can occur as the solution to a fractional inequality. A fractional inequality also can have as its solution a **finite interval**. For any two real numbers c and d, where $c < d$, there are four types of finite intervals, as illustrated in the following chart.

2.2. Linear Inequalities

Type of Interval	Graph	Inequality Notation	Interval Notation
open	$\underset{c\quad\quad d}{(\rule{2em}{0pt})}$	$c < x < d$	(c, d)
closed	$\underset{c\quad\quad d}{[\rule{2em}{0pt}]}$	$c \leq x \leq d$	$[c, d]$
half-open or half-closed	$\underset{c\quad\quad d}{[\rule{2em}{0pt})}$	$c \leq x < d$	$[c, d)$
half-open or half-closed	$\underset{c\quad\quad d}{(\rule{2em}{0pt}]}$	$c < x \leq d$	$(c, d]$

Inequalities with Fractional Expressions. A fraction is positive if its numerator and denominator are both positive or both negative. For instance,

$$\frac{7}{8} > 0 \quad \text{and} \quad \frac{-7}{-8} > 0.$$

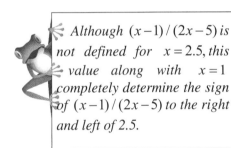

Although $(x-1)/(2x-5)$ is not defined for $x = 2.5$, this value along with $x = 1$ completely determine the sign of $(x-1)/(2x-5)$ to the right and left of 2.5.

The same principle applies to rational expressions in x. Furthermore, as x takes on values from $-\infty$ to ∞, a rational expression, such as $(x - 1)/(2x - 5)$, can change sign only if either the numerator or the denominator changes sign, and either of these can change sign only by going through 0. Now $x - 1 = 0$ when $x = 1$, and $2x - 5 = 0$ when $x = 2.5$. As illustrated in Figure 5, $(x - 1)/(2x - 5) > 0$ for all x in either $(-\infty, 1)$ or $(2.5, \infty)$, and $(x - 1)/(2x - 5) < 0$ for all x in the interval $(1, 2.5)$. Also, as shown in Figure 6, both $x - 1$ and $2x - 5$ are negative on $(-\infty, 1)$, and both are positive on $(2.5, \infty)$.

Figure 5

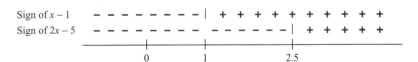

Figure 6

Therefore, to solve a rational inequality, first mark off the point(s) on the x-axis for which the numerator is 0 and those for which the denominator is 0. These points partition the x-axis into intervals on which the corresponding rational expression does not change sign. By testing a point in each interval, we can solve the inequality.

Example 7 Solve $\dfrac{2x+1}{x-2} \leq 0$.

Solution: $2x + 1 = 0$ if $x = -0.5$, and $x - 2 = 0$ if $x = 2$. The points $-0.5, 2$ partition the x-axis into the intervals $(-\infty, -0.5), (-0.5, 2)$, and $(2, \infty)$. Choose a point in each of these intervals, say $-1, 1$, and 3, and evaluate $(2x+1)/(x-2)$ at each of these points, getting $1/3, -3$, and 7. Then, $(2x + 1)/(x - 2) < 0$ on the interval $(-0.5, 2)$, as illustrated in Figure 7. Finally, since $2x + 1 = 0$ when $x = -0.5$, the solution to $(2x + 1)/(x - 2) \leq 0$ is the half-closed interval $[-0.5, 2)$. ■

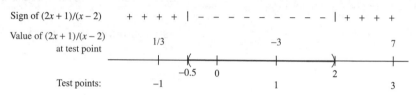

Figure 7

If a rational inequality does not have 0 on one side, we can first perform some algebraic steps to transform the inequality into one with 0 on one side, and then apply the previous method. This process is illustrated in the following example.

Example 8 Solve the inequality $\dfrac{x+8}{x+3} > 2$.

Solution: We first transform the given inequality into one with 0 on the right side as follows:

$$\dfrac{x+8}{x+3} > 2 \qquad \text{Given}$$

$$\dfrac{x+8}{x+3} - 2 > 0 \qquad \text{Subtract 2 from both sides.}$$

$$\dfrac{x+8}{x+3} - \dfrac{2(x+3)}{x+3} > 0 \qquad \text{Obtain a common denominator.}$$

$$\dfrac{x+8-2(x+3)}{x+3} > 0 \qquad \text{Subtract fractions with a common denominator.}$$

$$\dfrac{-x+2}{x+3} > 0 \qquad \text{Simplify.}$$

$$\dfrac{x-2}{x+3} < 0. \qquad \text{Multiply both sides by } -1 \text{ and reverse the direction of the inequality.}$$

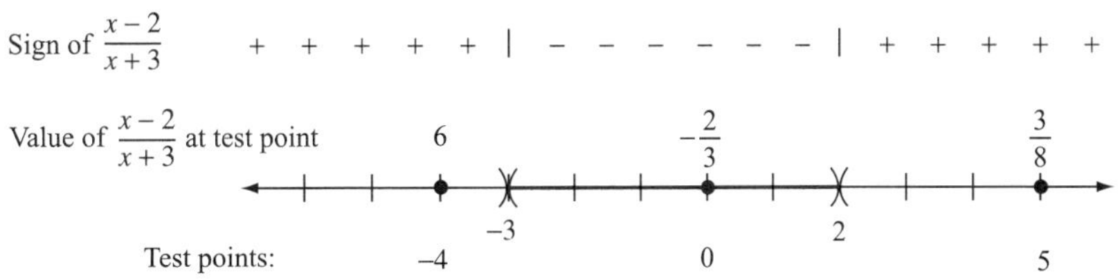

2.2. Linear Inequalities 113

Sign of $\frac{x-2}{x+3}$ + + + + + | − − − − − − − | + + + + +

Value of $\frac{x-2}{x+3}$ at test point   6 $-\frac{2}{3}$ $\frac{3}{8}$

Test points: −4   −3   0   2   5

Figure 8

We now solve the inequality $(x - 2)/(x + 3) < 0$, which is equivalent to the given one. The points 2 and -3, which are the respective roots of $x - 2 = 0$ and $x + 3 = 0$, break up the real line into the intervals $(-\infty, -3)$, $(-3, 2)$, and $(2, \infty)$. In each of these, the sign of $x - 2$ and $x + 3$ does not change. Therefore, the sign of $(x - 2)/(x + 3)$ does not change in each of these intervals. As illustrated in Figure 8, we can compute the value of $(x - 2)/(x + 3)$ at one convenient test point in each interval to determine the sign $(x - 2)/(x + 3)$ over the entire interval

We choose the test points $-4, 0$, and 5, and conclude that $(x-2)/(x+3)$ is negative on the interval $(-3, 2)$. Hence, $(-3, 2)$ is also the solution of the given inequality $(x + 8)/(x + 3) > 2$. ∎

Figure 9 $|x| > 2$

Inequalities with Absolute Values. We have seen that $|x|$ is equal to the distance between 0 and x on the real number line. Therefore, the solution of the inequality $|x| > 2$ consists of all real numbers whose distance from 0 is greater than 2, namely all numbers to the right of 2 *or* to the left of -2 on the number line. That is, the solution of $|x| > 2$ is the set $(-\infty, -2) \cup (2, \infty)$, which is illustrated in Figure 9. Hence, the single inequality $|x| > 2$ is equivalent to the two *separate* inequalities $x > 2$ or $x < -2$. This special case leads to the following general rule, in which M is any algebraic expression in x and c is any real number.

> The solution of $|M| > c$ consists of all real numbers x that satisfy *either* $M > c$ or $M < -c$. In short,
>
> $$|M| > c \Leftrightarrow M > c \quad \text{or} \quad M < -c.$$

Note that if c is negative, then $|M| > c$ for *all* real numbers for which M is defined. Therefore, we are mainly concerned with nonnegative c's. Also, the general rule holds if $>$ and $<$ are replaced by $\geq$ and $\leq$, respectively.

Example 9 Solve $|5x - 1| > 6$.

Solution:

$$5x - 1 > 6 \quad \text{or} \quad 5x - 1 < -6 \qquad \text{Each inequality is solved separately.}$$

$$5x > 7 \qquad\qquad 5x < -5$$

$$x > \frac{7}{5} \qquad\qquad x < -1$$

Thus, the solution is $(-\infty, -1) \cup (7/5, \infty)$, as show in Figure 10. ∎

Figure 10

The solution of $|x| < 2$ consists of all real numbers x whose distance from 0 is less than 2, namely all x in the interval $(-2, 2)$ (Figure 11). Hence, the single inequality $|x| < 2$ is equivalent to the *simultaneous* inequalities $x < 2$ *and* $x > -2$; that is, $-2 < x < 2$.

This special case leads to the following general rule.

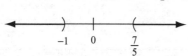

Figure 11 $|x| < 2$

> The solution of $|M| < c$ consists of all real numbers x that satisfy *both* $M < c$ and $M > -c$. In short,
>
> $$|M| < c \Leftrightarrow -c < M < c.$$

Note that if c is negative or zero, then the solution of $|M| < c$ is the empty set ∅. Therefore, we are mainly concerned with positive c's. Also, the general rule applies when $<$ and $>$ are replaced by $\leq$ and $\geq$, respectively.

Example 10 Solve $|2x + 7| < 9$.

Solution:

$$-9 < 2x + 7 < 9 \qquad \text{The two inequalities are solved simultaneously.}$$

$$-16 < \quad 2x \quad < 2 \qquad \text{Subtract 7 from each expression.}$$

$$-8 < \quad x \quad < 1 \qquad \text{Divide each expression by 2.}$$

Therefore, the solution is $(-8, 1)$, as shown in Figure 12. ∎

Figure 12

Example 11 Solve $|4x - 3| \leq 8$.

Solution:

$$-8 \leq 4x - 3 \leq 8$$

$$-5 \leq \quad 4x \quad \leq 11 \qquad \text{Add 3 to each expression.}$$

$$\frac{-5}{4} \leq \quad x \quad \leq \frac{11}{4} \qquad \text{Divide each expression by 4.}$$

Therefore, the solution is $[-5/4, 11/4]$, as shown in Figure 13. ∎

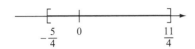

Figure 13

Exercises 2.2

Fill in the blanks to make each statement true.

1. A statement that one algebraic expression is not equal to another is called an _____.
2. For algebraic expressions M and N, the four basic inequalities are M _____ N, M _____ N, M _____ N, and M _____ N.
3. If $M < N$ and $c < 0$, then $M + c$ _____ $N + c$, and Mc _____ Nc.
4. The solution of $|M| > c$ consists of all real numbers x that satisfy either _____ or _____.
5. The solution of $|M| < c$ consists of all real numbers x that satisfy both _____ and _____.

Write true or false for each statement.

6. If $M + 5 < N + 4$, then $M < N$.
7. If $-5M < -5N$, then $M < N$.
8. If $M \leq N$, then $M + 1 < N + 2$.
9. If $M < 5$, then $|M| < 5$.
10. If $|M| < 5$, then $M < 5$.

Linear Inequalities

Solve the following inequalities. Graph the solutions and express the solutions in interval notation.

11. $2x - 4 > 0$
12. $3x + 5 > 0$
13. $-2x + 7 \geq 0$
14. $-4x + 1 \geq 0$
15. $2x - 7 < 0$
16. $4x + 2 < 0$
17. $-3x - 1 \leq 0$
18. $-4x - 5 \leq 0$
19. $5x + 1 > 3x + 4$
20. $3x - 1 > x + 2$
21. $2(x + 5) \geq 3x - 4$
22. $3(x - 2) \geq 2x + 5$
23. $5x - 3 < 2(3x + 1)$
24. $2x + 2 < 5(x - 1)$
25. $3(2x - 4) \leq 6x + 5$
26. $4x + 7 \leq 2(2x + 1)$

Solve the following inequalities for x in terms of the remaining constants and variables.

27. $ax + b > 0 \quad (a < 0)$
28. $ax - by \leq 0 \quad (a > 0)$
29. $\dfrac{x}{a} + \dfrac{y}{b} < 1 \quad (a > 0, b \neq 0)$
30. $\dfrac{x}{a} - \dfrac{y}{b} \geq 1 \quad (a < 0, b \neq 0)$
31. $ax + b > cx + d \quad (a > c)$
32. $ax + b \leq cx + d \quad (a < c)$

Inequalities with Fractional Expressions

Solve the following inequalities. Express the solutions in interval notation and graph.

33. $\dfrac{3}{4x + 1} > 0$
34. $\dfrac{2}{2x - 5} < 0$
35. $\dfrac{x}{x - 1} \leq 0$
36. $\dfrac{x}{x + 2} \geq 0$
37. $\dfrac{x + 3}{2x - 1} < 0$
38. $\dfrac{x - 2}{4x + 5} > 0$
39. $\dfrac{1}{x - 1} \geq 2$
40. $\dfrac{1}{x - 1} \leq -2$
41. $\dfrac{4}{x + 3} < 2$
42. $\dfrac{3}{x - 5} > -2$
43. $\dfrac{x + 1}{x - 1} \geq 5$
44. $\dfrac{2x - 3}{x - 3} \leq 1$
45. $\dfrac{x + 2}{x + 3} > 1$
46. $\dfrac{x - 5}{x - 4} < 2$
47. $\dfrac{2x + 7}{x + 3} \geq -2$
48. $\dfrac{3x - 9}{x - 7} \leq -1$

Solve each of the following for x in terms of the remaining constants and variables. Express your answers in interval notation.

49. $\dfrac{1}{x-a} > 1 \quad (x > a)$

50. $\dfrac{1}{x-a} < -1 \quad (x < a)$

51. $\dfrac{x}{a} \geq \dfrac{y}{a} \quad (a > 0)$

52. $\dfrac{x}{a} \geq \dfrac{y}{a} \quad (a < 0)$

53. $\dfrac{1}{x-a} + \dfrac{1}{x-1} < 0 \quad (a < x < 1)$

54. $\dfrac{1}{x-a} + \dfrac{1}{x-1} < 0 \quad (x < a < 1)$

Inequalities with Absolute Values

Solve and graph the following inequalities. Express your answers in interval notation.

55. $|x - 1| > 2$
56. $|x - 3| > 5$
57. $|x + 2| \geq 1$
58. $|x + 4| \geq 5$
59. $|2x - 3| < 6$
60. $|3x + 4| < 3$
61. $|4x + 5| \leq 2$
62. $|2x + 3| \leq 4$
63. $|5x - 1| > 0$
64. $|2x - 7| \leq 0$
65. $|5x - 4| \geq -1$
66. $|2x - 1| < -1$

Miscellaneous

Combine the techniques in this section to solve each of the following inequalities. Express your answers in interval notation and graph.

67. $5x + 3(x - 1) > 2(x + 4) - 7$
68. $3(2x + 1) + 5 > 5(4x + 2) - 3x$
69. $-8x + 1 + 2(x - 3) \geq -4(x + 2) + 5(x - 1)$
70. $-3x + 7 + 2(4x + 1) \geq 6(x + 2) - (2x + 3)$
71. $-4x + 3[2 - (x - 4)] < 5[x - (1 - 2x)]$
72. $-5x + 3 + 2(x + 7) < 4 + 3[7 - (x - 2)]$
73. $3x + 2(3x - 7) \leq 5 + 3(3x - 1)$
74. $2(x + 7) + 3(2x - 4) \leq 4(2x + 12)$
75. $\dfrac{1}{x-1} > \dfrac{2}{x-1}$
76. $\dfrac{2}{x+2} > \dfrac{1}{x+2}$
77. $1 + \dfrac{1}{x} \geq 3 - \dfrac{2}{x}$
78. $2 - \dfrac{5}{x+1} \geq 1 + \dfrac{4}{x+1}$
79. $\dfrac{1}{x+2} - 3 < 3 - \dfrac{1}{x+2}$
80. $\dfrac{4}{x+4} + 5 < \dfrac{-4}{x+4} + 5$
81. $\dfrac{1}{2x-1} - 3 \leq \dfrac{1}{2x-1} - 3$
82. $4 - \dfrac{3}{x-2} \leq -4 - \dfrac{3}{x-2}$
83. $\dfrac{2}{|x+1|} > 1$
84. $\dfrac{4}{|x-2|} > 1$
85. $\dfrac{1}{|2x+3|} \geq 2$
86. $\dfrac{1}{|3x-4|} \geq 5$
87. $\dfrac{2}{|3x+1|} < 3$
88. $\dfrac{4}{|5x-1|} - 5 < 0$
89. $\dfrac{1}{|2x+7|} - 1 \leq 0$
90. $\dfrac{5}{|3x-2|} - 6 \leq 0$

2.3 Applications of Linear Equations and Inequalities

Percent
Motion
Mixtures
Work
Miscellaneous Applications

In an algebra class, one question that can always be counted on is "Will there be any word problems on the test?"

People solve linear equations every day, even though they are not always aware of it. Whenever they make change, compute the sales tax on an item, or glance at a clock to determine how much work or class time is left, they are solving linear equations. Whenever people budget their time to maintain a schedule, they are working with inequalities.

2.3. Applications of Linear Equations and Inequalities

Several standard applications of linear equations and inequalities are discussed in this section. To solve a word problem, use the following procedure.

1. Read the problem very carefully, identify the quantities involved, and assign letters to the unknown quantities.
2. Express any relationships that exist among the quantities in the form of algebraic equations and/or inequalities. Draw diagrams whenever possible.
3. Solve for the unknown quantities.
4. Interpret the algebraic solution in terms of the original problem.

The above four-step transformation of a word problem to its mathematical model and back to its solution is illustrated in the following flow chart.

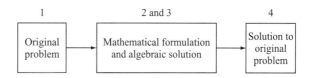

We now consider word problems involving several topics, including percent, motion, mixtures, and work.

Percent. Percent means "hundredths," and $r\%$ of a quantity Q means $(r/100)Q$. For instance, 15% of 250 means $(15/100)250 = (.15)250 = 37.5$.

Example 1 The retail price of a stereo is $225, which includes a 50% markup on the wholesale cost. (a) What is the wholesale cost of the stereo? (b) What percent of the retail price is the markup?

Solution:

(a) Let C = wholesale cost in dollars. Then

$$C + .5C = 225 \quad \text{Cost + markup = retail price}$$
$$1.5C = 225$$
$$C = \frac{225}{1.5}$$
$$= 150.$$

Hence, the wholesale cost of the stereo is $150.

(b) The markup in dollars is $225 - 150 = 75$. As a percent of retail price, the markup is

$$\left(\frac{75}{225} \cdot 100\right)\% = \left(\frac{1}{3} \cdot 100\right)\% = 33\frac{1}{3}\%. \quad \blacksquare$$

Caution When working with percents, it is important to specify the quantity on which the percent is computed. In Example 1, the $75 markup is 50% of the wholesale cost but only 33 1/3% of the retail price.

Example 2 The sticker price of a new car is $9995, which includes a $7545 base price plus $2450 in optional equipment. The base price includes a 14% markup on the base cost, and the options price includes an 18% markup on the options cost. What is the dealer's cost of the car?

Solution: Let B = the base cost to the dealer and let E = the options cost to the dealer, both in dollars. Then

$$B + .14B = 7545 \quad \text{and} \quad E + .18E = 2450$$
$$1.14B = 7545 \quad\quad\quad\quad 1.18E = 2450$$
$$B = \frac{7545}{1.14} \quad\quad\quad\quad E = \frac{2450}{1.18}$$
$$= 6618.42 \quad\quad\quad\quad = 2076.27.$$

Therefore, the dealer's cost of the car is

$$\$6618.42 + \$2076.27 = \$8694.69. \quad \blacksquare$$

Example 3 The workers at an electronics plant want their work week reduced from 40 hours to 35 hours but with the same weekly pay. If their current hourly rate is x dollars an hour, what will the new hourly rate be? To what percent increase is the new hourly rate equivalent?

Solution: The current weekly income is $40x$ dollars. If the new hourly rate is r dollars per hour, then the new weekly pay is $35r$ dollars. For the new weekly pay to be equal to the old, we must have $35r = 40x$. Therefore,

$$r = \frac{40x}{35} = \frac{8x}{7} = x + \frac{1}{7}x.$$

Hence, the new hourly rate is an increase of one seventh over the old. As a percent, the increase is $(1/7 \cdot 100)\%$, or approximately 14.3%. $\quad \blacksquare$

Example 4 A clothing retailer has a 35% markup over cost on a certain line of sport shirts. What percentage discount on the selling price will guarantee at least a 20% profit over cost?

Solution: Let C = the cost of a shirt to the retailer, and let x = the discount rate on the selling price.

2.3. Applications of Linear Equations and Inequalities 119

The current selling price of a shirt is

$$C + .35C = 1.35C,$$

and the discount price is to be at least

$$C + .2C = 1.2C.$$

That is,

$$1.35C - (1.35C)x \geq 1.2C \qquad \text{\textit{Selling price} $-$ \textit{discount} $\geq$ 20\% \textit{over cost}}$$
$$1.35C(1-x) \geq 1.2C$$
$$1 - x \geq \frac{1.2}{1.35} \approx .89 \qquad \text{\textit{Divide both sides by 1.35C.}}$$
$$1 - .89 \geq x$$
$$.11 \geq x.$$

Hence, a discount rate of 11% or less will insure a profit of at least 20%. Note that the result does not depend on the actual value of C. ∎

Motion. The most basic formula for motion is

$$\boldsymbol{d = rt},$$

where
$d =$ the distance traveled by an object along a given path,

$r =$ the average rate of speed of the object,

and $t =$ the time of travel.

The formula contains three quantities: $d, r,$ and t. If any two of these are known, the formula can be solved for the remaining one.

Example 5 Paul drove 150 miles in 3 1/2 hours. His average speed for the first 90 miles was 45 miles per hour. What was his average speed for the last 60 miles?

Solution: The formula can be applied to each part of the trip as needed.

Distance (mi)	Rate (mph)	Time (hr)
150		3.5
90	45	90/45 = 2
60	x	t

In the table we have set

$$x = \text{rate for last 60 miles,}$$
$$and \quad t = \text{time for last 60 miles.}$$

Now,

$$x = \frac{60}{t}. \quad Rate = \frac{distance}{time}$$

Also,

$$2 + t = 3.5 \quad \text{\textit{Time for first 90 miles + time for last}}$$
$$t = 1.5. \quad \text{\textit{60 miles = time for entire 150 miles.}}$$

Therefore,

$$x = \frac{60}{1.5}$$
$$= 40.$$

The average speed for the last 60 miles was 40 miles per hour. ∎

Example 6 At noon, cyclists Dee and Donna are 50 miles apart and pedaling toward each other. If Dee pedals at 10.5 miles per hour and Donna at 14.5 miles per hour, at what point between them and at what time do they meet?

Solution: We can apply the formula to each cyclist. If they meet in t hours then Dee has traveled x miles and Donna $50-x$ miles, as indicated in Figure 14 and the table.

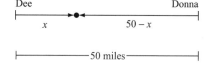

Figure 14

	Distance	Rate	Time
Dee	x	10.5	t
Donna	$50 - x$	14.5	t

Now,

$$\frac{x}{10.5} = \frac{50 - x}{14.5} \quad \text{\textit{Dee's time = Donna's time}}$$
$$14.5x = 10.5(50 - x)$$
$$14.5x = 525 - 10.5x$$
$$25x = 525$$
$$x = 21.$$

Therefore, they meet at a point 21 miles from Dee's noontime position. The time of travel is

$$t = \frac{21}{10.5} \quad \text{\textit{Time}} = \frac{\text{\textit{distance traveled by Dee}}}{\text{\textit{Dee's rate of speed}}}$$
$$= 2.$$

Hence, they meet at 2 P.M. ∎

2.3. Applications of Linear Equations and Inequalities

Example 7 An airplane traveling against a steady wind takes 2 hours to make a 500-mile trip. The return trip in the direction of the wind takes only 1 hour and 40 minutes. What is the speed of the plane in still weather, and what is the speed of the wind?

Solution: Let $r =$ the speed of the plane in still weather, and let $x =$ the speed of the wind. Then $r - x$ is the speed against the wind, and $r + x$ is the speed with the wind, as indicated in the table.

	Distance (mi)	Rate (mph)	Time (hr)
Against the wind	500	$r - x$	2
With the wind	500	$r + x$	$1\frac{2}{3} = \frac{5}{3}$

Now,

$$r - x = \frac{500}{2} \quad \text{and} \quad r + x = \frac{500}{\frac{5}{3}} \qquad Rate \;=\; \frac{distance}{time}$$

$$r - x = 250 \qquad r + x = 300.$$

Therefore,

$$r - x + r + x = 250 + 300 \quad \text{Add } r - x \text{ and } r + x \text{ to cancel } x.$$

$$2r = 550$$

$$r = 275,$$

giving $\qquad x = 25.$ $\qquad$ Use either $\; r - x = 250$
$\qquad\qquad\qquad\qquad\qquad\qquad\qquad$ or $r + x = 300.$

The speed of the plane in still weather is 275 miles per hour, and the speed of the wind is 25 miles per hour. ∎

Mixtures. A typical mixing problem involves combining two quantities having different concentrations to form a third quantity that has an intermediate concentration. For instance, if 10 liters of 30% acid are mixed with 15 liters of 60% acid, the result is 25 liters of acid with concentration x, where

$$\underbrace{10(.3)}_{\substack{\text{Liters of acid in}\\\text{30\% solution}}} + \underbrace{15(.6)}_{\substack{\text{Liters of acid in}\\\text{60\% solution}}} = \underbrace{25x.}_{\substack{\text{Total amount}\\\text{of acid}}} \qquad (1)$$

Therefore,

$$3 + 9 = 25x$$

$$12 = 25x$$

$$.48 = x.$$

The mixture is 48% acid.

In any mixing problem, each part to be combined contains a concentrate in some medium. In the above example, the concentrate is acid, which is measured in percent, and the medium is the solution, which is measured in liters. *We compute the amount of concentrate in each part and add these amounts to get the total amount of concentrate.* For convenience, we can list the given data for the above example in the following table.

	Part 1	Part 2	Mixture
Concentrate	30%	60%	x
Medium	10 liters	15 liters	25 liters
Amount of Concentrate	10(.3)	15(.6)	25x

Equation (1) is obtained from the last row of the table by adding the amount of concentrate in each part and equating the sum to the amount of concentrate in the mixture.

Example 8 Coffee worth $2.85 a pound is mixed with coffee worth $2.50 a pound to obtain 5 pounds of coffee worth $2.64 a pounds. How much of each type of coffee goes into the blend?

Solution: Here the concentrate is the value, which is measured in dollars per pound, and the medium is coffee, which is measured in pounds. Let $x =$ the number of pounds of $2.85 coffee required; then $5 - x =$ the number of pounds of $2.50 coffee needed. We have the following table of data.

	Part 1	Part 2	Mixture
Concentrate	$2.85/lb	$2.50/lb	$2.64/lb
Medium	x lbs	$5 - x$ lb	5 lbs
Amount of Concentrate	2.85x	2.50(5 − x)	2.64(5)

From the last row of the table, we add the amount of concentrate in each part to obtain

$$2.85x + 2.50(5 - x) = 2.64(5)$$
$$2.85x + 12.50 - 2.50x = 13.20$$
$$.35x = .70$$
$$x = 2.$$

Hence, 2 pounds of $2.85 coffee must be mixed with 3 pounds of $2.50 coffee to obtain the desired 5 pounds of $2.64 coffee. ■

As illustrated in the following example, we can also apply inequalities to mixing problems.

2.3. Applications of Linear Equations and Inequalities   123

Example 9   What is the least number of liters of 30% alcohol that must be mixed with 6 liters of 55% alcohol to make a solution that is at most 40% alcohol?

Solution: Here the concentrate is alcohol, which is measured in percent, and the medium is the solution, which is measured in liters. Let $x =$ the number of liters of 30% alcohol added. We have the following data.

	Part 1	Part 2	Mixture
Concentrate	30%	55%	$\leq 40\%$
Medium	x liters	6 liters	$x + 6$ liters
Amount of Concentrate	$.30x$	$.55(6)$	$\leq .40(x+6)$

Since the mixture is to be less than or equal to 40% alcohol, we add the amount of concentrate in each part and set the sum less than or equal to the amount of concentrate in a 40% mixture. That is,

$$.30x + .55(6) \leq .40(x+6)$$
$$.3x + 3.3 \leq .4x + 2.4$$
$$.9 \leq .1x$$
$$9 \leq x.$$

Therefore, at least 9 liters of 30% alcohol must be added.   ∎

Work.   In a typical work problem, we are given the time required for each member of a group to do a certain job when working individually, and we are asked to determine the time required when the members work together. We will consider situations in which the following basic principle is a reasonable assumption.

> If it takes n time units to do a given job, then $1/n$ of the job is completed per time unit.

Example 10   Andy can paint a fence in 4 hours, and Ben can do the job in 6 hours. If both work together, how long will it take to paint the fence?

Solution: Let $x =$ the number of hours required to paint the fence with both working. Then, according to the above basic principle, $1/x$ of the fence is painted per hour. Andy's contribution per hour is $1/4$, and Ben's is $1/6$. Therefore,

$$\frac{1}{4} + \frac{1}{6} = \frac{1}{x} \qquad \text{\textit{Andy's contribution per hour plus Ben's contribution}}$$
$$\text{\textit{per hour equals the total amount done per hour.}}$$
$$\frac{5}{12} = \frac{1}{x}$$
$$x = \frac{12}{5}.$$

Hence, it takes 2 2/5 hours, or 2 hours and 24 minutes, for Andy and Ben to paint the fence. ∎

Example 11 By typing together, secretaries A and B can finish a report in 2 hours. It would take B, typing alone, 5 hours to do the same report. How long would it take A?

Solution: Let $x =$ the number of hours it would take A to type the report. When A and B type together, $1/2$ of the report is finished in 1 hour. B's contribution in one hour is $1/5$ of the report, and A's is $1/x$. Therefore,

$$\frac{1}{5} + \frac{1}{x} = \frac{1}{2} \qquad \text{\textit{B's contribution per hour plus A's contribution}}$$
$$\text{\textit{per hour equals the total amount done per hour.}}$$
$$\frac{1}{x} = \frac{1}{2} - \frac{1}{5}$$
$$\frac{1}{x} = \frac{3}{10}$$
$$x = \frac{10}{3}.$$

Hence, it takes A 3 1/3 hours, or 3 hours and 20 minutes, to type the report. ∎

Example 12 One machine at a stationery company can produce 1000 envelopes in t minutes. The company wants to add another machine so that the two, working together, can produce 1000 envelopes in at most $t/3$ minutes, that is, at least $1000/(t/3)$ per minute. How fast must the new machine be?

Solution: Let $x =$ the number of minutes required for the new machine to produce 1000 envelopes. In one minute the old and the new machine, working together, produce

$$\underbrace{\frac{1000}{t}}_{\text{Old machine}} + \underbrace{\frac{1000}{x}}_{\text{New machine}}$$

envelopes, and we want this number to be at least

$$\frac{1000}{\frac{t}{3}}.$$

That is,

$$\frac{1000}{t} + \frac{1000}{x} \geq \frac{1000}{\frac{t}{3}}$$

$$\frac{1}{t} + \frac{1}{x} \geq \frac{3}{t}$$

$$x + t \geq 3x \qquad \textit{Multiply both sides by the positive quantity tx.}$$

$$t \geq 2x$$

$$\frac{t}{2} \geq x.$$

Therefore, the new machine must be at least twice as fast as the old one. ∎

Miscellaneous Applications. A variety of other situations give rise to linear equations and inequalities.

Example 13 A student has grades of 75, 80, and 87 on the first three math tests in a course. What grade must be obtained on the next test in order to give an average grade of at least 84 for the four tests?

Solution: Let $x =$ the grade needed on the next test. Then

$$\frac{75 + 80 + 87 + x}{4} \geq 84 \qquad \textit{The average of n numbers equals their sum divided by n.}$$

$$75 + 80 + 87 + x \geq 336$$

$$242 + x \geq 336$$

$$x \geq 94.$$

Hence, a grade of 94 better will bring the student's average up to at least 84. ∎

Example 14 Find three consecutive even integers such that twice the first plus three times the third is twenty more than four times the second.

Solution: Let x be the first of the three even integers. Then the next two are $x + 2$ and $x + 4$, and

$$2x + 3(x+4) = 4(x+2) + 20$$

Twice the first plus   Twenty more than
three times the third   four times the second

$$5x + 12 = 4x + 8 + 20$$
$$x = 16.$$

Therefore, the numbers are 16, 18, and 20. ■

Exercises 2.3

Percent

1. A clock radio that regularly sells for $24.95 is on sale at a 20% discount. What is the sale price?
2. A color TV is selling at a discount of 25%. The sale price is $375. What was the price of the TV before the sale?
3. During Goodwear Tire Company's annual sale, anyone who buys three tires at the regular price gets a fourth one free. To what percent discount on all four tires is this sale equivalent?
4. Alex bought gloves at a 50% discount for $10, a shirt at a 20% discount for $15, and slacks at a 25% discount for $30. What overall discount did Alex receive on his purchases?
5. The regular price of a sweater is $30, which includes a 50% markup over cost. If a retailer wants to clear out the inventory of sweaters and at least break even, what is the maximum discount that can be offered?
6. Rita's Appliances had 100 kerosene heaters to sell. It sold 75 of them at a 40% profit. What discount can it offer on the remaining 25 and still make a total profit of at least 35% on the 100 heaters?
7. A company's sales decreased by 18% from 1982 to 1983, and they increased by 20% from 1983 to 1984. What was the percent change from 1982 to 1984?
8. Newspaper A charges $6.50 a line for classified advertising, and Newspaper B charges $5.00 a line. What percent greater than B's charge is A's? What percent less than A's charge is B's?
9. Toni invested cash in a bank on December 31, 1982. On December 31, 1983 she received 10% interest and reinvested it in the same bank. On December 31, 1984 she received another 10% interest, and her grand total was $285.56 How much did Toni invest originally?
10. On June 1st a customer has a charge account balance of $500. At the end of each month a 1.5% finance charge is added to the previous month's balance, and then the customer makes a payment of 10% of the total amount. If no further purchases are made, what are the customer's payments on June 30, July 31, August 31, and September 30?

Motion

11. Frank drove 100 mi. For the first 55 mi he averaged 50 mph, and for the last 45 mi he averaged 30 mph. What was his average speed for the entire trip?
12. A bus driver wants to average at least 40 mph on a 70-mi trip. If the first 35 mi take the driver 1 hr, what should be the lowest average speed for the second 35 mi?
13. From a distance 102 mi apart, two cars travel toward each other, starting at 10 A.M. One car travels at 45 mph, and the other at 40 mph. At what point between them and at what time do they meet?
14. Drivers A and B are 50 mi apart, and they plan to meet at 7 P.M. at a point between them that is 30 mi from A. Both drive at 50 mph. What time should each driver start out?
15. At 3 P.M. Jean leaves her house and drives south on Interstate 95 at 50 mph. At 3:03 P.M. her sister Tammy leaves the house and follows Jean at 55 mph. When does Tammy catch up to Jean?

16. At 1 P.M. a person starts jogging, and from the same point a second person starts jogging at 1:05 P.M. The second catches up to the first at 1:25 P.M. How much faster than the first does the second jog?
17. Runner A can run at least one and one-half times as fast as runner B. In a 100-meter race, how many meter's head start can A give B and still not finish after B?
18. Runner A can run three-fourths as fast as Runner B. In a 5-mi race, B gives A a 1.5-mi head start. Who will win the race?
19. With the aid of the current, Patrick can row a canoe 3 mi in 12 min. Against the current, he requires 18 min to row the same distance. How fast does Patrick row in still water, and how fast is the current?
20. Kristi is walking alongside a merry-go-round. When she walks in the direction of the rotation, it takes 10 sec for a given point of the merry-go-round to return to her. When she walks against the rotation, it takes only 5 sec. How much faster than Kristi is the edge of the merry-go-round moving?

Mixtures

21. If 3 liters of 30% alcohol are mixed with 2 liters of 50% alcohol, what will be the concentration of the mixture?
22. Two liters of 40% alcohol are mixed with 4 liters of another alcohol to form a 50% alcohol solution. What is the strength of the added alcohol?
23. Five liters of 55% acid are mixed with a 40% acid to form a solution that is at most 50% acid. At least how many liters of 40% acid were used?
24. At most how much water can be added to 4 liters of a solution that is 60% acid to obtain a solution that is at least 50% acid?
25. How many pounds of hamburger worth $1.89 a pound must be mixed with 2 lb of hamburger worth $1.59 a pound to make hamburger worth $1.69 a pound?
26. If 2 lb of hamburger worth $1.49 a pound are mixed with 3 lb of hamburger worth $1.79 a pound, what should be the price per pound of the mixture?
27. A 15-qt car radiator is filled with a 10% antifreeze solution. If at most 4 qt are drained off and replaced with pure antifreeze, what is the minimum strength of the new solution?
28. A 10-qt car radiator is filled with a 20% antifreeze solution. At least how many quarts must be drained off and replaced with pure antifreeze in order to bring the strength up to at least 50% antifreeze?
29. If two gasolines with octane ratings of 89 and 92 are mixed 2 parts 89 to 3 parts 92, what is the octane rating of the mixture?
30. If three gasolines with octane ratings of 89, 92, and 94, respectively, are mixed 1 part 89 to 2 parts 92 to 3 parts 94, what is the octane rating of the mixture?

Work

31. It takes Charles 1 hr to cut and trim the grass, whereas Richard can do the same job in 1/2 hr. How long will it take with both of them working?
32. One snow blower can clear the snow from a driveway in 20 min. It takes another one 30 min to do the same job. How long would it take both of them working together?
33. Machine A processes 100 forms in 4 hr. Machines A and B together can process 100 forms in 1.5 hr. How long does it take B to process 100 forms working alone?
34. Bank tellers A and B can process 25 customers in 60 minutes. A can process 25 in 150 minutes. How long does it take B?
35. If machine A works twice as fast as B, how much faster are the two of them than each one?
36. A company has three machines that manufacture computer chips. They can process a given order in 2, 3, and 4 hours, respectively. How fast can all three machines working together process the order?
37. A company has two machines than can turn in 1000 items in 4 and 12 minutes, respectively. By adding a third machine they can turn out the 1000 items in 1 minute. How fast can the new machine do the job?
38. Plants A and B produce n_1 and n_2 autos per day, respectively. In t_3 days the two plants produce as many autos as A does in t_1 days and B does in t_2 days. That is, $n_1 t_1 = n_2 t_2 = (n_1 + n_2) t_3$ (production rate · time = total production). Show that $1/t_1 + 1/t_2 = 1/t_3$.

39. In a parallel circuit, electrical current of $I_1 + I_2$ amperes splits into I_1 amperes across a resistor of R_1 ohms and I_2 amperes across a resistor of R_2 ohms. According to Ohm's law, the total resistance in the parallel circuit is R_3 ohms, where
$$I_1 R_1 = I_2 R_2 = (I_1 + I_2) R_3. \quad \text{(current· resistance)} \\ = \text{voltage})$$
Show that $1/R_1 + 1/R_2 = 1/R_3$.

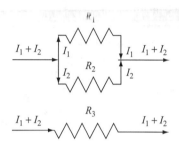

40. Compare the work equations in Exercise 38 with those in Exercise 39, and insert the corresponding electrical terms in the following chart.

Work	Parallel Circuit
Production Rate	
Time	
Total production	

Miscellaneous

41. A math professor gives four tests each worth 1/6 of the final grade and a comprehensive exam worth 2/6 of the final grade. If Tina gets 76, 80, 82, and 88 on the four tests, what is the lowest grade she can receive on the final exam and still have a final grade of at least 85?

42. A student averaged 79 on the first two tests in a course and scored 91 on the third test. The student concluded that the average for the three tests is $(79 + 91)/2 = 85$. Explain why this conclusion is wrong, and find the correct average.

43. Show that for any three consecutive integers, the sum of the first and third is twice the second.

44. The square of the sum of two numbers is 20 more than the square of their difference. One of the numbers is 1. What is the other number?

45. One number is 5 more than another, and the difference of their squares is 30. What are the two numbers?

46. The sum of three numbers is 79. The second number is 2 more than 3 times the first, and the third is 3 more than 2 times the second. What are the numbers?

47. Fifty micrograms of Vitamin B-12 is 83.3% of the U.S. recommended daily allowance for adults. How many micrograms constitute 100% of the recommended allowance?

48. A 10-oz jar of instant coffee costs $2.12. What is the equivalent price per pound?

49. If the sum of two numbers is greater than or equal to 5, and the difference is greater than or equal to 7, what is the smallest possible value of the large number?

50. If the sum of two numbers is less than or equal to 5 and the difference is greater than or equal to 7, what is the largest possible value of the smaller number?

2.4 The Coordinate Plane

Algebra and geometry come together in the coordinate plane.

Coordinates
The Distance Formula
The Midpoint Formula

Coordinates. We used the real number line to indicate solutions of linear equations and inequalities in one variable x. For equations and inequalities in two variables x and y, a two-dimensional plane is needed. To determine points in a plane, we position two perpendicular number lines to intersect at 0 on their respective number scales. We assume that

2.4. The Coordinate Plane

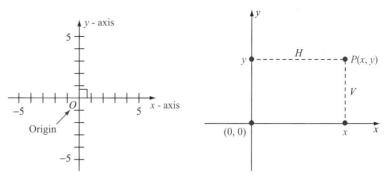

Figure 15   **Figure 16**

one line, called the ***x*-axis**, is horizontal, with its positive direction to the right, and the other, called the ***y*-axis**, is vertical, with its positive direction upward (Figure 15).

Now let P be a point in the plane. Through P draw a vertical line segment V and a horizontal line segment H as shown in Figure 16. The point x of intersection of V with the x-axis is called the ***x*-coordinate** or **abscissa** of P, and the point y of intersection of H with the y-axis is called the ***y*-coordinate** or **ordinate** of P. The ordered pair of numbers (x, y), with x to the left of y, are called the **rectangular coordinates**, or simply the **coordinates** of P. In this manner we establish a one-to-one correspondence between the points in the plane and ordered pairs of real numbers. This correspondence enables us to identify points with their coordinates, and we will use the notation P or $P(x, y)$ or simply (x, y) to denote a point in the plane. In particular, $(0, 0)$, which is the point of intersection of the x-axis and y-axis, is called the **origin** of the coordinate system.

Rectangular coordinates are also called Cartesian coordinates in honor of René Descartes (1596-1650), one of the founders of analytic geometry.

The notation (x, y) is also used to denote an open interval. We will rely on context to distinguish coordinates form open intervals.

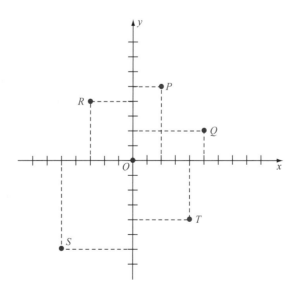

Figure 17

Example 1 Find the coordinates of the points O, P, Q, R, S, and T in Figure 17.

Solution: $O(0,0)$, $P(2,5)$, $Q(5,2)$, $R(-3,4)$, $S(-5,-6)$, and $T(4,-4)$ ∎

Example 2 Locate the points with the following coordinates: $(2,-5)$, $(-5,2)$, $(-3,-7)$, $(4,4)$, $(6,0)$, $(0,5)$.

Solution: To locate the point (x,y), start at the origin, go x directed units along the x-axis and then y directed units parallel to the y-axis. The endpoint has coordinates (x,y) (Figure 18). ∎

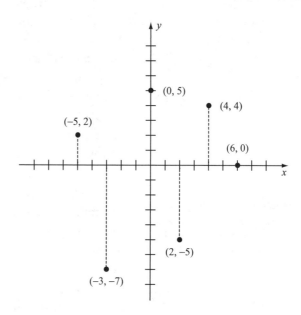

Figure 18

The coordinate axes partition the plane into four quarter planes called quadrants (Figure 19). For every point in quadrant I, both coordinates are positive. In quadrant II, the abscissa is negative and the ordinate positive. In III both coordinates are negative, and in IV the abscissa is positive and the ordinate negative. The coordinate axes form the boundaries of the four quadrants.

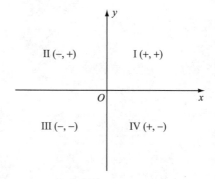

Figure 19

2.4. The Coordinate Plane

Example 3 Determine the quadrants of each of the following points: (a) $(-2, 100)$; (b) $(-10, -50)$; (c) $(a, -a)$ if $a > 0$; (d) $(b, -b)$ if $b < 0$; (e) $(-c, -c)$ if $c < 0$.

Solution: (a) II   (b) III   (c) IV   (d) II   (e) I   ■

The Distance Formula. Let x_1 and x_2 be any two points on a number line, and let d be the distance between them. If $x_2 > x_1$, there are three possibilities for their positions, as indicated in Figure 20.

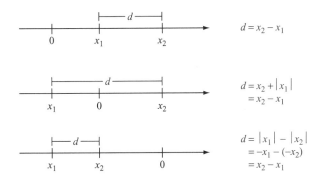

Figure 20

Hence, if $x_2 > x_1$, then $d = x_2 - x_1 = |x_2 - x_1|$ regardless of the positions of x_1 and x_2 relative to 0. Similarly, if $x_1 > x_2$, then $d = x_1 - x_2 = -(x_2 - x_1) = |x_2 - x_1|$. Hence, for any two points x_1 and x_2 on a number line,

$$d = |x_2 - x_1|. \qquad \text{distance formula along a number line}$$

Example 4 In each of the following cases, locate the two points on a number line and compute the distance between them.

(a) $x_1 = 2; x_2 = 5$ (b) $x_1 = -2; x_2 = 5$
(c) $x_1 = 2; x_2 = -5$ (d) $x_1 = -2; x_2 = -5$

Solution: The given points in each case are shown in Figure 21.   ■

Now let $P_1(x_1, y_1)$ and $P_2(x_2, y_2)$ be two points in the plane where the line segment P_1P_2 from P_1 to P_2 is neither horizontal nor vertical (Figure 22). Then P_1P_2 is the hypotenuse of a right triangle whose vertices are P_1, P_2, and $P_3(x_1, y_2)$.

By our previous formula for distance along a number line, the base P_3P_2 of the triangle has length $|x_2 - x_1|$ and the height P_3P_1 has length $|y_2 - y_1|$. Therefore, by the **Pythagorean Theorem**, the distance d between P_1 and P_2 satisfies the equation

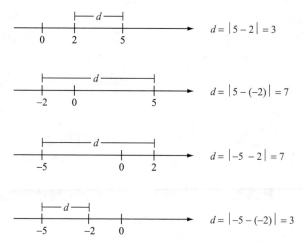

Figure 21

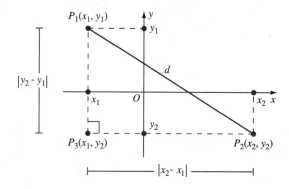

Figure 22 Distance in the Plane

$$d^2 = |x_2 - x_1|^2 + |y_2 - y_1|^2$$
$$= (x_2 - x_1)^2 + (y_2 - y_1)^2.$$

Hence,

$$d = \sqrt{(x_2 - x_1)^2 + (y_2 - y_1)^2}. \qquad \text{Distance Formula in the plane}$$

Example 5 Compute the distance between $P_1(5, 3)$ and $P_2(7, -1)$.

Solution:

$$d = \sqrt{(7-5)^2 + (-1-3)^2}$$
$$= \sqrt{4 + 16}$$
$$= \sqrt{20}$$
$$= 2\sqrt{5}$$

If P_1P_2 is a horizontal line, then $y_1 = y_2$ and $d = |x_2 - x_1| = \sqrt{(x_2 - x_1)^2 + 0^2}$. Also, if P_1P_2 is vertical, then $x_1 = x_2$ and $d = |y_2 - y_1| = \sqrt{0^2 + (y_2 - y_1)^2}$. Therefore, the distance formula applies to *any* two points in the plane. We also use the notation $d(\boldsymbol{P_1}, \boldsymbol{P_2})$ to denote the distance between P_1 and P_2, especially when more than two points are under consideration.

Question 1 *In the distance formula, does it matter which point is labeled P_1 and which is labeled P_2?*

Answer: No. If P_1 and P_2 are interchanged, then $x_2 - x_1$ becomes $x_1 - x_2$ and $y_2 - y_1$ becomes $y_1 - y_2$, but $(x_1 - x_2)^2 = (x_2 - x_1)^2$ and $(y_1 - y_2)^2 = (y_2 - y_1)^2$. Therefore, d remains the same. ∎

Example 6 Show that the points $P_1(2, 3)$, $P_2(8, 3)$, and $P_3(5, 3 + 3\sqrt{3})$ are the vertices of an equilateral triangle.

Solution:

$$d(P_1, P_2) = \sqrt{(8 - 2)^2 + (3 - 3)^2} = 6$$
$$d(P_1, P_3) = \sqrt{(5 - 2)^2 + (3 + 3\sqrt{3} - 3)^2} = 6$$
$$d(P_2, P_3) = \sqrt{(5 - 8)^2 + (3 + 3\sqrt{3} - 3)^2} = 6$$

The triangle is shown in Figure 23. Since its three sides are of equal length, it is equilateral. ∎

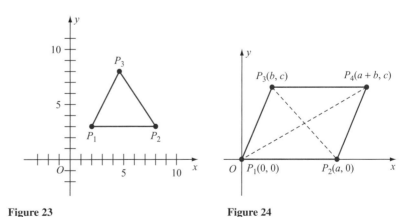

Figure 23

Figure 24

Example 7 Show that the sum of the squares of the sides of a parallelogram is equal to the sum of the squares of the diagonals.

Solution: Any parallelogram can be positioned as shown in Figure 24. The lengths of the sides are

$$d(P_1, P_2) = a = d(P_3, P_4)$$
and
$$d(P_1, P_3) = \sqrt{b^2 + c^2} = d(P_2, P_4).$$

The lengths of the diagonals are
$$d(P_1, P_4) = \sqrt{(a+b)^2 + c^2} = \sqrt{a^2 + 2ab + b^2 + c^2}$$
and
$$d(P_2, P_3) = \sqrt{(b-a)^2 + c^2} = \sqrt{b^2 - 2ab + a^2 + c^2}.$$

Therefore,
$$d^2(P_1, P_2) + d^2(P_3, P_4) + d^2(P_1, P_3) + d^2(P_2, P_4) = 2(a^2 + b^2 + c^2),$$

and
$$d^2(P_1, P_4) + d^2(P_2, P_3) = a^2 + 2ab + b^2 + c^2 + b^2 - 2ab + a^2 + c^2$$
$$= 2(a^2 + b^2 + c^2),$$

which proves the desired result. ∎

The Midpoint Formula. Let x_1 and x_2 be any two points on a number line, and let x be the midpoint of the line segment from x_1 to x_2, as shown in Figure 25.

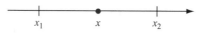

Figure 25

Then
$$x - x_1 = x_2 - x$$
$$2x = x_1 + x_2$$
$$x = \frac{x_1 + x_2}{2}. \quad \text{Midpoint Formula along a number line.}$$

Therefore, the midpoint is located at the **average value** of x_1 and x_2. Although x_2 is to the right of x_1 in Figure 25, the location of the midpoint does not depend on which value is larger. Therefore, the formula is valid regardless of the relative positions of x_1 and x_2.

Example 8   Find the midpoint of the two points in each of the following cases.

(a) $x_1 = 3$; $x_2 = 7$ (b) $x_1 = -3$; $x_2 = 7$
(c) $x_1 = 3$; $x_2 = -7$ (d) $x_1 = -3$; $x_2 = -7$

Solution: Each case is illustrated in Figure 26 ∎

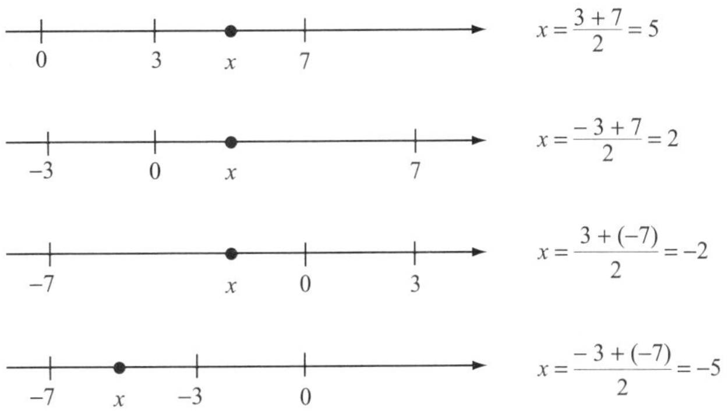

2.4. The Coordinate Plane

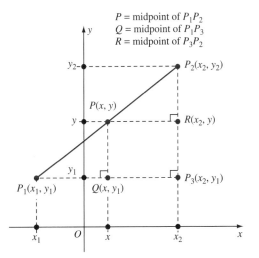

Figure 26

If we let $P_1(x_1, y_1)$ and $P_2(x_2, y_2)$ be any two points in the plane and $P(x, y)$ be the midpoint of the line segment $P_1 P_2$ (Figure 27), then $Q(x, y_1)$ is the midpoint of the horizontal line segment from P_1 to $P_3(x_2, y_1)$, and $R(x_2, y)$ is the midpoint of the vertical line segment from P_3 to P_2 (*why?*). Therefore, by the midpoint formula for points along a number line, the coordinates of P are

$$x = \frac{x_1 + x_2}{2} \quad \text{and} \quad y = \frac{y_1 + y_2}{2}. \qquad \text{midpoint formula in the plane}$$

Figure 27

Example 9 Find the distance between the point $(5,0)$ and the midpoint of the line segment from $(1,2)$ to $(7,-8)$.

Solution: As shown in Figure 28, the midpoint P has coordinates

$$\left(\frac{1+7}{2}, \frac{2-8}{2} \right) = (4, -3).$$

The distance between (5,0) and (4, −3) is

$$d = \sqrt{(4-5)^2 + (-3-0)^2} = \sqrt{10}.$$ ∎

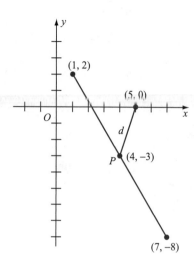

Figure 28

Example 10 Show that the diagonals of a parallelogram bisect each other.

Solution: With reference to Figure 24 in Example 7, the midpoint of diagonal $P_1 P_4$ has coordinates

$$\left(\frac{a+b}{2}, \frac{c}{2} \right),$$

and these are also the coordinates of the midpoint of diagonal $P_2 P_3$. Since both diagonals have the same midpoint, it follows that the diagonals bisect each other. ∎

Example 11 Show that the midpoints of the sides of any quadrilateral are the vertices of a parallelogram.

Solution: As shown in Figure 29, let P_1, P_2, P_3, P_4 be the vertices of a quadrilateral. The midpoints are M_1, M_2, M_3, M_4, where

$$M_1 = \left(\frac{x_1 + x_2}{2}, \frac{y_1 + y_2}{2} \right) \qquad M_3 = \left(\frac{x_3 + x_4}{2}, \frac{y_3 + y_4}{2} \right).$$

$$M_2 = \left(\frac{x_2 + x_3}{2}, \frac{y_2 + y_3}{2} \right) \qquad M_4 = \left(\frac{x_1 + x_4}{2}, \frac{y_1 + y_4}{2} \right).$$

We prove that M_1, M_2, M_3, M_4 is a parallelogram by showing that opposite side are equal in length. Now

$$d(M_1, M_2) = \sqrt{\left(\frac{x_3 - x_1}{2}\right)^2 + \left(\frac{y_3 - y_1}{2}\right)^2} = d(M_3, M_4)$$

and $$d(M_2, M_3) = \sqrt{\left(\frac{x_4 - x_2}{2}\right)^2 + \left(\frac{y_4 - y_2}{2}\right)^2} = d(M_1, M_4)$$

Therefore, M_1, M_2, M_3, M_4 is a parallelogram. ∎

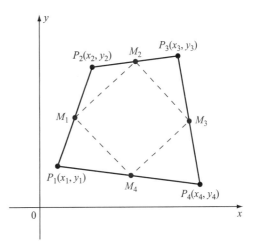

Figure 29

Comment Until the seventeenth century, algebra and geometry developed for the most part as separate subjects. Algebra was concerned mainly with solving equations, while plane geometry studied figures constructed with ruler and compass, the conic sections, and other loci. When the two subjects did meet, it was usually the case of a geometric construction to solve an algebraic equation. That is, the relationship was more geometric algebra than algebraic geometry. However, in the seventeenth century, Pierre Fermat (1601-65) and René Descartes (1596-1650) began to systematically apply the methods of algebra to geometry. Their new idea was the "coordinate plane." which enabled them to identify a curve with the locus of points that satisfy an equation in two variables x and y. This blend of algebra and geometry forged a major new branch of mathematics called **analytic geometry**. Analytic geometry, which is also called **coordinate geometry**, provides the perfect setting for the development of calculus, and it is an indispensable tool in applied mathematics and science. Much of the material in the remainder of this book is analytic geometry.

Exercises 2.4

Fill in the blanks to make each statement true.

1. The x-coordinate of a point P is also called the _____ of P, and the y-coordinate is also called the _____ of P.
2. The x-coordinate of every point on the y-axis is _____.
3. If $a > 0$ and $b > 0$, then the point $(-a, b)$ is in quadrant _____, and $(a, -b)$ is in quadrant _____.
4. If (a, b) is in the second quadrant, then $(-a, -b)$ is in quadrant _____, and (b, a) is in quadrant _____.
5. The distance formula in the plane is equivalent to the _____ theorem from geometry.

Write true or false for each statement.

6. If the ordinate of a point P is 0, then P lies on the y-axis.
7. The distance from $P_1(x_1, y_1)$ to $P_2(x_2, y_2)$ is $|x_2 - x_1| + |y_2 - y_1|$.
8. If $d(P_1, P_2) = d(P_2, P_3)$, where $P_1, P_2,$ and P_3 are in the plane, then P_2 is the midpoint of the line segment $P_1 P_3$.
9. If $d(P_1, P_2) + d(P_2, P_3) = d(P_1, P_3)$, where P_1, P_2 and P_3 are in the plane, then P_2 is on the line segment $P_1 P_3$.
10. The points (a, b) and (b, a) are always in the same quadrant.

Coordinates

Accompany each of the following with an appropriate diagram.

11. On a coordinate plane, locate each of the following points.

    (a) $(5, 2)$ (b) $(-4, 3)$
    (c) $(-6, -5)$ (d) $(5, -8)$

12. If $a < 0$ and $b > 0$, determine the quadrant of each of the following points.

    (a) (a, b) (b) $(-a, b)$
    (c) $(a, -b)$ (d) $(-a, -b)$

13. Three vertices of rectangle are $(-1, 2), (4, 2),$ and $(-1, -3)$. Find the fourth vertex.

14. Three vertices of a parallelogram are $(-3, 2), (1, 7),$ and $(12, 2)$. Find three choices for the fourth vertex.
15. Find two points P_1 and P_2 such that $P_1, P_2, (-4, -2),$ and $(5, 3)$ are the vertices of a rectangle whose sides are parallel to the coordinate axes.
16. Find two points P_1 and P_2 such that $P_1, P_2, (2, 0),$ and $(2, 6)$ are the vertices of a square.
17. Find a point P for which $Q = (2, 3), R = (10, 3)$ and P are the vertices of an isosceles triangle (two sides equal) with base QR and height 6.
18. Find a point P for which $Q = (3, -1), R = (3, 7),$ and P are the vertices of a right triangle with area 20 and height QR.
19. Find the coordinates of another point on the line through $(1, 3)$ and $(5, 6)$.
20. Find the point at which the line through $(1, 8)$ and $(2, 6)$ intersects the x-axis.

Distance

Include a coordinate diagram for each of the following.

21. Compute $d(P_1, P_2)$ for each of the following pairs of points.

    (a) $P_1(2, 1), P_2(-3, 4)$ (b) $P_1(-4, -2), P_2(4, 2)$
    (c) $P_1(5, 0), P_2(3, 6)$ (d) $P_1(8, -5), P_2(-5, 8)$

22. Find the perimeter of the triangle with vertices $P_1(3, 1), P_2(-4, 2),$ and $P_3(2, -3)$.
23. Find the lengths of the diagonals of the quadrilateral with vertices $P_1(2, 0), P_2(5, 0), P_3(7, 4),$ and $P_4(0, 6)$.
24. Given $P_1(2, -3), P_2(4, 1),$ and $P(x, 5)$, find the value of x for which $d(P, P_1) = d(P, P_2)$.
25. Given $P_1(3, 4), P_2(-1, 6),$ and $P(2, y)$, find the value of y for which $d(P, P_1) = d(P, P_2)$.
26. Show that $P_1(2, 1), P_2(4, -1),$ and $P_3(7, 4)$ are the vertices of an isosceles triangle (two sides equal in length).
27. Show that $P_1(0, 0), P_2(1, \sqrt{3}),$ and $P_3(-1, \sqrt{3})$ are the vertices of an equilateral triangle.

28. Show that $P_1(-1, 4)$, $P_2(2, 7)$, $P_3(5, 4)$ and $P_4(2, 1)$ are the vertices of a square.
29. Are $P_1(-4, 4)$, $P_2(1, 10)$, $P_3(6, 4)$ and $P_4(1, -2)$ the vertices of a square? Why or why not?
30. Show that $P_1(2, 5)$, $P_2, (4, -1)$, and $P_3(-7, 2)$ are the vertices of a right triangle.
31. A diameter of a circle has endpoints $P_1(-1, 4)$ and $P_2(10, 1)$. Is the point $P(8, 7)$ on the circle? Why or why not?
32. Show that the points $P_1(5, -1)$, $P_2(6, 0)$, and $P_3(-1, 7)$ lie on a circle with center $P(2, 3)$.
33. Show that the points $P_1(-3, 4)$, $P_2(2, 1)$, and $P_3(7, -2)$ are collinear (lie on the same straight line).
34. Jason walks 2 mi east, then 4 mi north, and then $5\sqrt{2}$ mi northeast. How far is Jason from the starting point?
35. Lynn walks 1 mi east and the 2 mi northeast. She returns by walking west and then south. How long was the entire walk?

Midpoints

Work each of the following exercises with the aid of a diagram.

36. Find the midpoint of the line segment P_1P_2 in each of the following cases.

    (a) $P_1(4, 5)$, $P_2(-3, 3)$ (b) $P_1(8, 0)$, $P_2(0, 8)$
    (c) $P_1(-6, 1)$, $P_2(-2, 7)$ (d) $P_1(2, 7)$, $P_2(-2, -7)$

37. Find the midpoints of the sides of the quadrilateral with vertices $P_1(1, 5)$, $P_2(-2, 7)$, $P_3(-3, -6)$, $P_4(4, -8)$.
38. $P(2, 3)$ is the midpoint of line segment P_1P_2, where P_1 has coordinates $(-4, 5)$. Find the coordinates of P_2.
39. $P(x, 4)$ is the midpoint of the line segment from $P_1(2, 5)$ to $P_2(-7, y_2)$. Find x and y_2.
40. The points $P_1(-1, 2)$ and $P_2(3, 2)$ are on a circle with center $(1, 6)$. Find two other points on the circle.
41. $P(3, -2)$ is the center of a circle of radius 5. Find four points on the circle.
42. One diagonal of a parallelogram has endpoints $P_1(4, 5)$ and $P_2(-1, -3)$. The other diagonal is a vertical line segment of length 6. Find the other two vertices of the parallelogram.
43. $P_1(-1, 2)$ is a vertex of a parallelogram. One diagonal of the parallelogram is a horizontal segment of length 6 with one endpoint $P_2(5, 4)$. Find the other two vertices of the parallelogram (two solutions).
44. Show that the line segments joining the midpoints of opposite sides of a quadrilateral bisect each other.
45. Show that the larger of any two points x_1, x_2 on the x-axis is $(x_1 + x_2)/2 + |x_2 - x_1|/2$, and the smaller is $(x_1 + x_2)/2 - |x_2 - x_1|/2$.

Miscellaneous

46. Show that if $P(x, y)$ is the point on the line segment P_1P_2 for which

    $$\frac{d(P_1, P)}{d(P_1, P_2)} = r \quad (0 \le r \le 1),$$

    then $\quad x = x_1 + r(x_2 - x_1)$
    and $\quad y = y_1 + r(y_2 - y_1)$.   **ratio formula**
    Show that the midpoint formula is the special case of the ratio formula corresponding to $r = 1/2$.

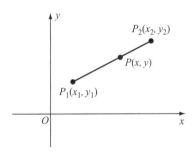

47. Use the ratio formula to show that the medians of a triangle intersect in a point whose distance from a given vertex is two thirds the length of the corresponding median. (A median of a triangle is a line segment from a vertex to the midpoint of the opposite side.)
48. Show that the sum of the squares of the medians of a triangle is equal to three fourths the sum of the squares of the sides.
49. Show that the midpoint of the hypotenuse of a right triangle is equidistant from the three vertices.
50. Show that the length of the line segment joining the midpoints of two sides of a triangle is one half the length of the remaining side.

2.5 Equations of Straight Lines

Slope of a Line
Point-Slope Equation
Slope-Intercept Equation
General Linear Equation

Single points have no length, but if you place enough of then side by side, you can fill out a straight line!

Until now we have considered equations in one unknown x whose solutions are points on the real line. We now begin a study of equations in two unknowns x and y whose solutions are sets of points in the plane. The set of all points satisfying a given equation is called the **graph** of the equation. Our study will focus on the following two problems:

1. given an equation, find its graph;
2. given a geometric figure, find an equation whose graph is the given figure.

These problems will play a central role throughout the rest of the book, a role that carries over into calculus. We start here with their application to straight lines.

Slope of a Line. Two points $P_1(x_1, y_1)$ and $P_2(x_2, y_2)$ determine a line L. Suppose L is not a vertical line, and let $P(x, y)$ be another point on L. By equating ratios of corresponding sides of the similar right triangles P_1RP and P_1QP_2 (Figure 30), we conclude that

$$\frac{y - y_1}{x - x_1} = \frac{y_2 - y_1}{x_2 - x_1}. \tag{1}$$

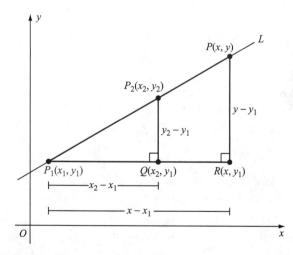

Figure 30 Similar Triangles

Conversely, if (1) is satisfied for some pair (x, y), then the point $P(x, y)$ lies on L.

2.5. Equations of Straight Lines

Example 1 Show that the three points $P_1(-1, 2)$, $P_2(3, 5)$, and $P_3(7, 8)$ are collinear (lie on the same line).

Solution:

$$\frac{y_2 - y_1}{x_2 - x_1} = \frac{5 - 2}{3 - (-1)} = \frac{3}{4} \quad \text{and} \quad \frac{y_3 - y_1}{x_3 - x_1} = \frac{8 - 2}{7 - (-1)} = \frac{3}{4}.$$

Hence, P_3 is on the line determined by P_1 and P_2. ∎

With reference to Figure 30, the quantity

$$m = \frac{y_2 - y_1}{x_2 - x_1} \tag{2}$$

is called the **slope** of L. The slope of a nonvertical line L is a geometric parameter whose value can be computed by means of any two points P_1 and P_2 on L. As Figure 31 illustrates, m is a measure of the

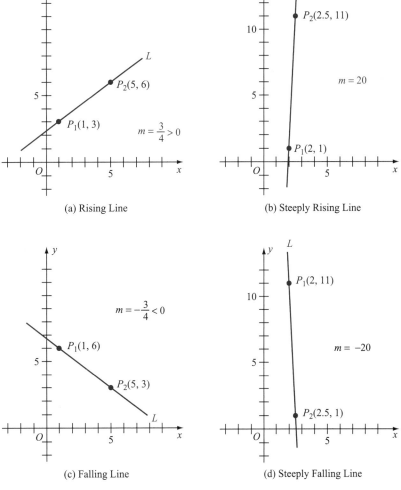

(a) Rising Line

(b) Steeply Rising Line

(c) Falling Line

(d) Steeply Falling Line

Figure 31

steepness and *direction* of L. Lines that rise from left to right have *positive* slopes [Figures 31(a) and 31(b)], and lines that fall have *negative* slopes [Figures 31(c) and 31(d)]. Also, the steeper a line is, the larger its slope is in *magnitude*.

Question 1 *In computing the slope of L by means of two points, does it matter which of the two is labeled P_1 and which is P_2?* ■

Answer: No. If the points are interchanged, then $x_2 - x_1$ becomes 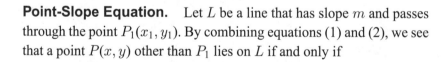 but

$$\frac{y_1 - y_2}{x_1 - x_2} = \frac{y_2 - y_1}{x_2 - x_1}.$$

Therefore, the slope remains the same. Hence,

$$m = \frac{\text{change in } y}{\text{change in } x}$$

in going from *any* point on L to *any other* point on L. Note that if $x_2 - x_1 = \pm 1$, then $y_2 - y_1 = \pm m$. That is, if x changes by ± 1 unit, then y changes by $\pm m$ units. For example, if L has slope 3 and $(5, 4)$ lies on L, then $(5 + 1, 4 + 3) = (6, 7)$ and $(5 - 1, 4 - 3) = (4, 1)$ also lie on L (Figure 32).

Two extreme cases regarding steepness are *horizontal* and *vertical* lines (Figure 33). A horizontal line has slope 0 because the numerator $y_2 - y_1$ in equation (2) is 0. For a vertical line, m is *not defined* because the denominator $x_2 - x_1$ in equation (2) is 0. To put it another way, there is no real number large enough to be the slope of a vertical line.

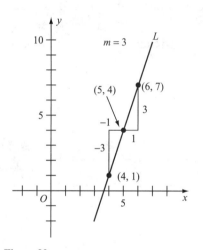

Figure 32

Point-Slope Equation. Let L be a line that has slope m and passes through the point $P_1(x_1, y_1)$. By combining equations (1) and (2), we see that a point $P(x, y)$ other than P_1 lies on L if and only if

$$\frac{y - y_1}{x - x_1} = m,$$

that is,

$$y - y_1 = m(x - x_1). \tag{3}$$

Equation (3) has line L for its graph and is called a **point-slope equation** of L. We note that although equation (3) was derived under the assumption that $P(x, y)$ was a point on L other than $P_1(x_1, y_1)$, the equation is also satisfied when $x = x_1$ and $y = y_1$. That is, *every* point on L satisfies (3).

2.5. Equations of Straight Lines   143

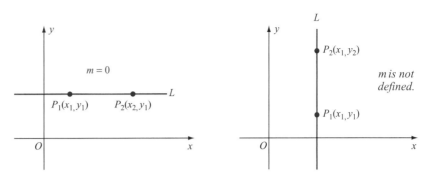

Figure 33 Horizontal Line and Vertical Line

Example 2   Find a point-slope equation of the line L that has slope 2 and passes through the point $(1, 3)$, and then graph L.

Solution: From (3), the corresponding point-slope equation of L is

$$y - 3 = 2(x - 1).$$

One point on L is, of course, $(1, 3)$. Another is $(2, 5)$ (*why?*). These two points can be used to graph L (Figure 34). ■

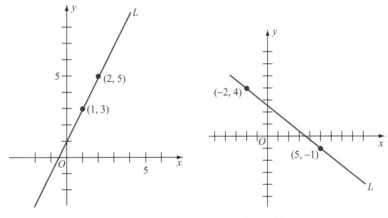

Figure 34   **Figure 35**

Example 3   Graph the line L that passes through the points $(-2, 4)$ and $(5, -1)$. Find a point-slope equation of L.

Solution: The graph of L through the two given points is shown in Figure 35. From (2), the slope of L is

$$m = \frac{(-1) - 4}{5 - (-2)} = -\frac{5}{7}.$$

Therefore, a point-slope equation of L can be written as either

144   Chapter 2. Algebra and Graphs of Linear Expressions

$$y - 4 = -\frac{5}{7}(x + 2) \quad \text{Using the point } (-2, 4)$$

or $\quad y + 1 = -\frac{5}{7}(x - 5). \quad \text{Using the point } (5, -1)$

Note that both of these equations are equivalent to

$$y = -\frac{5}{7}x + \frac{18}{7}. \quad \blacksquare$$

A horizontal line L has slope $m = 0$. If L passes through the point (a, b) as in Figure 36, then the corresponding point-slope equation of L is

$$y - b = 0(x - a),$$

or $\quad$ **$y = b.$** $\qquad\qquad$ Horizontal Line

For example, the equation of the horizontal line through the point $(-2, 3)$ is $y = 3$, and the equation of the x-axis is $y = 0$. In the plane, we can interpret the equation $y = b$ as saying "y must equal the constant value b, but x can equal any value."

A vertical line L has undefined slope and therefore no point-slope equation. However, if L passes through the point (a, b) as in Figure 37, then for every point on L, the x-coordinate is equal to a. Therefore, L is the graph of the equation

$$\boldsymbol{x = a.} \qquad\qquad \text{Vertical Line}$$

For example, the equation of the vertical the through $(-2, 3)$ is $x = -2$, and the equation of the y-axis is $x = 0$. In the plane, the equation $x = a$ says "x must have the constant value a, but y can have any value."

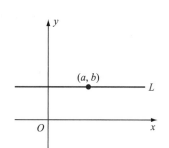

Figure 36

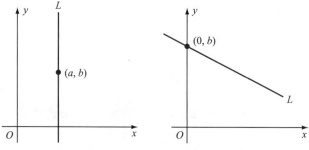

Figure 37 $\qquad\qquad$ Figure 38

Slope-Intercept Equation. A nonvertical line L will intersect the y-axis at a point whose coordinates are $(0, b)$ for some value of b (Figure 38). The point $(0, b)$, or simply b itself, is called the **y-intercept** of L. By using $(0, b)$ for (x_1, y_1) in the point-slope equation (3), we obtain

2.5. Equations of Straight Lines

$$y - b = m(x - 0),$$
$$y = mx + b, \quad (4)$$

or

where m is the slope of L. Equation (4) is called the **slope-intercept equation** of L.

Example 4 Let L be the line with point-slope equation $y - 5 = 3(x - 1)$. Find the y-intercept and the slope-intercept equation of L, and then graph L.

Solution: By substituting $x = 0$ in the point-slope equation, we obtain $y = 2$ as the y-intercept. Therefore, the slope-intercept equation of L is

$$y = 3x + 2.$$

The points $(1,5)$ and $(0,2)$ can be used to graph L (Figure 39).

Alternative Solution:

$$y - 5 = 3(x - 1) \qquad \textit{Point-slope equation of L.}$$
$$y = 3x - 3 + 5 \qquad \textit{Solve for y in terms of x.}$$
$$y = 3x + 2 \qquad \textit{Slope-intercept equation of L.} \quad \blacksquare$$

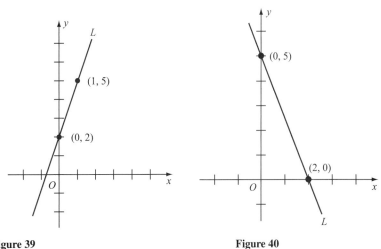

Figure 39   Figure 40

We note that a nonvertical line L of slope m has exactly one y-intercept b. Therefore, the slope-intercept equation of L is *unique*, whereas a point-slope equation of L depends on which point is chosen.

Example 5 Find the slope-intercept equation of the line L that crosses the x-axis at $(2,0)$ and the y-axis at $(0,5)$, and then graph L.

Solution: The slope of L is $(5-0)/(0-2) = -5/2$, and the y-intercept is 5. Therefore, the slope-intercept equation of L is

$$y = -\frac{5}{2}x + 5.$$

The graph of L is shown in Figure 40. ■

Example 6 Show that the point-slope equations $y-2 = 3(x+1)$ and $y-8 = 3(x-1)$ each have the same graph.

Solution:

$$y - 2 = 3(x + 1) \quad y - 8 = 3(x - 1)$$
$$y - 2 = 3x + 3 \quad y - 8 = 3x - 3$$
$$y = 3x + 5 \quad y = 3x + 5$$

Each equation has for its graph the line with slope 3 and y-intercept 5 (Figure 41). ■

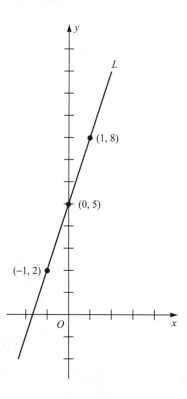

Figure 41

General Linear Equation. The equation

$$Ax + By = C, \tag{5}$$

where either $A \neq 0$ or $B \neq 0$, is called the **general linear equation**. As the name implies, the graph of (5) is a straight line. There are three possible cases.

2.5. Equations of Straight Lines

1. If $A = 0$ and $B \neq 0$, then $y = C/B$, a horizontal line. For example, the graph of $2y = 3$ is the horizontal line $y = 3/2$.
2. If $A \neq 0$ and $B = 0$, then $x = C/A$, a vertical line. For example, the graph of $-5x = 7$ is the vertical line $x = -7/5$.
3. If $A \neq 0$ and $B \neq 0$, then $y = -(A/B)x + C/B$, a line with slope $-A/B$ and y-intercept C/B. For example, $-2x + 3y = 5$ is equivalent to $y = (2/3)x + 5/3$, whose graph is the straight line with slope $2/3$ and y-intercept $5/3$.

Hence, the general linear equation covers all possible lines, including vertical ones.

Example 7 Graph the line $L : 3x - 5y = 15$.

Solution: Any two points that satisfy the equation will determine the graph. For instance, if $x = 0$, then $y = -3$. Also, if $y = 0$, then $x = 5$. Hence, L passes through the points $(0, -3)$ and $(5, 0)$ (Figure 42). ∎

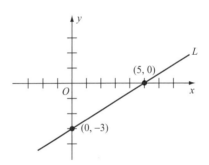

Figure 42

Example 8 Line L passes through the points $(3, -5)$ and $(-4, 6)$. Write an equation of L in (a) point-slope form, (b) slope-intercept form, and (c) general linear form.

Solution: First find the slope of L:

$$m = \frac{6 - (-5)}{-4 - 3} = -\frac{11}{7}.$$

(a) $y + 5 = -\dfrac{11}{7}(x - 3)$ *Use point* $(3, -5)$.

$y - 6 = -\dfrac{11}{7}(x + 4)$ *Use point* $(-4, 6)$.

(b) $y = -\dfrac{11}{7}x - \dfrac{2}{7}$ *Obtained from either equation in (a) by solving for y in terms of x.*

(c) $11x + 7y = -2$ *Obtained from (b).* ∎

Question 2 What is the graph of $Ax + By = C$ if $A = 0$ and $B = 0$?

Answer: The graph of $0x + 0y = C$ depends on C. If $C = 0$, then *every* point (x, y) satisfies the equation, and the graph is the *whole plane*. If $C \neq 0$, then *no* point (x, y) satisfies the equation, and the graph is the *empty set*. To avoid these extreme cases, we insist that either $A \neq 0$ or $B \neq 0$ in the general linear equation. ∎

Comment In Section 2.1, where we considered equations in one variable, the equation $ax + b = 0$ was called a linear equation. In this section, $Ax + By = C$ is called a linear equation. Later in the text, when we consider three variables x, y, and z, the equation $Ax + By + Cz = D$ will be called a linear equation. Of the three, only the second one actually has a line for its graph. However, all three equations have the same *form* with respect to the number of variables considered. In general, for n variables $x_1, x_2, \ldots, x_n$ a **linear equation** is of the form

$$A_1 x_1 + A_2 x_2 + \cdots + A_n x_n = C.$$

Exercises 2.5

Fill in the blanks to make each statement true.

1. The set of all points in the plane satisfying a given equation is called the _____ of the equation.
2. The _____ of a line is a geometric parameter that measures the steepness and direction of the line.
3. If a line L has slope 4, and the point $(2, 1)$ is on L, then the points $(3, \text{\_\_\_\_})$ and $(1, \text{\_\_\_\_})$ are also on L.
4. Three forms of equations of a straight line are _____ , _____, and _____.
5. The horizontal line through the point $(3, -4)$ has equation _____, and the vertical line through the same point has equation _____.

Write true or false for each statement.

6. Two different lines can have the same slope.
7. Two different lines can have the same y-intercept.
8. Two different lines can have the same slope and y-intercept.
9. A line with slope -10 is steeper than a line with slope 5.

10. If $(2, 1)$ and $(3, 5)$ are on line L, $(4, 9)$ is also on L.

Slope

Compute, if possible, the slope of the line through each of the pairs of points P_1 and P_2 in Exercises 11-16.

11. $P_1(4, 3), P_2(3, 4)$
12. $P_1(2, -5), P_2(6, -7)$
13. $P_1(8, 1), P_2(0, 3)$
14. $P_1(0, 0), P_2(4, 5)$
15. $P_1(3, 7), P_2(3, 5)$
16. $P_1(4, -1), P_2(4, 3)$

Find the value of each indicated variable in Exercises 17-20.

17. L has slope 1, and points $P_1(2, 5)$ and $P_2(4, y)$ are on L. Find y.
18. L has slope -3, and points $P_1(1, 4)$ and $P_2(x, 2)$ are on L. Find x.

19. L has slope 2, and points $P_1(0,5)$ and $P_2(x,0)$ are on L. Find x.
20. L has slope -1, and points $P_1(3,0)$ and $P_2(0,y)$ are on L. Find y.

Equations

Find a point-slope equation of each line L in Exercises 21-26, and then graph L.

21. L passes through point $P_1(-2,4)$ and has slope $m = 3$.
22. L passes through point $P_1(5,-5)$ and has slope $m = -2$.
23. L passes through the points $P_1(3,0)$ and $P_2(0,-3)$.
24. L passes through the points $P_1(6,-3)$ and $P_2(-1,1)$.
25. L has slope $m = 2$ and passes through the midpoint of the line segment from $(-3,4)$ to $(3,4)$.
26. L passes through $P(-1,2)$ and the midpoint of the line segment from $(4,7)$ to $(2,3)$.

Find the slope-intercept equation of each line L in Exercises 27-32, and then graph L.

27. L passes through point $P_1(3,5)$ and has slope $m = 2$.
28. L passes through point $P_1(-4,3)$ and has slope $m = -2$.
29. L passes through the points $P_1(4,4)$ and $P_2(5,7)$.
30. L passes through the points $P_1(-2,-3)$ and $P_2(3,2)$.
31. L passes through the midpoint of the line segment from $P_1(4,-1)$ to $P_2(4,5)$ and the midpoint of the segment from $Q_1(2,3)$ to $Q_2(4,3)$.
32. L passes through $P(2,5)$ and the midpoint of the line segment from $(6,2)$ to $(-8,8)$.

Graph each of the following. Also, where possible, determine the slope and the y-intercept.

33. $4x - 3y = 7$
34. $-2x + 5y = 5$
35. $4x = 3$
36. $-2y = 5$
37. $x + y = 1$
38. $x - y = 0$

Miscellaneous

The graph of the equation

$$\frac{x}{a} + \frac{y}{b} = 1 \quad \text{intercept equation}$$

is a straight line L with y-intercept b and x-intercept a [i.e., L intersects the x-axis at the point (a,0)]. Find the intercept equation of the line L determined in each of the following cases, and then graph L.

39. L has x-intercept 4 and y-intercept 3.
40. L has x-intercept 2 and y-intercept -5.
41. L passes through the points $P_1(2,4)$ and $P_2(8,-4)$.
42. L passes through the points $P_1(-3,5)$ and $P_2(2,0)$.
43. L passes through point $P_1(-2,-6)$ and has slope $m = 1$.
44. L passes through the point $P_1(7,4)$ and has slope $m = -1$.

The ratio formula of Exercise 46 in Section 2.4 gives the coordinates of a point $P(x,y)$ on the line segment from P_1 to P_2. That is, the equations

$$x = x_1 + r(x_2 - x_1)$$
$$y = y_1 + r(y_2 - y_1)$$

*have for their graph, as the parameter r varies from 0 to 1, the line segment $P_1(x_1, y_1)$ to $P_2(x_2, y_2)$. If the parameter r ranges through all real numbers, the same equations have for their graph the entire line L through P_1 and P_2. They are called **parametric equations** of L. Find parametric equations for each line L determined in Exercises 45-50, and then graph L.*

45. $P_1(2,-5)$, $P_2(5,-2)$
46. $P_1(4,3)$, $P_2(-4,-3)$
47. $P_1(6,2)$, $P_2(6,-2)$
48. $P_1(8,6)$, $P_2(6,8)$
49. L passes through point $P_1(7,2)$ and has slope $m = 3$.
50. L passes through point $P_1(-3,5)$ and has slope $m = -3$.
51. Use the distance formula in the plane to determine the equation of the perpendicular bisector of the line segment from $P_1(x_1, y_1)$ to $P_2(-x_1, -y_1)$.

52. The graph of the equation relating temperature in Fahrenheit (F) to temperature in Celsius (C) is a straight line L. With C as abscissa and F as ordinate, find the equation of L, given that water freezes at $0°C$ and $32°F$, and it boils at $100°C$ and $212°F$.

53. After school, Ted walks part of the way home and jogs the rest, both along the same straight line. One half hour after starting to jog, Ted is 3 mi from school, and after three fourths of an hour, he is 4 1/4 mi from school.

   (a) How fast does Ted jog?
   (b) How far from school was Ted when he started to jog?
   (c) Let $x =$ time spent jogging and $y =$ distance from school. Express y in terms of x and graph the equation.

54. A gas station charges a certain rate per gallon for the first 10 gal and a reduced rate for every gallon over 10. Amy pays $12.00 for 11 gal and $16.00 for 15 gal.

   (a) What is the rate per gallon for the first 10 gal?
   (b) What is the rate per gallon after the first 10 ?
   (c) Let $y =$ total amount paid for gas and let $x =$ the number of gallons over 10. Express y in terms of x, and graph the equation.

55. If P_0 dollars are invested at a monthly simple interest rate of r, the accumulated amount P after t months is

$$P = P_0 + P_0 rt.$$

Lou invests P_0 dollars on December 31. On March 31 Lou has accumulated $155.53, and on October 21 the amount is $166.10.

   (a) How much did Lou invest?
   (b) What was the monthly interest rate?
   (c) Graph the equation relating P to t.

2.6 Systems of Two Linear Equations in Two Unknowns

Two points determine a line, and in turn, two lines determine a point, possibly at infinity.

Parallel Lines
Perpendicular Lines
Intersecting Lines
Determinants
Applications

Two distinct lines in the plane either intersect in a single point or are parallel. Intersecting lines may or may not be perpendicular to each other. In this section we investigate these and other geometric properties of lines by means of their algebraic equation.

Parallel Lines. Distinct lines L_1 and L_2 are parallel if and only if they are both vertical or their slopes m_1 and m_2 are equal:

$$\boldsymbol{m_1 = m_2}. \quad \text{Parallel lines (nonvertical)} \qquad (1)$$

Example 1 Which of the following pairs of equations represent parallel lines?

(a) $2x - 3y = 4$
$\quad\;\, -4x + 6y = 5$

(b) $y = 3x - 1$
$\quad\;\, y = -3x + 1$

(c) $x = -3$
$\quad\;\, x = 3$

(d) $y - 1 = 2(x + 1)$
$\quad\;\, y - 5 = 2(x - 1)$

2.6. Systems of Two Linear Equations in Two Unknowns

Solution:

(a) The equations are equivalent to $y = (2/3)x - 4/3$ and $y = (2/3)x + 5/6$, respectively. Both of the corresponding lines have slope $2/3$, so they are parallel.

(b) The lines have unequal slopes, so they are not parallel.

(c) The corresponding graphs are parallel vertical lines.

(d) Both equations are equivalent to $y = 2x + 3$, and therefore the lines coincide. Coincident lines are not considered to be parallel; by definition, a line is not parallel to itself.

■

Example 2 Find an equation of the line L through the point $(-2, 1)$ and parallel to the line $2x - 5y = 7$.

Solution: The line $2x - 5y = 7$ has slope $2/5$ (why?). Therefore, a point-slope equation of L is

$$y - 1 = \frac{2}{5}(x + 2) \quad \text{or, equivalently,} \quad 2x - 5y = -9.$$

■

If L_1 and L_2 have general linear equations

$$L_1 : A_1 x + B_1 y = C_1$$
$$L_2 : A_2 x + B_2 y = C_2$$

then $m_1 = -A_1/B_1$ and $m_2 = -A_2/B_2$, assuming $B_1 \neq 0$ and $B_2 \neq 0$. Hence, the equation $m_1 = m_2$ is equivalent to $-A_1/B_1 = -A_2/B_2$, or

$$A_1 B_2 - A_2 B_1 = 0. \quad \text{Parallel lines} \qquad (2)$$

Furthermore, if L_1 and L_2 are vertical lines, and therefore parallel, then $B_1 = 0$ and $B_2 = 0$ and equation (2) still holds. Therefore, equation (2) is the algebraic condition for any two lines, vertical or not, to be parallel.

Caution As stated in Example 1 (d), the designation "parallel" does not apply to lines that coincide. Parallel lines are *distinct* lines in the plane that *do not intersect*. Coincident lines fail on both counts. To determine whether nonvertical lines coincide, write their respective equations in slope-intercept form by solving for y in terms x. If these equations are identical, the lines coincide. Vertical lines are coincident if their equations $x = c$ are identical.

152   Chapter 2. Algebra and Graphs of Linear Expressions

Example 3   Determine which of the following pairs of equations represent parallel lines and which represent coincident lines.

(a) $2x - 3y = 1$
      $-8x + 12y = -4$

(b) $5x - 2y = 3$
      $-10x + 4y = 6$

(c) $3x = 2$
      $6x = 4$

(d) $2y = 5$
      $10y = 15$

Solution:

(a) Each equation is equivalent to $y = (2/3)x - 1/3$. Therefore, the lines are coincident.

(b) The first equation is equivalent to $y = (5/2)x - 3/2$ and the second to $y = (5/2) + 3/2$. The lines are parallel.

(c) Each equation is equivalent to $x = 2/3$. The graphs are coincident vertical lines.

(d) The first equation is equivalent to $y = 5/2$ and the second to $y = 3/2$. The graphs are parallel horizontal lines. ∎

Perpendicular Lines.   Now suppose that lines L_1 and L_2 are perpendicular and that $P_0(x_0, y_0)$ is their point of intersection. If one of the lines is vertical, then the other is horizontal. If neither line is vertical, then one of the lines will have a positive slope and the other will have a negative slope *(why?)*. In this case let L_1 have positive slope m_1 and let L_2 have negative slope m_2. As indicated in Figure 43, if x changes from x_0 to $x_0 + 1$, then along L_1, P_0 changes to $P_1(x_0 + 1, y_0 + m_1)$, and along L_2, P_0 changes to $P_2(x_0 + 1, y_0 + m_2)$ (Along a straight line, when x increases by 1 unit, y changes by m units.)

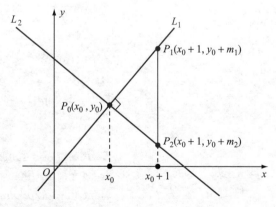

Figure 43   Perpendicular Lines

By the Pythagorean theorem applied to right triangle $P_0 P_1 P_2$, we obtain

2.6. Systems of Two Linear Equations in Two Unknowns   153

$$d(P_1, P_2)^2 = d(P_0, P_1)^2 + d(P_0, P_2)^2,$$

which becomes

$$[y_0 + m_2 - (y_0 + m_1)]^2 = (x_0 + 1 - x_0)^2 + (y_0 + m_1 - y_0)^2$$
$$+ (x_0 + 1 - x_0)^2 + (y_0 + m_2 - y_0)^2$$

or $(m_2 - m_1)^2 = 1 + m_1^2 + 1 + m_2^2$

and simplifies to

$$\boldsymbol{m_1 m_2} = -1. \quad \text{Perpendicular lines (neither one vertical)} \tag{3}$$

If the equation of L_1 is $A_1 x + B_1 y = C_1$ and that of L_2 is $A_2 x + B_2 y = C_2$, then $m_1 = -A_1/B_1$ and $m_2 = -A_2/B_2$, and equation (3) is equivalent to

$$\boldsymbol{A_1 A_2 + B_1 B_2} = 0. \quad \text{Perpendicular lines} \tag{4}$$

Conversely, if (3) or (4) is satisfied, then L_1 is perpendicular to L_2. Furthermore, equation (4) also includes the case where one of the lines is vertical (B_1 or $B_2 = 0$) and the other is horizontal (A_2 or $A_1 = 0$).

Example 4   Which of the following pairs of equations represent perpendicular lines?

(a) $y = (1/3)x - 1$ (b) $y - 2 = 4(x - 1)$
    $y = -3x + 2$ $y + 1 = -4(x + 2)$

(c) $y = 2$ (d) $4x - 3y = -1$
    $x = 5$ $3x + 4y = 7$

Solution:

(a) Slope $m_1 = 1/3$ and $m_2 = -3$; the lines are perpendicular.
(b) Slope $m_1 = 4$ and $m_2 = -4$; the lines are not perpendicular.
(c) One lines are horizontal and one is vertical; therefore, they are perpendicular.
(d) Slope $m_1 = 4/3$ and $m_2 = -3/4$; the lines are perpendicular. Alternatively, using (4), $A_1 A_2 + B_1 B_2 = 4 \cdot 3 + (-3)4 = 0$. ∎

Example 5   Find an equation of the perpendicular bisector of the line segment from $(-2, 3)$ to $(6, 7)$.

Solution: The midpoint of the line segment is

$$\left(\frac{-2 + 6}{2}, \frac{3 + 7}{2} \right) = (2, 5),$$

and the slope of the segment is

$$\frac{7-3}{6+2} = \frac{1}{2}.$$

Therefore, the perpendicular bisector passes through the point (2,5) and has slope -2 (Figure 44). Its corresponding point-slope equation is $y - 5 = -2(x - 2)$ ∎

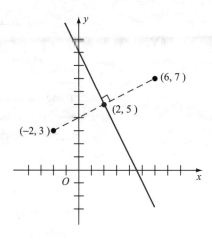

Figure 44

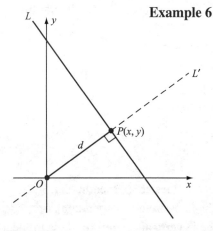

Figure 45

Example 6 Show that the (shortest) distance from the origin to the line L having equation $Ax + By = C$ is

$$\frac{|C|}{\sqrt{A^2 + B^2}}.$$

Solution: Let L' be the line through the origin and perpendicular to L. The shortest distance from the origin to L is the length of segment OP, where P is the point of intersection of L and L' (Figure 45). An equation of L' is $Bx - Ay = 0$. (why?), and the coordinates of P can be obtained as follows:

$$\begin{aligned} A^2x + ABy &= AC \qquad &\text{Multiply equation of } L \text{ by } A.\\ B^2x - ABy &= 0 \qquad &\text{Multiply equation of } L' \text{ by } B.\\ \hline (A^2 + B^2)x &= AC \qquad &\text{Add the two equations.} \end{aligned}$$

$$x = \frac{AC}{A^2 + B^2}.$$

Similarly, by multiplying L by B and L' by A and subtracting, we obtain

$$y = \frac{BC}{A^2 + B^2}.$$

Therefore, by the distance formula, the length of OP is

$$d = \sqrt{\left(\frac{AC}{A^2+B^2}\right)^2 + \left(\frac{BC}{A^2+B^2}\right)^2}$$

$$= \sqrt{\frac{A^2C^2 + B^2C^2}{(A^2+B^2)^2}}$$

$$= \sqrt{\frac{(A^2+B^2)C^2}{(A^2+B^2)^2}} = \frac{|C|}{\sqrt{A^2+B^2}}.$$

■

Intersecting Lines. In example 6 we determined the point of intersection of the perpendicular lines $Ax + By = C$ and $Bx - Ay = 0$. A similar procedure can be used to determine the point $P(x, y)$ of intersection of any two lines. Let L_1 and L_2 have the equations

$$L_1 : A_1x + B_1y = C_1 \qquad (5a)$$

$$L_2 : A_2x + B_2y = C_2. \qquad (5b)$$

We obtain the x-coordinate of P as follows:

$$\begin{aligned}
A_1B_2x + B_1B_2y &= C_1B_2 &&\text{Multiply } L_1 \text{ by } B_2. \\
A_2B_1x + B_2B_1y &= C_2B_1 &&\text{Multiply } L_2 \text{ by } B_1. \\
(A_1B_2 - A_2B_1)x &= C_1B_2 - C_2B_1 &&\text{Subtract.}
\end{aligned}$$

$$x = \frac{C_1B_2 - C_2B_1}{A_1B_2 - A_2B_1}. \qquad (6a)$$

Similarly, by multiplying L_1 by A_2 and L_2 by A_1 and subtracting, we have

$$y = \frac{A_1C_2 - A_2C_1}{A_1B_2 - A_2B_1}. \qquad (6b)$$

These formulas for x and y need not be memorized. In any given problem, the point of intersection can be obtained by repeating the above procedure. However, the formulas show that there is a unique pair (x, y) of real numbers that satisfies both equations (5a) and (5b) if and only if $A_1B_2 - A_2B_1 \neq 0$. This algebraic statement is equivalent to the evident geometric property that two distinct lines in the plane will intersect in a unique point if and only if the lines are not parallel [see equation (2)].

156   Chapter 2. Algebra and Graphs of Linear Expressions

Example 7   Find the point of intersection of the line $L_1 : 3x - 4y = 7$ and $L_2 : 2x + 5y = -3$

Solution:

$$
\begin{aligned}
15x - 20y &= 35 &&\text{Multiply } L_1 \text{ by 5.} \\
\underline{8x + 20y} &= \underline{-12} &&\text{Multiply } L_2 \text{ by 4.} \\
23x &= 23 &&\text{Add to eliminate } y. \\
x &= 1
\end{aligned}
$$

We can obtain y by a similar procedure or by substituting $x = 1$ into either L_1 or L_2. For instance, from L_1 we obtain

$$3 \cdot 1 - 4y = 7$$
$$y = -1$$

Therefore, the point of intersection of L_1 and L_2 is $(1, -1)$, as shown in Figure 46 ∎

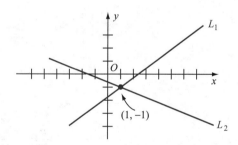

Figure 46

An alternate method for solving a **system of two linear equations in two unknowns**, such as equations (5a) and (5b), is illustrated in the following example.

Example 8   Solve the system of equations

$$-2x + 3y = 18$$
$$3x - 2y = -17.$$

Solution: From the first equation, express y in terms of x:

$$y = 6 + \frac{2}{3}x.$$

Substitute this expression for y into the second equation and solve for x.

$$3x - 2\left(6 + \frac{2}{3}x\right) = -17$$

$$3x - 12 - \frac{4}{3}x = -17$$

$$\frac{5}{3}x = -5$$

$$x = -3$$

Now obtain the value of y by substituting $x = -3$ in the above expression for y.

$$y = 6 + \frac{2}{3}(-3)$$

$$y = 4$$

Hence, the solution is $x = -3$ and $y = 4$. ∎

The procedure for solving the equations in Example 7 is called the method of **elimination**, and the procedure in Example 8 is called the method of **substitution**. Equations (6a), and (6b), which were derived by the method of elimination, provide formulas for obtaining the solution directly. These formulas are recast in the next section in a way that makes them easier to remember.

Determinants. The expression

$$A_1 B_2 - A_2 B_1 \tag{7}$$

has emerged as an important quantity in our investigation of the relationship between the lines

$$L_1 : A_1 x + B_1 y = C_1$$

and

$$L_2 : A_2 x + B_2 y = C_2.$$

For any four numbers A_1, B_1, A_2, and B_2, we denote quantity (7) by

$$\begin{vmatrix} A_1 & B_1 \\ A_2 & B_2 \end{vmatrix} \tag{7'}$$

which is called a **determinant of order two**. For example,

$$\begin{vmatrix} 3 & 5 \\ 2 & 6 \end{vmatrix} = 3 \cdot 6 - 2 \cdot 5 = 8.$$

From equations (6a), and (6a), we see that the coordinates of the **point $P(x, y)$ of intersection of L_1 and L_2** can be expressed in determinants

158   Chapter 2. Algebra and Graphs of Linear Expressions

as follows:

$$x = \frac{\begin{vmatrix} C_1 & B_1 \\ C_2 & B_2 \end{vmatrix}}{\begin{vmatrix} A_1 & B_1 \\ A_2 & B_2 \end{vmatrix}} \quad \text{and} \quad y = \frac{\begin{vmatrix} A_1 & C_1 \\ A_2 & C_2 \end{vmatrix}}{\begin{vmatrix} A_1 & B_1 \\ A_2 & B_2 \end{vmatrix}}. \qquad (8)$$

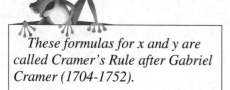
These formulas for x and y are called Cramer's Rule after Gabriel Cramer (1704-1752).

Of course, for x and y to be defined, the denominators in equations (8) cannot equal 0, which is another way of saying that distinct lines L_1 and L_2 will intersect if and only if they are not parallel [see equation (2)]

Example 9 Use determinants to find the point of intersection of the lines

$$-4x + 5y = 31$$
$$3x + 7y = 9.$$

Solution: By substituting in (8), we obtain

$$x = \frac{\begin{vmatrix} 31 & 5 \\ 9 & 7 \end{vmatrix}}{\begin{vmatrix} -4 & 5 \\ 3 & 7 \end{vmatrix}} = \frac{31 \cdot 7 - 9 \cdot 5}{(-4)7 - 3 \cdot 5} = \frac{172}{-43} = -4$$

and

$$y = \frac{\begin{vmatrix} -4 & 31 \\ 3 & 9 \end{vmatrix}}{\begin{vmatrix} -4 & 5 \\ 3 & 7 \end{vmatrix}} = \frac{(-4)9 - 3 \cdot 31}{(-4)7 - 3 \cdot 5} = \frac{-129}{-43} = 3.$$

The point of intersection is $(-4, 3)$. ∎

Applications. The following example illustrates how two equations in two unknowns can be used to solve practical problems.

Example 10 In order to start a small business, a young couple borrowed $25,000, some from parents at 5% simple interest and the rest from a bank at 11% interest. If their interest payment is $2450, how much did they borrow from each lender?

Solution: Let $x =$ the number of dollars borrowed from parents, and $y =$ the number of dollars borrowed from the bank. Then

$$x + y = 25{,}000$$
$$.05x + .11y = 2450.$$

Equivalently,

2.6. Systems of Two Linear Equations in Two Unknowns 159

$$5x + 5y = 125,000$$
$$5x + 11y = 245,000.$$

By subtracting the first equation from the second, we get

$$6y = 120,000 \quad \text{or} \quad y = 20,000,$$

and therefore $x = 5000.$

Hence, the couple borrowed $5000 from parents and $20,000 from the bank. ∎

The following example combines the notions of distance, slope, equations of a line, perpendicular lines, point of intersection, and determinants.

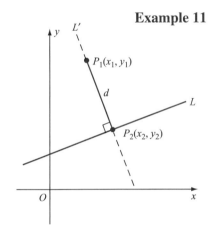

Figure 47

Example 11 Show that the (shortest) distance from a point $P_1(x_1, y_1)$ to a line L with equation $Ax + By = C$ is

$$d = \frac{|Ax_1 + By_1 - C|}{\sqrt{A^2 + B^2}}.$$

Solution: Let L' be the line through P_1 and perpendicular to L. If $P_2(x_2, y_2)$ is the point of intersection of L' and L, then $d(P_1, P_2)$ is the shortest distance from P_1 to L as indicated in Figure 47. If L is not vertical, then its slope is $m = -A/B$ and therefore the slope of L' is $m' = B/A$. A point-slope equation of L' is

$$L' : y - y_1 = \frac{B}{A}(x - x_1),$$

which is equivalent to the equation

$$L' : Bx - Ay = Bx_1 - Ay_1.$$

We find the point $P_2(x_2, y_2)$ of intersection of L' and L by the method of determinants [equation (8)]:

$$x_2 = \frac{\begin{vmatrix} C & B \\ Bx_1 - Ay_1 & -A \end{vmatrix}}{\begin{vmatrix} A & B \\ B & -A \end{vmatrix}} = \frac{-AC - B(Bx_1 - Ay_1)}{-A^2 - B^2}$$

$$= \frac{AC + B^2 x_1 - ABy_1}{A^2 + B^2}$$

$$y_2 = \frac{\begin{vmatrix} A & C \\ B & Bx_1 - Ay_1 \end{vmatrix}}{\begin{vmatrix} A & B \\ B & -A \end{vmatrix}} = \frac{A(Bx_1 - Ay_1) - BC}{-A^2 - B^2}$$

$$= \frac{BC + A^2 y_1 - ABx_1}{A^2 + B^2}.$$

By substituting these values for x_2 and y_2 in the distance formula, we get

$$d(P_1, P_2) = \sqrt{\left(\frac{AC + B^2 x_1 - ABy_1}{A^2 + B^2} - x_1\right)^2 + \left(\frac{BC + A^2 y_1 - ABx_1}{A^2 + B^2} - y_1\right)^2}.$$

We leave it for the reader to verify that the above expression simplifies to

$$d(P_1, P_2) = \sqrt{\frac{(A^2 + B^2)(Ax_1 + By_1 - C)^2}{(A^2 + B^2)^2}} = \frac{|Ax_1 + By_1 - C|}{\sqrt{A^2 + B^2}}.$$

In Exercise 58 you will be asked to verify the same result for the case in which L is a vertical line. ■

Exercises 2.6

Fill in the blanks to make each statement true.

1. The lines $y = m_1 x + b_1$ and $y = m_2 x + b_2$ ($b_1 \neq b_2$) are parallel if and only if _____.
2. If $L_1 : A_1 x + B_1 y = C_1$ is parallel to $L_2 : A_2 x + B_2 y = C_2$, then _____.
3. The lines $y = m_1 x + b_1$ and $y = m_2 x + b_2$ are perpendicular if and only if _____.
4. If $L_1 : A_1 x + B_1 y = C_1$ is perpendicular to $L_2 : A_2 x + B_2 y = C_2$ if and only if _____.
5. Lines $A_1 x + B_1 y = C_1$ and $A_2 x + B_2 y = C_2$ intersect in a unique point if and only if _____.

Write true or false for each statement. Assume L_1, L_2 and L_3 are distinct lines in the plane.

6. If L_1 is parallel to L_2 and L_2 is parallel to L_3, then L_1 is parallel to L_3.
7. If L_1 is perpendicular to L_2 and L_2 is perpendicular to L_3, then L_1 is perpendicular to L_3.
8. If L_1 is parallel to L_2 and L_2 is perpendicular to L_3, then L_1 is perpendicular to L_3.
9. If $L_1 : A_1 x + B_1 y = C_1$ and $L_2 : A_2 x + B_2 y = C_2$ satisfy $A_1 B_2 - A_2 B_1 = 0$, then L_1 is parallel to L_2.
10. If $L_1 : A_1 x + B_1 y = C_1$ and $L_2 : A_2 x + B_2 y = C_2$ satisfy $A_1 B_2 - A_2 B_1 = 0$ and $A_1 = 0$, then $A_2 = 0$.

Parallel and Perpendicular Lines

Determine which pairs of lines in Exercise 11-22 are parallel, which are perpendicular, and which are coincident.

11. $y = 2x + 3$
    $y = 2x - 5$

12. $y = 3x - 4$
    $y = 2x + 3$

13. $y = x - 2$
    $y = -x + 2$

14. $y = x - 4$
    $y = x + 4$

15. $y - 2 = -\frac{3}{2}(x - 1)$
    $y - 1 = \frac{2}{3}(x + 4)$

16. $y - 5 = -3(x + 2)$
    $y - 5 = 3(x + 2)$

17. $y - 3 = 5(x - 1)$
    $y + 7 = 5(x + 1)$

18. $y + 4 = 2(x + 1)$
    $y - 4 = 2(x + 1)$

19. $4x - 2y = 3$
    $-2x + y = 1$

20. $2x - 3y = 1$
    $-4x + 6y = -2$

21. $3x = 4$
    $5y = 2$

22. $5x - 4y = 2$
    $4x + 5y = 1$

23. Find a general linear equation of the line through the point (3, 5) and parallel to the line $4x - 7y = 8$.

24. Find a general linear equation of the line through the point $(-4, 6)$ and perpendicular to the line $5x + 2y = 7$.

25. Show that a general linear equation of the line through the point (x_0, y_0) and parallel to the line $L : Ax + By = C$ is $Ax + By = Ax_0 + By_0$. Assume (x_0, y_0) is not on L.

26. Show that a general linear equation of the line through the point (x_0, y_0) and perpendicular to the line $Ax + By = C$ is $Bx - Ay = Bx_0 - Ay_0$.

27. Show that the line L_1 through the points (0, 0) and (3, 4) is perpendicular to the line L_2: $3x + 4y = 1$.

28. Show that the line through the points (0, 0) and (5, 2) is parallel to the line $-2x + 5y = 1$.

29. Show that the line through the points (0, 0) and (A, B) is perpendicular to the line $Ax + By = C$.

30. Show that the line through the points (0, 0) and $(B, -A)$ is parallel to the line $Ax + By = C$. Assume $C \neq 0$.

31. Given that $L_1 : A_1x + B_1y = C_1$ and $L_2 : A_2x + B_2y = C_2$ are parallel, show that the line through the points (A_1, B_1) and (A_2, B_2) is perpendicular to both L_1. and L_2. Assume $A_1 \neq A_2$.

32. Given that $L_1 : A_1x + B_1y = C_1$ and $L_2 : A_2x + B_2y = C_2$ are parallel, show that the line through the points (B_1, A_2) and (B_2, A_1) is parallel to both L_1. and L_2. Assume neither point is on L_1 or L_1.

Intersecting Lines

Use the method of elimination to solve each of the following systems of equations.

33. $2x - y = 4$
    $3x + 5y = 8$

34. $4x - 5y = 0$
    $x + 2y = 6$

35. $-2x + 7y = 4$
    $3x - 4y = 7$

36. $7x - 5y = -1$
    $3x + 2y = 12$

37. $3x - 7 = 2y + 4$
    $x + 4 = y - 3$

38. $2x - 6y - 5 = 0$
    $4x - 3y + 6 = 0$

Use the method of substitution to solve each of the following systems of equations.

39. $2x + y = 5$
    $-4x + 3y = 10$

40. $x - 2y = 4$
    $6x + 5y = 2$

41. $4x + 5y = 7$
    $3x + 4y = -2$

42. $6x - 4y = 0$
    $3x - 5y = 7$

43. $7x - 3y = x + y + 2$
    $2x + 3y = 3x + 2y + 4$

44. $3x - y = x - 3y + 1$
    $3x + 2y = 2x + 3y - 2$

Determinants

Use the method of determinants to solve each of the following systems of equations.

45. $4x + 5y = 3$
    $2x + 4y = 7$

46. $3x + y = 9$
    $2x + 8y = 7$

47. $3x - 2y = 4$
    $7x + 6y = -3$

48. $-2x + 6y = 1$
    $2x - 4y = -3$

49. $2x = 7$
    $4x - 3y = 1$

50. $5x - 7y = 4$
    $2y = 3$

Applications

51. If a chemist has one solution of 20% acid and another of 50% acid, how many milliliters of each are needed to obtain 18 ml of 30% acid?

52. The cost of 40 lb of sugar and 16 lb of flour is $26, as is the cost of 30 lb of sugar and 25 lb of flour. Find the cost per pound of each item.

53. There is a fixed charge for a telegram of up to 15 words and an additional charge for each word after the fifteenth. A telegram of 20 words costs $32.90, and one of 25 words costs $37.90. What are the fixed charge and the rate for each word after the fifteenth?

54. Flying with a tail wind, a plane covers 1800 mi in 2 hr, but returns against the wind in 3 hr and 36 min. Find the speed of the plane and the speed of the wind.

55. A company makes two styles of blouses. The first style requires 12 min for cutting and 15 min for sewing, while the second needs 18 min for cutting and 10 min for sewing. How many of each type are produced if a total of 850 hr are spent on cutting and 750 hr on sewing?

Miscellaneous

56. Find the distance between the origin and the line $5x + 4y = -3$.

57. Find the distance between the point $(-5, 2)$ and the line $3x - 4y = 7$.

58. Show that the result in Example 11 holds when L is a vertical line, that is, when $B = 0$ in the equation of L.

In Exercises 59-61, L_1 and L_2 are lines, not necessarily distinct, with the general equations

$$L_1 : A_1 x + B_1 y = C_1$$
$$L_2 : A_2 x + B_2 y = C_2.$$

59. Show that if $A_1 B_2 - A_2 B_1 = 0$, then L_1 and L_2 are either parallel or coincident.
    [*Hint:* show that the only possibilities are
    (a) $B_1 = 0$ and $B_2 = 0$ or
    (b) $B_1 \neq 0$ and $B_2 \neq 0$.]

60. Show that if $A_1C_2 - A_2C_1 = 0$, then either L_1 and L_2 are parallel horizontal lines, or they intersect at some point on the x-axis, or they are coincident.
[*Hint:* consider the following four cases separately:

(a) $A_1 = 0, A_2 = 0$;
(b) $A_1 = 0, A_2 \neq 0$;
(c) $A_1 \neq 0, A_2 = 0$;
(d) $A_1 \neq 0, A_2 \neq 0$.]

61. Show that if $B_1C_2 - B_2C_1 = 0$, then either L_1 and L_2 are parallel vertical lines, or they intersect at some point on the y-axis, or they are coincident.
[*Hint:* consider the following four cases separately:

(a) $B_1 = 0, B_2 = 0$;
(b) $B_1 = 0, B_2 \neq 0$;
(c) $B_1 \neq 0, B_2 = 0$;
(d) $B_1 \neq 0, B_2 \neq 0$.]

62. Given $L_1 : A_1x + B_1y = C_1$ and $L_2 : A_2x + B_2y = C_2$, use the result of Exercises 59, 60, and 61 to classify each of the following cases as either intersecting lines, parallel vertical lines, parallel horizontal lines, parallel lines that are neither vertical nor horizontal, or coincident lines.

(a) $\begin{vmatrix} A_1 & B_1 \\ A_2 & B_2 \end{vmatrix} \neq 0$

(b) $\begin{vmatrix} A_1 & B_1 \\ A_2 & B_2 \end{vmatrix} = 0, \begin{vmatrix} C_1 & B_1 \\ C_2 & B_2 \end{vmatrix} \neq 0, \begin{vmatrix} A_1 & C_1 \\ A_2 & C_2 \end{vmatrix} \neq 0$

(c) $\begin{vmatrix} A_1 & B_1 \\ A_2 & B_2 \end{vmatrix} = 0, \begin{vmatrix} C_1 & B_1 \\ C_2 & B_2 \end{vmatrix} \neq 0, \begin{vmatrix} A_1 & C_1 \\ A_2 & C_2 \end{vmatrix} = 0$

(d) $\begin{vmatrix} A_1 & B_1 \\ A_2 & B_2 \end{vmatrix} = 0, \begin{vmatrix} C_1 & B_1 \\ C_2 & B_2 \end{vmatrix} = 0, \begin{vmatrix} A_1 & C_1 \\ A_2 & C_2 \end{vmatrix} \neq 0$

(e) $\begin{vmatrix} A_1 & B_1 \\ A_2 & B_2 \end{vmatrix} = 0, \begin{vmatrix} C_1 & B_1 \\ C_2 & B_2 \end{vmatrix} = 0, \begin{vmatrix} A_1 & C_1 \\ A_2 & C_2 \end{vmatrix} = 0$

2.7 Systems of Three Linear Equations in Three Unknowns

Linear Systems in Three Variables
Reduction Method
Geometric Meaning of Solutions
Applications

We can make a point by intersecting three planes.

In the previous section we solved systems of two linear equations in two unknowns. Geometrically such systems represent two lines in a plane that are parallel, coincide, or intersect in a unique point. To solve the systems, three algebraic methods were used: elimination, substitution, and determinants.

Here we consider systems with three linear equations in three unknowns, whose solutions may be interpreted as intersections of planes in 3-space. The method of substitution, also called the reduction method, will be used to solve such systems.

Linear Systems in Three Variables. Our concern in this section is to solve a system of three linear equations in three unknowns x, y, and z:

$$\begin{aligned} G_1 &: a_1x + b_1y + c_1z = d_1 \\ G_2 &: a_2x + b_2y + c_2z = d_2 \\ G_3 &: a_3x + b_3y + c_3z = d_3. \end{aligned} \quad (1)$$

No knowledge of three-dimensional geometry is required, if it is accepted that a linear equation in $x, y,$ and z represents a plane in 3-space.

Geometrically, the solution of system (1) is the intersection of the three planes $G_1, G_2,$ and G_3 in space. If we assume that the planes are distinct and that no two of them are parallel, then G_1 and G_2 will intersect in a line L, and G_1 and G_3 will intersect in a line L'. The lines L and L' are both in the plane G_1; hence, if they are not parallel or coincident, they will intersect in a point $P(x_1, y_1, z_1)$ (Figure 48).

We can solve the system algebraically for $x_1, y_1,$ and z_1 by several methods. To facilitate our discussion, we introduce the following terminology. The real numbers $a_i, b_i,$ and c_i ($i = 1, 2, 3$) in system (1) are called the **coefficients**, $x, y,$ and z are called the **variables or unknowns**, and the d_i ($i = 1, 2, 3$) are called the **constant terms**.

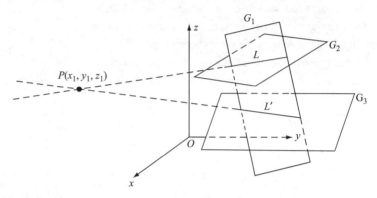

Figure 48 Intersection of Three Planes

Reduction Method. The basic idea in our first method is to express one of the unknowns, for example z, in terms of the remaining ones (x and y) and then by substitution to reduce a system of three equations to a system of two equations. Hence, the procedure is called the method of **reduction by substitution** or simply the **reduction method.**

Example 1 Solve the following system by the method of reduction.

$$3x - 2y + z = 7$$
$$2x + 5y - 4z = -3$$
$$4x + y - z = 6$$

Solution: From the first equation,

$$z = -3x + 2y + 7.$$

Substitute this expression for z into the last two equations.

$$2x + 5y - 4(-3x + 2y + 7) = -3$$
$$4x + y - (-3x + 2y + 7) = 6$$

2.7. Systems of Three Linear Equations in Three Unknowns

Equivalently,

$$14x - 3y = 25$$
$$7x - y = 13,$$

Whose solution is $x = 2$ and $y = 1$. Therefore, $z = -3 \cdot 2 + 2 \cdot 1 + 7 = 3$, and the solution of the original system is $x = 2$, $y = 1$, and $z = 3$. Geometrically, the given system represents three planes that intersect in a unique point (2, 1, 3). ∎

Example 2 Apply the reduction method to the following system.

$$x - 2y - 3z = 2$$
$$x - 4y - 13z = 14$$
$$-3x + 5y + 4z = 2$$

Solution: In the first equation, it is easiest to solve for x in terms of y and z. We get

$$x = 2y + 3z + 2.$$

Substituting this expression for x in the last two equations of the given system gives

$$(2y + 3z + 2) - 4y - 13z = 14$$
$$-3(2y + 3z + 2) + 5y + 4z = 2.$$

or equivalently

$$-y - 5z = 6$$
$$-y - 5z = 8,$$

which is a system that has no solution for y and z (*why?*). Hence, the given system has *no* solution for x, y, and z. ∎

Example 3 Apply the reduction method to the following system.

$$x - 2y - 3z = 2$$
$$x - 4y - 13z = 14$$
$$-3x + 5y + 4z = 0$$

Solution: The first equation gives

$$x = 2y + 3z + 2,$$

and substitution in the remaining two equations yields

$$(2y + 3z + 2) - 4y - 13z = 14$$
$$-3(2y + 3z + 2) + 5y + 4z = 0.$$

The problem is therefore reduced to solving the system

$$-y - 5z = 6$$
$$-y - 5z = 6,$$

which says that $y = -5z - 6$, and then by substitution, $x = -7z - 10$. We may write the solution of the system as $x = -7z - 10$, $y = -5z - 6$, and $z = $ any real number. Thus, by varying the value of z, we have and *infinite* number of triples (x, y, z) that satisfy the equation. ∎

Geometric Meaning of Solutions. Example 1 shows that the linear system (1) may have a unique solution, which means that (1) represents three planes G_1, G_2, and G_3 intersecting in a unique point as shown in Figure 48. But Examples 2 and 3 illustrate that the linear system (1) may not have a unique solution; geometrically, the planes G_1, G_2, G_3 may not intersect in a unique point. This could happen in any of the following ways.

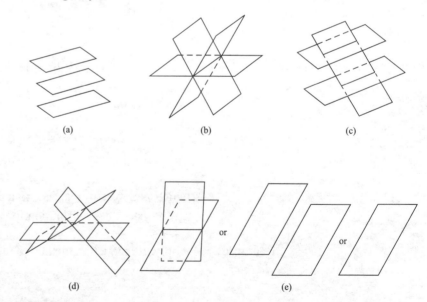

Figure 49

(a) The planes are parallel [Figure 49(a)].
(b) All three planes intersect in a single line [Figure 49(b)].
(c) Two of the planes are parallel, and the third plane intersects each of these in a line, resulting in two parallel lines [Figure 49(c)].

2.7. Systems of Three Linear Equations in Three Unknowns 167

(d) Each pair of planes intersects in a line, resulting in three parallel lines [Figure 49(d)].

(e) Two or all three of the planes are coincident [Figure 49(e)].

In each of these five cases, there is a fourth plane G that is perpendicular to G_1, G_2, and G_3, and it can be shown that the converse is also true (see Exercises 31-32). That is, if there is a plane G perpendicular to G_1, G_2, and G_3, then the system (1) has no solution [cases (a), (c), (d), and (e)] or an infinite number of solutions [cases (b) and (e)].

Although we have limited our discussion in this section to three equations in three unknowns, the reduction method can be applied to any number of linear equations in any number of unknowns. For example, given four equations in five unknowns, we first solve one of the equations for one of the unknowns in terms of the remaining four. We then substitute this expression into the remaining three equations, and thereby reduce the original system to three equations in four unknowns. We can repeat this process until we arrive at a solution.

Applications. Many of the decisions that must be made in business and in science require the solution of linear systems of many equations in large numbers of unknowns. The most common approach to such problems is to use the method of Gaussian elimination, which will be discussed in section 6.4. Here we limit ourselves to an example involving three linear equations in three unknowns.

Example 4 A zoo dietician must produce 100 pounds of an animal feed that contains 20% protein and 10% fat. Three types of food are available for mixing. Type A is 15% protein and 10% fat, type B is 15% protein and 7.5% fat, and type C is 25% protein and 11% fat. How many pounds of each type should be used to produce the desired feed?

Solution: Let x, y, and z be the number of pounds of food types A, B, and C, respectively, to be mixed, The mixture must be 100 pounds and contain 20 pounds of protein and 10 pounds of fat. Therefore,

Type A	Type B	Type C		
$x\ +$	$y\ +$	$z = 100$		*Total pounds*
$.15x\ +$	$.15y\ +$	$.25z =$	20	*Pounds of protein*
$.1x\ +$	$.075y\ +$	$.11z =$	$10.$	*Pounds of fat*

This linear system has the unique solution $x = 30$, $y = 20$, and $z = 50$. Hence, the dietician should mix 30 pounds of type A, 20 pounds of type B, and 50 pounds of type C. ∎

Exercises 2.7

Fill in the blanks to make each statement true.

1. A system of three linear equations in three unknowns may have _____, _____, or _____ solutions.
2. If a system of three linear equations in three unknowns has a unique solution (x_1, y_1, z_1), then geometrically the system represents three _____ intersecting in a _____.
3. If a system of three linear equations in three unknowns geometrically represents three planes that have a line in common, then algebraically an _____ number of solutions is obtained.
4. The reduction method applied to a linear system in three variables involves substitution for one variable so that the reduced system involves only _____ variables.
5. If a linear system in three variables has the solution $x = y = z = 0$, then all of the _____ terms in the equations are zero.

Write true of false for each statement.

6. The reduction (substitution) method can be applied to a linear system involving any number of variables.
7. A system of linear equations can be solved only if the number of equations equals the numbers of variables.
8. A linear system of three equations in three variables represents three planes that either intersect in one point or are parallel.
9. A system of four linear equations in four unknowns may be reduced to a system of three equations in three unknowns by substitution of one variable as a linear expression in the others.
10. If a system of three equations in three unknowns has no solution, then at least two of the equations represent parallel planes.

Reduction Method

In Exercises 11-16, solve the linear systems by the reduction method.

11. $x - 2y + z = -2$
     $2x + y + 5z = 3$
     $4x - 5y - 4z = 6$

12. $x - y + z = 2$
     $2x + y - z = 1$
     $2x - y - 3z = -9$

13. $x + y - z = 4$
     $x - 3y + z = -2$
     $3x - y - 2z = 6$

14. $2x + 5y - 3z = 4$
     $4x - y - 2z = 2$
     $5x - 2y + z = 15$

15. $x - 2y + 3z = 5$
     $2x + 5y - 2z = -5$
     $4x + y + z = -4$

16. $3x + 2y - 7z = 10$
     $2x - 5y + 3z = -1$
     $5x + 6y - z = 6$

17. Show that there is no solution for the system

$$x - 2y + 3z = 2$$
$$2x - 3y + 5z = 7$$
$$5x + 4y + z = 1$$

by reducing to a system in x and y only.

18. Solve Exercise 17 by reducing to a system involving y and z only.

19. Solve Exercise 17 by reducing to a system involving x and z only.

In Exercise 20-25, use the reduction method to show that the system has infinitely many solutions.

20. $2x - y = 7$
     $8x - 4y = 28$

21. $x - y + 4z = 0$
     $2x + y + 2z = 0$
     $3x + 2y + 2z = 0$

22. $x + y = 2$
     $x - z = 1$
     $2x + y - z = 3$

23. $x + 4y - z = 12$
     $3x + 8y - 2z = 4$

24. $2x + y - 4z = 3$
     $2x + 3y + 2z = -1$
     $y + 3z = -2$

25. $x + z = 2$
     $y + z = 6$
     $y + t = 0$
     $x + y + z + t = 2$

Solve the systems in Exercise 26-27 by the reduction method.

26. $2x - 3y - z = 2$
    $3y - 2z - 4t = -17$
    $x + y - 2t = -4$
    $x + z + 2t = 8$

27. $x + y + 2z = 13$
    $2x - y = 1$
    $4x + 3y - t = 19$
    $x + y + z - t = 11$

Geometric Meaning of Solutions

28. Each of the systems in Exercises 11-16 has a unique solution. What does that mean geometrically?
29. The system of equations given in Exercise 17 represents three distinct planes, no two of which are parallel, but the system has no solution. Explain how this is possible geometrically.
30. The linear system in Exercise 21 represents three planes passing through (0, 0, 0) (*why?*) and has an infinite number of solutions. Given that the three planes are distinct, explain how this is possible geometrically.

Exercises 31 and 32 illustrate that when three planes do not intersect in a unique point, there is a fourth plane perpendicular to all three.

31. Given the linear system

    $$G_1 : x - 2y + 3z = 2$$
    $$G_2 : 2x - 3y + 5z = 7$$
    $$G_3 : 5x + 4y + z = 1,$$

    do each of the following.

    (a) Show that the system has no solution.
    (b) Find a plane $G : Ax + By + Cz = D$ that is perpendicular to G_1, G_2, and G_3. (*Hint*: G is perpendicular to G_1, G_2, and G_3 if and only if A, B, and C satisfy the system

    $$A - 2B + 3C = 0$$
    $$2A - 3B + 5C = 0$$
    $$5A + 4B + C = 0.)$$

32. Given the linear system

    $$G_1 : x - 2y + 3z = 2$$
    $$G_2 : 2x - 3y + 5z = 7$$
    $$G_3 : 5x - 8y + 13z = 16,$$

    do each of the following.

    (a) Show that the system has an infinite number of solutions.
    (b) Find a plane $Ax + By + Cz + D = 0$ that is perpendicular to G_1, G_2, and G_3. (*Hint*: see Exercise 31.)

Applications

Solve Exercises 33-40 by using a system of three linear equation in three unknowns.

33. A couple has $5000 to invest. They decide to divide it into three parts, putting the first part in a certificate of deposit at 4% annual interest, the second part in bonds paying 5% per year, and the third part in a mutual fund paying 6% per year. The third investment equals the sum of the first two. Find the amount of each of the three investments, if at the of one year they receive $265 income from them.
34. A movie theater charges $5.00 admission for adults, $3.00 for senior citizens, and $2.00 for children. If 75 tickets are sold for a total of $285, with three times as many tickets for senior citizens as for children, how many tickets of each type are sold?
35. A store purchased 100 dresses of three qualities at $45, $30, and $25 each for a total of $3050. The store sold the dresses for $90, $65, and $50, respectively, for a total of $6250. How many of each type were sold?
36. A manufacturer makes three kinds of suits. One week he sold a total of 1050 suits. He sold the first, second, and third kinds to a small retail store at prices of $150, $100, and $80 per suit, respectively, for a total of $22,000. He also five sold five times as many of each kind to a large discount store at prices of $125, $90, and $75, respectively, for a total of $94,375. How many suits of each kind were sold to the small retail store?

37. A company packages one-quart cartons of three kinds of blended fruit juices. Each carton of type 1 contains 20 oz of orange juice, 10 oz of grapefruit juice, and 2 oz of pineapple juice. Type 2 has 8 oz of orange, 20 oz of grapefruit, and 4 oz of pineapple per carton, while type 3 has 12 oz of orange and 10 oz each of grapefruit and pineapple per carton. How many cartons of each type are packaged if a total of 14,400 oz of orange, 14,000 oz of grapefruit, and 3,600 oz of pineapple juice are used?

38. A company makes three kinds of calculators, each calculator of the first kind costs $10 for parts, $20 for labor, and $1 for advertising. For the second kind those costs are $5, $15, and $.50, respectively, and for the third kind, $4, $10, and $.25, respectively. How many of each type are made of the company spends a total of $13,700 on parts, $30,500 on labor, and $1325 on advertising?

39. Find an equation of the circle passing through points $P_1(2, 1)$, $P_2(-1, 3)$, and $P_3(3, -2)$. (*Hint*: a circle has an equation of the form $Ax^2 + Ay^2 + Cx + Dy = E$.)

40. Find an equation of the parabola $y = ax^2 + bx + c$ passing through $P_1(1, 4)$, $P_2(2, 1)$, and $P_3(-1, 8)$.

Chapter 2 Review Outline

2.1 Linear Equations

Definitions

An equation in x is a statement that two algebraic expressions in x are equal. A solution is a value of x for which the statement is true.

Two equations are equivalent if one can be transformed into the other by the algebraic properties of equality. (Equivalent equations have the same solution.)

A linear equations in x is one equivalent to $ax + b = 0$, where $a \neq 0$.

Linear Equations

The following properties of equality are used to reduce a given linear equation to an equivalent one in which all terms involving x are on one side.

(1) $M = N \Leftrightarrow M \pm c = N \pm c$
(2) If $c \neq 0$, then $M = N \Leftrightarrow Mc = Nc$ and $M/c = N/c$.

Equations with Fractions

First clear the fractions and then proceed as above. *Check answers for extraneous roots*.

Equations with Radicals

Eliminate one radical at a time by isolating it on one side and squaring both sides. *Check answers for extraneous roots*.

Equations with Absolute Values

First eliminate one absolute value $|M|$ by replacing the equation with two auxiliary equations, one with M in place of $|M|$ and the other with $-M$ in place of $|M|$. If $|N|$ is in the auxiliary equations, repeat the process for N in each auxiliary equation. Continue until all absolute values have been eliminated. Solve each of the final auxiliary equations for x, and *check answers for extraneous roots in the original equation*.

2.2 Linear Inequalities

Inequalities in which M and N are linear expressions in x can be solved by applying the following properties:

(1) $M > N \Leftrightarrow M \pm c > N \pm c$
(2) If $c > 0$, then $M > N \Leftrightarrow Mc > Nc$ and $M/c > N/c$
(3) If $c < 0$, then $M > N \Leftrightarrow Mc < Nc$ and $M/c < N/c$.

Any inequality with fractional expressions can be transformed to one of the form $M/N > 0, \geq 0, < 0,$ or ≤ 0, which can be solved by the method of testing points. The properties for solving inequalities with absolute values are as follows:

(1) $|M| > c \Leftrightarrow M > c$ or $M < -c$
(2) $|M| < c \Leftrightarrow -c < M < c$.

2.3 Applications of linear equations and inequalities

In a word problem, first translate the wording into equations and/or inequalities; then solve algebraically and interpret the results in terms of the original problem.

Percent: $r\%$ of $Q = \dfrac{r}{100} \cdot Q$

Motion: distance = rate · time

Mixtures: In any mixing problem, each quantity to be combined contains a certain concentrate in some mixture. Compute the amount of concentrate in each quantity to be combined, and add these amounts to get the total amount of concentrate.

Work: If it takes n time units to do a given job, then we assume that $1/n$ of the job is completed in each time unit.

2.4 The Coordinate Plane

Every Point P in the plane corresponds to an ordered pair (x, y) of real numbers.

Distance Formula:
$d(P_1, P_2) = \sqrt{(x_2 - x_1)^2 + (y_2 - y_1)^2}$

midpoint $(P_1, P_2) = \left(\dfrac{x_1 + x_2}{2}, \dfrac{y_1 + y_2}{2} \right)$

2.5 Equations of straight lines

The slope m of a nonvertical line through the points $P_1(x_1, y_1)$ and $P_2(x_2, y_2)$ is defined as

$$m = \dfrac{y_2 - y_1}{x_2 - x_1} \quad (x_1 \neq x_2).$$

A rising (falling) line has positive (negative) slope. A horizontal line has slope 0; the slope is not defined for a vertical line. The steeper the line, the larger the magnitude of its slope. When x changes by ± 1 unit along a line, y changes by $\pm m$ units.

Equations of a line

Point-Slope: $y - y_1 = m(x - x_1)$
Slope-Intercept: $y = mx + b$
Vertical Line: $x = c$
Horizontal Line: $y = c$
General Linear: $Ax + By = c$ (A or $B \neq 0$)

2.6 Systems of Two Linear Equations

Given two distinct lines

$$L_1 : A_1 x + B_1 y = C_1$$
$$L_2 : A_2 x + B_2 y = C_2$$

then

(1) $L_1 \parallel L_2 \Leftrightarrow m_1 = m_2$ (nonvertical) $\Leftrightarrow$ $A_1 B_2 - A_2 B_1 = 0$
(2) $L_1 \perp L_2 \Leftrightarrow m_1 m_2 = -1$ (neither line vertical) $\Leftrightarrow$ $A_1 A_2 + B_1 B_2 = 0$.
(3) If L_1 and L_2 are not parallel, their point of intersection may be found by the elimination or substitution methods, or by the formulas

$$x = \dfrac{\begin{vmatrix} C_1 & B_1 \\ C_2 & B_2 \end{vmatrix}}{\begin{vmatrix} A_1 & B_1 \\ A_2 & B_2 \end{vmatrix}}, \quad y = \dfrac{\begin{vmatrix} A_1 & C_1 \\ A_2 & C_2 \end{vmatrix}}{\begin{vmatrix} A_1 & B_1 \\ A_2 & B_2 \end{vmatrix}},$$

where $\begin{vmatrix} A & B \\ C & D \end{vmatrix} = AD - BC$

for any real numbers $A, B, C,$ and D.
The distance from a point $P(x_1, y_1)$ to a line $L: Ax + By = C$ is

$$d = \dfrac{|Ax_1 + By_1 - C|}{\sqrt{A^2 + B^2}}.$$

2.7 System of Three Linear Equations

To solve a system of three linear equations in three unknowns by the reduction method, one equation is solved for one variable in terms of the other two; then, by substitution, the remaining two equations involve only two variables and are solved by the methods in Section 2.6. A linear system of three equations in three unknowns represents three planes in space. It has a unique solution if the three planes intersect in one point, but such a system may have no solutions or infinitely many solutions.

Chapter 2 Review Exercises

Solve each equation or inequality in Exercises 1-14.

1. $3x - 2 = 7x + 12$
2. $4.5x - 3 = 1.5(x - 2) + 7.25$
3. $3x - 2 < 7x + 12$
4. $5x + 4 \geq 4x + 5$
5. $\dfrac{3}{2x - 4} - \dfrac{1}{2} = \dfrac{1}{2 - x}$
6. $\dfrac{3x}{x + 1} = 1 - \dfrac{3}{x + 1}$
7. $\sqrt{x^2 + 25} = x + 5$
8. $\sqrt{x + 7} = 1 + \sqrt{x}$
9. $\sqrt{x + 1} + \sqrt{x - 1} + 2 = 0$
10. $|2x + 4| < 8$
11. $|3x - 5 \geq 4|$
12. (a) $|4x - 7| > -2$ (b) $|4x - 7| \leq -2$
13. $|x - 2| = |2x - 7|$
14. $|3x - 2| + |x - 5| = 11$
15. Solve the equation $a^2 x - 4 = a + 16x$ for x in terms of a.
16. Solve $2/(x + a) + 3/x = 5/(x^2 + ax)$ for x in terms of a.
17. Solve the inequality $(x - 2)/(2x + 1) > 1$.
18. Solve $(x - 2)/(2x + 1) \leq 1$, and express the solution in interval notation.
19. Solve $1/(x + a) + 1/(x - a) > 0$ for x, given that $a > 0$ and $-a < x < a$.
20. Let $x_1 = 6$ and $x_2 = -7$ be two points on the real line. Find their midpoint and the distance between them.
21. For the points $(6, -2)$ and $(-7, 3)$ in the plane, find their midpoint and the distance between them.
22. Show that $P_1(8, -2)$, $P_2(-1, -1)$, and $P_3(3, -6)$ are vertices of a right triangle by using the following procedures.
    (a) distance formula   (b) slopes
23. Find a value of x for which the point $P(x, 7)$ is equidistant from $P_1(-2, 4)$ and $P_2(3, 6)$.
24. If $P(3, 4)$ is the midpoint of the line segment from $P_1(-1, 5)$ to $P_2(x, y)$, find x and y.
25. Show that all points $P(x, y)$ equidistant from the two points $P_1(5, 2)$ and $P_2(2, 6)$ lie on a line L. What is an equation of L?
26. Find the equation of the horizontal line through the point $P(6, 4)$ and also the equation of the vertical line through P.
27. Find the slope-intercept equation of the line through $(3, -2)$ and $(-7, 4)$.
28. Find the slope-intercept equation for the line $3x - 2y = 4$.
29. Find a general linear equation of the perpendicular bisector of the line segment from $P_1(5, 5)$ to $P_2(2, 6)$. Compare with Exercise 25.
30. Find an equation of the line that passes through $(-1, 5)$ and is parallel to $2x - 3y = 7$.
31. By the method of elimination, solve the system
$$\begin{cases} 3x - 2y = 18 \\ 4x + 3y = 7. \end{cases}$$
32. Solve the system in Exercise 31 by the method of substitution.
33. Solve the system in Exercise 31 by the method of determinants.
34. Use any method to solve the system $\begin{cases} 5x - 6y = 18 \\ y = -x + 6. \end{cases}$
35. In the plane, graph the solution of the system $x \geq 3$, $y \geq 2$, $x + y \leq 10$.
36. In the plane, graph the solution of the system
$$\begin{cases} y - 2x \leq 4 \\ x + y < 3. \end{cases}$$

In Exercises 37-40, determine whether the system has a unique solution, no solution, or an infinite number of solutions.

37. $x + y + z = 3$
    $x - 2y + 2z = 1$
    $2x + 3y + z = 4$

38. $x + y - 3z = 1$
    $x - y + z = 1$
    $3x + y - 5z = 3$

39. $x - y + z = 5$
    $2x + 3y + z = 7$
    $3x + 2y + 2z = 4$

40. $x + 2y - z = 0$
    $2x - y + z = 0$
    $5x + 5y - 2z = 0$

Graphing Calculator Exercises

Verify your answers to the indicated review exercises of this section by using a graphing calculator for the given functions and windows $[x_{\min}, x_{\max}]$, $[y_{\min}, y_{\max}]$

41. Exercise 3: $Y_1 = 7x + 12 - (3x - 2)$, $[-7, 2.4]$, $[-15, 15]$
42. Exercise 5: $Y_1 = 3/(2x - 4) - 1/2 - 1/(2 - x)$, $[0, 18.8]$, $[-1, 1]$
43. Exercise 7: $Y_1 = \sqrt{x^2 + 25} - (x + 5)$, $[-4, 5.4]$, $[-5, 5]$
44. Exercise 8: $Y_1 = \sqrt{x + 7} - (1 + \sqrt{x})$, $[0, 18.8]$, $[-1, 1]$
45. Exercise 13: $Y_1 = |x - 2| - |2x - 7|$, $[0, 9.4]$, $[-5, 5]$
46. Exercise 31: $Y_1 = 3x/2 - 9$, $Y_2 = -4x/3 + 7/3$. $[0, 9.4]$, $[-5, 5]$
47. Exercise 34: $Y_1 = 5x/6 - 18/6$, $Y_2 = -x + 6$, $[0, 9.4]$, $[-5, 5]$

3 Algebra and Graphs of Quadratic Expressions

3.1 Quadratic Equations

3.2 Complex Numbers

3.3 Equations Reducible to Quadratic Equations

3.4 Quadratic Inequalities

3.5 Applications of Quadratic Equations and Inequalities

3.6 Circles and Parabolas

3.7 Translation of Axes and Intersections

3.8 Ellipses and Hyperbolas

Hey Students!
You will benefit more from the classroom lectures and have a better understanding of the material if you take the time to view the Gilbert Review Videos @
www.CollegeAlgebraBySchiller.com
before attending the class lecture for this chapter.

Chapter 3. Algebra and Graphs of Quadratic Expressions

Chapter 2 was devoted to the algebra and geometry of linear expressions. Now we consider **quadratic expressions,** which are polynomials of degree 2 in one or more variables. We will restrict our attention to the cases

$$(1) \quad ax^2 + bx + c \quad (a \neq 0)$$
$$\text{and} \quad (2) \quad ax^2 + by^2 + cx + dy + e \quad (a \neq 0 \text{ or } b \neq 0),$$

where the coefficients a, b, c, d, and e are constant real numbers, and x and y are the variables. In seeking the zeros of (1), we are led to the quadratic formula and to the complex number system. The zeros of (2) can be any of a variety of curves in the plane called **conic sections.**

3.1 Quadratic Equations

People have been solving quadratic equations since the time of ancient Babylonia.

The Quadratic Form
Solutions by Factoring
The Quadratic Formula
The Discriminant
Factoring by Roots

The Quadratic Form. The standard form of a **quadratic equation** is

$$\boldsymbol{ax^2 + bx + c = 0}, \tag{1}$$

where the coefficients a, b, and c are real numbers and a is not zero. Some examples of quadratic equations and their solutions are shown below.

Equation	Solution (roots)
(a) $x^2 + 2x - 3 = 0$	$-3, 1$
(b) $2x^2 - 5x + 2 = 0$	$1/2, 2$
(c) $x^2 - 10x + 25 = 0$	5

We note that a quadratic equation can have *two* roots, while a linear equation always has exactly one. We now consider the problem of finding the roots of a quadratic equation.

Solutions by Factoring. When the left side of the quadratic equation (1) is factorable by integers, we can obtain its roots by setting each factor equal to zero. This method is based on the zero-product property, which states that a product of real numbers is zero if and only if at least one of its factors is zero.

Example 1 Find the roots of equations (a), (b), and (c) by factoring.

Solution:

(a) $x^2 + 2x - 3 = 0$

$(x+3)(x-1) = 0$

$x + 3 = 0 \quad \text{or} \quad x - 1 = 0$

$x = -3 \qquad\qquad x = 1$

Therefore, the roots are -3 and 1.

(b) $2x^2 - 5x + 2 = 0$

$(2x - 1)(x - 2) = 0$

$2x - 1 = 0 \quad \text{or} \quad x - 2 = 0$

$x = \dfrac{1}{2} \qquad\qquad x = 2$

The roots are $1/2$ and 2.

(c) $x^2 - 10x + 25 = 0$

$(x - 5)(x - 5) = 0$

$x = 5$

The only root is 5. ■

When a quadratic equation has two distinct roots, as in equations (a) and (b) above, each root is called a **simple root** or a **root of order 1**. In the case where a quadratic equation has only one distinct root (i.e., two equal roots) as in (c) above, the root is called a **double root** or a **root of order 2**.

If we are given one or two real numbers, we can construct a quadratic equation having these numbers for roots as illustrated in the following example.

Example 2 In each of the following cases, find a quadratic equation having the given roots.

(a) $2, 3$
(b) $3/2, -5$
(c) 6 only
(d) $2 + \sqrt{3}, 2 - \sqrt{3}$

Solution: For each simple root a_k, we construct the linear factor $x - a_k$, and for a double root a_k, we construct *two* linear factors $x - a_k$. We then multiply the factors and set the product equal to zero.

(a) $(x - 2)(x - 3) = 0$

$x^2 - 5x + 6 = 0$

(b) $\left(x - \dfrac{3}{2}\right)(x + 5) = 0$

$x^2 + \dfrac{7}{2}x - \dfrac{15}{2} = 0$

or $2x^2 + 7x - 15 = 0$ *Multiply both sides by 2*

(c) $(x - 6)(x - 6) = 0$

$x^2 - 12x + 36 = 0$

(d) $[x - (2 + \sqrt{3})][x - (2 - \sqrt{3})] = 0$

$[(x - 2) - \sqrt{3}][(x - 2) + \sqrt{3}] = 0$

$(x - 2)^2 - 3 = 0$

$x^2 - 4x + 1 = 0$ ∎

The Quadratic Formula. The method of factoring to find roots works well for quadratics that are easily factored. A more general method can be used to solve *any* quadratic equation. First, from the equation

$$x^2 + bx + \dfrac{b^2}{4} = \left(x + \dfrac{b}{2}\right)^2$$

we can deduce the following result, which is used in the process called **completing the square.**

If $\dfrac{b^2}{4}$ is added to $x^2 + bx$, then a perfect square $\left(x + \dfrac{b}{2}\right)^2$ is obtained.

Note that the coefficient of x^2 must be 1 in the expression $x^2 + bx$ that is to be completed, and the quantity added is $b^2/4 = (b/2)^2$, the square of one-half the coefficient of x. We now apply completing the square to the quadratic equation $2x^2 - 5x + 1 = 0$ in which the left side cannot be factored by integers.

$2x^2 - 5x + 1 = 0$ *Given.*

$x^2 - \dfrac{5}{2}x + \dfrac{1}{2} = 0$ *Divide both sides by 2.*

$x^2 - \dfrac{5}{2}x = -\dfrac{1}{2}$ *Subtract $\dfrac{1}{2}$ from both sides.*

$x^2 - \dfrac{5}{2}x + \dfrac{25}{16} = -\dfrac{1}{2} + \dfrac{25}{16}$ *Complete the square on left by adding $\left(-\dfrac{5}{4}\right)^2$ to both sides.*

3.1. Quadratic Equations

$$\left(x - \frac{5}{4}\right)^2 = \frac{17}{16} \qquad \text{Add on right side with LCD.}$$

$$x - \frac{5}{4} = \pm\frac{\sqrt{17}}{4} \qquad \text{Take square roots on both sides.}$$

$$x = \frac{5 \pm \sqrt{17}}{4} \qquad \text{Add } \frac{5}{4} \text{ to both sides.}$$

We could use the method of completing the square to solve any quadratic equation, but it is more efficient to use it to derive the quadratic formula, which can then be used to solve any quadratic equation without the need to repeat the steps in the above example. To obtain the quadratic formula, we apply the method of completing the square to the general quadratic equation $ax^2 + bx + c = 0$ as follows:

$$ax^2 + bx + c = 0 \qquad \text{Given.}$$

$$x^2 + \frac{b}{a}x + \frac{c}{a} = 0 \qquad \text{Divide both sides by } a.$$

$$x^2 + \frac{b}{a}x = -\frac{c}{a} \qquad \text{Subtract } \frac{c}{a} \text{ from both sides.}$$

$$x^2 + \frac{b}{a}x + \frac{b^2}{4a^2} = -\frac{c}{a} + \frac{b^2}{4a^2} \qquad \text{Complete the square on left by adding } \left(\frac{b}{2a}\right)^2 \text{ to both sides.}$$

$$\left(x + \frac{b}{2a}\right)^2 = \frac{b^2 - 4ac}{4a^2} \qquad \text{Add on right side with LCD} = 4a^2.$$

$$x + \frac{b}{2a} = \pm\frac{\sqrt{b^2 - 4ac}}{2a} \qquad \text{Take square roots on both sides.}$$

$$x = \frac{-b \pm \sqrt{b^2 - 4ac}}{2a} \qquad \text{Add } -\frac{b}{2a} \text{ to both sides.}$$

Hence, we obtain the **quadratic formula**:

The roots of the quadratic equation $ax^2 + bx + c = 0$ $(a \neq 0)$, are

$$a_1 = \frac{-b + \sqrt{b^2 - 4ac}}{2a} \quad \text{and} \quad a_2 = \frac{-b - \sqrt{b^2 - 4ac}}{2a}.$$

Example 3 Find the roots of the equations in Example 1 by using the quadratic formula.

Solution:

(a) $x^2 + 2x - 3 = 0 \quad (a = 1, b = 2, c = -3)$

$$x = \frac{-2 \pm \sqrt{4 - (-12)}}{2}$$
$$= \frac{-2 \pm \sqrt{16}}{2}$$
$$= \frac{-2 \pm 4}{2}$$
$$= -1 \pm 2$$

The roots are $a_1 = -1 + 2 = 1$ and $a_2 = -1 - 2 = -3$.

(b) $2x^2 - 5x + 2 = 0 \quad (a = 2, b = -5, c = 2)$

$$x = \frac{-(-5) \pm \sqrt{(-5)^2 - 16}}{4}$$
$$= \frac{5 \pm \sqrt{9}}{4} = \frac{5 \pm 3}{4}$$

The roots are $a_1 = (5+3)/4 = 2$ and $a_2 = (5-3)/4 = 1/2$.

(c) $x^2 - 10x + 25 = 0 \quad (a = 1, b = -10, c = 25)$

$$x = \frac{-(-10) \pm \sqrt{(-10)^2 - 100}}{2}$$
$$= \frac{10 \pm \sqrt{0}}{2} = 5$$

Here the only root is the double root $a_1 = 5$. ∎

Example 4 Solve each equation by using the quadratic formula.

(a) $\frac{1}{2}x^2 + 2x + 1 = 0$

(b) $4x^2 + 8x + 8 = 0$

The quadratics in Example 3 are factorable by integers, but these in Example 4 are not.

Solution:

(a) We first multiply both sides by 2 in order to clear fractions and thereby obtain the equivalent equation $x^2 + 4x + 2 = 0$. We now use the quadratic formula with $a = 1, b = 4$, and $c = 2$ to get

$$x = \frac{-4 \pm \sqrt{16 - 8}}{2}$$
$$= \frac{-4 \pm \sqrt{8}}{2}$$
$$= \frac{-4 \pm 2\sqrt{2}}{2}$$
$$= -2 \pm \sqrt{2}.$$

The roots are $a_1 = -2 + \sqrt{2}$ and $a_2 = -2 - \sqrt{2}$.

(b) We can first divide both sides by 4 to obtain the equivalent equation $x^2 + 2x + 2 = 0$. We now use the quadratic formula with $a = 1$, $b = 2$ and $c = 2$ to get

$$x = \frac{-2 \pm \sqrt{4-8}}{2}$$

$$= \frac{-2 \pm \sqrt{-4}}{2}.$$

However, $\sqrt{-4}$ has no meaning in the real number system. Therefore, the quadratic equation $x^2 + 2x + 2 = 0$ has *no real roots;* that is, it has *no solution in the real number system.* ■

Example 5 Put each of the following quadratic equations in standard form and solve by any method.

(a) $2x^2 - x = 5x + 3$
(b) $3x^2 + 7x - 8 = x^2 + 2x + 4$
(c) $4x^2 + 6x + 18 = 3x^2 - 6x - 18$
(d) $x^2 + 5x + 4 = x - 2$

The standard form of a quadratic is $ax^2 + bx + c$

Solution:

(a) Subtract $5x + 3$ from both sides to obtain $2x^2 - 6x - 3 = 0$. By the quadratic formula,

$$x = \frac{6 \pm \sqrt{36 + 24}}{4}$$

$$= \frac{6 \pm \sqrt{60}}{4}$$

$$= \frac{6 \pm 2\sqrt{15}}{4}$$

$$= \frac{2(3 \pm \sqrt{15})}{4}$$

$$= \frac{3 \pm \sqrt{15}}{2}.$$

Therefore, the roots are $a_1 = \dfrac{3 + \sqrt{15}}{2}$ and $a_2 = \dfrac{3 - \sqrt{15}}{2}$.

(b) Subtract $x^2 + 2x + 4$ from both sides to obtain $2x^2 + 5x - 12 = 0$, which can be factored by integers into $(2x - 3)(x + 4) = 0$. Therefore, the roots are $a_1 = 3/2$ and $a_2 = -4$.

(c) Subtract $3x^2 - 6x - 18$ from both sides to obtain $x^2 + 12x + 36 = 0$, which can be written as $(x+6)^2 = 0$. The only root is the double root $a_1 = -6$.

(d) Subtract $x - 2$ from both sides to obtain $x^2 + 4x + 6 = 0$. By the quadratic formula,

$$x = \frac{-4 \pm \sqrt{16 - 24}}{2}$$
$$= \frac{-4 \pm \sqrt{-8}}{2}.$$

Since $\sqrt{-8}$ has no meaning as a real number, there are no roots in the real number system. ∎

The discriminant. As the previous examples show, a quadratic equation can have two distinct real roots, two equal roots, or no real root at all. To discover which of these possibilities is the case, we look at the sign of the quantity

$$b^2 - 4ac,$$

which is called the **discriminant** of $ax^2 + bx + c$. By the quadratic formula,

> the quadratic equation $ax^2 + bx + c = 0$ will have
>
> (a) two distinct real roots $\Leftrightarrow b^2 - 4ac > 0$,
> (b) two equal real roots $\Leftrightarrow b^2 - 4ac = 0$,
> (c) no real roots $\Leftrightarrow b^2 - 4ac < 0$.

Example 6 Without actually solving for the roots, determine the exact number of distinct real roots in each of the following cases.

(a) $x^2 + 50x + 750 = 0$
(b) $3x^2 + 70x - 125 = 0$
(c) $4x^2 - 50x + 156.25 = 0$

Solution:

(a) $a = 1$, $b = 50$, $c = 750$

$$b^2 - 4ac = 2500 - 3000$$
$$= -500$$

Since -500 is less than zero, there are *no* real roots.

(b) $a = 3$, $b = 70$, $c = -125$

$$b^2 - 4ac = 4900 + 1500$$
$$= 6400$$

There are *two* distinct real roots.

(c) $a = 4$, $b = -50$, $c = 156.25$

$$b^2 - 4ac = 2500 - 2500$$
$$= 0$$

The equation has *one* distinct real root (two equal real roots). ∎

Factoring by Roots. We have seen that if $ax^2 + bx + c$ factors into $a(x - a_1)(x - a_2)$, then a_1 and a_2 are the roots of $ax^2 + bx + c = 0$. We now use the quadratic formula to show that the converse is also true.

Let a_1 and a_2 be the roots of $ax^2 + bx + c = 0$ obtained by the quadratic formula. Now

$$a(x - a_1)(x - a_2) = a(x^2 - a_1 x - a_2 x + a_1 a_2)$$
$$= a[x^2 - (a_1 + a_2)x + a_1 a_2]. \quad (2)$$

By expressing a_1 and a_2 in terms of a, b, c as given in the quadratic formula, we obtain the following formulas for the sum and product of the roots of $ax^2 + bx + c = 0$ (see Exercise 86):

$$\boldsymbol{a_1 + a_2 = -\frac{b}{a}} \quad \text{and} \quad \boldsymbol{a_1 a_2 = \frac{c}{a}}. \quad (3)$$

We now substitute from (3) into (2) to obtain

$$a(x - a_1)(x - a_2) = a[x^2 - (a_1 + a_2)x + a_1 a_2]$$
$$= a\left[x^2 + \frac{b}{a}x + \frac{c}{a}\right]$$
$$= ax^2 + bx + c.$$

Hence, we have the following result for factoring a quadratic by its zeros.

If a_1 and a_2 are the roots of $ax^2 + bx + c = 0$, then

$$\boldsymbol{ax^2 + bx + c = a(x - a_1)(x - a_2)}. \quad (4)$$

We note, however, that if $b^2 - 4ac < 0$, then a_1 and a_2 are not real numbers and therefore $ax^2 + bx + c$ cannot be factored by real numbers. In this case we say that $ax^2 + bx + c$ is **irreducible** in the real number system.

Example 7 Use result (4) to factor each of the following quadratics by their real zeros.

(a) $x^2 - 6x + 7$
(b) $4x^2 - 20x + 25$
(c) $6x^2 - 5x + 1$
(d) $x^2 + 4x + 7$

Solution:

(a) The roots of $x^2 - 6x + 7 = 0$ are $a_1 = 3 + \sqrt{2}$ and $a_2 = 3 - \sqrt{2}$. Therefore, $x^2 - 6x + 7 = [x - (3 + \sqrt{2})][x - (3 - \sqrt{2})]$.

(b) The equation $4x^2 - 20x + 25 = 0$ has $a_1 = 5/2$ as a double root. Therefore $4x^2 - 20x + 25 = 4(x - 5/2)^2$.

(c) The roots of $6x^2 - 5x + 1 = 0$ are $a_1 = 1/2$ and $a_2 = 1/3$. Therefore, $6x^2 - 5x + 1 = 6(x - 1/2)(x - 1/3)$.

(d) The discriminant of $x^2 + 4x + 7$ is the negative quantity -12. Therefore, $x^2 + 4x + 7$ is irreducible in the real number system. ∎

Question *Can two different quadratic equations have the same roots?*

Answer: Yes, but only if one of the quadratic expressions is a constant multiple of the other. Suppose $ax^2 + bx + c = 0$ and $Ax^2 + Bx + C = 0$ have the same roots, a_1 and a_2. Then, by result (4),

$$ax^2 + bx + c = a(x - a_1)(x - a_2)$$

and
$$Ax^2 + Bx + C = A(x - a_1)(x - a_2).$$

Therefore,

$$Ax^2 + Bx + C = \frac{A}{a}[a(x - a_1)(a - a_2)] = \frac{A}{a}(ax^2 + bx + c). \quad \blacksquare$$

Comment In Chapter 2 we saw that every linear equation has exactly one root in the real number system. However, to a Greek mathematician in 250 B.C., the linear equation $2x + 1 = x$ would have *no* solution, since only *positive* numbers had a geometric interpretation and therefore meaning in ancient Greek mathematics. If we want every linear equation to have a solution, then we must extend the set of positive real numbers to the larger system of *all* real numbers, positive, negative, and zero. In a similar way, if we want every quadratic equation to have a solution, then we must extend the real number system to a larger system in which it is permissible to take a square root of a negative quantity. Mathematicians began to do this in the seventeenth century, but it was not until the latter part of the nineteenth century that this larger system, called the **complex number system,** was completely understood and fully accepted by all mathematicians.

Exercises 3.1

Fill in the blanks to make each statement true.

1. The standard form of a quadratic equation is _____, where the coefficients a, b, and c are _____ and _____, $\neq 0$.
2. A quadratic equation can have as many as _____ roots, whereas a linear equation always has exactly _____ root(s).
3. Three methods for finding the roots of a quadratic equation are _____, _____, and _____.
4. By the quadratic formula, the roots of $ax^2 + bx + c = 0$ are _____.
5. The quadratic equation $ax^2 + bx + c = 0$ will have two distinct real roots if _____, two equal real roots if _____, and no real roots if _____.

Write true or false for each statement.

6. If the quadratic expression $ax^2 + bx + c$ is factorable by integers, then the roots of $ax^2 + bx + c = 0$ are rational numbers.
7. All the roots of a given quadratic equation can be obtained by the quadratic formula.
8. If a_1 and a_2 are the roots of $ax^2 + bx + c = 0$, then $ax^2 + bx + c = a(x - a_1)(x - a_2)$.
9. If $ax^2 + bx + c = a(x - a_1)(x - a_2)$, then the roots of $ax^2 + bx + c = 0$ are a_1 and a_2.
10. If $a > 0$ and $c < 0$, then the quadratic equation $ax^2 + bx + c = 0$ has two distinct real roots.

Factoring

Solve each quadratic equation by the method of factoring.

11. $x^2 - 7x + 12 = 0$
12. $x^2 - 7x + 10 = 0$
13. $x^2 - 6x - 16 = 0$
14. $x^2 - 2x - 35 = 0$
15. $x^2 - 8x + 16 = 0$
16. $x^2 - 25 = 0$
17. $4x^2 - 9 = 0$
18. $4x^2 + 12x + 9 = 0$
19. $3x^2 + 5x + 2 = 0$
20. $6x^2 - x - 1 = 0$
21. $x^2 + (a + b)x + ab = 0$
22. $ax^2 + bx = 0$

Find a quadratic equation with integral coefficients that has the given roots in each of the following cases.

23. $3, -3$
24. $\dfrac{2}{5}, -\dfrac{2}{5}$
25. $\dfrac{3}{2}, \dfrac{4}{3}$
26. $5, 2$
27. $\dfrac{7}{2}$ only
28. -8 only
29. $3 + \sqrt{2}, 3 - \sqrt{2}$
30. $\dfrac{1 + \sqrt{5}}{2}, \dfrac{1 - \sqrt{5}}{2}$

The Quadratic Formula

Use the quadratic formula to solve each of the following equations.

31. $x^2 - x - 1 = 0$
32. $x^2 + x - 1 = 0$
33. $x^2 + 4x + 2 = 0$
34. $x^2 + 4x - 2 = 0$
35. $2x^2 - 6x + 3 = 0$
36. $2x^2 + 6x - 3 = 0$
37. $3x^2 - 10x + 4 = 0$
38. $3x^2 + 10x - 4 = 0$
39. $2x^2 - 5x - 2 = 0$
40. $2x^2 - 5x + 2 = 0$
41. $x^2 - (a + b)x + ab = 0$
42. $x^2 - (a + 1)x + a = 0$

Chapter 3. Algebra and Graphs of Quadratic Expressions

Use factoring or the quadratic formula to find all real roots of each of the following.

43. $x^2 - 3x + 2 = 3x + 2$
44. $x^2 - 4 = x - 2$
45. $3x^2 - 2x + 4 = x^2 - 5x + 3$
46. $5x^2 + 6x - 4 = x^2 - 2x + 1$
47. $2x^2 - 4x + 5 = -2x + 6$
48. $x^2 + 4x + 2 = 2x^2 + 1$
49. $4x^2 + 3x + 3 = -2x^2 + 4x + 4$
50. $3x^2 + 7x + 8 = -3x^2 - 6x + 2$

The Discriminant

Without solving the equation, find the number of distinct real roots in each of the following.

51. $x^2 - 100x + 275 = 0$
52. $x^2 - 100x - 275 = 0$
53. $5x^2 + 120x + 720 = 0$
54. $5x^2 - 120x + 720 = 0$
55. $x^2 + 25x + 175 = 0$
56. $x^2 - 25x - 175 = 0$
57. $321x^2 + 225x - 720 = 0$
58. $321x^2 + 225x + 720 = 0$
59. $36x^2 - 204x + 289 = 0$
60. $36x^2 - 204x - 289 = 0$

Factoring by Roots

If possible, factor each of the following quadratics $ax^2 + bx + c$ as $a(x - a_1)(x - a_2)$, where a_1 and a_2 are the real roots of $ax^2 + bx + c = 0$.

61. $2x^2 - 13x + 20$
62. $6x^2 + 13x + 6$
63. $x^2 - 10x + 22$
64. $x^2 - 6x + 4$
65. $4x^2 - 12x + 9$
66. $9x^2 + 30x + 25$
67. $2x^2 + 5x + 5$
68. $3x^2 + 4x + 2$
69. $4x^2 - 4x - 1$
70. $9x^2 - 6x - 1$

Miscellaneous

Solve each of the following quadratic equations by completing the square. (Go through each step used in the derivation of the quadratic formula.)

71. $x^2 - 5x + 5 = 0$
72. $x^2 + 6x + 4 = 0$
73. $2x^2 + 4x - 3 = 0$
74. $3x^2 + 6x + 1 = 0$
75. $4x^2 + 4x - 1 = 0$
76. $2x^2 + 4x - 3 = 0$
77. $3x^2 - 2x - 1 = 0$
78. $4x^2 - 10x + 5 = 0$

The process of completing the square can be used to write any quadratic expression as a constant times a perfect square plus a constant. For example,

$$2x^2 + 5x + 7 = 2\left(x^2 + \frac{5}{2}x + \underline{}\right) + (7 - \underline{})$$

$$= 2\left(x^2 + \frac{5}{2}x + \frac{25}{16}\right) + \left(7 - 2 \cdot \frac{25}{16}\right)$$

↑
(*Why the factor 2?*)

$$= 2\left(x + \frac{5}{4}\right)^2 + \frac{31}{8}.$$

Apply this procedure to each of the quadratic expressions in Exercises 79-84.

79. $2x^2 + 12x + 3$
80. $3x^2 - 3x + 1$
81. $4x^2 + 2x - 5$
82. $x^2 - 18x - 17$
83. $x^2 + x + 1$
84. $2x^2 + 4x$

85. **Perfect Square.** Show that $ax^2 + bx + c$ ($a > 0$) is a perfect square $\Leftrightarrow b^2 - 4ac = 0$. (*Hint*: complete the square for $ax^2 + bx + c$.)

86. **Sum and Product of Roots.** Given that a_1 and a_2 are the roots of $ax^2 + bx + c$, use the quadratic formula to show that

$$a_1 + a_2 = -\frac{b}{a} \quad \text{and} \quad a_1 a_2 = \frac{c}{a}.$$

(In the case in which $b^2 - 4ac = 0$, take $a_1 = a_2$.)

87. Use the result from Exercise 86 to find the sum and product of the roots in Exercises 51, 53, 57, and 59.

88. Use the result from Exercise 86 to find the sum and product of the roots in Exercises 52, 54, 56, and 60.

89. **The Case of the Missing Root.** Consider the linear and quadratic equations

(i) $bx + c = 0$ and (ii) $ax^2 + bx + c = 0$
where $a > 0$, $b > 0$, and $c < 0$. Now (ii) has two distinct roots (*why?*), and (i) has the single root $-c/b$. This suggests that if the value of the coefficient a approaches 0 (written: $a \to 0$), then one of the roots of (ii) approaches the value $-c/b$. Verify that such is the case, and also determine the fate of the other root of (ii) as $a \to 0$.
(*Hint:* the two roots of (ii) are

$$a_1 = \frac{-b + \sqrt{b^2 - 4ac}}{2a} \text{ and } a_2 = \frac{-b - \sqrt{b^2 - 4ac}}{2a}.$$

Now complete the following steps:

1. Multiply numerator and denominator of a_1 by $-b - \sqrt{b^2 - 4ac}$.

2. Multiply numerator and denominator of a_2 by $-b + \sqrt{b^2 - 4ac}$.)

3.2 Complex Numbers

Mathematicians will do anything to solve a problem, even if it requires taking square roots of negative numbers!

The Complex Number System

Complex Roots of a Quadratic Equation

Geometric Representation of Complex Numbers

The Complex Number System. Square roots of negative numbers arise when we formally apply the quadratic formula to certain quadratic equations, such as $x^2 + 2x + 2 = 0$. In order that such quadratic equations have roots, we extend the real number system to a larger one in which square roots of negative numbers exist. The larger system is called the **complex number system,** and a number in this system is defined as follows.

Note that the definition of i implies that $i^2 = -1$.

A **complex number** is a number

$$z = a + bi,$$

where a and b are real numbers and $i = \sqrt{-1}$; a is called the **real part** of z, and b is called the **imaginary part** of z.

For example, $4 + 3i$ and $2 - \sqrt{5}\, i$ are complex numbers. Also, the real number 4 is a complex with zero imaginary part, and the number $3i$ is a complex number with zero real part. Complex numbers with zero real part, such as $3i$, are called **pure imaginary numbers.**

Two complex numbers are said to be equal if and only if their real parts are equal and their imaginary parts are equal. That is,

$$a + bi = c + di \quad \Leftrightarrow \quad a = c \text{ and } b = d. \qquad \text{Equality of Complex Numbers}$$

The operations of addition, subtraction, multiplication, and division of complex numbers are defined in such a way that the basic properties (1) through (9) of Chapter R continue to hold. Hence, by definition,

$$(a + bi) + (c + di) = (a + c) + (b + d)i \qquad \text{Addition of Complex Numbers}$$

and

$$(a + bi) - (c + di) = (a - c) + (b - d)i. \qquad \text{Subtraction of Complex Numbers}$$

For example, $(4+3i)+(2+5i) = 6+8i$ and $(4+3i)-(2+5i) = 2-2i$.

To arrive at the definition for multiplication, we formally apply the usual rules for multiplying two binomials.

$$(a + bi)(c + di) = ac + adi + bci + bdi^2$$
$$= (ac - bd) + (ad + bc)i \quad i^2 = -1$$

Therefore, we define

$$(a + bi)(c + di) = (ac - bd) + (ad + bc)i. \qquad \text{Multiplication of Complex Numbers}$$

For example, $(4 + 3i)(2 + 5i) = (8 - 15) + (20 + 6)i$
$$= -7 + 26i.$$

To obtain a rule for dividing complex numbers, we recall that for real numbers the division $a/b = c$ means $bc = a$. Similarly, for complex numbers,

$$\frac{a + bi}{c + di} = x + yi \text{ means } (c + di)(x + yi) = a + bi.$$

Therefore, by the above rule for multiplication,

$$(cx - dy) + (cy + dx)i = a + bi,$$

and by the rule for equality of complex numbers,

$$cx - dy = a$$
$$\text{and} \quad dx + cy = b,$$

which are two equations in two unknowns x and y. By the method of determinants from Section 2.6,

$$x = \frac{\begin{vmatrix} a & -d \\ b & c \end{vmatrix}}{\begin{vmatrix} c & -d \\ d & c \end{vmatrix}} \quad \text{and} \quad y = \frac{\begin{vmatrix} c & a \\ d & b \end{vmatrix}}{\begin{vmatrix} c & -d \\ d & c \end{vmatrix}}$$

$$= \frac{ac+bd}{c^2+d^2} \qquad\qquad\qquad = \frac{bc-ad}{c^2+d^2}.$$

Hence,

$$\frac{a+bi}{c+di} = \frac{ac+bd}{c^2+d^2} + \frac{bc-ad}{c^2+d^2}i. \qquad \text{Division of Complex Numbers}$$

For example,

$$\frac{4+3i}{5+2i} = \frac{20+6}{25+4} + \frac{15-8}{25+4}i$$

$$= \frac{26}{29} + \frac{7}{29}i.$$

In the complex number system, zero is $0 + 0i$ and division by zero is prohibited just as in the real number system.

The rules for multiplication and division of complex numbers need not be memorized. To multiply two complex numbers, we can just apply the usual rules for multiplying two binomials, keeping in mind that $i^2 = -1$. To divide two complex numbers, the following technique can be used.

$$\frac{a+bi}{c+di} = \frac{a+bi}{c+di} \cdot \frac{c-di}{c-di} \qquad \textit{Multiply numerator and denominator by } c - di.$$

$$= \frac{ac - adi + bci - bdi^2}{c^2 - cdi + dci - d^2i^2}$$

$$= \frac{(ac+bd) + (bc-ad)i}{c^2+d^2} \qquad \textit{The denominator is a real number.}$$

$$= \frac{ac+bd}{c^2+d^2} + \frac{bc-ad}{c^2+d^2}i.$$

For example,

$$\frac{4+3i}{5+2i} = \frac{4+3i}{5+2i} \cdot \frac{5-2i}{5-2i}$$

$$= \frac{(20+6) + (-8+15)i}{25+4}$$

$$= \frac{26}{29} + \frac{7}{29}i.$$

Chapter 3. Algebra and Graphs of Quadratic Expressions

The number $c - di$ is called the **complex conjugate** of $c + di$, and

$$(c + di)(c - di) = c^2 + d^2.$$

Hence, to divide two complex numbers, multiply numerator and denominator by the complex conjugate of the denominator.

Example 1 Compute each of the following.

(a) $(2 - 3i)(1 + i)$

(b) $(4 + i)i$

(c) $\dfrac{5 + i}{1 + i}$

(d) $\dfrac{1}{i}$

(e) $(2i)^2$

Solution:

(a) $(2 - 3i)(1 + i) = 2 + 2i - 3i - 3i^2 \qquad i^2 = -1$
$= 5 - i$

(b) $(4 + i)i = 4i + i^2$
$= -1 + 4i$

(c) $\dfrac{5 + i}{1 + i} = \dfrac{(5 + i)(1 - i)}{(1 + i)(1 - i)} \qquad$ *Multiply numerator and denominator by the conjugate of the denominator.*

$= \dfrac{5 - 5i + i - i^2}{1 + 1}$

$= \dfrac{6 - 4i}{2}$

$= 3 - 2i$

(d) $\dfrac{1}{i} = \dfrac{1(-i)}{i(-i)} \qquad$ *The conjugate of i is $-i$.*

$= \dfrac{-i}{-i^2}$

$= \dfrac{-i}{-(-1)}$

$= -i \qquad$ *The reciprocal of i is $-i$.*

(e) $(2i)^2 = (2i)(2i) = 4i^2 = 4(-1) = -4$ ∎

From part (e) of Example 1, we see that $(2i)^2 = -4$. Similarly, $(-2i)^2 = -4$, so that both $2i$ and $-2i$ are square roots of -4. We use the notation $\sqrt{-4}$ to denote $2i$. In general, if p is any positive number, then by definition,

$$\sqrt{-p} = \sqrt{p}\sqrt{-1} = \sqrt{p}\,i \quad (p > 0),$$

where $\sqrt{p}$ denotes the positive square root of p.

Complex Roots of a Quadratic Equation. If $ax^2 + bx + c = 0$ cannot be solved by factoring with real numbers, then we use the quadratic formula to find its roots in the complex number system.

Example 2 Find the roots of each of the following quadratic equations.

(a) $x^2 + 2x + 2 = 0$
(b) $x^2 + 4x + 6 = 0$

Solution:

(a) By the quadratic formula, with $a = 1, b = 2$, and $c = 2$, we have

$$x = \frac{-2 \pm \sqrt{4-8}}{2}$$
$$= \frac{-2 \pm \sqrt{-4}}{2}$$
$$= \frac{-2 \pm 2i}{2} \qquad \sqrt{-4} = \sqrt{4}\sqrt{-1} = 2i$$
$$= -1 \pm i.$$

(b) Here, $a = 1, b = 4$, and $c = 6$. Therefore,

$$x = \frac{-4 \pm \sqrt{16-24}}{2}$$
$$= \frac{-4 \pm \sqrt{-8}}{2}$$
$$= \frac{-4 \pm \sqrt{8}\,i}{2} = \frac{-4 \pm 2\sqrt{2}\,i}{2} = -2 \pm \sqrt{2}\,i. \quad \blacksquare$$

In Example 2, the roots $-1+i$ and $-1-i$ of part (a) are complex conjugates. Similarly, in (b) the roots $-2 + \sqrt{2}i$ and $-2 - \sqrt{2}i$ are complex conjugates. In general, if $a + bi$ is a complex root of a quadratic equation with real coefficients, then so is its complex conjugate $a - bi$ (*why?*). We can therefore reformulate our previous result (Section 3.1), which related the roots of the equation $ax^2 + bx + c = 0$ to its discriminant as follows.[1]

---

[1] *The relationship between the roots and the discriminant of a quadratic equation was discovered by Sir Isaac Newton (1642-1727).*

> The quadratic equation $ax^2 + bx + c = 0$, where a, b, and c are real numbers, has
>
> (a) two distinct real roots $\Leftrightarrow b^2 - 4ac > 0$,
> (b) two equal real roots $\Leftrightarrow b^2 - 4ac = 0$,
> (c) two complex conjugate roots $\Leftrightarrow b^2 - 4ac < 0$.

The result $ax^2 + bx + c = a(x - a_1)(x - a_2)$, where a_1 and a_2 are the roots of $ax^2 + bx + c = 0$, applies when a_1 and a_2 are complex numbers, as is illustrated in the next example.

Example 3 Factor each of the following quadratics by their complex zeros.

(a) $x^2 + 4$
(b) $2x^2 + 2x + 1$

Solution:

(a) The quadratic equation $x^2 + 4 = 0$ has roots $a_1 = 2i$ and $a_2 = -2i$. Therefore,

$$x^2 + 4 = (x - 2i)[x - (-2i)]$$
$$= (x - 2i)(x + 2i).$$

(b) The roots of $2x^2 + 2x + 1 = 0$ are $a_1 = -\dfrac{1}{2} + \dfrac{i}{2}$ and $a_2 = -\dfrac{1}{2} - \dfrac{i}{2}$. Therefore,

$$2x^2 + 2x + 1 = 2\left[x - \left(-\dfrac{1}{2} + \dfrac{i}{2}\right)\right]\left[x - \left(-\dfrac{1}{2} - \dfrac{i}{2}\right)\right]$$
$$= 2\left(x + \dfrac{1}{2} - \dfrac{i}{2}\right)\left(x + \dfrac{1}{2} + \dfrac{i}{2}\right). \blacksquare$$

Question *Does the quadratic formula apply to quadratic equations whose coefficients are complex numbers?*

Answer: Yes. The algebraic properties of real numbers needed to derive the quadratic formula in the previous section also hold for complex numbers. However, we will not encounter quadratic equations with complex coefficients in this course. ∎

Geometric Representation of Complex Numbers. The complex number $z = a + bi$ can be represented in the plane by the directed line segment from the origin to the point with coordinates (a, b). The real numbers a and b are called the **rectangular coordinates of z** (Figure 1).

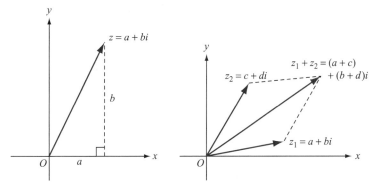

Figure 1 Complex Number z

Figure 2 Addition of Complex Numbers

Addition of complex numbers follows the **parallelogram rule.** As shown in Figure 2, the sum of two complex numbers z_1 and z_2 is the diagonal (from the origin) of the parallelogram formed by z_1 and z_2 (see Exercise 55).

Example 4 In each of the following cases, compute $z_1 + z_2$ and illustrate the result with a corresponding parallelogram.

(a) $z_1 = 3 + 2i$; $z_2 = 1 + 4i$
(b) $z_1 = 3 + 2i$; $z_2 = 3 - 2i$

Solution: Each result is indicated in Figure 3. ∎

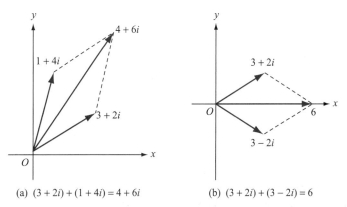

(a) $(3 + 2i) + (1 + 4i) = 4 + 6i$ (b) $(3 + 2i) + (3 - 2i) = 6$

Figure 3

Although addition of complex numbers is easily represented geometrically in the rectangular coordinate system of the plane, a convenient representation of multiplication requires the use of trigonometry (see also Exercises 56 and 57).

194   Chapter 3. Algebra and Graphs of Quadratic Expressions

Comment As mentioned earlier, complex numbers were developed by mathematicians in order to solve equations such as $x^2 + 2 = 1$, just as negative numbers were introduced to solve equations such as $x + 2 = 1$. The acceptance of both types of numbers was a very gradual process. However, our geometric interpretation of real numbers as points on a directed line and complex numbers as points in the plane, along with a geometric interpretation of addition and multiplication, serve to dispel any mystery about either type of number (at least among mathematicians).

Exercises 3.2

Fill in the blanks to make each statement true.

1. A complex number is a number of the form $a + bi$, where a and b are _____ and $i =$ _____.
2. The complex conjugate of $a + bi$ is _____.
3. A real number has imaginary part _____.
4. A complex number with zero real part is called a _____ number.
5. Geometrically, the sum of two complex numbers z_1 and z_2 is a _____ of the parallelogram with sides z_1 and z_2.

Write true or false for each statement.

6. A real number is not a complex number.
7. A complex number is a point in the plane; points along the x-axis correspond to real numbers, and points along the y-axis correspond to pure imaginary numbers.
8. If $3 + 2i$ is a root of a quadratic equation with real coefficients, then $3 - 2i$ is also a root of the equation.
9. Two complex numbers are equal if and only if their rectangular coordinates are equal.
10. To divide two complex numbers, multiply the numerator and denominator by the conjugate of the numerator.

The Complex Number System

Perform the indicated operations in Exercises 11-26.

11. $(2 + 3i) + (5 - 7i)$ 12. $(3 - 5i) - (-3 + 4i)$

13. $(3 + 2i) + 4i$
14. $6 + (4 - 2i)$
15. $(4 + i)(3 - 2i)$
16. $(-3 + 5i)(2 + 7i)$
17. $(4 + 3i)(4 - 3i)$
18. $(-2 + 5i)(-2 - 5i)$
19. $3i(1 - i)$
20. $(4 + 2i)5i$
21. $\dfrac{3 + 2i}{1 + i}$
22. $\dfrac{2 - i}{2 + i}$
23. $\dfrac{i}{5 - 2i}$
24. $\dfrac{6i}{2 - 3i}$
25. $\dfrac{7 + 4i}{i}$
26. $\dfrac{-5 + 6i}{-3i}$

27. Show that $i^4 = 1$ and use this result to compute each of the following.

    (a) i^5 (c) i^{15}
    (b) i^{10} (d) i^{20}

28. Any positive integer n can be written as $4q + r$, where q and r are nonnegative integers and $r \leq 3$. Show that $i^n = i^r$.

Complex Roots of a Quadratic Equation

Solve each of the following quadratic equations.

29. $x^2 + 9 = 0$
30. $4x^2 + 9 = 0$
31. $x^2 - 4x + 5 = 0$
32. $x^2 - 2x + 5 = 0$
33. $x^2 + 8x + 25 = 0$
34. $x^2 + 10x + 26 = 0$
35. $2x^2 - 2x + 1 = 0$

36. $2x^2 + 2x + 1 = 0$
37. $2x^2 - 4x + 3 = 0$
38. $2x^2 + 4x + 3 = 0$

Factor each of the following quadratics $ax^2 + bx + c$ as $a(x - a_1)(x - a_2)$, where a_1 and a_2 are the complex roots of $ax^2 + bx + c = 0$.

39. $x^2 + 9$
40. $x^2 - 4x + 5$
41. $x^2 + 8x + 25$
42. $x^2 + x + 1$
43. $x^2 + 10x + 26$
44. $2x^2 + 2x + 1$
45. $2x^2 + 4x + 3$
46. $3x^2 + 10x + 13$

Geometric Representation of Complex Numbers

In each of the following cases, compute $z_1 + z_2$ and illustrate the result by means of the parallelogram rule.

47. $z_1 = 3 + 2i$; $z_2 = 2 + 3i$
48. $z_1 = 4 - 2i$; $z_2 = 2 + 5i$
49. $z_1 = -3 + 4i$; $z_2 = 1 + 2i$
50. $z_1 = 5 - 3i$; $z_2 = -3 - i$
51. $z_1 = 4 + 2i$; $z_2 = -4 + 2i$
52. $z_1 = 2 - 3i$; $z_2 = 2 + 3i$
53. $z_1 = -3$; $z_2 = 5 + 2i$
54. $z_1 = 2i$; $z_2 = -4 + 2i$
55. *Parallelogram Rule for Addition.* Prove the parallelogram rule for addition of complex numbers as follows:
    (a) Use congruent triangles to show that the coordinates of Q are $(a + c, b)$.
    (b) Use part (a) to show that the coordinates of P_3 are $(a + c, b + d)$.

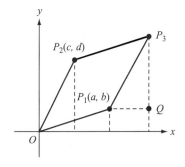

56. *Multiplication by i.* If the complex number $c + di$ is multiplied by i, we get

$$i(c + di) = -d + ci.$$

As indicated in the figure, the directed line segment $-d + ci$ is perpendicular to $c + di$. Also, both line segments have the same length, $\sqrt{c^2 + d^2}$, and when $c + di$ is in quadrant I, II, III, or IV, $-d + ci$ is in II, III, IV, or I, respectively. Therefore, *the effect of multiplying by i is to rotate $c + di$ through $90°$.* Draw a figure to illustrate each of the following products.

(a) $i(3 + 4i)$
(b) $i(-3 + 4i)$
(c) $i \cdot i$
(d) $i \cdot 5$

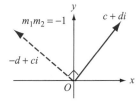

57. *Construction of $(a + bi)(c + di)$.* As indicated in the figure, we can construct the product of two complex numbers as the sum of two perpendicular directed line segments. Now

$$(a + bi)(c + di) = a(c + di) + bi(c + di)$$
$$= a(c + di) + b(-d + ci).$$

As in Exercise 56, $-d + ci$ is obtained by rotating $c + di$ through $90°$. The effect of the real factor a is to multiply the length of the line segment $c + di$ by $|a|$, and if a is negative, to reverse its direction. The effect of the factor b is similar. Draw a figure to illustrate each of the following products.

(a) $(1 + i)(3 + 4i)$
(b) $(1 - i)(3 + 4i)$
(c) $(-1 + i)(3 + 4i)$

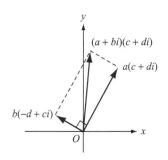

3.3 Equations Reducible to Quadratic Equations

The equations in this section are just quadratic equations in disguise.

Quadratics Obtained by Substitution
Equations with Fractional Expressions
Equations with Radicals

Quadratics Obtained by Substitution. It is sometimes possible to use substitution to transform a nonquadratic equation into a quadratic equation. Some examples are listed below.

	Original Equation	Substitution	Quadratic Equation
(a)	$x^4 - 6x^2 + 5 = 0$	$w = x^2$	$w^2 - 6w + 5 = 0$
(b)	$(x-1)^6 - 7(x-1)^3 - 8 = 0$	$w = (x-1)^3$	$w^2 - 7w - 8 = 0$
(c)	$2x + 3\sqrt{x} - 2 = 0$	$w = \sqrt{x}$	$2w^2 + 3w - 2 = 0$

The resulting quadratic equation in each case can be solved for w either by factoring or by the quadratic formula. Then, by reversing the substitution to express x in terms of w, we obtain the roots of the original equation.

Caution Some substitutions place a restriction on the w-roots. For instance, in equations (a) and (c) above, only *nonnegative* values of w are acceptable in the real number system. All values of x obtained by a substitution process should be checked for acceptability in the original equation.

Example 1 Find all real solutions of the above equations (a), (b), and (c) by means of the substitutions given there.

Solution:

(a) The roots of $w^2 - 6w + 5 = 0$ are $w_1 = 1$ and $w_2 = 5$. The substitution $w = x^2$ implies that $x = \pm\sqrt{w}$. Therefore, the real roots of $x^4 - 6x^2 + 5 = 0$ are $a_1 = 1$, $a_2 = -1$, $a_3 = -\sqrt{5}$, and $a_4 = -5$.

Check each root in $x^4 - 6x^2 + 5 = 0$:
$a_1:\ 1^4 - 6 \cdot 1^2 + 5 = 1 - 6 + 5 = 0$
$a_2:\ (-1)^4 - 6(-1)^2 + 5 = 1 - 6 + 5 = 0$
$a_3:\ (\sqrt{5})^4 - 6(\sqrt{5})^2 + 5 = 25 - 30 + 5 = 0$
$a_4:\ (-\sqrt{5})^4 - 6(-\sqrt{5})^2 + 5 = 25 - 30 + 5 = 0.$

Note that $x^4 - 6x^2 + 5 = 0$ has four roots. In general, if $P(x)$ is a polynomial of degree n, then the equation $P(x) = 0$ has n roots, including both real and complex roots as well as multiple roots.

(b) The roots of $w^2 - 7w - 8 = 0$ are $w_1 = -1$ and $w_2 = 8$. Now $w = (x-1)^3$ implies that $x - 1 = \sqrt[3]{w}$ or $x = 1 + \sqrt[3]{w}$. Hence, the real roots of $(x-1)^6 - 7(x-1)^3 - 8 = 0$ are $a_1 = 1 + \sqrt[3]{-1} = 0$ and $a_2 = 1 + \sqrt[3]{8} = 3$.

Check each root in $(x-1)^6 - 7(x-1)^3 - 8 = 0$:
$a_1:\ (0-1)^6 - 7(0-1)^3 - 8 = 1 + 7 - 8 = 0$
$a_2:\ (3-1)^6 - 7(3-1)^3 - 8 = 64 - 56 - 8 = 0.$

(c) The roots of $2w^2 + 3w - 2 = 0$ are $w_1 = -2$ and $w_2 = 1/2$. The root $w_1 = -2$ is unacceptable since the substitution $w = \sqrt{x}$ implies that w is nonnegative. From the root $w_2 = 1/2$, we obtain $x = w_2{}^2 = 1/4$ as the only real root of the original equation.

Check this root in $2x + 3\sqrt{x} - 2 = 0$:
$2(1/4) + 3\sqrt{1/4} - 2 = 1/2 + 3/2 - 2 = 0.$ ■

Equations with Fractional Expressions. If an equation contains fractional expressions in x, we proceed as in Section 2.1. That is, we first clear the fractions by multiplying both sides of the equation by the factors of the denominators. If the resulting equation is quadratic, we can then solve it by factoring or by the quadratic formula.

Caution As stated in Section 2.1, any value of x obtained as a solution after clearing fractions must be tested in the original equation. If the value makes a denominator equal to zero, it is an extraneous root and cannot be accepted.

Example 2 Solve $\dfrac{x}{x-1} + \dfrac{6}{x} = \dfrac{7}{x-1}$.

Solution:

$$\dfrac{x \cdot x(x-1)}{\cancel{x-1}} + \dfrac{6 \cdot x(x-1)}{\cancel{x}} = \dfrac{7 \cdot x(x-1)}{\cancel{x-1}} \quad \textit{Multiply both sides by } x(x-1).$$

$$x^2 + 6(x-1) = 7x$$
$$x^2 - x - 6 = 0$$
$$(x+2)(x-3) = 0$$
$$x = -2 \quad \text{or} \quad x = 3$$

Since neither of these values of x makes any denominator of the original equation equal to zero, the solutions of the original equation are -2 and 3. ■

Example 3 Solve $\dfrac{x^2}{x-5} + 3 = \dfrac{5x}{x-5}$.

Solution:

$$\dfrac{x^2 \cdot (x-5)}{x-5} + 3 \cdot (x-5) = \dfrac{5x \cdot (x-5)}{x-5} \quad \textit{Multiply both sides by } x-5.$$

$$x^2 + 3x - 15 = 5x$$
$$x^2 - 2x - 15 = 0$$
$$(x+3)(x-5) = 0$$
$$x = -3 \quad \text{or} \quad x = 5$$

The value 5 is an extraneous root of the original quadratic equation *(why?)*, but -3 is an acceptable root. Therefore, the solution to the original equation is -3. ∎

Equations with Radicals. As stated in Section 2.1, the general procedure for solving equations with radicals is to eliminate one radical at a time by squaring both sides. Again, because squaring may result in extraneous roots, all values obtained by this method must be checked for acceptability by substitution in the original equation.

Example 4 Solve $2x - 1 = \sqrt{x+1}$.

Solution:

$$4x^2 - 4x + 1 = x + 1 \quad \text{\textit{Square both sides.}}$$
$$4x^2 - 5x = 0$$
$$x(4x - 5) = 0$$
$$x = 0 \quad \text{or} \quad x = \frac{5}{4}$$

By substituting these values into the original equation, we see that 0 is an extraneous root *(why?)*, but 5/4 is a acceptable. Therefore, the solution to the original equation is 5/4. ∎

Example 5 Solve $2\sqrt{3x+1} = \dfrac{2}{\sqrt{3x+1}} - 3$.

Solution: First clear the fraction by multiplying both sides by $\sqrt{3x+1}$. Then eliminate the radical by squaring.

$$2\sqrt{3x+1} \cdot \sqrt{3x+1} = \frac{2\sqrt{3x+1}}{\sqrt{3x+1}} - 3\sqrt{3x+1} \quad \text{\textit{Clear fraction.}}$$
$$2(3x+1) = 2 - 3\sqrt{3x+1}$$
$$2x = -\sqrt{3x+1}$$
$$4x^2 = 3x + 1 \quad \text{\textit{Square both sides.}}$$
$$4x^2 - 3x - 1 = 0$$
$$x = 1 \quad \text{or} \quad x = -\frac{1}{4}$$

By substituting these values into the original equation, we see that 1 is an extraneous root *(why?)*, but $-1/4$ is acceptable. Hence, the solution to the original equation is $-1/4$. ∎

Exercises 3.3

Fill in the blanks to make each statement true.

1. A nonquadratic equation can sometimes be transformed into a quadratic equation by means of a _____.

2. An equation containing fractions cannot have as a solution any value that makes one of its denominators equal to _____.

3. A value obtained after squaring both sides of an equation must be tested for acceptability by substitution into _____.

4. If an equation is solved for w after making the substitution $w = x^{2/3}$, then only _____ values of w are acceptable in the real number system.

5. The equation $(x+1)^3 + (x+1)^{3/2} - 2 = 0$ becomes a quadratic equation in w if we let $w =$ _____.

Write true or false for each statement.

6. If an equation in x is transformed into a quadratic equation in w by means of a substitution, then the original equation cannot have more than two solutions for x.

7. If an equation in x becomes a quadratic equation in w by means of the substitution $w = (x-1)^2$, then any real solution x of the original equation must satisfy $x \geq 1$.

8. An equation obtained by clearing fractions in a given equation is always equivalent to the original equation.

9. If an equation in x becomes a quadratic equation in x after clearing fractions, both solutions of the quadratic equation may be solutions of the original equation.

10. An equation with radicals always has an extraneous root.

Substitution

Find all real solutions of the following equations by the substitution process.

11. $x^4 - 3x^2 + 2 = 0$
12. $x^6 - 6x^3 + 5 = 0$
13. $4x - 5\sqrt{x} - 9 = 0$
14. $3(x-1) + 2\sqrt{x-1} - 8 = 0$
15. $\left(x + \dfrac{1}{x}\right)^2 - 2\left(x + \dfrac{1}{x}\right) - 3 = 0$
16. $\left(x - \dfrac{1}{x}\right)^2 - 13\left(x - \dfrac{1}{x}\right) + 36 = 0$
17. $(x^2 - 1)^2 - 8(x^2 - 1) + 16 = 0$
18. $(x^2 + 1)^2 - 12(x^2 + 1) + 36 = 0$
19. $(2x + 3)^{-2} - 9(2x + 3)^{-1} + 20 = 0$
20. $(3x + 5)^{-4} + 4(3x - 5)^{-2} + 3 = 0$

Fractional Expressions

Find all solutions of the following equations.

21. $x + \dfrac{6}{x - 5} = 0$
22. $x + \dfrac{2}{x + 3} = 0$
23. $\dfrac{x}{x + 1} + \dfrac{9}{x(x + 1)} = \dfrac{5}{x}$
24. $\dfrac{1}{x - 6} + \dfrac{1}{x - 5} + \dfrac{12}{x(x - 5)(x - 6)} = 0$
25. $\dfrac{1}{x - 2} + \dfrac{3}{x(x - 5)} = \dfrac{9}{x(x - 2)(x - 5)}$
26. $\dfrac{x}{x - 3} + \dfrac{3}{(x - 3)(x - 4)} = \dfrac{3}{(x - 4)}$
27. $\dfrac{1}{x + 2} + \dfrac{x + 1}{3x - 2} = 1$
28. $\dfrac{x + 4}{x + 5} + \dfrac{x + 3}{2x + 7} = 1$
29. $\dfrac{2x}{x - 7} - \dfrac{1}{x - 5} = \dfrac{1}{(x - 7)(x - 5)}$
30. $\dfrac{x^2}{(x + 2)(x + 4)} + \dfrac{2x}{x + 4} = \dfrac{2}{x + 2}$

Equations with Radicals

Find all real solutions of the following equations.

31. $x - 3 = \sqrt{3x - 11}$
32. $2(x - 1) = \sqrt{2x^2 - x - 1}$
33. $x + 1 = \sqrt{\dfrac{x + 5}{3}}$
34. $x - 2 = \sqrt{16 - 3x}$

35. $\sqrt{2x+1} - \sqrt{x+4} = 1$

36. $\sqrt{3x+4} - \sqrt{x} = 2$

37. $\sqrt{x-1} + \sqrt{x-4} = \sqrt{x+4}$

38. $\sqrt{2x-6} + \sqrt{2x-1} = \sqrt{5x}$

39. $\sqrt{\dfrac{x+2}{3x+7}} - \dfrac{1}{\sqrt{2x+6}} = 0$

40. $\dfrac{1}{\sqrt{x+6}} - \dfrac{\sqrt{x+3}}{\sqrt{2-x}} = 0$

3.4 Quadratic Inequalities

The sign of $ax^2 + bx + c$ depends on the sign of a and the roots of $ax^2 + bx + c = 0$.

Quadratics with Distinct Real Roots
Quadratics with Equal Real Roots
Quadratics with Complex Roots

In Section 2.2 we solved the four basic inequalities

$$M > 0, \ M \geq 0, \ M < 0, \text{ and } M \leq 0,$$

where M was a linear expression $ax + b$. We now consider the same inequalities where M is a quadratic expression $ax^2 + bx + c$. We consider separately the cases in which $ax^2 + bx + c = 0$ has distinct real roots, equal real roots, and complex roots. These cases correspond to $b^2 - 4ac > 0$, $b^2 - 4ac = 0$, and $b^2 - 4ac < 0$, respectively.

Quadratics with Distinct Real Roots. In the case of distinct real roots, we can factor $ax^2 + bx + c$ and proceed as we did for fractional inequalities in Section 2.2. That is, the roots partition the x-axis into three intervals, and on each interval the sign of $ax^2 + bx + c$ does not change.

Example 1 Solve $-x^2 + 3x + 10 \geq 0$.

Solution: By first multiplying both sides by -1, we obtain the equivalent inequality

$$x^2 - 3x - 10 \leq 0.$$

Note that multiplying by -1 changes $\geq$ to $\leq$.

The quadratic $x^2 - 3x - 10$ factors into $(x+2)(x-5)$, which has zeros at -2 and 5. These points partition the x-axis into the three intervals $(-\infty, -2)$, $(-2, 5)$, and $(5, \infty)$, as illustrated in Figure 4. By evaluating $(x+2)(x-5)$ at an arbitrarily chosen point in each of these intervals, we find that $(x+2)(x-5) < 0$ on the open interval $(-2, 5)$, and therefore $(x+2)(x-5) \leq 0$ on the closed interval $[-2, 5]$ ∎

Example 2 Solve $6x^2 - 7x - 20 \geq 0$.

3.4. Quadratic Inequalities   201

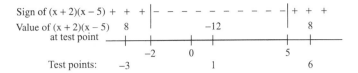

Figure 4

Solution: The quadratic $6x^2 - 7x - 20$ factors into

$$(3x+4)(2x-5) = 6\left(x + \frac{4}{3}\right)\left(x - \frac{5}{2}\right).$$

Here the equation $6x^2 - 7x - 20 = 0$ has distinct real roots $-4/3$ and $5/2$.

As shown in Figure 5, $-4/3$ and $5/2$ break up the number line into the intervals $(-\infty, -4/3)$, $(-4/3, 5/2)$, and $(5/2, \infty)$. In each of these, the sign of $x + 4/3$ and of $x - 5/2$ does not change. Hence, the sign of $6(x+4/3)(x-5/2)$ does not change in each of these intervals.

We choose three convenient test points, -2, 0 and 3 (one in each interval), and evaluate $6x^2 - 7x - 20$ at each of the test points. As indicated in Figure 5, we find that $6x^2 - 7x - 20$ is positive on $(-\infty, -4/3)$ and on $(5/2, \infty)$. Finally, since $6x^2 - 7x - 20 = 0$ for $x = -4/3$ and $x = 5/2$, we conclude that the solution of $6x^2 - 7x - 20 \geq 0$ consists of all x in either $(-\infty, -4/3]$ or $[5/2, \infty)$. ∎

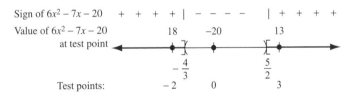

Figure 5

Quadratics with Equal Real Roots. The quadratic $4x^2 - 12x + 9$ factors into

$$(2x-3)^2 = 4\left(x - \frac{3}{2}\right)^2.$$

Here the equation $4x^2 - 12x + 9 = 0$ has a double root $3/2$. The factored form is zero when $x = 3/2$ and is positive for all other values of x. Therefore, $4x^2 - 12x + 9 \geq 0$ for all x.

In general, if a_1 is a double root of $ax^2 + bx + c = 0$, then

$$ax^2 + bx + c = a(x - a_1)^2.$$

Therefore, $ax^2 + bx + c = 0$ for $x = a_1$ and has the same sign as a for all other values of x.

Example 3 Solve $-4x^2 + 4x - 1 \geq 0$.

Solution: We first multiply both sides by -1 to obtain the equivalent inequality

$$4x^2 - 4x + 1 \leq 0.$$

Now the equation $4x^2 - 4x + 1 = 0$ has a double root $1/2$, and the factored form

$$(2x - 1)^2 = 4\left(x - \frac{1}{2}\right)^2$$

is positive for all other values of x. Hence, $4x^2 - 4x + 1 \leq 0$ is satisfied for $x = 1/2$ only. Therefore, the solution of the given inequality consists of the single point $1/2$ on the number line (Figure 6). ■

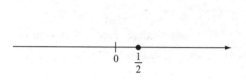

Figure 6

Quadratics with Complex Roots. The equation $x^2 - 2x + 2 = 0$ has complex roots $1 + i$ and $1 - i$. (Note that the discriminant $b^2 - 4ac = -4$ is negative.) Therefore,

$$\begin{aligned} x^2 - 2x + 2 &= [x - (1 + i)][x - (1 - i)] \\ &= [(x - 1) - i][(x - 1) + i] \\ &= (x - 1)^2 + 1. \end{aligned}$$

This same result could be obtained by completing the square in x. Now $(x - 1)^2 + 1 > 0$ for all real x, and therefore $x^2 - 2x + 2 > 0$ for all real numbers x.

In general, if the equation $ax^2 + bx + c = 0$ has complex roots $a_1 + b_1 i$, $a_1 - b_1 i$, then

$$ax^2 + bx + c = a[(x - a_1)^2 + b_1^2].$$

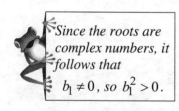

Since the roots are complex numbers, it follows that $b_1 \neq 0$, so $b_1^2 > 0$.

Since $(x - a_1)^2 + b_1^2 > 0$ for all real x, it follows that $ax^2 + bx + c$ has the same sign as a for all real numbers x. That is, *a quadratic with complex roots is either positive for all real numbers x or negative for all real numbers x.*

Example 4 Solve $x^2 + 2x + 3 > 0$.

Solution: The equation $x^2 + 2x + 3 = 0$ has complex roots since the discriminant $b^2 - 4ac = 4 - 12 = -8 < 0$. Therefore, $x^2 + 2x + 3$ is either positive for all real values of x or is negative for all real x. Now when $x = 0$, we have $0^2 + 2 \cdot 0 + 3 = 3 > 0$. Thus, $x^2 + 2x + 3 > 0$ for all real numbers x, and the solution is the entire real line $(-\infty, \infty)$. ∎

Example 5 Solve $-2x^2 + 5x - 5 \geq 0$.

Solution: The equation $-2x^2 + 5x - 5 = 0$ has complex roots, since $b^2 - 4ac = 25 - 40 = -15 < 0$. Therefore, $-2x^2 + 5x - 5$ has the same sign for all x. For $x = 0$, we get $-2 \cdot 0^2 + 5 \cdot 0 - 5 = -5 < 0$. Thus, $-2x^2 + 5x - 5 < 0$ for all real numbers x, and the solution of the given inequality is the empty set. ∎

We can summarize our results for quadratic inequalities as follows:

---

1. If $ax^2 + bx + c = 0$ has distinct real roots a_1 and a_2 ($a_1 < a_2$), then $ax^2 + bx + c = a(x - a_1)(x - a_2)$, which does not change sign within each of the intervals $(-\infty, a_1)$, (a_1, a_2), and (a_2, ∞).
2. If $ax^2 + bx + c = 0$ has a double root a_1, then $ax^2 + bx + c = a(x - a_1)^2$, which has the same sign as a for all values of x other than a_1.
3. If $ax^2 + bx + c = 0$ has complex roots, then $ax^2 + bx + c$ is irreducible in the real number system and has the same sign as a for all real values of x.

Exercises 3.4

Fill in the blanks to make each statement true.

1. The solution of a quadratic inequality $ax^2 + bx + c > 0$ is completely determined by the sign of _____ and the _____ of $ax^2 + bx + c = 0$.

2. If $a > 0$ and $ax^2 + bx + c = 0$ has roots a_1 and a_2 ($a_1 < a_2$), then $ax^2 + bx + c = a(x - a_1)(x - a_2)$ is _____ 0 for x in the interval (a_1, a_2).

3. If $b^2 - 4ac < 0$, then the solution to $ax^2 + bx + c > 0$ is either _____ or _____.

4. The quadratic inequality $a(x - a_1)^2 > 0$ has the empty set as its solution if _____.

5. If $ax^2 + bx + c = 0$ has roots a_1 and a_2 ($a_1 < a_2$), and $ax^2 + bx + c > 0$ for $x = (a_1 + a_2)/2$, then the solution to $ax^2 + bx + c \leq 0$ is _____.

Write true or false each statement

6. If $b^2 - 4ac = 0$, the solution of $ax^2 + bx + c > 0$ is a single point on the line.
7. If $b^2 - 4ac < 0$, then the inequality $ax^2 + bx + c > 0$ has no solution in the real number system.
8. If $b^2 - 4ac > 0$, and a_1 and a_2 ($a_1 < a_2$) are the roots of $ax^2 + bx + c = 0$, then the solution of $ax^2 + bx + c > 0$ is either $(-\infty, a_1) \cup (a_2, \infty)$ or (a_1, a_2).

9. If we know the roots of $ax^2 + bx + c = 0$, then we know the solution of $ax^2 + bx + c < 0$.
10. If we know the sign of a, then we know the solution of $ax^2 + bx + c \geq 0$.

Quadratics with Distinct Real Roots
Solve each of the following inequalities, and graph the solution on the real number line.

11. $x^2 - 4x + 3 \geq 0$
12. $x^2 + 7x + 10 > 0$
13. $x^2 + 2x - 8 < 0$
14. $x^2 - 3x - 4 \leq 0$
15. $2x^2 + 7x - 3 \leq 0$
16. $-3x^2 + 17x - 10 < 0$
17. $3x^2 - 10x - 8 \leq 0$
18. $4x^2 - 5x - 6 < 0$
19. $-6x^2 + 13x + 8 \leq 0$
20. $-6x^2 + 7x + 20 < 0$

Quadratics with Equal Real Roots
Solve each of the following inequalities.

21. $x^2 - 4x + 4 \geq 0$
22. $x^2 + 10x + 25 > 0$
23. $x^2 + 6x + 9 < 0$
24. $x^2 - 2x + 1 \leq 0$
25. $-4x^2 + 12x - 9 > 0$
26. $-9x^2 + 12x - 4 \geq 0$
27. $32x^2 - 16x + 2 \leq 0$
28. $50x^2 + 40x + 8 < 0$
29. $-48x^2 + 72x - 27 < 0$
30. $-20x^2 + 20x - 5 \geq 0$

Quadratics with Complex Roots
Solve each of the following inequalities in the real number system.

31. $x^2 + x + 1 \geq 0$
32. $x^2 + 2x + 4 < 0$
33. $-x^2 + 5x - 7 \leq 0$
34. $-2x^2 + 6x - 5 > 0$
35. $3x^2 - 4x + 2 \leq 0$
36. $5x^2 - 10x + 7 > 0$

Miscellaneous
Solve each of the inequalities in Exercises 37-48.

37. $12x^2 - 27 \geq 0$
38. $75 - 12x^2 \leq 0$
39. $(2x + 1)(x - 3) < 0$
40. $(3x - 2)(x + 1) > 0$
41. $x^2 + 3x - 4 < 4 - 3x - 4x^2$
42. $3x^2 + 12x - 3 > 2 - 6x^2$
43. $(x - 5)(x - 1) > x(5 - x)$
44. $(x - 3)(x + 1) \leq x(3 - x)$
45. $\dfrac{x^2 + 10}{x^2 + 4} > 2$
46. $\dfrac{x + 1}{x^2 + 1} > 1$
47. $\dfrac{1}{x^2 + 2} + 1 \leq \dfrac{4x}{x^2 + 2}$
48. $\dfrac{x}{x^2 + 1} \geq \dfrac{x + 2}{x^2 + x + 1}$

49. Find all points on the x-axis whose distance from the point $P(8, 2)$ is at least twice its distance from $Q(2, 1)$.

50. Find all points on the x-axis whose distance from the $P(3, 4)$ is at least 5.

51. A ball is tossed upward. After t sec its height is y ft, where $y = -16t^2 + 32t$. For what values of t is the ball at least 7 ft above the ground?

52. For what values of t is the ball in Exercise 51 less than 12 ft above the ground?

3.5 Applications of Quadratic Equations and Inequalities

Percent
Motion Based on Average Speed
Motion Due to Gravity
Work
Miscellaneous Applications

Why is that when a quadratic equation is used to solve a problem, usually only one of the roots makes sense?

In this section we investigate word problems whose mathematical interpretation results in a quadratic equation or inequality. In addition to the topics of percent, average speed, and work, which were introduced in Section 2.3, we also discuss motion due to gravity and some miscellaneous geometric and numeric problems. Although the expressions here are quadratic, they are arrived at in the same manner as in the linear case

3.5. Applications of Quadratic Equations and Inequalities

of Section 2.3, and you should follow the guidelines and flow chart given there.

Percent. When a percentage increase or decrease is applied to a quantity, a linear equation results. If a second increase or decrease is applied, a quadratic equation may result.

Example 1 After graduating in June with an engineering degree, James gets a job starting at \$24,000 per year. In December he receives an $r\%$ increase, and six months later he gets a $2r\%$ increase, making his new salary \$27,720. What is r?

Solution: Let x be the decimal equivalent of $r\%$: $x = r/100$. We have the following:

$$\text{salary in December} = \text{starting salary} + (\text{starting salary})x$$
$$= 24{,}000 + 24{,}000 \cdot x$$
$$= 24{,}000(1 + x);$$
$$\text{salary six months later} = \text{December salary} + (\text{December salary})2x$$
$$= 24{,}000(1 + x) + 24{,}000(1 + x)2x$$
$$= 24{,}000(1 + x)(1 + 2x)$$
$$= 24{,}000(1 + 3x + 2x^2)$$

Therefore,

$$24{,}000(1 + 3x + 2x^2) = 27{,}720$$
$$1 + 3x + 2x^2 = 1.155 \qquad 27{,}720/24{,}000 = 1.155$$
$$2x^2 + 3x - .155 = 0.$$

The roots are $x = .05$ and $x = -1.55$, but only the positive value is meaningful here. Hence, $x = .05$ and $r = 5$, so that the first increase is 5% and the second is 10%. ∎

Example 2 A clothing store wants to sell 100 suits originally priced at \$225 each. Seventy five of them are sold at an $r\%$ discount, and the remaining 25 are marked down by another $r\%$. All 100 suits are sold for a total of \$14,568.75. What is r?

Solution: As in Example 1, let $x = r/100$, the decimal equivalent of $r\%$. Then

$$\text{selling price of first 75 suits} = \text{original price} - (\text{original price})x$$
$$= 225 - 225 \cdot x$$
$$= 225(1 - x);$$

selling price of last 25 suits = selling price of first 75 suits
$$- \text{(selling price of first 75 suits)}x$$
$$= 225(1-x) - 225(1-x)x$$
$$= 225(1-x)(1-x)$$
$$= 225(1-x)^2.$$

Therefore,

$$75 \cdot 225(1-x) + 25 \cdot 225(1-x)^2 = 14{,}568.75, \qquad (1)$$

Income from first 75 units   *Income from second 25 units*   *Total Income*

Equation (1) is a quadratic in $w = 1 - x$. That is,

$$25 \cdot 225 w^2 + 75 \cdot 225 w - 14{,}568.75 = 0$$
$$\text{or} \qquad w^2 + 3w - 2.59 = 0$$

By the quadratic formula, $w = .7$ and $w = -3.7$. Hence, $x = .3$ and $x = 4.7$. Only the root $x = .3$ has a meaningful interpretation (*why?*). Therefore, $r = 30\%$. ∎

Motion Based on Average Speed. Quadratic equations often arise in basic motion problems in which an event taking t time units consists of two parts of t_1 and t_2 time units, respectively. The event is described by the equation

$$t_1 + t_2 = t,$$
$$\text{or} \qquad \frac{d_1}{r_1} + \frac{d_2}{r_2} = t,$$

where d_1 and d_2 are the distances and r_1 and r_2 are the average rates for the first and second parts, respectively.

Example 3 A bus travels 60 miles at a certain average rate of speed and another 75 miles at an average rate 10 miles per hour faster. The entire trip takes 3 hours. What are the two rates of speed?

Solution: Let $x =$ the rate of speed in miles per hour for the first 60 miles. Then $x + 10 =$ the rate of speed for the last 75 miles. Also, $t_1 = 60/x$ is the time (in hours) for the first 60 miles, and $t_2 = 75/(x+10)$ is the time for the last 75 miles. We have

3.5. Applications of Quadratic Equations and Inequalities

$$t_1 + t_2 = 3,$$

or $$\frac{60}{x} + \frac{75}{x+10} = 3. \qquad (2)$$

After clearing fractions in equation (2), we obtain the quadratic equation

$$x^2 - 35x - 200 = 0$$

whose roots are $x = 40$ and $x = -5$. Only the positive root $x = 40$ is meaningful. Therefore, the average rate of speed for the first 60 miles is 40 miles per hour, and that for the last 75 miles is 50 miles per hour. ∎

Example 4 Herb and Ray row at the same speed. Herb rows 3 miles with the aid of a 1-mile-per-hour current and then back 3 miles against the same current. Ray rows 6 miles in still water. If Herb's trip takes 4 hours, how long does Ray's take?

Solution:

Let $x =$ the rate (in miles per hour) at which both Herb and Ray row,

$t_1 =$ the time for Herb's trip downstream, and

$t_2 =$ the time for Herb's trip upstream. Then

$$t_1 + t_2 = 4,$$

or $$\frac{3}{x+1} + \frac{3}{x-1} = 4. \qquad \text{Herb's speed downstream is } x + 1, \text{ and upstream it is } x - 1.$$

When fractions are cleared in the above equation, the result is the quadratic equation

$$2x^2 - 3x - 2 = 0$$

whose roots are $x = 2$ and $x = -1/2$. Only the positive root $x = 2$ is meaningful. Therefore, both Herb and Ray row at 2 miles per hour in still water. Hence, the time for Ray's trip is $6/2 = 3$ hours. ∎

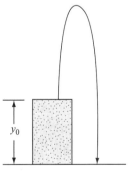

Figure 7

Motion Due to Gravity. If an object is projected with a vertical velocity of v_0 feet per second from a point y_0 feet above the ground (Figure 7), then until it hits the ground its **height y** is given by

$$y = -16t^2 + v_0 t + y_0. \qquad (3)$$

The height y is measured in feet, and the time t is measured in seconds. Formula (3) can be proven using calculus, and is based on the assumption that gravity is the only force acting on the object and that the acceleration due to gravity is a constant -32 ft/sec^2. Also, the **vertical velocity** v of the object at time t is given by the formula

$$v = -32t + v_0. \tag{4}$$

The velocity is measured in feet per second; it is positive when the object is rising and negative when the object is falling.

Example 5 An object is projected upward from ground level with an initial velocity of 128 feet per second. For what period of time is the object at least 192 feet above the ground?

Solution: We are given $y_0 = 0$ and $v_0 = 128$. Therefore, formula (3) becomes

$$y = -16t^2 + 128t.$$

Set $y \geq 192$ and solve for t:

$$-16t^2 + 128t \geq 192$$
$$-t^2 + 8t - 12 \geq 0$$
$$t^2 - 8t + 12 \leq 0$$
$$(t-2)(t-6) \leq 0.$$

By the method of Section 3.4, the solution of the inequality $(t-2)(t-6) \leq 0$ is $2 \leq t \leq 6$. Therefore, the object is at least 192 feet above the ground from 2 seconds to 6 seconds after it leaves the ground. ∎

Example 6 Jacqui drops a coin into a deep wishing well. She hears the splash 2.86 seconds later. How far down is the water level of the well? (Sound travels at 1100 feet per second.)

Solution: As indicated in Figure 8, let $d =$ the depth of the water level in feet, $t_1 =$ the time in seconds for the coin to hit the water, and $t_2 =$ the time of the sound to return to the top of the well. The motion downward is due to gravity with zero initial velocity (since the coin was just dropped) and an initial height of d feet. Therefore, form the formula (3),

$$-16t_1^2 + d = 0$$

or $\quad t_1 = \dfrac{\sqrt{d}}{4}.$

The motion upward is sound traveling at 1100 ft/sec. Therefore,

$$d = 1100 t_2,$$

or $\quad t_2 = \dfrac{d}{1100}.$

We are given that

$$t_1 + t_2 = 2.86.$$

Therefore, $\quad \dfrac{\sqrt{d}}{4} + \dfrac{d}{1100} = 2.86.$ \hfill (5)

After clearing the fractions and substituting $w = \sqrt{d}$, equation (5) becomes

$$w^2 + 275w - 3146 = 0,$$

whose roots are $w = 11$ and $w = -286$. (Only $w = 11$ is acceptable.) Since $d = w^2$, the water level of the well is $11^2 = 121$ feet down. ■

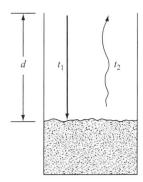

Figure 8

Work. Here, as in the linear case, we consider work problems in which the following principle applies.

If it takes n time units to do a given job, then $1/n$ of the job is completed per time unit.

Example 7 It takes secretary A 3 hours more than secretary B to prepare a certain report. Working together, the two can complete the job in 2 hours. How long does it take each secretary to finish the report when working alone?

Solution: Let $x =$ the number of hours required for B to prepare the report. Then $x + 3 =$ the number of hours required for A. When the two work together, 1/2 of the job is completed per hour. B's contribution in one hour is $1/x$ and A's is $1/(x+3)$. Therefore,

$$\frac{1}{x} + \frac{1}{x+3} = \frac{1}{2}.$$

We clear fractions by multiplying both sides by $2x(x+3)$ to obtain

$$2(x+3) + 2x = x(x+3)$$
$$\text{or} \quad x^2 - x - 6 = 0,$$

whose roots are $x = 3$ and $x = -2$. Only the positive root is meaningful here. Therefore, it takes B 3 hours to prepare the report when working alone, and it takes A 6 hours for the same report. ∎

Example 8 Machine A can do a job in 3 1/3 hours. Machines A and B do the same job in 3 hours less time than it takes B working alone. How long does it take B to do the job?

Solution: Let $x =$ number of hours for B to do the job. Then $x - 3 =$ number of hours for A and B working together. Each hour A does 3/10 of the job, B contributes $1/x$, and A and B together contribute $1/(x-3)$. Therefore,

$$\frac{3}{10} + \frac{1}{x} = \frac{1}{x-3}.$$

which, after clearing fractions, becomes

$$3x(x-3) + 10(x-3) = 10x$$
$$3x^2 - 9x - 30 = 0$$
$$3(x^2 - 3x - 10) = 0$$
$$3(x-5)(x+2) = 0$$
$$x = 5 \quad \text{or} \quad x = -2.$$

Therefore, it takes B 5 hours to do the job. ∎

Miscellaneous Applications. Here we treat some geometric and numeric problems that can be solved by means of quadratic equations.

Example 9 A string 12 inches long is cut into two pieces. Each piece is formed into a circle, and the sum of the areas of the two circles is $61/(2\pi)$ square inches. What is the length of each piece?

Solution: Let $x =$ the length of one of the pieces. Then $12 - x =$

3.5. Applications of Quadratic Equations and Inequalities 211

the length of the other piece (Figure 9). We can express the area of each circle (Figure 10) in terms of x as follows:

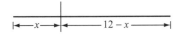

Figure 9

$$\text{circumference} = 2\pi \cdot \textit{radius} \qquad \text{area} = \pi(\textit{radius})^2$$

$$x = 2\pi \cdot r_1 \qquad\qquad A_1 = \pi r_1^2$$

$$r_1 = \frac{x}{2\pi} \qquad\qquad = \pi\left(\frac{x}{2\pi}\right)^2$$

$$= \frac{x^2}{4\pi}$$

$$\text{circumference} = 2\pi \cdot \textit{radius} \qquad \text{area} = \pi(\textit{radius})^2$$

$$12 - x = 2\pi \cdot r_2 \qquad\qquad A_2 = \pi r_2^2$$

$$r_2 = \frac{12 - x}{2\pi} \qquad\qquad = \pi\left(\frac{12-x}{2\pi}\right)^2$$

$$= \frac{144 - 24x + x^2}{4\pi}$$

We are given that

$$A_1 + A_2 = \frac{61}{2\pi}.$$

Therefore, $\quad\dfrac{x^2}{4\pi} + \dfrac{144 - 24x + x^2}{4\pi} = \dfrac{61}{2\pi},$

which simplifies to

$$x^2 - 12x + 11 = 0.$$

The roots are $x = 11$ and $x = 1$. In the first case $12 - x = 1$, and in the second case $12 - x = 11$. Therefore, one piece of the cut string is 11 inches, and the other piece is 1 inch. ∎

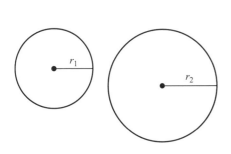

Figure 10

Example 10 It can be shown that for any positive integer n,

$$1^2 + 2^2 + 3^3 + \ldots + n^2 = \frac{n(n + 1)(2n + 1)}{6}.$$

What is n if the sum of the squares from 1 to n is $20n$?

Solution:

$$\frac{\cancel{n}(n + 1)(2n + 1)}{6} = 20\cancel{n} \qquad\textit{Divide both sides by } n.$$

$$2n^2 + 3n + 1 = 120 \qquad\textit{Multiply both sides by 6.}$$

$$2n^2 + 3n - 119 = 0$$

By the quadratic formula,

$$n = \frac{-3 \pm \sqrt{9 + 952}}{4}$$

$$= \frac{-3 + \pm\sqrt{961}}{4}$$

$$= \frac{-3 \pm 31}{4},$$

$$n = 7 \quad \text{or} \quad -\frac{34}{4},$$

Since $-34/4$ is not a positive integer, the only answer is $n = 7$.

Check: $1^2 + 2^2 + 3^2 + 4^2 + 5^2 + 6^2 + 7^2 = 1 + 4 + 9 + 16 + 25 + 36 + 49 = 140 = 20 \cdot 7$. ∎

Exercises 3.5

Most of the quadratic equations generated in the following exercises can be solved by factoring. However, in some cases, especially in the exercises involving percent, it will be helpful to use the quadratic formula with the aid of a calculator.

Percent

1. A union contract calls for an $r\%$ salary increase the first year and another $r\%$ increase the second year. As a result, a worker's salary of $24,000 will increase to $26,460 after two years. What is r?

2. An IRA earned $r\%$ interest the first year and $r\%$ the second. If an investment of $2000 grew to $2420 at the end of the second year, what is r?

3. In a closeout sale, ARA Audio cuts prices by $r\%$ the first week and another $r/2\%$ the second. If a stereo that sold for $500 before the sale sells for $240 in the second week of the sale, what is r?

4. A stock selling for $50 goes up $r\%$ one day and $2r\%$ the next, resulting in a price of $51.51. What is r?

5. A $1000 investment grew to $1875 in two years. The percent appreciation for the second year was double that of the first year. What was the percent appreciation for each year?

6. Stock in the MICRO Computer Company appreciated 75% in one year. If the percent appreciation over the second half of the year was triple that of the first half, what was the appreciation in each 6-month period?

7. A sum of $1000 was invested in a mutual fund. After 6 months $500 was withdrawn, and at the end of the year $609.90 remained. If the money earned $r\%$ the first 6 months and another $r\%$ the second, what is r?

8. Trish deposited $1200 in a savings account. Eight months later she deposited another $600, and at the end of the year she had $1993.86. If her investment earned r percent the first 8 months and $r/2$ percent the last 4 months, what is r?

Motion Based on Average Speed

9. Jean rides a bicycle 5 mi to school each morning. Her roommate Margo rides home from school at a rate 5 mph faster than Jean. The total time for both trips is 50 min. How fast does each woman travel?

10. When driving through a 10-mi construction area, a car's speed was 15 mph slower than for the remaining 60 mi of the trip. If the entire trip took 1 hr and 54 min, what was the car's speed through the construction area?

11. An airplane flies 990 mi against a 30-mph wind and 990 mi back in the direction of the same wind. The round trip takes 6 2/3 hr. What is the speed of the plane in still weather?

12. Cindy and Lynne row at the same speed in still water. Cindy rows 2 mi against a steady current and back 2 mi in the direction of the current. Lynne makes the same trip in still water. Whose trip takes more time?

13. A Jogger runs 5 mi in one hour. If he runs x mph faster than his average for the first 3 mi and x mph slower for the last 2 mi, what is x?
14. For what values of x in Exercise 13 will the jogger run the five miles in less than one hour?

Motion Due to Gravity

15. From ground level an object is fired upward with an initial velocity of 144 ft/sec.
    (a) When does the object reach its highest point? (*Hint*: when the highest point is reached, the vertical velocity of the object is zero.)
    (b) What is the highest point reached by the object?
    (c) When does the object return to ground level?
16. An object is tossed upward from a point 16 ft above the ground with an initial velocity of 96 ft/sec. For what period of time is the object at least 144 ft above the ground?
17. If a baseball is hit from a point 4 ft above the ground with an initial vertical velocity of 64 ft/sec, then its height y above the ground at time t is
    $$y = -16t^2 + 64t + 4.$$
    If the ball's horizontal velocity is a constant 90 ft/sec and there is a 10-ft-high fence 330 ft away, will the ball go over the fence?

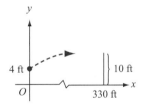

18. A football is place-kicked from ground level at the 50-yard line with an initial vertical velocity of 80 ft/sec. If at time r the horizontal distance traveled before hitting the ground is
    $$x = 15t - t^2,$$
    will the ball clear 10-ft-high goal posts on the goal line?

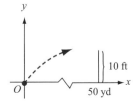

19. In a test of strength at a carnival, a hammer drives a metal weight up a rod to ring a bell at the top. If the bell is 24 ft above the starting position of the weight, what is the minimum initial vertical velocity needed to ring the bell? (*Hint*: when the weight reaches its highest point, its vertical velocity is zero.)
20. Formula (3) in this section is valid under the conditions stated in the text. Is it possible, under these conditions, for an object to be launched with an initial velocity large enough so that the object never returns to earth?

Work

21. It takes one mail carrier 1 hour more than another to deliver mail along a certain route. Working together, the two can deliver the mail in 1 hr and 12 min. How long does it take each carrier to deliver the mail?
22. Don and Rich together can set up all the banquet tables in a VFW hall in 20 min less time than Don can and 45 min less than Rich can, each working alone. How long does it take the two of them to set up the tables?
23. Machine A takes 4 min longer and machine B takes 9 min longer than it takes both A and B together to do a certain job. How long does it take each to do the job?
24. Tom can deliver papers in 15 min less time than his sister, and his father can deliver them twice as fast as Tom. With all three working, the morning delivery takes 12 min. How long does it take each person?
25. Printer A can do a job in the same time as printers B and C working together. If C takes 4 min longer than B, and it takes all three 2 min and 24 sec, how long does it take each printer to do the job working alone?
26. Machine A does a job in the same time as B and C working together, and B does the job in the same time as C and D working together. If D takes 1 hr longer than C and all four can do the job in 20 min, what is the time required for each machine?

Miscellaneous

27. One positive number is 5 more than another, and their product is 14. Find the numbers.
28. The difference between a number and its reciprocal is 1/2. Find all such pairs of numbers and check your answer.

29. A gardener wants to use 350 yd of fencing to enclose a rectangular garden with an area of 7500 sq yd. What should be the dimensions of the garden?

30. In Exercise 29, the gardener now wants to enclose his garden and also divide the total area in half with a piece of fence in the middle. Show that this is impossible with only 350 yd of fencing.

31. A page in a book is 7 wide and 9 in long. It must contain 35 sq in of printed material and have a margin of the same width on top, bottom, and both sides. How wide should the margin be?

32. If the page of the book in Exercise 31 must have a margin at top and bottom that is twice as wide as the margin at the sides, how wide should the margin be?

3.6 Circles and Parabolas

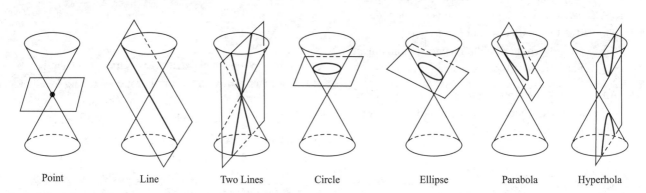

Second-Degree Equations
Circles
Parabolas
Applications

Second-Degree Equations. In Section 2.5 we studied the first-degree (linear) equation

$$Ax + By = C \quad (A \text{ or } B \neq 0), \tag{1}$$

and now we investigate the **second-degree equation**[1]

$$Ax^2 + By^2 + Cx + Dy = E \quad (A \text{ or } B \neq 0). \tag{2}$$

The graph of equation (1) is always a straight line, but equation (2) can have for its graph a point, a line, two lines, a circle, an ellipse, a parabola, a hyperbola, or the empty set. All of these curves, except for parallel lines and the empty set, can be obtained as the intersection of a cone and a plane (see diagram above). Hence, they are called **conic sections**. However, the term *conic section* is usually restricted to the circle, parabola, ellipse, and hyperbola. We consider the circle and parabola in this section, and the ellipse and hyperbola in Section 3.8.

[1] *The most general second-degree equation is $Ax^2 + By^2 + Cxy + Dx + Ey = F$. However, the Cxy term can be eliminated by a rotation of axes. The equations for the rotation require trigonometry.*

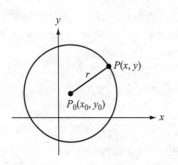

Figure 11

Circles. The set of all points $P(x, y)$ in the plane at a fixed distance $r > 0$ from a fixed point $P_0(x_0, y_0)$ is a **circle** with **center** (x_0, y_0) and **radius** r, as shown in Figure 11. By the distance formula in the plane

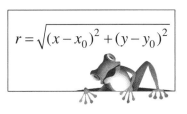

(Section 2.4), a point $P(x, y)$ will lie on the circle of Figure 11 if and only if x and y satisfy

$$(x - x_0)^2 + (y - y_0)^2 = r^2, \qquad (3)$$

which is the **standard equation of a circle**. If the center is at the origin $(0, 0)$, then equation (3) simplifies to

$$x^2 + y^2 = r^2. \qquad (4)$$

Example 1 In each of the following cases, find the standard equation of the circle with the given center and radius. Also, graph the circle.

(a) $(0, 0)$, 3
(b) $(3, 1)$, 2
(c) $(-4, 0)$, 1

Solution:

(a) With the center at the origin and $r = 3$ [Figure 12(a)], we get

$$x^2 + y^2 = 9.$$

(b) With center $(3, 1)$ and radius 2 [Figure 12(b)], the equation is

$$(x - 3)^2 + (y - 1)^2 = 4.$$

(c) Here [Figure 12(c)] we get

$$[x - (-4)]^2 + y^2 = 1^2 \text{ or } (x + 4)^2 + y^2 = 1. \quad \blacksquare$$

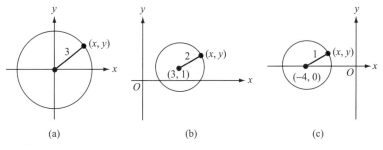

(a) (b) (c)

Figure 12

Example 2 Find the standard equation of the circle that has center $P_0(-2, 3)$ and passes through the point $P_1(1, 2)$. Also, graph the circle.

Solution: The radius of the circle is

$$d(P_0, P_1) = \sqrt{(1+2)^2 + (2-3)^2} = \sqrt{10}.$$

Therefore, the equation is

$$(x+2)^2 + (y-3)^2 = 10,$$

whose graph is given in Figure 13. ∎

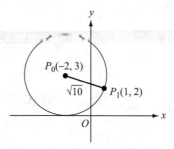

Figure 13

Example 3 Use the result of Example 6 in Section 2.6 to find the equation of the circle with center at the origin and tangent to the line $3x + 4y = 10$.

Solution: By Example 6 in Section 2.6, the shortest distance from the origin $(0, 0)$ to the line $3x + 4y = 10$ is $10/\sqrt{3^2 + 4^2} = 2$. This distance is achieved by the point of intersection of a *radial line* through $(0, 0)$ and the line $3x + 4y = 10$ (Figure 14). Hence, the radius is 2 and equation (4) of the circle becomes

$$x^2 + y^2 = 4. \quad \blacksquare$$

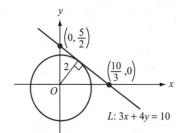

Figure 14

If we start with any second-degree equation (2) in which $A = B \neq 0$, we can proceed to an equation $(x - x_0)^2 + (y - y_0)^2 = c$ by completing the square in x and in y as will be shown in Example 4. Example 4 illustrates that there are three possible graphs: a circle ($c > 0$,), a single point $c = 0$, and the empty set ($c < 0$).

Example 4 In each of the following, complete the square in x and in y to determine the graph.

(a) $x^2 + y^2 - 6x - 2y = -6$
(b) $x^2 + y^2 + 2x - 2y = -2$
(c) $x^2 + y^2 - 2x - 4y = -6$

Solution:

(a) $x^2 + y^2 - 6x - 2y = -6$ *Given*
  $x^2 - 6x + y^2 - 2y = -6$ *Combine x terms and y terms.*

$$x^2 - 6x + 9 + y^2 - 2y + 1 = -6 + 9 + 1 \quad \text{Complete the square}$$
$$(x-3)^2 + (y-1)^2 = 4 \quad \text{in x and in y.}$$

The graph is a circle with center $(3, 1)$ and radius 2.

(b)
$$x^2 + y^2 + 2x - 2y = -2 \quad \text{Given}$$
$$x^2 + 2x + y^2 - 2y = -2 \quad \text{Combine x terms and y terms.}$$

$$x^2 + 2x + 1 + y^2 - 2y + 1 = -2 + 1 + 1 \quad \text{Complete the square}$$
$$(x+1)^2 + (y-1)^2 = 0 \quad \text{in x and in y.}$$

Here the graph is the single point $(-1, 1)$ (*why?*).

(c)
$$x^2 + y^2 - 2x - 4y = -6 \quad \text{Given}$$
$$x^2 - 2x + y^2 - 4y = -6 \quad \text{Combine x terms and y terms.}$$

$$x^2 - 2x + 1 + y^2 - 4y + 4 = -6 + 1 + 4 \quad \text{Complete the square}$$
$$(x-1)^2 + (y-2)^2 = -1 \quad \text{in x and in y.}$$

The graph is the empty set (*why?*). ■

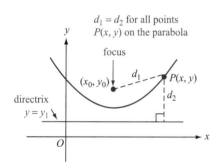

Figure 15 Parabola

Parabolas. The set of all points $P(x, y)$ equidistant from a fixed point $P_0(x_0, y_0)$ and a fixed line L not containing P_0 is called a **parabola** with **focus** (x_0, y_0) and **directrix** L (Figure 15). With reference to Figure 15, the condition $d_1 = d_2$ becomes the equation

$$\sqrt{(x-x_0)^2 + (y-y_0)^2} = |y - y_1|,$$

which simplifies (see Exercise 57) to

$$y = ax^2 + bx + c, \qquad (5)$$

where,

$$a = \frac{-1}{2(y_1 - y_0)} \neq 0, \quad b = \frac{x_0}{y_1 - y_0}, \quad \text{and} \quad c = \frac{y_1^2 - y_0^2 - x_0^2}{2(y_1 - y_0)}. \quad (6)$$

Equation (5) is called the **standard equation of a parabola.**

Example 5 In each of the following cases, find the standard equation of the parabola with the given focus and directrix.

(a) $(3, 0), \quad L : y = -1$
(b) $(3, 0), \quad L : y = 1$

Solution:

(a) From the equation $d_1 = d_2$ [Figure 16(a)] we have

$$\sqrt{(x-3)^2 + y^2} = |y + 1|.$$

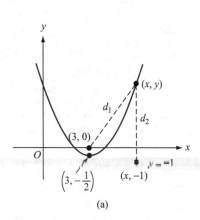

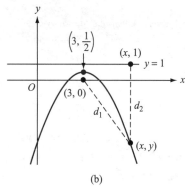

Figure 16

After squaring both sides and simplifying, we obtain

$$y = \frac{1}{2}x^2 - 3x + 4$$

as the equation of the parabola.

(b) Here, from $d_1 = d_2$ [Figure 16(b)] we have

$$\sqrt{(x-3)^2 + y^2} = |y - 1|.$$

If we again square both sides, we get

$$y = -\frac{1}{2}x^2 + 3x - 4$$

as the equation of the parabola. ∎

If $a > 0$ in equation (5), then the parabola opens upward as in Example 5(a); if $a < 0$, it opens downward as in Example 5(b). Either the low point $(3, -1/2)$ in Example 5(a) or the high point $(3, 1/2)$ in Example 5(b) is called the **vertex** of the parabola. In general, the vertex occurs at

$$\left(x_0, \frac{y_0 + y_1}{2}\right),$$

which is the midpoint of the vertical line segment from the focus to the directrix [see Exercise 60(a)].

The part of the parabola to the right of the vertical line $x = x_0$ is a reflection of the part to the left. This symmetry property and the location of the vertex are immediate consequences of the condition $d_1 = d_2$ [see Exercise 60(b)]. From the first two of equations (6), we get $x_0 = -b/2a$. Therefore, the **axis of symmetry** of the parabola (5) is the vertical line

$$x = -\frac{b}{2a}. \tag{7}$$

The vertex lies on the axis of symmetry. Hence, we can obtain the y-coordinate of the vertex by substituting $x = -b/2a$ in the equation $y = ax^2 + bx + c$.

Example 6 Find the axis of symmetry and the vertex for the parabola $y = x^2 - 4x + 3$. Also, graph the parabola.

Solution: Since $a = 1$ and $b = -4$, the axis of symmetry is the vertical line $x = -b/2 = 2$. Also, when $x = 2$, $y = 2^2 - 4 \cdot 2 + 3 = -1$. Therefore, the vertex is the point $(2, -1)$. To obtain other points on the graph, we select a few values of x on each side of the vertex and compute the corresponding values of y from the equation of the parabola. We then draw a smooth curve through the determined points (Figure 17). ∎

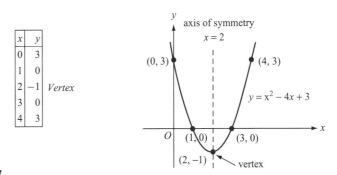

Figure 17

We point out that *although the focus and directrix appear in the definition of a parabola, the two most important items for graphing a parabola are the vertex and the axis of symmetry*. Also, it is not necessary to memorize $x = -b/2a$ for the axis of symmetry of the parabola $y = ax^2 + bx + c$. For instance, in Example 6, the parabola crosses the x-axis at 1 and 3, which are the roots of the quadratic equation $x^2 - 4x + 3 = 0$, and the axis of symmetry $x = 2$ is the vertical line through the *midpoint* of the line segment joining the two roots. For a general parabola $y = ax^2 + bx + c$, if a_1 and a_2 are the roots of $ax^2 + bx + c = 0$, then the **axis of symmetry** is the vertical line

$$x = \frac{a_1 + a_2}{2}, \tag{7'}$$

which reduces to $x = a_1$ if $a_1 = a_2$. [See Exercise 59, which shows that the axis of symmetry is given by (7'), even if a_1 and a_2 are complex numbers.]

Example 7 Graph each of the following parabolas, showing the axis of symmetry, vertex, and, if possible, the points where the graph intersects the x-axis.

(a) $y = x^2 + 2x - 3$
(b) $y = -4x^2 + 20x - 25$
(c) $y = \frac{1}{2}x^2 - 3x + 5$

Solution:

(a) The roots of $x^2 + 2x - 3 = 0$ are $a_1 = -3$ and $a_2 = 1$. Therefore, the axis of symmetry is the vertical line

$$x = \frac{-3 + 1}{2} = -1.$$

When $x = -1$, $y = (-1)^2 + 2(-1) - 3 = -4$. Hence, the vertex is $(-1, -4)$. The points where the graph crosses the x-axis are $(-3, 0)$ and $(1, 0)$ [Figure 18(a)].

(b) The equation $-4x^2 + 20x - 25 = 0$ has 5/2 as a double root. Setting $a_1 = 5/2 = a_2$, and proceeding as in part (a), we see that the axis of symmetry is $x = 5/2$ and the vertex is $(5/2, 0)$. Some other points on the graph are $(1, -9)$, $(2, -1)$, $(3, -1)$, and $(4, -9)$ [Figure 18(b)].

(c) The equation $(1/2)x^2 - 3x + 5 = 0$ has the complex roots $a_1 = 3 + i$ and $a_2 = 3 - i$. Hence, the axis of symmetry is

$$x = \frac{a_1 + a_2}{2} = 3.$$

Also, if $x = 3$, then $y = 1/2$, so the vertex is $(3, 1/2)$. The graph does not cross the x-axis (*why?*), but some other points on the graph are $(0, 5)$, $(2, 1)$, $(4, 1)$, and $(6, 5)$ [Figure 18(c)] ∎

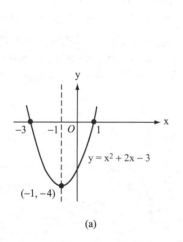
(a)

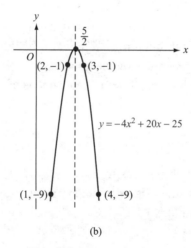
(b)

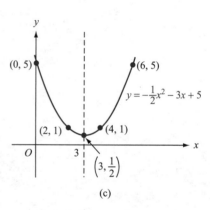
(c)

Figure 18

A third way to determine the vertex and axis of symmetry of a parabola is by **completing the square.** For instance, in Example 7(c), the axis $x = 3$ may be obtained by completing the square in x as follows.

$$y = \frac{1}{2}x^2 - 3x + 5 \qquad \textit{Given.}$$
$$2y = x^2 - 6x + 10 \qquad \textit{Multiply both sides by 2.}$$
$$2y = x^2 - 6x + 9 + 1 \qquad \textit{Complete the square in x.}$$
$$2y = (x - 3)^2 + 1$$
$$y = \frac{1}{2}(x - 3)^2 + \frac{1}{2} \qquad \textit{Divide both sides by 2.}$$

Since $(1/2)(x-3)^2 \geq 0$ for all values of x, it follows that $y \geq 1/2$ for all values of x. That is, the smallest value of y is $1/2$, and it occurs when $x = 3$. Hence, the lowest point or vertex of the graph is $(3, 1/2)$ and the axis of symmetry is $x = 3$.

If we interchange x and y in equation (5), we obtain

$$x = ay^2 + by + c,$$

whose graph is also a **parabola** (Figure 19) but with the roles of x and y interchanged. That is, the axis of symmetry is the *horizontal line* $y = -b/(2a) = (a_1 + a_2)/2$, where a_1 and a_2 are the roots of $ay^2 + by + c = 0$. Also, the vertex is the *leftmost point* [Figure 19(a)] or the *rightmost point* [Figure 19(b)] of the graph. The graph opens to the right if $a > 0$ [Figure 19(a)] and to the left if $a < 0$ [Figure 19(b)].

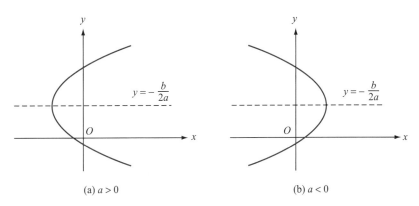

Figure 19 Parabola $x = y^2 + by + c$

Example 8 Find the vertex and axis of symmetry, and sketch the graph of the parabola $x = y^2 + 2y - 3$. [Compare with Example 7(a).]

Solution: The equation $y^2 + 2y - 3 = 0$ has roots $a_1 = -3$ and $a_2 = 1$. Therefore, the axis of symmetry is the horizontal line $y = (-3+1)/2 = -1$. Also, when $y = -1$, $x = (-1)^2 + 2(-1) - 3 = -4$. Hence, the vertex is $(-4, -1)$. We can determine several other points on the curve by assigning values to y and computing the corresponding values of x. For instance, when $y = -3, -2, 0, 1$, and 2, we get $x = 0, -3, -3, 0$, and 5, respectively. Hence, the points $(0, -3), (-3, -2), (-3, 0), (0, 1)$, and $(5, 2)$ are also on the graph (Figure 20) ∎

In general, the second-degree equation (2) represents a parabola when either $B = 0$, $A \neq 0$, and $D \neq 0$ (vertical axis of symmetry) or $A = 0$, $B \neq 0$, and $C \neq 0$ (horizontal axis of symmetry).

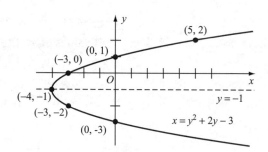

Figure 20

Question *If either $A = 0$ or $B = 0$ (but not both) in the second-degree equation (2), what graphs other than a parabola are possible?*

Answer: (We consider the case $B = 0$, $A \neq 0$, and $D = 0$. The other cases are similar.) Suppose $B = 0$, $A \neq 0$, and $D = 0$. Then equation (2) becomes $Ax^2 + Cx = E$ or $Ax^2 + Cx - E = 0$, which is a quadratic equation in x. Depending on the sign of the discriminant $C^2 + 4AE$, the quadratic equation could have (a) two distinct real roots a_1 and a_2, (b) one distinct real root a_1 or (c) two complex conjugate roots. In case (a) the graph consists of two vertical lines $x = a_1$ and $x = a_2$; in case (b) the graph is one vertical line $x = a_1$; and in case (c), the graph is the empty set. ■

Comment In graphing a parabola $y = ax^2 + bx + c$, as in Examples 6 and 7, for each value substituted for x we obtain exactly one corresponding value of y. Therefore, as discussed in Section 1.1, y is called a **function of x** (see Chapter 4 for a further discussion of functions). In particular, $y = ax^2 + bx + c$ $(a \neq 0)$ defines a **quadratic function of x** or a polynomial function of degree 2.

Applications. Parabolas appear in nature, science, and engineering. When a baseball or tennis ball is hit, a basketball is shot, or a football is kicked, the resulting path is a parabola. Optical, sonic, and electronic reflectors use parabolas because rays parallel to the axis of symmetry are reflected through the focus. Automobile headlights and some telescopes also use this principle. In the following examples, the vertex of a parabola is used to find the maximum or minimum value of a quadratic function.

Example 9 A paper airplane, in one of its dips, flies in a parabolic path whose equation is $y = x^2 - 4x + 5$, where x and y are measured in feet. What is the lowest point of the dip?

Solution: Since the coefficient $a = 1$ of x^2 is positive, the parabola opens upward, and the lowest point is at the vertex. The vertex is located at $x = -b/2a = -(-4/2) = 2$, and when $x = 2$ feet, $y = 2^2 - 4 \cdot 2 + 5 = 1$ foot (Figure 21). Hence, the lowest point of the dip is 1 foot above the ground. ∎

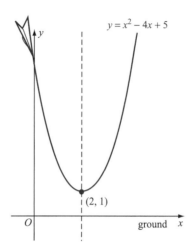

Figure 21

Example 10 A wholesaler sells dresses to a retail store for $150 each. If more than 100 are ordered, the price per dress is reduced by $1 for each dress over 100 ordered. What size order will give the largest amount of money to the wholesaler?

Solution: Let x be the number over 100 of dresses ordered. Then $100 + x$ dresses are ordered and the price per dress is $150 - x$ dollars. If y is the amount of money paid to the wholesaler, then

$$y = \text{(number of dresses)} \cdot \text{(price per dress)}$$
$$= (100 + x) \cdot (150 - x)$$
$$= -x^2 + 50x + 15{,}000.$$

Since $a = -1$ is negative, the graph is a parabola opening downward, and its highest point is its vertex. The roots of $-x^2 + 50x + 15{,}000 = 0$, as seen from the factored form, are $a_1 = -100$ and $a_2 = 150$. Hence, the axis of symmetry is $x = (a_1 + a_2)/2 = 25$, and when $x = 25$, $y = 15{,}625$. Therefore, an order for 125 dresses will give wholesaler $15,625$, the largest amount possible. ∎

Exercises 3.6

Fill in the blanks to make each statement true.

1. If the graph of $Ax^2 + By^2 + Cx + Dy = E$ is a circle, then _____ = _____.
2. The center and radius of a circle $x^2 + y^2 + Cx + Dy = E$ are usually found by the process of _____.
3. If the graph of $Ax^2 + By^2 + Cx + Dy = E$ is a parabola, then either _____ = 0 or _____ = 0.
4. If $y = ax^2 + bx + c$ is a parabola and $ax^2 + bx + c = 0$ has roots a_1 and a_1, then the line $x = (a_1 + a_2)/2$ is the _____ for the parabola.
5. The graph of $y = ax^2 + bx + c$ is a parabola that opens upward if _____ and downward if _____.

Write true or false for each statement.

6. The graph of $Ax^2 + By^2 + Cx + Dy = E$ can be a single point.
7. If $A = B \neq 0$, then the graph of $Ax^2 + By^2 + Cx + Dy = E$ is a circle.
8. If $b^2 - 4ac < 0$, then the parabola $y = ax^2 + bx + c$ does not cross the x-axis.
9. The parabola $y = ax^2 + bx + c$ has a horizontal directrix.
10. The parabola $x = ay^2 + by + c$ has a horizontal axis of symmetry.

Circles

Find the standard equation of the circle in each of the following cases.

11. center: (2, 1); radius: 3
12. center: $(-1, 3)$; radius: 5
13. center: $(2, -2)$; passes through the point (0, 0)
14. center: $(-1, -1)$; passes through the point (1, 1)
15. center: $(-1, 0)$; tangent to the y-axis
16. center: (0, 2); tangent to the x-axis
17. center: (0, 0); tangent to the line $x + y = 1$
18. center: (2, 1); tangent to the line $y = 4$
19. the circle having the line segment from $(-2, -1)$ to (6, 3) as a diameter
20. he circle having the diagonals of the square with vertices (0, 0), (3, 0), (3, 3), and (0, 3) as diameters

By completing the square in x and y, determine the nature of the graph (circle, point, or empty set) of each of the following second-degree equations. Sketch the graph, if possible.

21. $x^2 + y^2 - 4x + 4y = 1$
22. $x^2 + y^2 + 6x - 2y = 15$
23. $4x^2 + 4y^2 + 12x - 4y = -1$
24. $2x^2 + 2y^2 - 8x + 8y = -7$
25. $x^2 + y^2 + 6x - 2y = -10$
26. $x^2 + y^2 - x - y = -2$
27. $x^2 + y^2 + x - 4y = -5$
28. $2x^2 + 2y^2 - 4x - 2y = -3$
29. $3x^2 + 3y^2 + 3x + 6y = -4$
30. $3x^2 + 3y^2 + 3x + 6y = 21/4$

Parabolas

In Exercises 31-34, find the equation of the parabola with the given focus and directrix by setting $d_1 = d_2$.

31. focus: (0, 1); directrix: $y = 2$
32. focus: (0, 2); directrix: $y = 0$
33. focus: (2, 1); directrix: $y = 3$
34. focus: $(-1, 0)$; directrix: $y = 1$

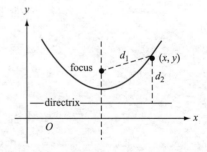

Find the vertex and axis of symmetry for each of the parabolas in Exercises 35-44. Graph each parabola.

35. $y = x^2$
36. $y = -\dfrac{1}{2}x^2$

37. $y = x^2 + 2$
38. $y = x^2 - 2$
39. $y = x^2 + 6x + 10$
40. $y = x^2 - 6x + 5$
41. $y = x^2 - 4x$
42. $y = -\frac{1}{2}x^2 + x - 1$
43. $2y = x^2 + 4x + 2$
44. $12y = x^2 - x - 1$

Find the vertex and axis of symmetry for each of the parabolas in Exercises 45-50. Graph each parabola.

45. $x = \frac{1}{2}y^2$
46. $x = \frac{1}{2}y^2 + 1$
47. $x = y^2 - 6y$
48. $x = y^2 + y$
49. $2x = y^2 - 3y - 4$
50. $2x = y^2 + 4y + 2$

Applications

51. A rectangular parking lot with a straight road as one side is to be fenced on the other three sides by 1000 ft of aluminum fencing (see the figure). If the area of the lot is to be maximized, what should be its length and width? (*Hint:* write the area of the lot as $ax^2 + bx + c$.)

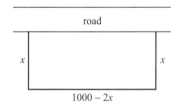

52. Find the dimensions and the area of the parking lot in Exercise 51 if the fourth side along the road is also to be fenced except for an entrance gate 10 ft wide.

53. A farmer has a crop of 600 bushels that he can sell now at $1 per bushel. He estimates that for the next few weeks, his crop will increase by 100 bushels each week but the price per bushel will decrease 10 cents each week. How many weeks should he wait in order to make the most money? Assume that the entire crop must be harvested and sold at one time.

54. A club charters a bus containing 50 seats for an outing. The bus company charges $30 for each passenger if the bus is filled, but increases the price by $1 per passenger for each empty seat. How many empty seats will produce the largest income for the bus company?

55. A toy rocket launched from origin O has a parabolic path as indicated in the figure. At time t sec from launch to impact, the coordinates (x, y) of the rocket satisfy the equations

$$x = 64t \quad \text{and} \quad y = -16t^2 + 128t.$$

(a) At what time t_1 will the rocket land?
(b) How far will the rocket travel horizontally by the time of impact?
(c) What is the maximum height of the rocket?
(d) Find the equation of the parabolic path by expressing y in terms of x.

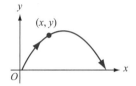

Miscellaneous

56. Each of the following is the graph of a parabola $y = ax^2 + bx + c$. Determine whether the discriminant is positive, negative, or zero. Also determine the sign of the coefficient a.

(a)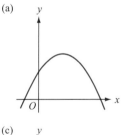

(b)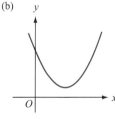

(c)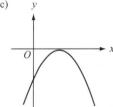

(d)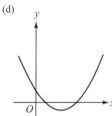

57. By squaring both sides and simplifying, show that the equation $\sqrt{(x - x_0)^2 + (y - y_0)^2} = |y - y_1|$ is equivalent to $y = ax^2 + bx + c$, where

$$a = \frac{-1}{2(y_1 - y_0)}, \quad b = \frac{x_0}{y_1 - y_0},$$

and

$$c = \frac{y_1^2 - y_0^2 - x_0^2}{2(y_1 - y_0)}.$$

58. Use the results in Exercise 57 to show that

$$x_0 = -\frac{b}{2a}, \qquad y_0 = \frac{4ac - b^2 + 1}{4a},$$

and

$$y_1 = \frac{4ac - b^2 - 1}{4a}.$$

59. A given parabola has the equation $y = ax^2 + bx + c$. Complete the square in x to show that the axis of symmetry is the vertical line $x = -b/(2a)$. Also, use the quadratic formula to show that if the roots of $ax^2 + bx + c = 0$ are a_1 and a_2 (real or complex), then

$$\frac{a_1 + a_2}{2} = -\frac{b}{2a},$$

and therefore, the line $x = (a_1 + a_2)/2$ is the axis of symmetry.

60. A given parabola has focus (x_0, y_0) and directrix $L: y = y_1$ (see the figure in Exercises 31-34).

(a) Use the property $d_1 = d_2$ to show that $(x_0, [y_0 + y_1]/2)$ is the point on the parabola that is closest to the directrix.

(b) Use the property $d_1 = d_2$ to show that if $(x_0 + h, y)$ is on the parabola, then so is $(x_0 - h, y)$ for any real number h.

3.7 Translation of Axes and Intersections

To simplify a second-degree equation, complete the square in x and / or y, and then make the corresponding substitutions.

Translation of Axes
Points of Intersection
Regions of Intersection

Translation of Axes. The process of completing the square used in Example 4 of Section 3.6 to put the equation of a circle in standard form has a geometric application that is useful for many graphs. if we start with the equation of a circle

$$(x - x_0)^2 + (y - y_0)^2 = r^2, \tag{1}$$

and make the **substitution**

$$x' = x - x_0 \quad \text{and} \quad y' = y - y_0, \tag{2}$$

we then obtain the equation

$$x'^2 + y'^2 = r^2. \tag{3}$$

Now equation (3) is the equation of a circle with center at the origin, provided we consider new $x'y'$-axes with the origin located at (x_0, y_0) (Figure 22).

As indicated in Figure 22, equations (1) and (3) represent the same circle, but the coordinates have changed from xy-coordinates to $x'y'$-coordinates. Geometrically, we can imagine the $x'y'$-axes as being obtained by moving the xy-axes parallel to themselves so that the xy-origin

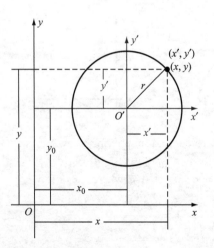

Figure 22 Translation of Axes

3.7. Translation of Axes and Intersections

$(0, 0)$ moves to the point (x_0, y_0). In this sense, the $x'y'$-axes are called a **translation** of the xy-axes. Every point in the plane has both xy-coordinates and $x'y'$-coordinates, which are related by the **substitution equations** (2). Similarly, every curve that has an equation in the variables x and y also has an equation in the variables x' and y' and conversely.

Example 1 The $x'y'$-axes are a translation of the xy-axes in which the point $(x_0, y_0) = (6, 3)$ becomes the $x'y'$-origin O'. Find the $x'y'$-coordinates of the $P : (x, y) = (8, 7)$, and find the xy-coordinates of $Q : (x', y') = (-5, 2)$.

Solution: With reference to Figure 23, the substitution equations (2) become

$$x' = x - 6 \quad \text{and} \quad y' = y - 3.$$

Hence, the $x'y'$-coordinates of P are

$$x' = 8 - 6 = 2 \quad \text{and} \quad y' = 7 - 3 = 4.$$

The xy-coordinates of Q must satisfy

$$-5 = x - 6 \quad \text{and} \quad 2 = y - 3.$$

Hence, $x = 1, y = 5$. ∎

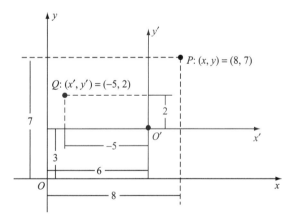

Figure 23

Example 2 A given line L has equation $y = 2x + 1$. Translate the xy-axes to $x'y'$-axes in such a way that the point $(1, 3)$ on L becomes the $x'y'$-origin. Find the equation of L in the variables x' and y', and graph L in both coordinate systems.

Solution: The substitution equations (2) are

$$x' = x - 1 \quad \text{and} \quad y' = y - 3$$
$$\text{or} \quad x = x' + 1 \quad \text{and} \quad y = y' + 3.$$

Hence, by substituting for x and y in $y = 2x + 1$, the equation of L in the $x'y'$-coordinates is

$$y' + 3 = 2(x' + 1) + 1$$
$$\text{or} \quad y' = 2x'.$$

The graph of L in both coordinate systems is shown in Figure 24. ∎

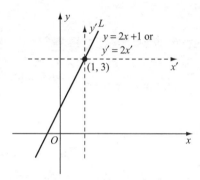

Figure 24

Example 3 A given parabola has equation $y = 2x^2 + 4x + 5$. Transform the xy-axes to $x'y'$-axes in such a way that the vertex of the parabola becomes the $x'y'$-origin. Find the equation of the parabola in terms of x' and y', and graph the parabola in both coordinate systems.

Solution: The axis of symmetry of a parabola $y = ax^2 + bx + c$ is the vertical line $x = -b/2a$. Here the axis is $x = -4/4 = -1$, and when $x = -1$, we have $y = 3$. Hence, the vertex is $(-1, 3)$, and the substitution equations (2) become

$$x' = x + 1 \quad \text{and} \quad y' = y - 3$$
$$\text{or} \quad x = x' - 1 \quad \text{and} \quad y = y' + 3.$$

The equation of the parabola in the $x'y'$-coordinates is

$$y' + 3 = 2(x' - 1)^2 + 4(x' - 1) + 5$$
$$= 2x'^2 - 4x' + 2 + 4x' - 4 + 5$$
$$\text{or} \quad y' = 2x'^2.$$

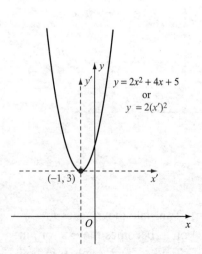

Figure 25

The graph of the parabola in both coordinate systems is given in Figure 25. Note that the simplified equation $y' = 2x'^2$ could also be obtained by completing the square in x in the given equation $y = 2x^2 + 4x + 5$ to

get $y - 3 = 2(x+1)^2$ and then substituting y' for $y - 3$ and x' for $x + 1$.
∎

In Examples 2 and 3, a translation of axes was performed in order to obtain a simpler equation of the given curve. The curve remained fixed, but the axes and variables changed. We can also obtain a simpler equation by keeping the axes fixed and moving the curve to a more convenient position. For example, if we start with the circle

$$(x - x_0)^2 + (y - y_0)^2 = r^2 \tag{1'}$$

and make the **replacements**

$$x - x_0 \text{ by } x \quad \text{and} \quad y - y_0 \text{ by } y, \tag{2'}$$

we obtain the equation

$$x^2 + y^2 = r^2. \tag{3'}$$

Equation (3′) is the equation of a circle with center (0, 0) and radius r. That is, circle (3′) is **congruent** to circle (1′). Here the axes and variables remain the same, but the circle is replaced by a congruent one with a simpler equation (Figure 26).

We call circle (3′) a **translation** of circle (1′). These two procedures, translation of axes by the substitution equations (2) and translation of curves by the replacements (2′), are equivalent ways to achieve more convenient equations. For a second-degree equation, values of x_0 and y_0 can be obtained by completing the square in x, y, or both, as noted in Example 3.

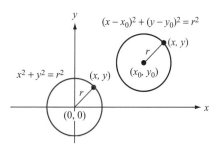

Figure 26 Translation of a Curve

Points of Intersection. In the previous chapter we discussed the intersection of two straight lines represented by linear equations. We now consider intersections involving lines, circles, and parabolas.

Example 4 Find all points of intersection of the line $x + y = 3$ and the circle $(x - 1)^2 + y^2 = 4$.

Solution: The intersections occur at those points $P(x, y)$ whose coordinates satisfy *both* equation $x + y = 3$ and equation $(x - 1)^2 + y^2 = 4$. We solve the first of these for y in terms of x, obtaining $y = 3 - x$, and then substitute this expression for y in the second equation. We get

$$(x - 1)^2 + (3 - x)^2 = 4$$
$$x^2 - 2x + 1 + 9 - 6x + x^2 = 4$$
$$2x^2 - 8x + 6 = 0$$
$$x^2 - 4x + 3 = 0$$
$$x = 1 \quad \text{or} \quad x = 3.$$

When $x = 1$, we have $y = 3 - x = 2$, and when $x = 3$, we get $y = 0$. Hence, the points of intersection are $(1, 2)$ and $(3, 0)$ (Figure 27). ∎

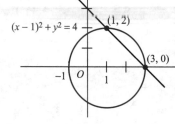

Figure 27

Example 5 Find all points of intersection of the parabolas $y = x^2 - 4$ and $y = -x^2 + 3x - 2$.

Solution: Here we equate the y values of the curves, obtaining

$$x^2 - 4 = -x^2 + 3x - 2$$
$$2x^2 - 3x - 2 = 0$$
$$(2x + 1)(x - 2) = 0$$
$$x = -1/2 \quad \text{or} \quad x = 2.$$

When $x = -1/2$, we obtain $y = x^2 - 4 = -15/4$, and when $x = 2$, we get $y = 0$. Hence, the curves intersect in the points $(-1/2, -15/4)$ and $(2, 0)$, as shown in Figure 28. ∎

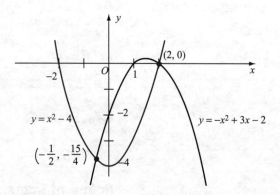

Figure 28

Comment Geometrically, it is evident that a straight line can intersect a circle or parabola in at most two points. Also, two parabolas that are both of the form $y = ax^2 + bx + c$ or both of the form $x = ay^2 + by + c$ can intersect in at most two points. Hence, it is not surprising that a quadratic equation results when the points of intersection are determined algebraically, as in Examples 4 and 5. We can also see geometrically that two circles with different centers can intersect in at most two points. In this case, the algebraic solution involves working with radicals, but eventually a quadratic equation results (see Exercises 49 and 50). On the other hand, a circle and a parabola can intersect in up to four points, as can a parabola of the type $y = ax^2 + bx + c$ with one of the type $x = ay^2 + by + c$. In these cases, an algebraic process to determine the points of intersection can result in a fourth-degree equation. For example, the system

$$x^2 + y^2 = 5$$
$$y = 3x^2 + 3x - 6$$

results in the fourth-degree equation

$$9x^4 + 18x^3 - 26x^2 - 36x + 31 = 0.$$

At present, we can solve only very special cases of fourth-degree equations, for example, those that are actually quadratics in x^2. Hence we cannot yet determine the points of intersection of arbitrary circles and parabolas.

Regions of Intersection. In Example 4, we saw that the line L with equation $x + y = 3$ intersects the circle C with equation $(x-1)^2 + y^2 = 4$ in the two points (1, 2) and (3, 0). The points *above* L satisfy $y > 3 - x$ and those *below* L satisfy $y < 3 - x$ (*why?*). Similarly, the points *outside* C satisfy $(x-1)^2 + y^2 > 4$ and those *inside* C satisfy $(x-1)^2 + y^2 < 4$. Hence, the system of inequalities

$$y > 3 - x$$
$$(x-1)^2 + y^2 < 4$$

is satisfied by all points in the region above L and inside C [Figure 29(a)], and the system

$$y < 3 - x$$
$$(x-1)^2 + y^2 < 4$$

is satisfied by all points in the region below L and inside C [Figure 29(b)].

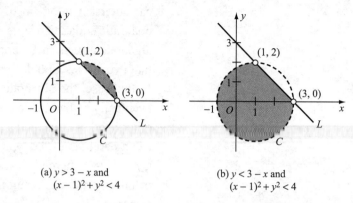

(a) $y > 3 - x$ and $(x-1)^2 + y^2 < 4$

(b) $y < 3 - x$ and $(x-1)^2 + y^2 < 4$

Figure 29

Example 6 Graph the region satisfied by the system of inequalities

$$y \geq x^2 - 4$$
$$y \leq -x^2 + 3x - 2.$$

Solution: As illustrated in Figure 28 in Example 5, parabolas, P_1 with equation $y = x^2 - 4$ and P_2 with equation $y = -x^2 + 3x - 2$ intersect in the two points $(2, 0)$ and $(-1/2, -15/4)$. The points satisfying $y \geq x^2 - 4$ lie on or above P_1, and those satisfying $y \leq -x^2 + 3x - 2$ lie on or below P_2. Hence, the points satisfying *both* of the given inequalities lie in the shaded region of Figure 30. ∎

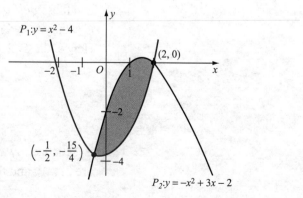

Figure 30 $y \geq x^2 - 4$ and $y \leq -x^2 + 3x - 2$

Exercises 3.7

Fill in the blanks to make each statement true.

1. If the xy-axes are translated to $x'y'$-axes so that the $x'y'$-origin is at the point (x_0, y_0), then the xy-coordinates of a point P are related to the $x'y'$-coordinates of P by the equations _____.

2. If (x_0, y_0) is the origin of the $x'y'$-coordinate system, then the equation of the circle $(x - x_0)^2 + (y - y_0)^2 = r^2$ in $x'y'$-coordinates is _____.

3. If the circle $(x - x_0)^2 + (y - y_0)^2 = r^2$ is translated so that its center moves to the point (x_0, y_0), then the equation of the translated circle is _____.

4. A second-degree equation sometimes can be simplified by the process of _____.

5. A parabola can intersect a circle in at most _____ points.

Write true or false for each statement.

6. If the xy-axes are translated to $x'y'$-axes so that the linear equation $y = mx + b$ becomes $y' = m'x' + b'$, then $m' = m$.

7. If the xy-axes are translated to $x'y'$-axes so that the equation $(x - x_0)^2 + (y - y_0)^2 = r^2$ becomes $(x')^2 + (y')^2 = (r')^2$, then $r' = r$.

8. It is possible to translate axes so that the parabola $y = ax^2 + bx + c$ has equation $y' = a(x')^2$ in the new coordinates.

9. It is possible to translate axes so that the line $y = x$ has equation $y' = -x'$.

10. It is possible to translate axes so that the circle $(x - 1)^2 + (y - 2)^2 = 25$ has the equation $(x' - 2)^2 + (y' - 1)^2 = 25$.

Translation of Axes

11. The $x'y'$-axes are a translation of the xy-axes in which the xy-point $(2, 5)$ becomes the $x'y'$-origin. Find the $x'y'$-coordinates of the following points.

    (a) $P: (x, y) = (3, -2)$
    (b) $Q: (x, y) = (0, 0)$
    (c) $R: (x, y) = (2, 5)$

12. For the translation of Exercise 11, find the xy-coordinates of each of the following points.

    (a) $P: (x', y') = (4, 3)$
    (b) $Q: (x'y') = (0, 0)$
    (b) $R: (x', y') = (2, 5)$

Find the substitution equations in each of the following cases.

13. The xy-origin is the $x'y'$-point $(2, -4)$.
14. The xy-point $(5, 1)$ is the $x'y'$-point $(1, 5)$.
15. The $x'y'$-point $(3, 0)$ is the xy-point $(2, 2)$.
16. The $x'y'$-point $(-1, 2)$ is the xy-point $(0, 0)$.
17. The $x'y'$-origin is the center of the circle $x^2 + y^2 + 2x - 4y = 5$.
18. The $x'y'$-origin is the vertex of the parabola $y = 2x^2 + 4x - 3$.
19. The xy-origin is the y'-intercept of the line $2x' + 3y' = 6$.
20. The xy-origin is the x'-intercept of the line $4x' - 5y' = 12$.

If the $x'y'$-axes are a translation of the xy-axes with the point (x_0, y_0) as the $x'y'$-origin, find the $x'y'$-equation of each of the following curves. Sketch a graph showing both coordinate systems.

21. $2x + 3y = 7$; $(x_0, y_0) = (4, -3)$
22. $y - 2 = 3(x + 5)$; $(x_0, y_0) = (-5, 2)$
23. $x^2 + y^2 = 5$; $(x_0, y_0) = (2, -4)$
24. $x^2 + y^2 + 2x = 4$; $(x_0, y_0) = (-1, 0)$
25. $y = x^2 - 10x + 5$; $(x_0, y_0) = (5, -15)$
26. $y = 3x^2 + 4x - 2$; $(x_0, y_0) = (-2, 1)$

Translate each of the following curves as indicated. Sketch both curves in xy-coordinates.

27. Translate the circle $(x - 3)^2 + (y + 2)^2 = 25$ so that its center is at the origin.

234   Chapter 3. Algebra and Graphs of Quadratic Expressions

28. Translate the circle $x^2 + 8x + y^2 - 8y = 4$ so that its center is at the point (1, 2).

29. Translate the parabola $y = x^2 + 6x + 5$ so that its vertex is at the point (0, 1).

30. Translate the parabola $x = y^2 - 2y + 2$ so that its vertex is at the origin.

Simplify the equation of each of the following curves by completing the square in x and/or y and making the corresponding $x'y'$-substitutions.

31. $y = x^2 + 3x + 1$
32. $y = 4x^2 - 2x + 8$
33. $x = 2y^2 + 2y - 4$
34. $x^2 + y^2 + x + 5y = 0$
35. $2x^2 + 2y^2 + 6x + 8y = 5$
36. $3x^2 + 3y^2 + 2x + y = 10$

In Exercises 37-40, first substitute $x' + x_0$ for x and $y' + y_0$ for y and then choose values of x_0 and y_0 that simplify the equation. Give the simplified equation.

37. $y = 2x^2 + 3x - 5$
38. $x = y^2 - y + 4$
39. $4x^2 + 4y^2 - 2x + 4y = 5$
40. $x^2 + y^2 - 3x - 6y = 0$

Points of Intersection

Find all points of intersections of each of the following pairs of curves. Sketch a graph for each.

41. $y = 4 - x^2$
    $y = 1 - 2x$

42. $x + 2y - 3 = 0$
    $x - y^2 = 0$

43. $5x - 3y = 0$
    $x^2 + y^2 = 34$

44. $x^2 + y^2 = 25$
    $y = x + 1$

45. $y = x^2 - 1$
    $y = -x^2 + 3$

46. $y = x^2 + 3x + 2$
    $y = -x^2 - x + 2$

47. $x = 4 - (y - 2)^2$
    $x = y^2 - 4y + 6$

48. $x = y^2 - 4$
    $2x = y^2 - 4$

49. $x^2 + y^2 = 4$
    $(x - 1)^2 + (y + 1)^2 = 6$

50. $(x - 2)^2 + y^2 = 5$
    $x^2 + (y - 1)^2 = 10$

Regions of Intersection

Use the results of Exercises 41-46 to graph the regions determined by each of the following systems of inequalities.

51. $y < 4 - x^2$
    $y > 1 - 2x$

52. $x < 3 - 2y$
    $x > y^2$

53. $3y \le 5x$
    $x^2 + y^2 \le 34$

54. $x^2 + y^2 \le 25$
    $y \le x + 1$

55. $y \ge x^2 - 1$
    $y \le -x^2 + 3$

56. $y > x^2 + 3x + 2$
    $y < -x^2 - x + 2$

3.8 Ellipses and Hyperbolas

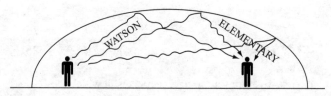

Words whispered at one focus of an elliptic "whispering chamber" can be heard loud and clear at the other focus.

Ellipses
Hyperbolas
Asymptotes

The ellipse and the hyperbola, like the parabola, appear in the physical universe, science, and engineering. Planets and satellites travel in ellip-

3.8. Ellipses and Hyperbolas

tical orbits. Elliptic and hyperbolic gears are used in various kinds of machinery. Hyperbolas are used in the design of telescopes because light from one focus reflects back along the line from the other focus. Electronic instruments on a ship locate its position at sea as the intersection of two hyperbolas.

Ellipses. Given two fixed points P_1 and P_2 in the plane and a positive number c, the set of all points P for which

$$d(P_1, P) + d(P_2, P) = c \tag{1}$$

is called an **ellipse**. Each of the points P_1 and P_2 is called a **focus**, and each of the distances $d_1 = d(P_1, P)$ and $d_2 = d(P_2, P)$ is called a **focal radius**. Figure 31 is an ellipse in which $P_1 = (c_1, 0)$ and $P_2 = (-c_1, 0)$. We note that the foci P_1 and P_2 and the positive quantity c are *fixed*, whereas the focal radii d_1 and d_2 can vary in value, but their *sum* is constant.

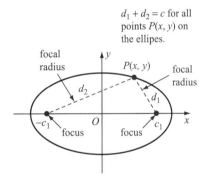

Figure 31 Ellipse

With reference to Figure 31, equation (1) becomes

$$\sqrt{(x - c_1)^2 + y^2} + \sqrt{(x + c_1)^2 + y^2} = c,$$

which simplifies (see Exercise 29) to

$$\frac{x^2}{a^2} + \frac{y^2}{b^2} = 1, \tag{2}$$

The x-intercepts of the ellipse are $\pm a$, and the y-intercepts are $\pm b$.

where $a = c/2$ and $b = \sqrt{a^2 - c_1^2}$ (Figure 32). Equation (2) is called the **standard equation of an ellipse**.

Example 1 Find the equation of the ellipse with foci $(2\sqrt{3}, 0)$ and $(-2\sqrt{3}, 0)$ and the sum of focal radii equal to 8. Also, graph the ellipse.

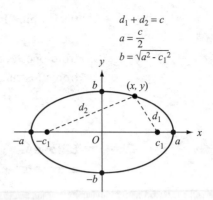

Figure 32 Ellipse $x^2/a^2 + y^2/b^2 = 1$

Solution: Here $c_1 = 2\sqrt{3}$ and $c = 8$. We could substitute $a = c/2 = 4$ and $b = \sqrt{a^2 - c_1^2} = 2$ into equation (2). However, in order to illustrate the way in which equation (2) follows from definition (1), we substitute the given data into (1) and simplify. That is, the condition $d_1 + d_2 = c$ becomes

$$\sqrt{(x - 2\sqrt{3})^2 + y^2} + \sqrt{(x + 2\sqrt{3})^2 + y^2} = 8,$$

which, in two squarings, simplifies to

$$x^2 + 4y^2 = 16.$$

Then, by dividing both sides by 16, we obtain

$$\frac{x^2}{16} + \frac{y^2}{4} = 1$$

for the equation of the ellipse. (Check this result with the equations in Figure 32). For graphing the ellipse, we can first plot the points where the graph crosses the coordinate axes, as shown in the table, and then draw a smooth curve through these points (Figure 33).

For more accuracy in the graph, several other points can be plotted. For instance, if $x = \pm 2$, then from the equation $x^2/16 + y^2/4 = 1$, we obtain $y = \pm\sqrt{3}$. Hence, the points $(2, \sqrt{3})$, $(2, -\sqrt{3})$, $(-2, \sqrt{3})$, and $(-2, -\sqrt{3})$ are also on the graph. ∎

For an ellipse [equation (2)] with its foci $(c_1, 0)$ and $(-c_1, 0)$ on the x-axis, we note that $a > b$ (*why?*). The line segment through the two foci from $-a$ to a along the x-axis is called the **major axis** of the ellipse, and the segment from $-b$ to b along the y-axis is called the **minor axis**.

If the foci are positioned at $(0, c_1)$ and $(0, -c_1)$ on the y-axis, then the equation of the ellipse remains the same as (2), but the roles of x and y

are interchanged, as are those of a and b. That is, $b > a$, the major axis is along the y-axis from $-b$ to b, and the minor axis is from $-a$ to a along the x-axis (Figure 34)

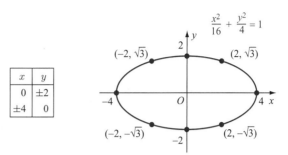

Figure 33

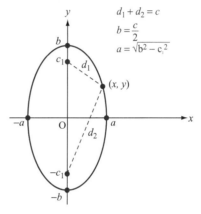

Figure 34 Ellipse $x^2/a^2 + y^2/b^2 = 1$ $(b > a)$

Example 2 Show that each of the following equations represents an ellipse. Graph each ellipse and identify its major and minor axis.

(a) $4x^2 + 5y^2 = 20$
(b) $25x^2 + 9y^2 = 16$

Solution:

(a) If both sides of the equation are divided by 20, we obtain

$$\frac{x^2}{5} + \frac{y^2}{4} = 1,$$

which is an ellipse with $a = \sqrt{5}$ and $b = 2$. Since $a > b$, the major axis is along the x-axis from $-\sqrt{5}$ to $\sqrt{5}$ [Figure 35(a)].

(b) By dividing both sides of the given equation by 16, we get

$$\frac{25x^2}{16} + \frac{9y^2}{16} = 1 \quad \text{or} \quad \frac{x^2}{\left(\frac{4}{5}\right)^2} + \frac{y^2}{\left(\frac{4}{3}\right)^2} = 1,$$

which is an ellipse with $a = 4/5$ and $b = 4/3$. Since $b > a$, the major axis is along the y-axis from $-4/3$ to $4/3$ [Figure 35(b)]. ∎

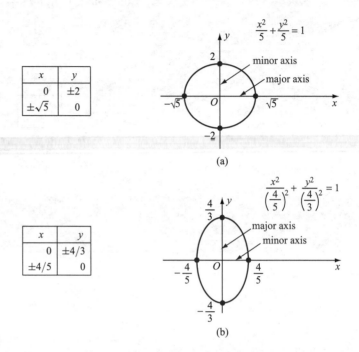

Figure 35

When an ellipse is given by equation (2), the origin $(0, 0)$ is called the **center** of the ellipse. We can translate the center to a point (x_0, y_0) by replacing x by $x - x_0$ and y by $y - y_0$ in (2). We then get

$$\frac{(x - x_0)^2}{a^2} + \frac{(y - y_0)^2}{b^2} = 1 \qquad (3)$$

as the equation of an **ellipse with center (x_0, y_0)** (Figure 36).

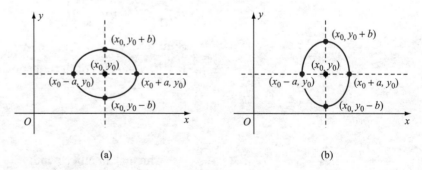

Figure 36 Ellipse $(x - x_0)^2/a^2 + (y - y_0)^2/b^2 = 1$

To determine whether a second-degree equation

$$Ax^2 + By^2 + Cx + Dy = E \qquad (4)$$

represents and ellipse, we complete the square in x and in y, as illustrated in the following example.

Example 3 Show that the equation $4x^2 + 9y^2 - 16x - 54y = -61$ represents an ellipse and sketch its graph.

Solution: As indicated above, we complete the square in x and in y.

$$4x^2 - 16x + 9y^2 - 54y = -16 \quad \text{\textit{Combine x terms and y terms.}}$$

$$4(x^2 - 4x) + 9(y^2 - 6y) = -61 \quad \text{\textit{Factor leading coefficient from x terms and y terms, and leave space for completing the square.}}$$

$$\begin{aligned}4(x^2 - 4x + 4) + 9(y^2 - 6y + 9) \\ = -61 + 4 \cdot 4 + 9 \cdot 9\end{aligned} \quad \text{\textit{Complete square in x and in y inside parentheses (Why } 4 \cdot 4 \text{ and } 9 \cdot 9 \text{ on right?).}}$$

$$4(x-2)^2 + 9(y-3)^2 = 36 \quad \text{\textit{Simplify.}}$$

$$\frac{(x-2)^2}{9} + \frac{(y-3)^2}{4} = 1 \quad \text{\textit{Divide both sides by 36.}}$$

The last equation is of type (3) with $x_0 = 2$, $y_0 = 3$, $a = 3$, and $b = 2$. Hence, the graph (Figure 37) is an ellipse with center $(2, 3)$ and with a horizontal major axis (*why horizontal?*). ∎

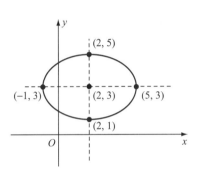

Figure 37

We saw in Section 3.6 that if a second-degree equation (4) represents a circle, then $A = B \neq 0$. Example 3 illustrates that if a second-degree equation represents an ellipse, then $A \neq B$ but A and B are either both positive or both negative ($AB > 0$). We also saw in Section 3.6 that the condition $A = B \neq 0$ does not *guarantee* a circle. Similarly, the conditions $A \neq B$ and $AB > 0$ do not guarantee an ellipse (see Exercises 23-28).

Hyperbolas. Given two points P_1 and P_2 in the plane and a positive number c, then the set of all points P for which

$$d(P_2, P) - d(P_1, P) = \pm c \qquad (5)$$

is called a **hyperbola**. A hyperbola consists of two branches, one for $+c$ and the other for $-c$. As in the case of an ellipse, points P_1 and P_2 are **foci**, and distances $d_1 = d(P_1, P)$ and $d_2 = d(P_2, P)$ are **focal radii**. Figure 38 is a hyperbola in which $P_1 = (c_1, 0)$ and $P_2 = (-c_1, 0)$. With reference to Figure 38, equation (5) becomes

$$\sqrt{(x+c_1)^2 + y^2} - \sqrt{(x-c_1)^2 + y^2} = \pm c,$$

which simplifies (see Exercise 41) to

$$\frac{x^2}{a^2} - \frac{y^2}{b^2} = 1, \qquad (6)$$

The x-intercepts are $\pm a$, there are no y-intercepts.

240   Chapter 3. Algebra and Graphs of Quadratic Expressions

where $a = c/2$ and $b = \sqrt{c_1^2 - a^2}$ (Figure 39). Equation (6) is called the **standard equation of a hyperbola**. Compare these equations with the corresponding ones for an ellipse in Figure 32.

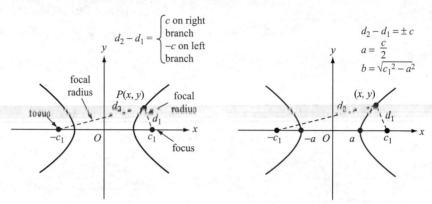

Figure 38 Hyperbola

Figure 39 Hyperbola $x^2/a^2 - y^2/b^2 = 1$

Example 4   Find the equation of the hyperbola with foci $(5, 0)$ and $(-5, 0)$ and difference of focal radii equal to ± 8. Also, graph the hyperbola.

Solution: From Figure 38, the condition $d_2 - d_1 = \pm 8$ becomes

$$\sqrt{(x+5)^2 + y^2} - \sqrt{(x-5)^2 + y^2} = \pm 8.$$

After two squarings to eliminate the radicals, the above equation simplifies to

$$9x^2 - 16y^2 = 144.$$

Then, by dividing both sides by 144, we obtain

$$\frac{x^2}{16} - \frac{y^2}{9} = 1$$

for the equation of the hyperbola. Here $a = 4$ and $b = 3$. (Check these values with the equations in Figure 39). To graph the hyperbola, we plot the points of intersection with the x-axis and several other convenient points, as indicated in the table, and draw smooth branches through these points (Figure 40). ∎

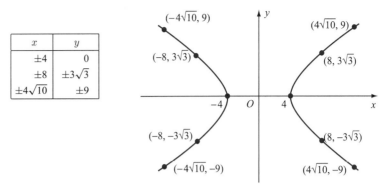

Figure 40 Hyperbola $x^2/16 - y^2/9 = 1$

If the foci of a hyperbola are positioned at $(0, c_1)$ and $(0, -c_1)$ on the y-axis, then the roles of x and y are interchanged, as are those of a and b. The equation becomes

$$\frac{y^2}{b^2} - \frac{x^2}{a^2} = 1, \tag{7}$$

where $b = c/2$ and $a = \sqrt{c_1^2 - b^2}$ (Figure 41).

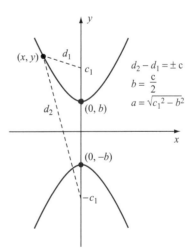

Figure 41 Hyperbola $y^2/b^2 - x^2/a^2 = 1$

When a hyperbola is given by equation (6) or (7), the origin (0, 0) is called the **center** of the hyperbola. We can translate the center to a point (x_0, y_0) by replacing x by $x - x_0$ and y by $y - y_0$ in (6) or (7). We then obtain

$$\frac{(x - x_0)^2}{a^2} - \frac{(y - y_0)^2}{b^2} = 1 \tag{8}$$

or

$$\frac{(y - y_0)^2}{b^2} - \frac{(x - x_0)^2}{a^2} = 1 \tag{9}$$

as the equation of a **hyperbola with center** (x_0, y_0) (Figure 42).

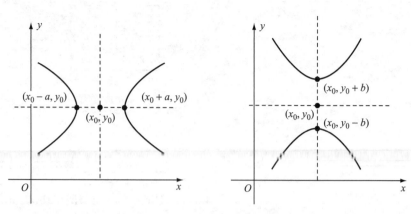

Figure 42 Hyperbola with Center (x_0, y_0)

As with the circle and ellipse, we can determine whether a second-degree equation represents a hyperbola by completing the square in x and in y, as illustrated in the following example.

Example 5   Show that the equation $3x^2 - 4y^2 - 6x - 16y = 1$ represents a hyperbola and sketch its graph.

Solution: We proceed as in Example 3.

$$3x^2 - 6x - 4y^2 - 16y = 1 \qquad \text{Combine x terms and y terms.}$$
$$3(x^2 - 2x) - 4(y^2 - 4y) = 1 \qquad \text{Factor out leading coefficients.}$$
$$3(x^2 - 2x + 1) - 4(y^2 + 4y + 4)$$
$$= 1 + 3 \cdot 1 - 4 \cdot 4 \qquad \text{Complete squares inside parentheses and balance equation.}$$
$$3(x-1)^2 - 4(y+2)^2 = -12 \qquad \text{Simplify.}$$
$$\frac{3(x-1)^2}{-12} - \frac{4(y+2)^2}{-12} = 1 \qquad \text{Divide both sides by } -12.$$
$$\frac{(y+2)^2}{3} - \frac{(x-1)^2}{4} = 1 \qquad \text{Simplify.}$$

The last equation is of type (9) with $y_0 = -2$, $x_0 = 1$, $b = \sqrt{3}$, and $a = 2$. Hence, the graph is a hyperbola with center at $(1, -2)$ (Figure 43). ∎

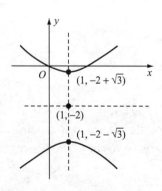

Figure 43 Hyperbola $3x^2 - 4y^2 - 6x - 16y = 1$

As illustrated in Example 5, if a second-degree equation (4) represents a hyperbola, then A and B have opposite signs ($AB < 0$). However, a second-degree equation could satisfy $AB < 0$ and have two lines for its graph (see Exercises 35-40).

Asymptotes.

As shown in Figure 44, the hyperbola

$$\frac{x^2}{a^2} - \frac{y^2}{b^2} = 1 \tag{10}$$

crosses the x-axis at $x = \pm a$. To obtain a geometric interpretation of the positive constant b, we first factor the left side of equation (10):

$$\left(\frac{x}{a} - \frac{y}{b}\right)\left(\frac{x}{a} + \frac{y}{b}\right) = 1.$$

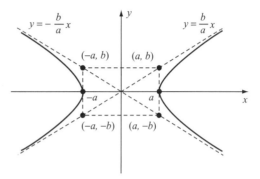

Figure 44 Asymptotes and Fundamental Rectangle

If x and y have the same sign (quadrants I and III) but get very large in magnitude, then the factor $x/a + y/b$ also gets very large in magnitude, and therefore, the factor $x/a - y/b$ must get very small in magnitude (since the product is 1 for all points on the hyperbola). That is,

$$\frac{x}{a} - \frac{y}{b} \to 0 \quad \text{or} \quad y \to \frac{b}{a}x.$$

Hence, the hyperbola approaches the line

$$y = \frac{b}{a}x \tag{11}$$

in quadrants I and III, as indicated in Figure 44. By similar reasoning, the hyperbola approaches the line

$$y = -\frac{b}{a}x \tag{12}$$

in quadrants II and IV. Lines (11) and (12) are called the **asymptotes** of hyperbola (10). The rectangle in Figure 44 with vertices (a, b), $(-a, b)$, $(-a, -b)$, $(a, -b)$, is called the **fundamental rectangle** for the hyperbola (10). By drawing the asymptotes through the diagonals of the fundamental rectangle, we can get an accurate graph of the hyperbola.

Example 6 Use the fundamental rectangle and asymptotes to graph each of the following hyperbolas.

(a) $4x^2 - 9y^2 = 36$
(b) $4y^2 - 25x^2 = 100$

Solution:

(a) By dividing both sides of the given equation by 36, we obtain

$$\frac{x^2}{9} - \frac{y^2}{4} = 1.$$

Here $a = 3$ and $b = 2$, and the asymptotes are the lines $y = (2/3)x$ and $y = -(2/3)x$. The fundamental rectangle, asymptotes, and graph of the hyperbola are shown in Figure 45(a).

(b) We divide both sides of the given equation by 100 to obtain

$$\frac{y^2}{25} - \frac{x^2}{4} = 1.$$

Here $a = 2$ and $b = 5$, and the asymptotes are $y = (5/2)x$ and $y = -(5/2)x$. These asymptotes, along with the fundamental rectangle and the graph of the hyperbola, are illustrated in Figure 45(b). ∎

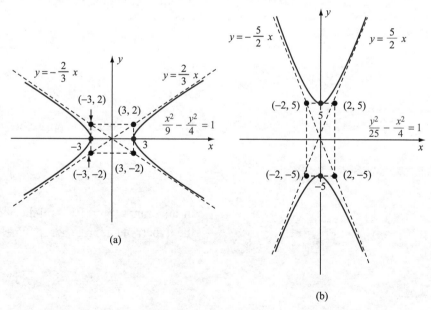

Figure 45

Comment The conic sections were discovered around 350 B.C. by the Greek geometer Menæchmus while working on a solution to the problem of "duplicating the cube," that is, constructing a cube whose volume is twice the volume of a given cube. About 120 years later, Apollonius made an exhaustive study of the ellipse, parabola, and hyperbola in the *Conics*, a collection of eight books. Apollonius also devised widely accepted system of *circular* planetary motion, apparently not realizing that the planets actually travel in elliptical orbits. The elliptical orbits of the planets were discovered empirically in 1609 by Johannes Kepler. About 70 years later, Sir Isaac Newton formulated his laws of gravitation and motion. From these laws it follows that planetary motion in the universe can be along any one of the three conics. Hence, the parabola, ellipse, and hyperbola, discovered in an investigation of a classical problem in geometry, have universal application.

Exercises 3.8

Fill in the blanks to make each statement true.

1. Let P_1 and P_2 two fixed points in the plane and let c be a positive number. The set of all points P in the plane for which _____ is called an ellipse, and the set of all P for which _____ is called a hyperbola.

2. Each of the fixed points P_1 and P_2 in the definition of an ellipse and hyperbola is called a _____.

3. The graph of $(x-3)^2/a^2 - (y+5)^2/b^2 = 1$ is a(n) _____ with center _____.

4. The length of the major axis of the ellipse $x^2/4 + y^2/10 = 1$ is _____ and that of the minor axis is _____.

5. If the graph of a second-degree equation $Ax^2 + By^2 + Cx + Dy = E$ is an ellipse or hyperbola, then $AB \neq$ _____. For an ellipse, AB _____ 0, and for a hyperbola, AB _____ 0.

Write true or false for each statement

6. A focal radius of an ellipse has constant length.
7. The major axis of an ellipse is always along the x-axis.
8. The lines $y = \pm(b/a)x$ are asymptotes for the hyperbola $x^2/a^2 - y^2/b^2 = 1$.
9. The lines $y = \pm(a/b)x$ are asymptotes for the hyperbola $y^2/b^2 - x^2/a^2 = 1$.
10. If $A \neq B$, then the graph of the second-degree equation $Ax^2 + By^2 + Cx + Dy = E$ is either an ellipse or a hyperbola.

Ellipses

In Exercises 11-14, find the equation of the ellipse with the given foci and focal radii.

11. foci $(1, 0)$ and $(-1, 0)$; $d_1 + d_2 = 4$
12. foci $(2, 0)$ and $(-2, 0)$; $d_1 + d_2 = 5$
13. foci $(0, 2)$ and $(0, -2)$; $d_1 + d_2 = 6$
14. foci $(0, 1)$ and $(0, -1)$; $d_1 + d_2 = 5$

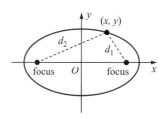

Graph each of the following ellipses.

15. $\dfrac{x^2}{16} + \dfrac{y^2}{25} = 1$

16. $\dfrac{x^2}{25} + \dfrac{y^2}{36} = 1$

17. $4x^2 + 9y^2 = 36$

18. $9x^2 + 16y^2 = 144$

19. $\dfrac{(x-2)^2}{16} + \dfrac{(y-5)^2}{25} = 1$

20. $\dfrac{(x+3)^2}{25} + \dfrac{(y+2)^2}{36} = 1$

21. $25(x-1)^2 + 4(y-2)^2 = 100$

22. $2(x+3)^2 + 8(y-3)^2 = 32$

By completing the square in x and y, determine the graph (ellipse, point, empty set) of each of the second-degree equations in Exercises 23-28.

23. $x^2 + 4y^2 + 4x - 16y = 80$

24. $2x^2 + y^2 + 4x + 10y = -30$

25. $3x^2 + 2y^2 + 6x + 8y = -11$

26. $2x^2 + 5y^2 - 4x - 10y = -7$

27. $4x^2 + 9y^2 + 16x - 18y = 11$

28. $2x^2 + 3y^2 + 2x = -1$

29. Eliminate the radicals in two squarings to show that the equation

$$\sqrt{(x-c_1)^2 + y^2} + \sqrt{(x+c_1)^2 + y^2} = c \ (c > 2c_1 > 0)$$

simplifies to

$$\dfrac{x^2}{a^2} + \dfrac{y^2}{b^2} = 1,$$

where $a = \dfrac{c}{2}$ and $b = \sqrt{a^2 - c_1^2}$.

30. The moon travels in an elliptical orbit about the earth as one of the foci. From the data in the figure, find the equation of the ellipse and the location of the other focus.

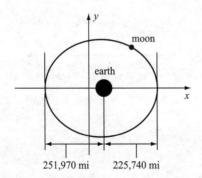

Hyperbolas

In Exercises 31-34, find the equation of the hyperbola with the given foci and focal radii.

31. foci $(1, 0)$ and $(-1, 0)$; $d_2 - d_1 = \pm 1$
32. foci $(2, 0)$ and $(-2, 0)$; $d_2 - d_1 = \pm 4$
33. foci $(0, 2)$ and $(0, -2)$; $d_2 - d_1 = \pm 2$
34. foci $(0, 3)$ and $(0, -3)$; $d_2 - d_1 = \pm 3$

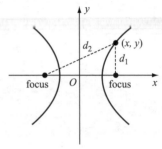

By completing the square in x and y, determine the graph (hyperbola or two lines) of each of the second-degree equations in Exercises 35-40.

35. $4x^2 - y^2 + 2x - y = 0$

36. $9x^2 - 16y^2 - 9x = 0$

37. $2x^2 - 5y^2 - 4x - 10y = -2$

38. $y^2 - x^2 + 3y - 3x = 1$

39. $8y^2 - x^2 - 8y = -1$

40. $y^2 - 8x^2 - 8x = -2$

41. Eliminate the radicals in two squarings to show that the equation

$$\sqrt{(x+c_1)^2 + y^2} - \sqrt{(x-c_1)^2 + y^2} = \pm c \quad (2c_1 > c > 0)$$

simplifies to

$$\dfrac{x^2}{a^2} - \dfrac{y^2}{b^2} = 1,$$

where $a = \dfrac{c}{2}$ and $b = \sqrt{c_1^2 - a^2}$.

Asymptotes

Graph each of the following hyperbolas, showing the fundamental rectangle and asymptotes.

42. $\dfrac{x^2}{16} - \dfrac{y^2}{25} = 1$

43. $\dfrac{x^2}{25} - \dfrac{y^2}{36} = 1$

3.8. Ellipses and Hyperbolas

44. $16y^2 - 9x^2 = 144$ **46.** $x^2 - 4y^2 = 16$

45. $y^2 - x^2 = 1$ **47.** $9x^2 - y^2 = 81$

48. $\dfrac{(x-2)^2}{9} - \dfrac{(y-5)^2}{16} = 1$

49. $\dfrac{(x-1)^2}{4} - \dfrac{(y-2)^2}{9} = 1$

Miscellaneous

50. *Eccentricity of an Ellipse.* If $a > b > 0$, the ellipse

$$\frac{x^2}{a^2} + \frac{y^2}{b^2} = 1$$

has its foci at $(\pm\sqrt{a^2 - b^2}, 0)$. The ratio

$$e = \frac{\sqrt{a^2 - b^2}}{a} \quad (0 < e < 1)$$

is called the **eccentricity** of the ellipse, and the vertical line

$$L : x = \frac{a}{e}$$

is called a **directrix**. In terms of eccentricity, the foci are at $(\pm ae, 0)$. With reference to the figure, show that a point $P(x, y)$ is on the ellipse if and only if

$$\frac{d_1}{d} = e.$$

(The line $x = -a/e$ is also called a directrix, and the above equation holds when $(ae, 0)$ and $x = a/e$ are replaced by $(-ae, 0)$ and $x = -a/e$, respectively.)

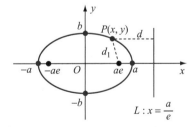

51. Discuss the change in shape of the ellipse $x^2/a^2 + y^2/b^2 = 1$ in each of the following cases.

(a) a is held fixed but the focus $(ae, 0)$ moves toward $(0, 0)$.

(b) a is held fixed but the focus $(ae, 0)$ moves toward $(a, 0)$.

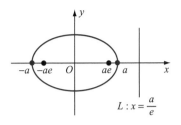

52. *Eccentricity of a Hyperbola.* The hyperbola

$$\frac{x^2}{a^2} - \frac{y^2}{b^2} = 1$$

has its foci at $(\pm\sqrt{a^2 + b^2}, 0)$. The ratio

$$e = \frac{\sqrt{a^2 + b^2}}{a} \quad (e > 1)$$

is called the **eccentricity** of the hyperbola, and the vertical line

$$L : x = \frac{a}{e}$$

is called a **directrix**. In terms of the eccentricity, the foci are at $(\pm ae, 0)$. With reference to the figure, show that a point $P(x, y)$ is on the hyperbola if and only if

$$\frac{d_1}{d} = e.$$

[As in the case of an ellipse, the line $x = -a/e$ is also called a directrix, and the above equation holds when $(ae, 0)$ and $x = a/e$ are replaced by $(-ae, 0)$ and $x = -a/e$, respectively.]

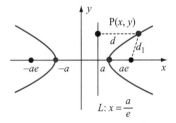

53. Discuss the change in shape of the hyperbola in each of the following cases. (*Hint:* draw the fundamental rectangle.)

(a) a is held fixed but ae approaches ∞.

(b) a is held fixed but ae approaches a.

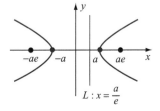

Chapter 3 Review Outline

3.1 Quadratic Equations

Quadratic Formula:
The roots of $ax^2 + bx + c = 0$ $(a \neq 0)$ are
$$\frac{-b \pm \sqrt{b^2 - 4ac}}{2a}.$$

Completing the Square:
If $b^2/4$ is added to $x^2 + bx$, the result is a perfect square $(x + b/2)^2$.

Factoring by Roots:
The roots of $ax^2 + bx + c = 0$ are a_1 and $a_1 \Leftrightarrow ax^2 + bx + c = a(x - a_1)(x - a_2)$.
If the discriminant $b^2 - 4ac < 0$, then $ax^2 + bx + c$ is irreducible (cannot be factored in the real number system).

3.2 Complex Numbers

A complex number is a number of the form $a + bi$, where a and b are real numbers and $i = \sqrt{-1}$.
$$a + bi = c + di \Leftrightarrow a = c \text{ and } b = d$$

Operations:
$$(a + bi) \pm (c + di) = (a \pm c) + (b \pm d)i$$
$$(a + bi)(c + di) = (ac - bd) + (ad + bc)i$$

To divide complex numbers, proceed as follows:
$$\frac{a + bi}{c + di} = \frac{a + bi}{c + di} \cdot \frac{c - di}{c - di}$$
$$= \frac{ac + bd}{c^2 + d^2} + \frac{bc - ad}{c^2 + d^2}i.$$

($c - di$ is the complex conjugate of $c + di$)

Classification of Roots:
The quadratic equation $ax^2 + bx + c = 0$ has

two distinct real roots $\Leftrightarrow b^2 - 4ac > 0$,

two equal real roots $\Leftrightarrow b^2 - 4ac = 0$,

two complex conjugate roots $\Leftrightarrow b^2 - 4ac < 0$.

3.3 Equations Reducible to Quadratic Equations

Some equations in x can be transformed into quadratic equations in w by substituting w for an expression in x.

Equations with fractional expressions or radicals can sometimes be transformed into quadratic equations by clearing fractions or squaring.
All solutions obtained by substitution, clearing fractions, or squaring must be tested for acceptability by substitution in the original equation.

3.4 Quadratic Inequalities

If $ax^2 + bx + c = 0$ has distinct real roots a_1 and a_2 $(a_1 < a_2)$, then $ax^2 + bx + c$ does not change sign in $(-\infty, a_1)$, (a_1, a_2) and (a_2, ∞).
If $ax^2 + bx + c = 0$ has a double root a_1, then $ax^2 + bx + c$ has the same sign as a for all other values of x.
If $ax^2 + bx + c = 0$ has complex roots, then $ax^2 + bx + c$ has the same sign as a for all values of x.

3.5 Applications of Quadratic Equations and Inequalities

The four-step approach to solving word problems and the basic principles for percent, average speed, and work problems are the same here as in Section 2.3.
The height y of an object moving under the influence of gravity only is
$$y = -16t^2 + v_0 t + y_0,$$
where v_0 is the object's vertical velocity at time $t = 0$, and y_0 is its height at time 0 (distance is measured in feet and time in seconds). The vertical velocity of the object is
$$v = -32t + v_0.$$

3.6 Circles and Parabolas

The graph of the second-degree equation
$Ax^2 + By^2 + Cx + Dy = E$ (A or $B \neq 0$) is a conic section or the empty set. Circles and parabolas are two types of conic sections. Circle: $(x - x_0)^2 + (y - y_0)^2 = r^2$, center (x_0, y_0), radius r Parabola: $y = ax^2 + bx + c$, axis of symmetry $x = -\frac{b}{2a} = \frac{a_1 + a_2}{2}$, where a_1 and a_2 are the roots of $ax^2 + bx + c = 0$ (With x and y interchanged, the equations are similar.)
A second-degree equation with $A = B$ is either a circle, a single point, or the empty set. To determine which is the case, complete the square in x and y.

A second-degree equation with either $B = 0, A \neq 0$, and $D \neq 0$ or $A = 0, B \neq 0$, and $C \neq 0$ is a parabola.

3.7 Translation of Axes and Intersections

If the $x'y'$-axes are a translation of the xy-axes in which (x_0, y_0) becomes the $x'y'$-origin, then a point in the plane has both xy-coordinates and $x'y'$-coordinates that are related by the substitution equations

$$x' = x - x_0 \quad \text{and} \quad y' = y - y_0.$$

To simplify a second-degree equation, complete the square in x and/or y and make the corresponding substitutions. A line can intersect a parabola or a circle in at most two points. Substituting from the linear equation into the second-degree equations leads to a quadratic equation. Two parabolas of the type $y = ax^2 + bx + c$ or of the type $x = ay^2 + by + c$ can intersect in at most two points.

Other pairs of parabolas, or a circle and a parabola, can intersect in up to four points; algebraically, a fourth-degree equation may result.

3.8 Ellipses and Hyperbolas

Two other conic sections are ellipses and hyperbolas.

Ellipse : $\dfrac{(x - x_0)^2}{a^2} + \dfrac{(y - y_0)^2}{b^2} = 1$, center (x_0, y_0)

Hyperbola : $\dfrac{(x - x_0)^2}{a^2} - \dfrac{(y - y_0)^2}{b^2} = 1$

or $\dfrac{(y - y_0)^2}{b^2} - \dfrac{(x - x_0)^2}{a^2} = 1,$

center (x_0, y_0)

For a hyperbola with center $(0, 0)$, the points (a, b), $(-a, b)$, $(-a, -b)$, $(a, -b)$ form the vertices of the fundamental rectangle of the hyperbola; the lines $y = (b/a)x$ and $y = -(b/a)x$ are the asymptotes of the hyperbola.

Chapter 3 Review Exercises

Solve each quadratic equation in Exercises 1-6.

1. $x^2 - 5x - 50 = 0$
2. $3x^2 - 10x - 8 = 0$
3. $3x^2 + 18x + 27 = 0$
4. $3x^2 + 2x + 2 = 0$
5. $3x^2 + 2x - 2 = 0$
6. $x^2 - \dfrac{5}{6}x + \dfrac{1}{3} = 0$

7. Factor the quadratic $8x^2 + 2x - 3$ in the form $a(x - a_1)(x - a_2)$.
8. Show that the quadratic $x^2 - 6x + 11$ is irreducible (has no factors) in the real number system.
9. Find a quadratic equation with integral coefficients whose roots are $2/3$ and $-3/4$.
10. Find all real roots of $x^4 - 6x^2 + 9 = 0$.
11. Solve $2\sqrt{x + 5} - \sqrt{2x + 1} = 3$.
12. Solve $x/(x + 1) + 2/x = 26/(5x)$.
13. Find the sum of $3 + 2i$, and $5 - 3i$.
14. Represent the numbers $3 + 2i$, $5 - 3i$, and their sum by directed line segments in the plane.
15. Compute $(4 + 3i)(2 - i)$ and $(4 + 3i)/(2 - i)$.

Solve each of the following inequalities.

16. $x^2 - 4x + 3 \geq 0$
17. $3x^2 + 10x - 8 > 0$
18. $4x^2 - 20x + 25 \leq 0$
19. $x^2 + 5x + 7 < 0$

20. Sketch the graph of $y = 3x^2 + 10x - 8$.
21. Sketch the graph of $x = y^2 + y - 12$.
22. Find the standard equation of the circle that has the line segment from $(3, -4)$ to $(5, 8)$ as its diameter.

Sketch the graph of each of the following. For hyperbolas, show the fundamental rectangle and the asymptotes.

23. $4x^2 + 4y^2 = 9$
24. $4(x - 2)^2 + 4(y + 3)^2 = 9$
25. $4(x - 2)^2 + 9(y + 3)^2 = 0$
26. $4x^2 + 9y^2 = 9$
27. $4x^2 - 9y^2 = 9$
28. $4y^2 - 9x^2 = 9$

Find the center and identify the type of each of the following curves.

29. $2x^2 + 2y^2 - 4x - 24y = -42$
30. $x^2 - y^2 + 10x + 8y = 0$

In each of the following, xy-axes are translated to $x'y'$-axes with the point (x_0, y_0) becoming the $x'y'$-origin. Find the $x'y'$-equation of the given curve.

31. (x_0, y_0) is the center of the circle
$x^2 + y^2 + 2x - 4y = 0$.
32. (x_0, y_0) is the center of the hyperbola
$4x^2 - y^2 - 20x = 10$.
33. (x_0, y_0) is the vertex of the parabola
$y = 2x^2 - 5x - 12$.

Find all points of intersections of each pair of curves in Exercises 34-35.

34. $x^2 + y^2 = 25$
$x - 3y = 5$
35. $y = x^2 - 2x - 3$
$2y = x - 3$

36. Graph the region in the plane determined by the system of inequalities.

$$y \geq x^2 - 9$$
$$y \leq -x^2 + 3x.$$

37. A car is priced at $10,000 on January 1. On May 1, the price is increased by r%, and on September 1 there is another r% increase, bringing the price up to $10,609 Find r.

38. Dick jogs 1/2 mph faster than Frank and covers 10 mi in 5 min less. Find the jogging speed of each man.

39. It takes Janet and Dan 6 min to wash a cars when working together. Working alone, it takes Janet 5 min more than Dan. How long does it take each when working alone?

40. Two thousand dollars are deposited in a bank account on January 1 and $1,000 more on February 1. If the monthly interest rate is r% and the deposits plus interest amount to $3,025.05 on March 1, what is r?

Graphing Calculator Exercises

Verify your answers to the indicated review exercises of this section by using a graphing calculator for the given functions and windows

$[x_{min}, x_{max}], [y_{min}, y_{max}]$.

41. Exercises 2: $Y_1 = 3x^2 - 10x - 8$, $[-3.4, 6]$, $[-20, 5]$
42. Exercises 3: $Y_1 = 3x^2 + 18x + 27$, $[-5.4, 4]$, $[-1, 30]$
43. Exercises 8: $Y_1 = x^2 - 6x + 11$, $[-3.4, 6]$, $[-1, 15]$
44. Exercises 12: $Y_1 = x/(x+1) + 2/x - 26/5x)$, $[-2.4, 7]$, $[-5, 5]$
45. Exercises 16: $Y_1 = x^2 - 4x + 3$, $[-2, 7.4]$, $[-2, 10]$
46. Exercises 19: $Y_1 = x^2 + 5x + 7$, $[-7, 2.4]$, $[-2, 5]$
47. Exercises 20: $Y_1 = 3x^2 + 10x - 8$, $[-5.4, 4]$, $[-20, 20]$
48. Exercises 26: $Y_1 = \sqrt{1 - 4x^2/9}$, $Y_2 = -Y_1$, $[-2.2, 2.5]$, $[-2, 2]$
49. Exercises 28: $Y_1 = 1.5\sqrt{1 + x^2}$, $Y_2 = -Y_1$, $[-4.4, 5]$, $[-5, 5]$
50. Exercises 35: $Y_1 = x^2 - 2x - 3$, $Y_2 = (x-3)/2$, $[-4, 5.4]$, $[-5, 5]$

4 Functions and their Graphs

4.1	Functions
4.2	Polynomial Functions
4.3	Rational Functions
4.4	Algebraic Functions
4.5	One-to-One and Inverse Functions
4.6	Variation

Hey Students!
You will benefit more from the classroom lectures and have a better understanding of the material if you take the time to view the Gilbert Review Videos @
www.CollegeAlgebraBySchiller.com
before attending the class lecture for this chapter.

Chapter 4. Functions and their Graphs

In Chapter 2 we saw that the graph of the equation

$$ax + by + c = 0 \quad (a \text{ or } b \neq 0)$$

is a straight line, and in Chapter 3 we saw that the graph of the equation

$$ax^2 + by^2 + cx + dy + e = 0 \quad (a \text{ or } b \neq 0)$$

is a conic section. In this chapter we first refine the idea of "equation" to that of "*function*," which is one of the most fundamental concepts in mathematics. We then proceed to investigate the graphs of algebraic functions, including polynomial and rational functions.

4.1 Functions

Functions are the machines of algebra.

The Concept of a Function
Graphs of Functions
Algebra of Functions
Composition of Functions

The Concept of a Function. Illustrations of the concept of a function occur in many aspects of our daily lives. For example,

(a) to each person in your math class, there corresponds a name on the professor's class list;
(b) to each item in the supermarket, there corresponds a price;
(c) to each of the fifty states in the U.S., there corresponds the name of its capital city;
(d) to each real number, there corresponds its square;
(e) to each nonvertical straight line in the coordinate plane, there corresponds its slope.

Notice that every example involves a rule for pairing each member of one set with exactly one member of a second set. These examples involve sets of people, names, prices, and lines as well as numbers, but in algebra we shall we most interested in rules of correspondence between sets of real numbers. Let us formalize the basic idea.

In general, a set can be any collection of objects, but we are mainly concerned with sets of real numbers

*The set Y, which contains the range, is often called the **codomain** of the function.*

A function from a set X to a set Y is a correspondence that assigns to each element of X exactly one element of Y. The set X is called the **domain** of the function, and the set of correspondents in Y is called the **range** of the function.

Example 1 Determine the domain and range of each of the functions (a) through (e).

Solution:

	Domain	Range
(a)	The set of students in your math class.	The set of names on the class list.
(b)	All items for sale in the supermarket.	All of their prices.
(c)	The fifty states in the U.S.	Their fifty capitals.
(d)	The set of real numbers.	The set of nonnegative real numbers.
(e)	All nonvertical lines in the plane.	All real numbers.

It is customary to denote functions by letters such as **f**, **g**, and **h**. If x is in the domain of a function f, then the corresponding element in the range of f is denoted by $f(x)$, read "f of x." We call x the **independent variable** and $y = f(x)$ the **dependent variable**. ∎

Example 2 Let $f(x) = x^3 - 1$ define a function whose domain is all real numbers. Find each of the following values of f.
(a) $f(0)$ (b) $f(2)$ (c) $f(-2)$ (d) $f(\sqrt[3]{5})$

Solution:
(a) $f(0) = 0^3 - 1 = -1$ (b) $f(2) = 2^3 - 1 = 7$
(c) $f(-2) = (-2)^3 - 1 = -9$ (d) $f(\sqrt[3]{5}) = (\sqrt[3]{5})^3 - 1 = 4$ ∎

We can think of a function as a machine that accepts an input x from its domain and produces an output $f(x)$ in its range (Figure 1). For instance, if 2 is fed into the "machine" $f(x) = x^3 - 1$, then the output is 7.

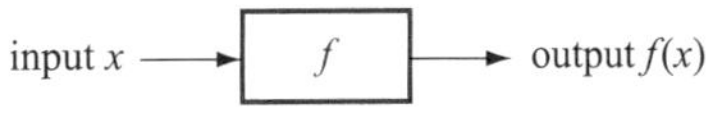

$$\text{input } x \longrightarrow \boxed{f} \longrightarrow \text{output } f(x)$$

Figure 1 A function as a Machine

Unless otherwise specified, the domain of a function f defined by an algebraic expression, such as $f(x) = \sqrt{x-3}$, is the set of all real numbers for which $f(x)$ is also a real number. We will call this set the **implied domain**. The implied or specified domain of f will be denoted by D_f and the range by R_f.

Example 3 Find the implied domain of each of the following functions.
(a) $f(x) = \sqrt{x-3}$. (b) $g(x) = 2x + 3$ (c) $h(x) = 1/(3x - 4)$

Solution:

(a) $\sqrt{x-3}$ is a real number if any only if $x - 3 \geq 0$. Therefore, $D_f = [3, \infty)$.

(b) $2x + 3$ is a real number for all real numbers x. Therefore $D_g = (-\infty, \infty)$.

(c) $1/(3x-4)$ is a real number if and only if $3x-4 \neq 0$, that is $x \neq 4/3$. Therefore, $D_h = (-\infty, 4/3) \cup (4/3, \infty)$. ∎

The correspondence between domain and range that defines a function can be denoted by various choices of letters. For example, $f(x) = x^2$ could also be denoted by $g(x) = x^2$ or by $f(t) = t^2$. All three expressions define the same function. On the other hand, we can modify $f(x)$ by replacing x with another algebraic expression, as illustrated in the following example.

Example 4 Let $f(x) = x^2 + 5x - 3$. Find each of the following.
(a) $f(2x)$ (b) $f(x^2)$ (c) $f(x+2)$

Solution:

(a) To determine $f(2x)$, we replace x by $2x$ in the expression for $f(x)$. That is, $f(2x) = (2x)^2 + 5(2x) - 3 = 4x^2 + 10x - 3$.

(b) Here we replace x by x^2 to obtain $f(x^2) = (x^2)^2 + 5x^2 - 3 = x^4 + 5x^2 - 3$.

(c) Proceeding as before, we get $f(x+2) = (x+2)^2 + 5(x+2) - 3 = x^2 + 4x + 4 + 5x + 10 - 3 = x^2 + 9x + 11$. ∎

Graphs of Functions. Each function f generates a set of ordered pairs $(x, f(x))$. For example, if $f(x) = x^2$, then

$$f(0) = 0, \quad f(1) = 1, \quad f(2) = 4, \quad f(-1) = 1,$$

and the corresponding ordered pairs are

$$(0,0), \quad (1,1), \quad (2,4) \quad (-1,1),$$

which may be interpreted as points in the coordinate plane. The set of all pairs (x, y) with x in the domain of f and $y = f(x)$ constitutes the **graph** of the function f (Figure 2). Notice that if all points (x, y) on the graph of f are projected vertically onto the x–axis, the resulting set of x's is the domain of f, and, similarly, if all (x, y) on the graph of f are projected horizontally onto the y–axis, the set of all y's is the range of f. For the example $f(x) = x^2$, these projections are the entire x–axis and the nonnegative y–axis, respectively.

4.1. Functions

Figure 2 Graph of $f(x) = x^2$

Question 1 *Why does the equation $y^2 = x + 4$ not define a function $y = f(x)$?*

Answer: Since $y = \pm\sqrt{x+4}$, it follows that for each $x > -4$, there are *two* corresponding values of y. Therefore, we should not write $f(x) = \pm\sqrt{x+4}$ because the definition of a function f requires exactly one number $f(x)$ for each x in the domain. ∎

> We could say that the equation $y^2 = x + 4$ defines the two branch functions $g(x) = \sqrt{x+4}$ and $h(x) = -\sqrt{x+4}$.

Let us now compare the graph of a function [Figure 3(a)] with one that is not of a function [Figure 3(b)]. As illustrated in Figure 3, we can determine whether a graph is that of a function by applying the following **vertical line test**.

If every vertical line intersects the graph in at most one point, the graph is that of a function. On the other hand, if some vertical line intersects the graph in more than one point, the graph is not that of a function.

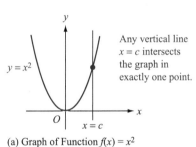
(a) Graph of Function $f(x) = x^2$

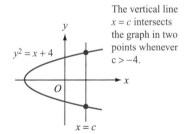
(b) Graph of Equation $y^2 = x + 4$

Figure 3

Example 5 Use the vertical line test to determine which of the graphs in Figure 4 are graphs of functions.

Solution:

(a) This is not the graph of a function, because the line $x = 1$ intersects the graph in two points.

(b) This is the graph of a function. Every vertical line $x = c$ ($c \neq 1$) intersects the graph in exactly one point, and the vertical line $x = 1$ does not intersect the graph at all.

256 Chapter 4. Functions and their Graphs

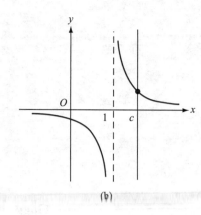

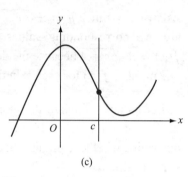

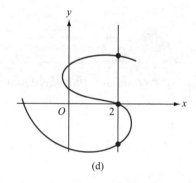

Figure 4

(c) This is the graph of a function. Every vertical line $x = c$ intersects the graph in exactly one point.

(d) This is not the graph of a function. The vertical line $x = 2$ intersects the graph in three points. ∎

Algebra of Functions. Functions may be added, subtracted, multiplied, and divided in much the same way as numbers. Starting with two functions f and g, these operations produce new functions $f + g$, $f - g$, fg, and f/g that are defined as follows.

$(f + g)(x) = f(x) + g(x)$	Sum Function
$(f - g)(x) = f(x) - g(x)$	Difference Function
$(fg)(x) = f(x)g(x)$	Product Function
$(f/g)(x) = f(x)/g(x)$	Quotient Function

To be in the domain of $f + g$, a real number must be in both the domain of f and the domain of g (why?). The same principle holds for $f - g$, fg, and f/g. For f/g, we must also have $g(x) \neq 0$. Therefore,

$$D_{f+g} = D_f \cap D_g = D_{f-g} = D_{fg}$$

$D_f \cap D_g$ is called the **intersection** of D_f and D_g and consists of all real numbers in both D_f and D_g.

and $\quad D_{f/g}$ = all x in $D_f \cap D_g$ for which $g(x) \neq 0$.

Example 6 In each of the following cases, determine $(f+g)(x)$, $(f-g)(x)$, $(fg)(x)$, $(f/g)(x)$, and the their respective domains.

(a) $f(x) = x^2 + 1, \quad g(x) = x - 2$
(b) $f(x) = \sqrt{x-3}, \quad g(x) = \sqrt{5-x}$

Solution:

(a) $(f+g)(x) = f(x) + g(x) = x^2 + 1 + x - 2 = x^2 + x - 1$
$(f-g)(x) = f(x) - g(x) = x^2 + 1 - (x-2) = x^2 - x + 3$
$(fg)(x) = f(x)g(x) = (x^2+1)(x-2) = x^3 - 2x^2 + x - 2$
$(f/g)(x) = f(x)/g(x) = \dfrac{x^2+1}{x-2}$
Now $D_f = D_g = (-\infty, \infty)$. Therefore, $D_{f+g} = (-\infty, \infty) = D_{f-g} = D_{fg}$.
Also, $D_{f/g}$ = all x except 2, that is, $(-\infty, 2) \cup (2, \infty)$.

(b) $(f+g)(x) = \sqrt{x-3} + \sqrt{5-x}$
$(f-g)(x) = \sqrt{x-3} - \sqrt{5-x}$
$(fg)(x) = \sqrt{x-3}\sqrt{5-x}$
$(f/g)(x) = \dfrac{\sqrt{x-3}}{\sqrt{5-x}}$

Now D_f = all $x \geq 3 = [3, \infty)$, and D_g = all $x \leq 5 = (-\infty, 5]$. Hence, $D_{f+g} = D_f \cap D_g = [3, \infty) \cap (-\infty, 5] = [3, 5] = D_{f-g} = D_{fg}$. For f/g, we cannot have $x = 5$ (*why?*). Therefore, $D_{f/g} = [3, 5)$. ■

The algebraic operations can be applied to more than two functions. For example, if $f(x) = x^2$, $g(x) = 2x+3$, $h(x) = x^2 - 1$, and $F(x) = x^2 + 1$, then

$$\dfrac{f(x)g(x) - h(x)}{F(x)} = \dfrac{x^2(2x+3) - (x^2-1)}{x^2+1} = \dfrac{2x^3 + 2x^2 + 1}{x^2+1}.$$

Caution When combining functions, it is important to keep in mind the restrictions placed on the domain of the combination. For example, if $f(x) = 2x^2$ and $g(x) = 1/x$, then $(fg)(x) = 2x$ for $x \neq 0$ but is *undefined* for $x = 0$. It is incorrect to simply say $(fg)(x) = 2x$. The graph of fg is the line $y = 2x$ with the origin $(0, 0)$ removed. Similarly, if $f(x) = 1 + 1/(x-3)$ and $g(x) = 1/(x-3)$, then $(f-g)(x) = 1$ for $x \neq 3$ but is undefined for $x = 3$. Here the graph of $f - g$ is the horizontal line $y = 1$ with the point $(3, 1)$ deleted.

Composition of Functions. Another way to combine functions involves following one by another. That is, *we let the output of one function serve as the input of another function*. For example, if $f(x) = x + 3$ and $g(x) = \sqrt{x}$, then f followed by g results in the function $\sqrt{x+3}$ (Figure 5).

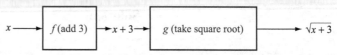

Figure 5 The Function $f(x) = x + 3$ followed by $g(x) = \sqrt{x}$

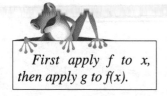

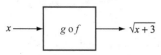

Figure 6 Composition $g \circ f$ for $f(x) = x + 3$ and $g(x) = \sqrt{x}$

In symbols, first $f(x) = x + 3$, then $g(f(x)) = g(x+3) = \sqrt{x+3}$. This combination of f and g is denoted by $\boldsymbol{g \circ f}$ and called the *composition of g with f* (Figure 6). In general, if f and g are any two functions for which the range (output) of f is contained in the domain (input) of g, then the **composition of $\boldsymbol{g}$ with $\boldsymbol{f}$** is defined by

$$(\boldsymbol{g \circ f})(x) = \boldsymbol{g(f(x))}.$$

Example 7 Find $(g \circ f)(x)$ in each of the following cases.

(a) $f(x) = x^2$, $g(x) = 3x + 5$
(b) $f(x) = 3x + 5$, $g(x) = 1/x$
(c) $f(x) = -(x^2 + 1)$, $g(x) = \sqrt{x}$

Solution:

(a) $(g \circ f)(x) = g(f(x)) = g(x^2) = 3x^2 + 5$

(b) $(g \circ f)(x) = g(f(x)) = g(3x + 5) = \dfrac{1}{3x+5}$

(c) The range of f is $R_f = (-\infty, -1]$, and the domain of g is $D_g = [0, \infty)$. Since R_f is not contained in D_g, the composition $g \circ f$ is not defined. [Note that if we formally apply the definition of composition, we obtain $(g \circ f)(x) = \sqrt{-(x^2+1)}$, which is meaningless in the real number system.] ■

Example 7(c) calls to our attention the importance of the domain in the definition of $g \circ f$. First x must be in D_f, then $f(x)$ must be in D_g. That is,

$$D_{g \circ f} \text{ consists of all } x \text{ in } D_f \text{ for which } f(x) \text{ is in } D_g.$$

In Example 7(c), $g \circ f$ is not defined because square roots of negative numbers do not exist in the real number system. Even when $g \circ f$ is defined, we must keep in mind any restrictions placed on its domain. For example, if $f(x) = \sqrt{x-1}$ and $g(x) = x^2 + 3$, then $(g \circ f)(x) = x + 2$, but the domain of $g \circ f$ is only $[1, \infty)$ (*why?*). The graph of $g \circ f$ consists of that part of the line $y = x + 2$ above the interval $[1, \infty)$.

Example 8 Given $f(x) = 2x + 3$ and $g(x) = \sqrt{x}$, find (a) $(g \circ f)(x)$, (b) $(f \circ g)(x)$, and their respective domains.

Solution:

(a) $(g \circ f)(x) = g(2x + 3) = \sqrt{2x + 3}$
Now $D_f = (-\infty, \infty)$ and $D_g = [0, \infty)$. Therefore, $D_{g \circ f}$ consists of all real numbers x for which $f(x) = 2x + 3 \geq 0$, that is, $x \geq -3/2$ or $[-3/2, \infty)$.

(b) $(f \circ g)(x) = f(\sqrt{x}) = 2\sqrt{x} + 3$
Since D_g consists of all nonnegative real numbers and D_f contains all real numbers, it follows that $D_{f \circ g}$ consists of all nonnegative real numbers, that is, $[0, \infty)$. ∎

Caution Example 8 shows that $g \circ f$ need not equal $f \circ g$; in other words, composition of functions is not a commutative operation.

Question 2 *Can $g \circ f = f \circ g$ for some functions f and g?*

Answer: Yes. For example, if $f(x) = 2x + 1$ and $g(x) = \dfrac{1}{2}(x - 1)$, then $(g \circ f)(x) = x = (f \circ g)(x)$. ∎

Exercises 4.1

Fill in the blanks to make each statement true

1. A function from set X to set Y assigns to each member of X _____ of set Y.
2. If f is a function from X to Y, then X is the _____ of f.
3. If $f(x) = x^3$, then $f(2) =$ _____ and $f(-2) =$ _____ are numbers in the _____ of f.
4. The domain of fg is the set of numbers that are _____.
5. To be in the domain of $g \circ f$, x must be in the domain of _____, and _____ must be in the domain of g.

Write true or false for each statement.

6. The domain and range of a function must be sets of real numbers.
7. If $f(x) = x^2 + 4$, then the domain of f is $(-\infty, \infty)$, and then range of f is $[0, \infty)$.
8. The point $(1, 5)$ belongs to the graph of $f(x) = x^2 + 4$.
9. Every vertical line $x = c$ intersects the graph of a function in exactly one point.
10. If x is in the domain of both f and g, then x is in the domain of f/g.

Functions

11. Given $f(x) = 2x^2 + 1$, find each value of f.
    (a) $f(0)$ (b) $f(2)$ (c) $f(-2)$
    (d) $f(\sqrt{2})$ (e) $f(-\sqrt{2})$ (f) $f(\pi)$
12. Given $g(x) = (3x + 1)/(5x - 2)$, find each value of g.
    (a) $g(6)$ (b) $g(2)$ (c) $g(-1)$
    (d) $g(1/3)$ (e) $g(1.5)$ (f) $g(-5/2)$
13. If $h(x) = \begin{cases} 1 \text{ for each rational number } x \\ -1 \text{ for each irrational number } x, \end{cases}$
    find each value of h.

(a) $h(3)$ (b) $h(-3)$ (c) $h(2/3)$
(d) $h(\sqrt{2}/3)$ (e) $h(\pi)$ (f) $h(3.14)$

14. If $F(x) = |x|/x$, find each value of F.
    (a) $F(2)$ (b) $F(10)$ (c) $F(-2)$
    (d) $F(-10)$ (e) $F(x)$ for $x > 0$ (f) $F(x)$ for $x < 0$

Find the implied domain of each of the following functions.

15. $f(x) = 2x - 5$
16. $g(x) = \dfrac{1}{2x - 5}$
17. $h(x) = \sqrt{2x - 5}$
18. $F(x) = \sqrt[3]{2x - 5}$
19. $G(x) = \dfrac{1}{\sqrt[3]{2x - 5}}$
20. $H(x) = \sqrt{x^2 + 1}$
21. $\alpha(x) = \sqrt{x^2 - 1}$
22. $\beta(x) = |x|$
23. $\gamma(x) = \dfrac{|x|}{x}$
24. $f_1(x) = \dfrac{1}{(x - 1)(x - 2)}$
25. $f_2(x) = \sqrt{(x - 1)(x - 2)}$

In Exercises 26-31, find all x in D_f satisfying $f(x) = 4$.

26. $f(x) = x - 4$
27. $f(x) = x^2$
28. $f(x) = x^2 - 5x$
29. $f(x) = x^2;\ D_f = [0, \infty)$
30. $f(x) = x^2;\ D_f = [-3, 3]$
31. $f(x) = x^2;\ D_f = [-1, 1]$

32. If $f(x) = x^2 + 2x + 3$, find each of the following.
    (a) $f(t)$ (b) $f(\sqrt{x})$ (c) $f(x^2)$
    (d) $f(x + 1)$ (e) $\dfrac{f(x) - f(1)}{x - 1}$ (f) $\dfrac{f(x + h) - f(x)}{h}$

33. If $g(x) = (x - 1)/(x + 1)$, find each of the following.
    (a) $g(2x)$ (b) $g(x^2)$ (c) $g(x + 1)$
    (d) $g(1/x)$ (e) $g(x + h)$ (f) $g(x - h)$

Decide whether each of the following equations can be put in the form $y = f(x)$, where f is a function. If it can be done, find $f(x)$.

34. $y - 3 = 4(x + 1)$
35. $y^2 - 3x - 1 = 0$
36. $x^2 + y^2 = 4$
37. $y^3 = x$
38. $|y| = x$
39. $xy - y = x^2$

Graphs of functions

Sketch the graph of each of the following equations and determine whether it is graph of a function $f(x)$.

40. $2x + 3y = 6$
41. $y - x^2 = 4$
42. $x^2 + y^2 = 9$
43. $x - y^2 = 4$
44. $x^2 - y^2 = 4$

Decide whether each of the following is the graph of a function $f(x)$.

45.
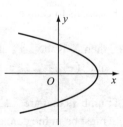

47.
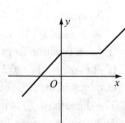

46.
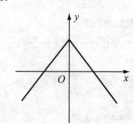

48.
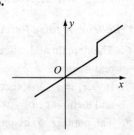

Algebra of Functions

In Exercise 49-56, find $f + g$, $f - g$, fg, and f/g, and determine their respective domains.

49. $f(x) = 2x$; $g(x) = 3x - 7$
50. $f(x) = 2x$; $g(x) = x^2 + 1$
51. $f(x) = x^2$; $g(x) = \sqrt{x}$
52. $f(x) = \dfrac{1}{x}$; $g(x) = \dfrac{1}{x-2}$
53. $f(x) = x^3$; $g(x) = \dfrac{1}{x^3}$
54. $f(x) = \sqrt{x+1}$; $g(x) = x^2$
55. $f(x) = \dfrac{x}{x^2+1}$; $g(x) = x^2 + 1$
56. $f(x) = x^2$; $g(x) = |x|$

Composition of Functions

For each of the following, fund $g \circ f$ and its domain.

57. $f(x) = 2x$; $g(x) = x^2$
58. $f(x) = x^2 + 1$; $g(x) = 2x$
59. $f(x) = x + 2$; $g(x) = x - 2$
60. $f(x) = \sqrt{x}$; $g(x) = x^2$
61. $f(x) = x^2$; $g(x) = \sqrt{x}$
62. $f(x) = \dfrac{1}{x}$; $g(x) = x$
63. $f(x) = x + 1$; $g(x) = \sqrt{x}$
64. $f(x) = x^2 + 1$; $g(x) = \sqrt{x}$

For each of the following, find $g \circ f$, $f \circ g$, and their respective domains.

65. $f(x) = x^2 - 4$; $g(x) = 1/x$
66. $f(x) = 3x - 2$; $g(x) = (x + 2)/3$
67. Let $f(x) = ax + b$ and $g(x) = cx + d$. Show that $(g \circ f)(x) = (f \circ g)(x) \Leftrightarrow ad + b = bc + d$.

Miscellaneous

In Exercises 68-77, write a formula for the indicated function.

68. $A(x) =$ the area of a square of side x
69. $A(r) =$ the area of a circle of radius r
70. $A(w) =$ the area of a rectangle whose length equals twice its width w
71. $V(s) =$ the volume of a cube of side s
72. $V(h) =$ the volume of a box whose width is half its height h and whose base is square
73. $C(x) =$ the cost of manufacturing x automobiles per day, if there is a fixed cost each day of \$5000 and the labor cost for each car is \$1500
74. $C(x) =$ the cost of renting a car per day, if the cost is \$20 per day plus 15 cents per mile for x miles
75. $S(x) =$ the surface area of a closed box whose base is a square of side x and whose height is twice its length
76. $A(x) =$ the area of a rectangular field to be fenced in with 3000 ft of fencing, if no fence is required on one side that is along a river and the other sides perpendicular to the river are each x ft long
77. $L(x) =$ the length of a fence around a rectangular field of length x ft and area 5000 sq ft
78. Given that $D(t) = 16t^2$ is the distance in feet that an object falls in t seconds, find each of the following.
    (a) distance the object falls in three seconds
    (b) distance the object falls in the third second
    (c) time it takes the object to reach the ground if it starts from 256 ft above the ground
79. For the distance function $D(t) = 16t^2$ given in Exercise 78, compute each of the following.
    (a) $\dfrac{D(4) - D(1)}{3}$
    (b) $\dfrac{D(1 + \Delta t) - D(1)}{\Delta t}$ $(\Delta t \neq 0)$
80. Give a physical interpretation of each of the answers in Exercise 79.

4.2 Polynomial Functions

The most basic numbers are integers, and the most basic functions are polynomials.

Definition of a Polynomial Function
Intercepts and Symmetry
Three-Step Procedure for Graphing Polynomial Functions

Definition of a Polynomial Function The simplest polynomial functions have already been discussed in Chapters 2 and 3, namely

$$f(x) = ax + b \qquad \text{Linear fuctions.}$$
and $$f(x) = ax^2 + bx + c \quad (a \neq 0). \quad \text{Quadratic functions.}$$

Their graphs are straight lines and parabolas, respectively. Here we discuss the general polynomial function and how to sketch its graph.

> The function f defined by
> $$f(x) = a_n x^n + a_{n-1} x^{n-1} + \ldots + a_1 x + a_0,$$
> where n is a nonnegative integer and $a_n \neq 0$, is called a **polynomial function of degree n**.

A constant polynomial $f(x) = a_0$ requires $a_0 \neq 0$ to have degree 0. Since $f(x) = 0$ does not satisfy this condition, we do not assign a degree to $f(x) = 0$.

The domain of any polynomial function is the set of all real numbers. As indicated above, polynomial functions of degree 1 are linear and those of degree 2 are quadratic. If $n = 0$, then $f(x) = a_0$ and f is a nonzero **constant function.** We include the constant function $f(x) = 0$ with the polynomials, but we do not assign a degree to it. In general, any algebraic expression in x generated by addition, subtraction, and multiplication determines a polynomial function.

Example 1 Which of the following define polynomial functions?
(a) $f(x) = 3x^4 - 2x^3 + x - 1$ (b) $g(x) = 1 - 5x$
(c) $h(x) = \sqrt{x} + 10$ (d) $F(x) = \dfrac{3}{x} + x^2$
(e) $G(x) = \dfrac{2x + 5}{x^3 + 2x^2 + 1}$ (f) $H(x) = 6$

Solution: The functions in (a), (b), and (f) are polynomial functions of degrees 4, 1, and 0, respectively. (Note that $H(x) = 6x^0$.) The functions $h(x) = x^{1/2} + 10$ and $F(x) = 3x^{-1} + x^2$ in (c) and (d) are not polynomials since they involve fractional and negative exponents, respectively. Also, $G(x)$ in part (e) is not a polynomial but rather the quotient of two polynomials. ∎

4.2. Polynomial Functions

Intercepts and Symmetry. The most elementary way to draw the graph of a function f is to find a set of coordinates (x, y) satisfying the equation $y = f(x)$, plot the corresponding points, and then, if possible, draw a smooth curve through those points. It can be shown by methods of calculus that the graph of any polynomial function is *continuous* (has no breaks in it) and *smooth* (has no sharp turns).

Example 2 Plot the graph of $g(x) = 4 - x^2$.

Solution: First a table of coordinates is obtained by computing $y = 4 - x^2$ for $x = 0, \pm 1, \pm 2, \pm 3$; then a smooth curve is drawn through the corresponding seven points. Since f is a quadratic function, its graph is a parabola, as shown in Figure 7. ■

Example 2 illustrates that it is desirable to include intercepts on the graph of a function. These are defined as follows.

If the graph of $f(x)$ crosses the x-axis at $(a, 0)$, then the point $(a, 0)$ or the number a is called an **x-intercept**. If the graph crosses the y-axis at $(0, b)$, then the point $(0, b)$ or the number b is called a **y-intercept**.

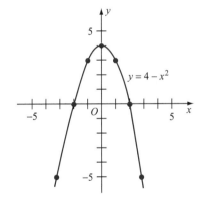

x	y
0	4
± 1	3
± 2	0
± 3	-5

Figure 7

If a graph has two y-intercepts, then two values of y correspond to $x = 0$, so the graph could not be that of a function.

We note that the graph of a function can have any number of x-intercepts, but at most one y-intercept. Example 2 also illustrates the simplification possible when a graph is symmetric with respect to the y-axis. Notice the pairs of points $(1, 3)$ and $(-1, 3)$, $(3, -5)$, and $(-3, -5)$ on the graph. In general, we have the following definition.

If for every point (x, y) on a graph, the point $(-x, y)$ is also on the graph, then the graph is **symmetric with respect to the y-axis.**

Figure 8 illustrates that when the graph of a function is symmetric with respect to the y-axis, then each point $P(x, y)$ on the graph can be paired with $Q(-x, y)$ also on the graph so that the y-axis is the perpendicular bisector of PQ.

Example 3 Sketch the graph of $F(x) = x^3 - 4x$.

Solution: First find intercepts. If $x = 0$, then $y = F(0) = 0$. Therefore, $(0, 0)$ is the y-intercept. If $y = 0$, then

$$x^3 - 4x = 0$$
$$x(x^2 - 4) = 0$$
$$x(x - 2)(x + 2) = 0$$
$$x = 0, 2, -2.$$

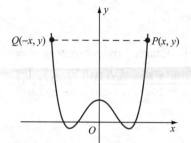

Figure 8 Symmetry with Respect to the y-axis

Therefore, $(0, 0)$, $(2, 0)$, and $(-2, 0)$ are the x-intercepts. ∎

Now construct a table of coordinate values including the intercepts and several other points on the graph. Then plot the points from the table and draw a smooth curve through them (Figure 9).

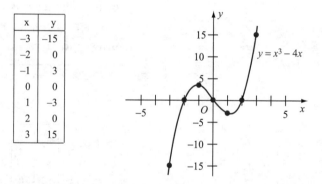

x	y
-3	-15
-2	0
-1	3
0	0
1	-3
2	0
3	15

Figure 9

The graph in Figure 9 illustrates symmetry with respect to the origin. Note that the pairs $(1, -3)$, $(-1, 3)$ and $(3, 15)$, $(-3, -15)$ are on the graph. In general, we have the following definition.

If for every point (x, y) on a graph, the point $(-x, -y)$ is also on the graph, then the graph is **symmetric with respect to the origin**.

Figure 10 illustrates that when the graph of a function is symmetric with respect to the origin, then each point $P(x, y)$ on the graph can be paired with $Q(-x, -y)$ also on the graph so that the origin is the midpoint of the line segment PQ.

Some graphs exhibit **symmetry with respect to the x-axis**. That is, for each point $P(x, y)$ on the graph, the point $Q(x, -y)$ is also on the graph.

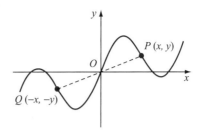

Figure 10 Symmetry with Respect to the Origin

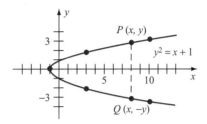

Figure 11 Symmetry with Respect to the x-axis

For example, the graph of $y^2 = x + 1$ (Figure 11) is symmetric with respect to the x-axis. But *the graph of a function $y = f(x)$ cannot be symmetric with respect to the x-axis unless the graph lies entirely on the x-axis (why?)*.

The previous examples illustrate how symmetry and intercepts can be utilized in graphing functions. We incorporate these two notions in the following systematic procedure for graphing any polynomial function.

Three-Step Procedure for Graphing Polynomial Functions.

1. *Test for symmetry.*

    (a) The graph of $y = f(x)$ is symmetric with respect to the y-axis if and only if replacing x by $-x$ does not change the value of y. That is, $f(-x) = f(x)$. Such functions are called **even functions.** A polynomial function f is even and therefore its graph is symmetric with respect to the y-axis precisely when f contains only *even* powers of x.

    (b) The graph of $y = f(x)$ is symmetric with respect to the origin if and only if replacing x by $-x$ changes y to $-y$; that is, $f(-x) = -f(x)$. Such functions are called **odd functions.** A polynomial function f is odd and therefore its graph is symmetric with respect to the origin precisely when f contains only *odd* powers of x.

2. *Solve for intercepts.*

    (a) To find the x-intercepts, set $y = 0$ and solve for x, if possible.

    (b) To find the y-intercept, set $x = 0$ and solve for y.

In addition to steps 1 and 2, the following step is useful in graphing polynomial functions.

3. *Determine the sign of $f(x)$.*

We saw in Chapter 3 that a quadratic function (parabola) can change sign only where its graph crosses the x-axis. More generally, if f is any polynomial function and hence has a continuous graph, then $f(x)$ can change sign only where $f(x) = 0$, that is, at the x-intercepts. These intercepts partition the x-axis into closed intervals, and inside each interval the sign of $f(x)$ does not change. (We use closed intervals $[a, b]$ to distinguish them from points (a, b). Of course, $f(x) = 0$ at the endpoints a and b.)

As an illustration of step 3, we look back at Example 3, where $y = F(x) = x^3 - 4x$. Here the x-intercepts are $-2, 0$, and 2, which partition the x-axis into the four intervals $(-\infty, -2], [-2, 0], [0, 2]$, and $[2, \infty)$. In each of these, the sign of y can be determined by selecting a point x in the interval and computing $y = F(x)$. The sign of $F(x)$ behaves as shown in the following table, which agrees with the graph (Figure 9) for Example 3.

Interval	$(-\infty, -2]$	$[-2, 0]$	$[0, 2]$	$[2, \infty)$
Selected x	-3	-1	1	3
Corresponding y	-15	3	-3	15
Sign of $F(x) = x^3 - 4x$ in interval	$-$	$+$	$-$	$+$

Question 1 *How can the fact that the graph in Example 3 (Figure 9) is symmetric with respect to the origin be used to shorten the above procedure for determining the sign of $F(x)$?*

Answer: Because its graph is symmetric with respect to the origin, the sign of $F(x)$ inside an interval $[a, b]$ is the opposite of the sign at corresponding points in $[-b, -a]$. Hence, we need only determine the sign inside $[0, 2]$ and $[2, \infty)$ to conclude what the sign is inside $[-2, 0]$ and $(-\infty, -2]$. ∎

Question 2 *Step 3 above says that if f changes sign at x, then $f(x) = 0$. Is the converse true; that is, if $f(x) = 0$, must f change signs at x?*

Answer: No. See Example 4 for an explanation. ∎

Example 4 Sketch the graph of $f(x) = x^3 - 3x + 2 = (x-1)^2(x+2)$.

Solution:

1. *Test for symmetry.*

$$f(-x) = (-x)^3 - 3(-x) + 2 = -x^3 + 3x + 2$$

Since $f(-x) \neq f(x)$, f is not even and the graph is *not* symmetric with respect to the y-axis.

Since $f(-x) \neq -f(x)$, f is not odd and the graph is *not* symmetric with respect to the origin.

2. *Solve for intercepts.*
   If $y = 0$, then $x^3 - 3x + 2 = (x - 1)^3(x + 2) = 0$. Therefore, $x = 1$ and $x = -2$ are the x-intercepts. If $x = 0$, then $y = 2$.

3. *Determine the sign of f(x).*
   The x-intercepts -2 and 1 determine three intervals on the x-axis, in each of which the sign of y is indicated in the following chart.

Interval	$(-\infty, -2]$	$[-2, 1]$	$[1, \infty)$
Selected x	-3	-1	2
Corresponding y	-16	4	4
Sign of $f(x)$	$-$	$+$	$+$

   We plot the intercepts $(1, 0)$, $(-2, 0)$, and $(0, 2)$, as well as the points $(-1, 4)$ and $(2, 4)$ from the chart, and connect them with a smooth curve (Figure 12). ∎

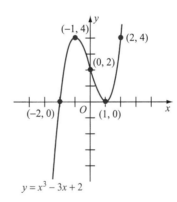

Figure 12

Example 5 Sketch the graph of $f(x) = x^6 - 4x^4 - x^2 + 4$.

Solution:

1. *Test for symmetry.*
   $$f(-x) = (-x)^6 - 4(-x)^4 - (-x)^2 + 4$$
   $$= x^6 - 4x^4 - x^2 + 4 = f(x)$$

   Therefore, the graph is symmetric with respect to the y-axis. Hence, it is sufficient to consider only $x \geq 0$ (*why?*).

2. *Solve for intercepts.*

Setting $y = 0$, we obtain

$$x^6 - 4x^4 - x^2 + 4 = x^4(x^2 - 4) - (x^2 - 4)$$
$$= (x^4 - 1)(x^2 - 4)$$
$$= (x^2 + 1)(x + 1)(x - 1)(x + 2)(x - 2) = 0$$

Therefore, $x = -1, 1, -2,$ and 2 are the x-intercepts. Setting $x = 0$, we obtain $y = 4$ as the y-intercept.

3. Determine the sign of $f(x)$.

We consider only the intervals determined by the x-intercepts for $x \geq 0$.

Interval	[0, 1]	[1, 2]	[2, ∞)
Selected x	.5	1.5	3
Corresponding y	≈ 3.5	≈ −7.1	400
Sign of $f(x)$	+	−	+

We use the intercepts $(1, 0)$, $(2, 0)$, and $(0, 4)$ and the points $(.5, 3.5)$ and $(1.5, -7.1)$ from the chart in order to plot the graph for $x \geq 0$, and then use the symmetry with respect to the y-axis to complete the graph for $x < 0$ (Figure 13). ∎

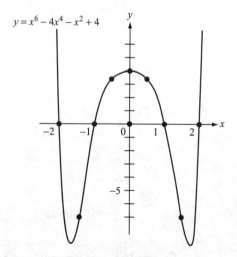

Figure 13

Comment On the graph of $f(x) = x^3 - 3x + 2$ (Example 4), the local "maximum" point $(-1, 4)$ and the local "minimum" point $(1, 0)$ were included because they happen to occur at integer values of x. Also, in the special case of quadratic functions (Example 2), the vertex of a parabola can be found by the methods of Section 3.6. But in general, such important points on graphs can be found only by calculus. That is the case in Examples 3 and 5. Also, when finding the x-intercepts of a polynomial function $y = f(x)$, we must solve the equation $f(x) = 0$ for

Exercises 4.2

Fill in the blanks to make each statement true.

1. A polynomial function of degree 4 is defined by $f(x) = a_4x^4 + a_3x^3 + a_2x^2 + a_1x + a_0$, provided _____.
2. The domain of any polynomial is _____.
3. If $f(-x) = f(x)$ for all x in the domain of f, then f is called an _____ function and its graph is _____.
4. For $f(x) = x(x+2)(x-3)(x^2+4)$, the x-intercepts are _____.
5. For $g(x) = x^3 - 3x^2 + 7x + 4$, the y-intercept is _____.

Write true or false for each statement.

6. The graph of every polynomial function has exactly one y-intercept.
7. The graph of every polynomial function has at least one x-intercept.
8. The graph of *a* quadratic polynomial is always symmetric with respect to a vertical line.
9. A graph that is symmetric with respect to the x-axis cannot be the graph of a function $y = f(x)$ unless $f(x) = 0$ for all x in D_f.
10. If $f(1) = 0$ and $f(3) = 0$, then $f(x)$ cannot change sign for $1 < x < 3$.

Polynomial Functions

11. Which of the following are polynomial functions?
    (a) $f(x) = x^{10}$
    (b) $g(x) = (x+1)^{10}$
    (c) $h(x) = x^2 + 2x + 3\sqrt{x} + 4$
    (d) $F(x) = 3x^2 - x^{-2}$
    (e) $G(x) = x^{2.1} + 5$
    (f) $H(x) = \dfrac{1}{x^{-3} + 3}$

12. Which of the following are *not* polynomial functions?
    (a) $f_1(x) = (x+3)^{1/3}$
    (b) $f_2(x) = (x+4)^3$
    (c) $f_3(x) = (x+4)^{-3}$
    (d) $f_4(x) = x^2(x+1)^3$
    (e) $f_5(x) = x^{-2}(x+1)^3$
    (f) $f_6(x) = 16$

Symmetry

13. Which of the following have graphs symmetric with respect to the y-axis
    (a) $f(x) = x^2 - 1$
    (b) $g(x) = x^3 - 1$
    (c) $h(x) = x^6 - 1$
    (d) $F(x) = 4$
    (e) $G(x) = x^4 - 2x^2 + 1$
    (f) $H(x) = x^5 - 2x^3 + x$

14. Which of the functions in Exercise 13 are even? Which are odd?

15. Which of the following are odd functions?
    (a) $F_1(x) = 2x^3 + x$
    (b) $F_2(x) = 2x^3 + x + 1$
    (c) $F_3(x) = \sqrt[3]{x}$
    (d) $F_4(x) = (x^3 + x)^3$
    (e) $F_5(x) = (x^3 + x)^2$
    (f) $F_6(x) = 6$

16. Which of the functions in Exercise 15 has a graph that is symmetric with respect to the origin?

Intercepts

17. Find the x- and y-intercepts for the graphs of the following functions.
    (a) $f(x) = x^2 + 2x + 1$
    (b) $g(x) = (x-1)^2$
    (c) $h(x) = 7$
    (d) $F(x) = (x-1)^{-1}$
    (e) $G(x) = x^2 - 4$
    (f) $H(x) = x^2 + 4$

18. Find x- and y-intercepts for the graphs of the following functions.

(a) $f_1(x) = x(x-3)(x^2-3)(x^3-1)$
(b) $f_2(x) = (x+3)(x^2+3)(x^3+3)$
(c) $f_3(x) = x^3 - 6x^2 + 11x - 6$
(d) $f_4(x) = x^5 + x^4 - 16x - 16$

21. $h(x) = x^3 - 3x^2$
22. $F(x) = x^3 - 3x$
23. $G(x) = (x^2-1)^2$
24. $H(x) = x^4 + 2x^3$
25. $p(x) = x^4 - 2x^2$
26. $q(x) = 3x^4 - 4x^3$
27. $r(x) = x^4 - 4$
28. $s(x) = x^5 + x^4 - 16x - 16$ (*Hint*: factor by grouping).
29. $t(x) = x^3 - 6x^2 + 11x - 6$ (*Hint*: $x-1$ is a factor).
30. $u(x) = x^3 - 3x^2 - 9x + 11$ (*Hint*: $x-1$ is a factor).

Graphs of Polynomial Functions

Using the three-step procedure, sketch the graphs of the following polynomial functions

19. $f(x) = 9 - x^2$ **20.** $g(x) = x^2 - 4x + 3$

4.3 Rational Functions

Rational functions are related to polynomials just as rational numbers are related to integers.

Definition of a Rational Function
Vertical Asymptotes
Horizontal Asymptotes
Four-Step Procedure for Graphing Rational Functions

Definition of a Rational Function. A **rational function** is the ratio of two polynomial functions. That is, f is a rational function if defined by

$$f(x) = \frac{P(x)}{Q(x)},$$

where $P(x)$ and $Q(x)$ are polynomials. Any algebraic expression in x generated by addition, subtraction, multiplication, and division determines a rational function. The class of rational functions includes the polynomial functions (*why?*).

Example 1 Which of the following are rational functions?

(a) $f(x) = \dfrac{x^3+6}{1-2x}$ (b) $g(x) = \dfrac{\sqrt{2}}{x^2+1}$

(c) $h(x) = \dfrac{x^{1/2} + x^{-1/2}}{x^2-1}$ (d) $F(x) = \dfrac{x^{-2} - x^{-3}}{x^2+x^3}$

Solution: The functions in (a), (b) and (d) are rational functions. Note that

$$F(x) = \frac{\dfrac{1}{x^2} - \dfrac{1}{x^3}}{x^2+x^3} = \frac{\dfrac{x-1}{x^3}}{x^2+x^3} = \frac{x-1}{x^5+x^6}.$$

4.3. Rational Functions

The function in (c) is not a rational function because the numerator, $x^{1/2} + x^{-1/2}$, has terms with fractional exponents.

While the domain of a polynomial function is $(-\infty, \infty)$, a rational function is undefined wherever its denominator is zero. ■

Example 2 Find the domain of each rational function in Example 1.

Solution:

If A and B are sets of real numbers, then $A \cup B$ is called the union of A and B and consists of all numbers in either A or B or both.

(a) $f(x) = (x^3 + 6)/(1 - 2x)$ is undefined at $x = 1/2$. $D_f = (-\infty, 1/2) \cup (1/2, \infty)$.
(b) $g(x) = \sqrt{2}/(x^2 + 1)$ is defined for every real number since the equation $x^2 + 1 = 0$ has no real solution. Hence, $D_g = (-\infty, \infty)$.
(c) $h(x) = (x^{1/2} + x^{-1/2})/(x^2 - 1)$ is not a rational function.
(d) $F(x) = (x - 1)/[x^5(x + 1)]$ is undefined at $x = 0$ and $x = -1$. Therefore, $D_F = (-\infty, -1) \cup (-1, 0) \cup (0, \infty)$. ■

Vertical Asymptotes. The techniques used in graphing polynomial functions [symmetry, intercepts, and sign of $f(x)$] are also useful for rational functions, but in addition, it is necessary to pay special attention to the values of x for which the denominator in zero. In the following definition, we assume that $P(x)$ and $Q(x)$ have no zeros in common.

> If $Q(a) = 0$, then the vertical line $x = a$ is called a **vertical asymptote** for the rational function $f(x) = P(x)/Q(x)$.

In the following example, we examine the behavior of the graph of a rational function near a vertical asymptote.

Example 3 For $f(x) = 2/(x - 1)$, find any vertical asymptote and sketch the graph of f near that line.

Solution: The vertical line $x = 1$ is the only vertical asymptote for $f(x)$. The table shows values of x approaching 1 and the corresponding values of y.

272 Chapter 4. Functions and their Graphs

x	$f(x)$	x	$f(x)$
0	-2	2	2
.9	-20	1.1	20
.99	-200	1.01	200
.999	-2000	1.001	2000
.9999	$-20{,}000$	1.0001	20,000
$\downarrow$	$\downarrow$	$\downarrow$	$\downarrow$
1^-	$-\infty$	1^+	∞

As x approaches 1 from the left (written: $x \to 1^-$), $f(x)$ is negative but $|f(x)|$ gets larger and larger [$f(x) \to -\infty$]. As x approaches 1 from the right ($x \to 1^+$), $f(x)$ is positive and gets larger and larger [$f(x) \to \infty$]. Hence, the graph of f falls as it approaches the line $x = 1$ from the left and rises as it approaches $x = 1$ from the right (Figure 14). ∎

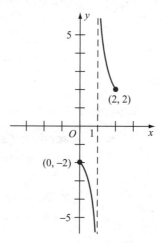

Figure 14 Graph of $f(x) = 2/(x-1)$ near Vertical Asymptote $x = 1$

Example 3 illustrates the following **characteristic property of a vertical asymptote**.

The line $x = a$ is a **vertical asymptote** for f if and only if $|f(x)|$ increases without bound as x approaches the finite value a.

Horizontal Asymptotes. We now consider the possibility that as $|x|$ increases without bound, $y = f(x)$ may approach a finite value b.

Example 3 Complete the graph of $f(x) = 2/(x-1)$.
(continued)

Solution: We consider values of x that approach ∞ and $-\infty$ and the corresponding values of y, as shown in the table.

4.3. Rational Functions   273

x	$y = f(x)$	x	$y = f(x)$
3	1	-1	-1
11	.2	-9	$-.2$
101	.02	-99	-0.2
1001	.002	-999	$-.002$
10,001	.0002	$-9,999$	$-.0002$
$\downarrow$	$\downarrow$	$\downarrow$	$\downarrow$
∞	0^+	$-\infty$	0^-

As x approaches ∞, $f(x)$ is positive and approaches 0 ($y \to 0^+$). As x approaches $-\infty$, y is negative and approaches 0 ($y \to 0^-$). Hence, the graph of f approaches the x-axis ($y = 0$) as x increases in magnitude (Figure 15). The line $y = 0$ is called a horizontal asymptote. ∎

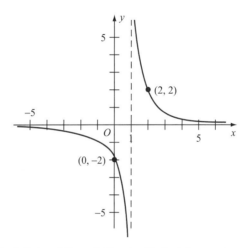

Figure 15 Graph of $f(x) = 2/(x - 1)$ Showing Vertical Asymptote $x = 1$ and Horizontal Asymptote $y = 0$

Example 3 illustrates the following definition and property of a horizontal asymptote.

> The horizontal line $y = b$ is a **horizontal asymptote** for f if and only if $f(x)$ approaches the finite value b as x approaches ∞ or as x approaches $-\infty$.

As in Example 3, we will see that $y = 0$ is a horizontal asymptote for a rational function whenever the degree of the numerator is *less than* the degree of the denominator. We now consider an example in which the degree of the numerator is *equal* to the degree of the denominator.

Example 4   Find any horizontal asymptote and sketch the graph for $f(x) = 2x/(x - 1)$.

Solution: Although we could construct a table similar to the one in Example 3, it is more efficient to write $f(x)$ in a different form by dividing the numerator and denominator by x, the highest power occurring in $f(x)$.

$$f(x) = \frac{\frac{2x}{x}}{\frac{x-1}{x}} = \frac{2}{1 - \frac{1}{x}} \quad (x \neq 0)$$

Now as $|x| \to \infty$, we have $1/x \to 0$ so that $f(x) \to 2/1 = 2$. Hence the line $y = 2$ is a horizontal asymptote. By observing that $x = 1$ is a vertical asymptote and plotting a few points as shown in the table, we can easily draw the graph (Figure 16).

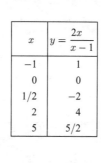

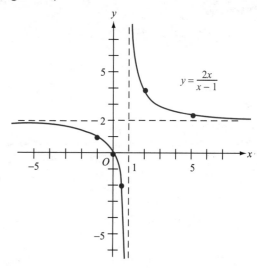

Figure 16

In our next example, the degree of the numerator is *greater than* that of the denominator.

Example 5 Sketch the graph of $f(x) = (x^2 + 1)/x$.

Solution:

$$f(x) = \frac{x^2 + 1}{x} = x + \frac{1}{x} \quad (x \neq 0)$$

Since $1/x \to 0$ as $|x| \to \infty$, we conclude that $f(x)$ is approximately equal to x for large values of x. The line $y = x$ is called an **oblique asymptote** for f. Hence, $f(x)$ does not approach any constant b as x increases in magnitude, and therefore $f(x)$ has *no* horizontal asymptote. By noting that the line $x = 0$ (the y-axis) is a vertical asymptote for $y = (x^2 + 1)/x$ and constructing the following table of values of x with corresponding values of y, we obtain the graph of f (Figure 17).

x	y	x	y
1/4	17/4	−1/4	−17/4
1/2	5/2	−1/2	−5/2
1	2	−1	−2
2	5/2	−2	−5/2
3	10/3	−3	−10/3
4	17/4	−4	−17/4

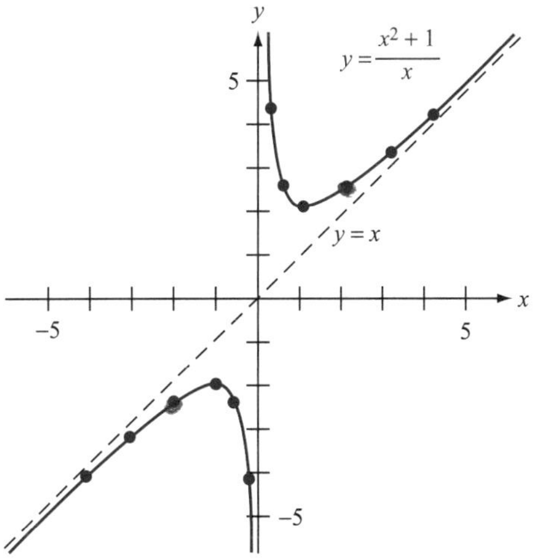

Figure 17

Note that only half of the table is really needed, since the graph is symmetric with respect to the origin. That is,

$$f(-x) = \frac{(-x)^2+1}{-x} = -\frac{x^2+1}{x} = -f(x). \quad \blacksquare$$

Examples 3, 4, and 5 illustrate the following result concerning **horizontal asymptotes**.

$x + \frac{1}{x}$ $\quad \frac{1x^2+1}{1x^1}$

Let $f(x) = \dfrac{a_m x^m + a_{m-1}x^{m-1} + \cdots + a_1 x + a_0}{b_n x^n + b_{n-1}x^{n-1} + \cdots + b_1 x + b_0}$ be a rational function, where $a_m \neq 0$ and $b_n \neq 0$. Then

1. if $m < n$, the line $y = 0$ (x-axis) is a horizontal asymptote for f;
2. if $m = n$, the line $y = a_m/b_n$ is a horizontal asymptote for f;
3. if $m > n$, f has no horizontal asymptote.

$\frac{Ax^m + C}{Bx^n}$

top exponent < bottom exponent $y = 0$

top exponent = bottom exponent $y = \frac{A}{B}$

top exponent > bottom exponent no H.A.

Note that a rational function of x can have at most one horizontal asymptote. The above result for horizontal asymptotes need not be memorized. For any particular rational function $f(x) = P(x)/Q(x)$, we can determine the existence of a horizontal asymptote by *first dividing $P(x)$ and $Q(x)$ by the highest power of x that is present in $f(x)$, and then examining the resulting expression as $|x| \to \infty$*. To see how this works, we will try a few examples.

1. $$\frac{3x^2+2x-4}{5x^3+1} = \frac{\frac{3}{x} + \frac{2}{x^2} - \frac{4}{x^3}}{5 + \frac{1}{x^3}}$$ *Divide numerator and denominator by* x^3

276 Chapter 4. Functions and their Graphs

As $|x| \to \infty$, the value of the resulting expression above approaches

$$\frac{0+0-0}{5+0} = 0,$$

So $y = 0$ is a horizontal asymptote.

2. $\dfrac{3x^2 + 2x - 4}{5x^2 + 1} = \dfrac{3 + \dfrac{2}{x} - \dfrac{4}{x^2}}{5 + \dfrac{1}{x^2}}$ *Divide numerator and denominator by x^2.*

As $|x| \to \infty$, the value of the resulting expression above approaches

$$\frac{3+0-0}{5+0} = \frac{3}{5},$$

so $y = 3/5$ is a horizontal asymptote.

3. $\dfrac{3x^2 + 2x - 4}{5x + 1} = \dfrac{3 + \dfrac{2}{x} - \dfrac{4}{x^2}}{\dfrac{5}{x} + \dfrac{1}{x^2}}$ *Divide numerator and denominator by x^2.*

As $|x| \to \infty$, the value of the resulting expression above approaches

$$\frac{3+0-0}{0+0},\qquad \text{Not defined.}$$

so there is no horizontal asymptote.

By including asymptotes along with the items previously considered for polynomials, we arrive at the following step-by-step method for graphing a rational function $f(x) = P(x)/Q(x)$. We assume that $P(x)$ and $Q(x)$ have no zeros in common.

Four-Step Procedure for Graphing Rational Functions.

1. *Test for symmetry.*
   Test for possible symmetry with respect to the y-axis or the origin by computing $f(-x)$. If either type of symmetry applies, it is sufficient to consider only $x \geq 0$.
2. *Solve for x-intercepts and vertical asymptotes.*
   Find the zeros of the numerator $P(x)$ to obtain the x-intercepts (if any). Find the zeros of the denominator $Q(x)$ to obtain the vertical asymptotes (if any).
3. *Determine the sign of $f(x)$.*
   The values of x determining x-intercepts and vertical asymptotes partition the x-axis into intervals, and inside each of these intervals the

4.3. Rational Functions

sign of $f(x)$ does not change. Determine the sign of $f(x)$ for a selected value of x in each interval.

4. *Solve for y-intercept and horizontal asymptote.*

   (a) Set $x = 0$ to find the y-intercept. (If 0 is not in D_f, then the line $x = 0$ is a vertical asymptote.)

   (b) Determine the horizontal asymptote, if any, by comparing the degree of $P(x)$ to that of $Q(x)$ or by considering the behavior of y as $|x| \to \infty$.

To draw the graph, we first draw broken lines for any asymptotes, horizontal or vertical. Next we plot all intercepts and the points determined in each interval by step 3. Finally, we join the points by a smooth curve broken only at the vertical asymptotes and approaching any horizontal asymptote as $|x| \to \infty$.

Example 6 Sketch the graph of $f(x) = x^2/(1 - x^2)$ by using the four-step procedure for graphing rational functions.

Solution:

1. *Test for symmetry.*

$$f(-x) = \frac{(-x)^2}{1 - (-x)^2} = \frac{x^2}{1 - x^2} = f(x)$$

Therefore, f is an even function and the graph is symmetric with respect to the y-axis.

2. *Solve for x-intercepts and vertical asymptotes.*
   If the numerator $x^2 = 0$, then $x = 0$. Hence, (0, 0) is the only x-intercept.
   If the denominator $1 - x^2 = 0$, then $x = 1$ or $x = -1$. Therefore, each of the lines $x = 1$ and $x = -1$ is a vertical asymptote.

3. *Determine the sign of $f(x)$.*
   Because of the symmetry with respect to the y-axis, it is necessary to consider only the intervals $[0, 1]$ and $[1, \infty)$ on the x-axis. (See the table). Symmetry with respect to the y-axis immediately yields the point $(-1/2, 1/3)$ in $[-1, 0]$ and the point $(-2, -4/3)$ in $(-\infty, -1]$.

4. *Solve for y-intercept and horizontal asymptote.*
   0 is in D_f and (0, 0) is the y-intercept (as well as the only x-intercept). Since the numerator x^2 and the denominator $1 - x^2$ are both of degree 2, the horizontal asymptote is $y = 1/(-1) = -1$. Alternatively, we note that

$$f(x) = \frac{x^2}{1 - x^2} = \frac{1}{\frac{1}{x^2} - 1} \to \frac{1}{0 - 1} = -1 \text{ as } |x| \to \infty.$$

To draw the graph (Figure 18), first draw broken lines for the vertical asymptotes $x = 1$ and $x = -1$ and for the horizontal asymptote $y = -1$.

Interval	[0, 1]	[1, ∞]
Selected x	1/2	2
Corresponding y	1/3	−4/3
Sign of $f(x)$	+	−

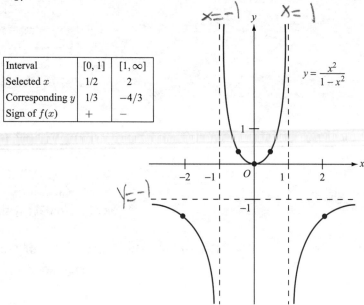

Figure 18

Next, plot the points (0, 0), (1/2, 1/3), (2, −4/3), (−1/2, 1/3), and (−2, −4/3). Finally, keeping in mind symmetry with respect to the y-axis and the sign of $f(x)$, join the points by a smooth curve broken at $x = 1$ and $x = -1$ and approaching $y = -1$ as $|x| \to \infty$. ∎

Comment As illustrated in Example 5, it is sometimes necessary to compute more points on the graph than those determined by "intercepts" and "sign of $f(x)$." Also, it is sometimes helpful to write the given expression for $f(x)$ in a different form. For instance, the function in Example 5 was written as

$$f(x) = x + \frac{1}{x}.$$

In this form we saw that for x large in magnitude, the term $1/x$ is relatively insignificant and the function behaves like

$$g(x) = x.$$

On the other hand, for x small in magnitude, x becomes the insignificant term and function behaves like

$$h(x) = \frac{1}{x}.$$

Similarly, we can decompose the function in Example 6 into a sum of simpler functions as follows:

$$f(x) = \frac{x^2 - 1 + 1}{1 - x^2} \qquad \text{Subtract and add } -1 \text{ in the numerator.}$$

$$= \frac{x^2 - 1}{1 - x^2} + \frac{1}{1 - x^2}$$

$$= -1 + \frac{1}{1 - x^2}$$

$$= -1 + \underbrace{\frac{1}{2} \cdot \frac{1}{1 - x}}_{\uparrow} + \underbrace{\frac{1}{2} \cdot \frac{1}{1 + x}}_{\uparrow} \cdot \qquad \text{See the section on partial fractions.}$$

↑ Most important term for x large in magnitude   ↑ Most important term for x close to 1   ↑ Most important term for x close to -1

In each example, the decomposition of $f(x)$ enables us to focus on a simple term that is the most significant contributor to the behavior of $f(x)$, depending on the location of x. From this point of view, go back and examine the graph in Example 6 and see if it becomes easier to understand.

Exercises 4.3

In Exercises 1-10, assume $P(x)$ and $Q(x)$ have no common zeros.
Fill in the blanks to make each statement true.

1. A rational function is the ratio of _____.
2. The line $x = a$ is a vertical asymptote for the graph of $f(x) = P(x)/Q(x)$ if _____.
3. The graph of $f(x) = (3x^2 + 2x - 1)/(2x^2 + 3x + 1)$ has the horizonal asymptote $y =$ _____.
4. The x-axis is a horizontal asymptote for the graph of a rational function $f(x) = P(x)/Q(x)$ if _____.
5. The graph of the rational function $f(x) = P(x)/Q(x)$ has no horizontal asymptote if _____.

Write true or false each statement.

6. A polynomial function is a rational function.
7. The domain of a rational function may be $(-\infty, \infty)$.
8. The graph of every rational function has at least one vertical asymptote.
9. The sign of a rational function $f(x) = P(x)/Q(x)$ always changes where $Q(x) = 0$.
10. For a rational function $f(x) = P(x)/Q(x)$, the roots of $P(x) = 0$ are x-intercepts.

Rational Functions

11. Which of the following are rational functions?

    (a) $f(x) = x^2 + 3x - 1$
    (b) $g(x) = \dfrac{x^2 + 3x - 1}{x^3 + 4}$
    (c) $h(x) = x^{-2} + 3x^{-1} + x^2$
    (d) $F(x) = (x^2 + 4)^{-3}$
    (e) $G(x) = (x^2 + 4)^{1/3}$
    (f) $H(x) = \sqrt[3]{2}$

12. Which of the following are not rational functions and why?

(a) $F_1(x) = x + \sqrt{2}$
(b) $F_2(x) = \sqrt{x} + 2$
(c) $F_3(x) = \dfrac{x + \sqrt{2}}{x - \sqrt{2}}$
(d) $F_4(x) = \dfrac{x^{-2}}{x^{-3} + 1}$
(e) $F_5(x) = \dfrac{x^{-1/2}}{x^{-3/2} + 1}$
(f) $F_6(x) = \dfrac{4}{x + 1}$

21. $f(x) = \dfrac{2x}{x + 1}$
22. $g(x) = \dfrac{2x}{x^2 + 1}$
23. $r(x) = \dfrac{3x^2 - 4x + 5}{1 - 2x^2}$
24. $h(x) = \dfrac{2x^2}{x + 1}$
25. $s(x) = 3x^2 - 4x + 5$
26. $t(x) = \dfrac{3x^2 - 4x + 5}{1 - 2x^3}$
27. $u(x) = \dfrac{3x^2 - 4x + 5}{1 - 2x}$
28. $v(x) = x^2 - \dfrac{1}{x + 1}$

Vertical Asymptotes

For each of the rational functions in Exercises 13-20, find the equations of the vertical asymptotes, if any.

13. $f(x) = \dfrac{x^2}{x + 1}$
14. $g(x) = \dfrac{x^2}{x^2 + 1}$
15. $h(x) = \dfrac{x^3}{x^2 - 1}$
16. $p(x) = \dfrac{x^2}{x^2 - 4x + 3}$
17. $q(x) = \dfrac{x^2}{x^2 + 4x + 3}$
18. $r(x) = \dfrac{x^2}{x^2 + 3x + 4}$
19. $s(x) = \dfrac{x^2}{x^3 - 1}$
20. $t(x) = \dfrac{x^3}{x^3 - 6x^2 + x - 6}$

Graphs of Rational Functions

Use the four-step procedure discussed in this section to sketch a graph for each of the following rational functions.

29. $f(x) = \dfrac{2}{x + 1}$
30. $g(x) = \dfrac{2x}{x + 1}$
31. $h(x) = \dfrac{1 - x}{x + 1}$
32. $F(x) = \dfrac{1}{x^2}$
33. $G(x) = \dfrac{1}{(x + 1)^2}$
34. $H(x) = \dfrac{1}{(x - 1)^2}$
35. $p(x) = \dfrac{x^2}{x^2 - 1}$
36. $q(x) = x - \dfrac{1}{x}$
37. $r(x) = \dfrac{2x}{x - 3}$
38. $s(x) = \dfrac{x}{x^2 + 1}$
39. $t(x) = \dfrac{x}{x^2 - 1}$
40. $u(x) = \dfrac{4 - x}{x - 2}$

Horizontal Asymptotes

For each of the rational functions in Exercises 21-28, find the equations of the horizontal asymptote, if any.

4.4 Algebraic Functions

Definition of an Algebraic Function
Graphing Algebraic Functions
Piecewise Algebraic Function

[1] More generally, if y satisfies an equation
$$P_n(x)y^n + P_{n-1}(x)y^{n-1} + \cdots + P_1(x)y + P_0(x) = 0,$$
where the coefficients $P_0(x), P_1(x), \ldots, P_{n-1}(x), P_n(x)$ are polynomials in x, then y is called an *algebraic function of x*.

Definition of an Algebraic Function. A function obtained by the algebraic operations of addition, subtraction, multiplication, division, and the taking of roots is called an **algebraic function**.[1] The rational functions are included among the algebraic ones, but some algebraic functions, such as $f(x) = \sqrt{x}$ and $g(x) = \sqrt[3]{(x-1)/(x+2)}$, are not rational.

4.4. Algebraic Functions

Example 1 Which of the following are algebraic functions?

(a) $f(x) = (x-3)^2$

(b) $g(x) = (x-3)^x$

(c) $h(x) = x^2 + \dfrac{2}{x} + \sqrt{\dfrac{x+2}{x-5}}$

(d) $F(x) = \sqrt[3]{x^{1/2} + x^{2/3} + 1}$

(e) $G(x) = x^{3.14} + x^{1.414}$

(f) $H(x) = x^\pi + x^{\sqrt{2}}$

Solution: First note that if m/n is a rational number, then $x^{m/n}$ is an algebraic expression because $x^{m/n} = \sqrt[n]{x^m}$. On the other hand, if r is an irrational number, then x^r is not an algebraic expression. Therefore, the functions (a), (c), (d), and (e) are algebraic functions, while those in (b) and (f) are not. ∎

Two limitations are placed on the domain of an algebraic function. First, if the function contains a term with a denominator, then the function is undefined wherever the denominator is zero. Second, if the function contains a term that is an even root, then the function is undefined wherever the radicand is negative.

Example 2 Find the domain of each of the following algebraic functions.

(a) $f(x) = \sqrt{x-2}$

(b) $g(x) = \sqrt[3]{x} + \dfrac{1}{\sqrt{x+4}}$

(c) $F(x) = x^{2/3} + x^{3/2}$

(d) $G(x) = x^{1.2} - x^{2.1}$

Solution:

(a) $\sqrt{x-2}$ is defined for $x - 2 \geq 0$ or $x \geq 2$. Therefore, $D_f = [2, \infty)$.

(b) $\sqrt[3]{x}$ is defined for all real number x. However, $1/\sqrt{x+4}$ is defined and its denominator is not zero only for $x + 4 > 0$ or $x > -4$. Therefore, $D_g = (-4, \infty)$.

(c) $x^{2/3}$ is defined for all real numbers x, and $x^{3/2}$ is defined for $x \geq 0$. Therefore, $D_F = [0, \infty)$.

(d) As a rational number, $1.2 = 12/10 = 6/5$. Hence $x^{1.2} = \sqrt[5]{x^6}$ is defined for all real numbers x. However, $2.1 = 21/10$, which means that $x^{2.1} = \sqrt[10]{x^{21}}$ is defined only for $x \geq 0$. Therefore, $D_G = [0, \infty)$. ∎

Graphing Algebraic Functions. We can apply the notions of symmetry, intercepts, asymptotes, and sign of $f(x)$ to the graphing of algebraic functions. However, special attention must be paid to the limitations placed on the domain of an algebraic function. Also, for algebraic functions, we define both vertical and horizontal asymptotes in terms of their characteristic properties.

> The vertical line $x = a$ is a **vertical asymptote** for an algebraic function $f(x)$ if $|f(x)| \to \infty$ as a $x \to a^+$ or a^-. The horizontal line $y = b$ is a **horizontal asymptote** for $f(x)$ if $f(x) \to b$ as $x \to \infty$ or $-\infty$.

These definitions are consistent with those given for rational functions, and our methods for finding asymptotes for rational function can be adapted to general algebraic functions. We now consider graphs of some algebraic functions, starting with a very simple case and working up to more complicated ones.

Example 3   Sketch the graph of $f(x) = \sqrt{x}$.

Solution: $\sqrt{x}$ is defined only for $x \geq 0$; also, $\sqrt{x} \geq 0$ so the graph is contained in the first quadrant. The only intercept is $(0, 0)$, and there are no asymptotes since $f(x) \to \infty$ if and only if $\to \infty$. The graph [Figure 19(a)] is the top half of the parabola $y^2 = x$ [Figure 19(b)].

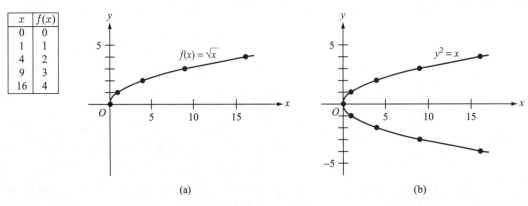

Figure 19

Example 4   Sketch the graph of $g(x) = \sqrt[3]{x}$.

Solution:
Domain: $(-\infty, \infty)$
Symmetry: $g(-x) = \sqrt[3]{-x} = -\sqrt[3]{x} = -g(x)$, so the graph is symmetric with respect to the origin.
Intercepts: $(0, 0)$
Asymptotes: none, since for $|g(x)| \to \infty \Leftrightarrow |x| \to \infty$.
Sign of $g(x)$: $g(x) > 0$ for $x > 0$, and $g(x) < 0$ for $x < 0$.
Since our analysis has given only one point, $(0, 0)$, we must first compute coordinates for a few more points, as shown in the table. Then, keeping

in mind symmetry with respect to the origin, we plot the graph as shown in Figure 20. ∎

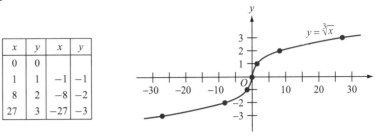

Figure 20

Example 5 Sketch the graph of $h(x) = \sqrt{(x-1)/(x+2)}$.

Solution: *Domain:* $\sqrt{(x-1)/(x+2)}$ is defined only for $(x-1)/(x+2) \geq 0$. The solution to this fractional inequality, by the method of Section 2.2, is $D_h = (-\infty, -2) \cup [1, \infty)$.
Symmetry: none, since $h(-x) \neq h(x)$ or $-h(x)$
Intercepts: The x-intercept is $(1, 0)$. There is no y-intercept, since 0 is not in D_h.
Asymptotes: $h(x) \to \infty$ as $x \to -2^-$, so the line $x = -2$ is a vertical asymptote. As $x \to \infty$ or ∞,

$$h(x) = \sqrt{\frac{1 - 1/x}{1 + 2/x}} \to \sqrt{\frac{1 - 0}{1 + 0}} = 1.$$

Therefore, the line $y = 1$ is a horizontal asymptote.
Sign of $h(x)$: By definition, $h(x)$ is nonnegative for all x in its domain. We can now graph the function h (Figure 21). It is worth nothing that to graph h only three points were actually computed. By analyzing the properties of h we were able to obtain its graph with a minimum number of computed points. ∎

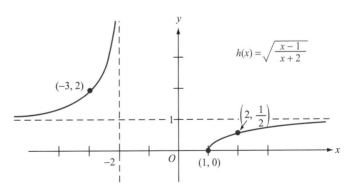

Figure 21

Comment In general, by analyzing domain, symmetry, intercepts, asymptotes, and the sign of $f(x)$, we have been able to graph some complicated functions without the need to compute a great number of points. However, for even more detailed information concerning the graph of a function, the power of calculus is needed. Calculus can give precise information on the following important properties of graphs. It can determine

(i) where the graph is rising and where it is falling.

(ii) the points at which the graph has its local maximums, and its local minimums.

(iii) where the graph is bending upward or and where it is bending downward or.

(iv) the points at which the graph changes its bend: or or or.

If we were given a cluster of points as in Figure 22(a), we would normally assume that the corresponding graph behaves as in Figure 22(b). However, a closer analysis, using items (i)–(iv) above, might show that the actual behavior is as in Figure 22(c).

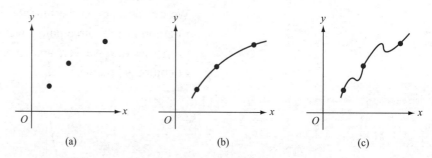

Figure 22 Graphing a Function from a Cluster of Points

Piecewise Algebraic Functions. The absolute value function $f(x) = |x|$ is an algebraic function since it can be written as $f(x) = \sqrt{x^2}$. More often, we write

$$f(x) = |x| = \begin{cases} x & \text{for } x \geq 0 \\ -x & \text{for } x < 0, \end{cases}$$

from which we can see that the graph of f is as in Figure 23. The graph of the absolute value function consists of two pieces, $y = x$ for $x \geq 0$ and $y = -x$ for $x < 0$. In general, if the graph of a function consists of several pieces, each one defined by a different algebraic expression, then the function is called a **piecewise algebraic function.** An important example is the **greatest integer function** $G(x) = [x]$, which is defined as the greatest integer less than or equal to x. That is,

$$G(x) = [x] = n \quad \text{for} \quad n \leq x < n+1,$$

where n is any integer. For instance, $[3] = 3$, $[3.95] = 3$, and $[-3.15] = -4$. The domain of G consists of the entire x-axis broken down into left-closed, right-open intervals $[n, n+1)$, and the value of $G(x)$ on $[n, n+1)$ is n, as shown in Figure 24, where points marked by closed dots are included, but those indicated by open dots are not.

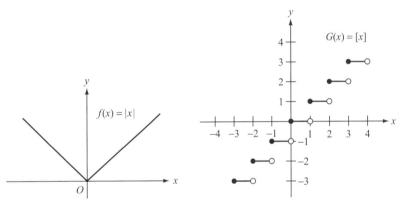

Figure 23   Figure 24

The graph of $G(x) = [x]$ shows why it is called a **step function.** Such functions occur often in applications, and one is described in Example 6 below.

Example 6 $P(x)$, the U.S. postage required in recent years for priority mailing of a letter weighing x ounces, has been defined as follows: the postage is 22 cents for a letter weighing one ounce or less plus 17 cents for each additional ounce or fraction thereof. Sketch the graph of the function P.

Solution: The graph of P is a step function with domain $x > 0$, as shown in Figure 25. ■

Example 7 Sketch the graph of H and give its domain and range if

$$H(x) = \begin{cases} x & \text{for } x \leq -1 \\ 1 & \text{for } -1 < x < 1 \\ x^2 & \text{for } x \geq 1. \end{cases}$$

286 Chapter 4. Functions and their Graphs

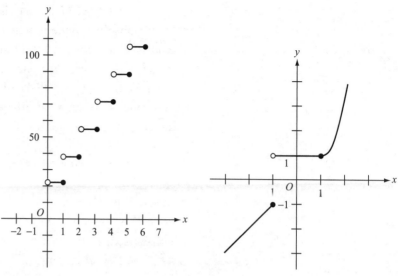

Figure 25

Figure 26

Solution: The definition of H shows that $D_H = (-\infty, \infty)$, and the graph of H (Figure 26) shows that $R_H = (\infty, -1] \cup [1, \infty)$. ∎

Comment The word *function* was first used in 1673 by Gottfried Wilhelm Leibniz to mean any quantity associated with the points on a curve. Later, function came to mean a relationship given by a single equation such as an algebraic function. The notation $f(x)$ was introduced by Leonhard Euler in 1734. The definition of a function as a correspondence that is not necessarily given by a single equation (for example, a piecewise algebraic function) is due to Peter Gustav Dirichlet (1805-1859). Today the function concept appears in every branch of mathematics.

Exercises 4.4

Fill in the blanks to make each statement true.

1. Rational functions are generated by the operations of ____, ____, ____, and ____. An algebraic function may also involve ____.

2. If an algebraic function contains a term that is an even root, then we must exclude from its domain all real numbers for which ____.

3. For $f(x) = \sqrt[n]{x}$ ($n = 3, 5, 7, \ldots$), the domain is ____.

4. If $F(x) = |x - 1|$, then $F(x) = x - 1$ for x ____ and $F(x) = 1 - x$ for x ____.

5. Included among the ____ functions are the rational functions, and included among the rational functions are the ____ functions.

Write true or false for each statement.

6. If $f(x)$ involves a square root, then its domain cannot be $(-\infty, \infty)$.

7. If $(-\infty, -1) \cup (1, \infty)$ is the implied domain of an algebraic function $f(x)$, then $f(x)$ must have a term that is an even root.

8. Every polynomial function is algebraic.

9. Some algebraic functions are not rational.
10. If f is a piecewise algebraic function, then its graph must be disconnected for some values of x.

Algebraic Functions

11. Which of the following algebraic functions are not rational?

    (a) $f(x) = \sqrt{3}$
    (b) $g(x) = \sqrt{x}$
    (c) $h(x) = (x+3)^{1/3}$
    (d) $F(x) = \dfrac{x^2 - 2x + 5}{x + 3}$
    (e) $G(x) = \dfrac{x^{1/2} - 2x + 5}{x + 3}$
    (f) $H(x) = \sqrt{\dfrac{x^2 - 2x + 5}{x + 3}}$

12. Which of the following are *not* algebraic functions?

    (a) $f(x) = x^2 + 2^x$
    (b) $g(x) = x^{\sqrt{3}} + \sqrt{3}$
    (c) $h(x) = (x)^{3/2} - x^{3.2}$
    (d) $F(x) = \sqrt[3]{x - 4}$
    (e) $G(x) = \dfrac{(x-4)^{1/3}}{x^2}$
    (f) $H(x) = \dfrac{x^3 + 3^x}{\sqrt{x}}$

13. Find the domain of each of the following algebraic functions.

    (a) $f(x) = \dfrac{x^3 - 3x^2 + 7x - 5}{x^2 - 2x + 1}$
    (b) $g(x) = \dfrac{x + 2}{x - 1}$
    (c) $h(x) = \sqrt{\dfrac{x + 2}{x - 1}}$
    (d) $F(x) = \sqrt[3]{\dfrac{x + 2}{x - 1}}$
    (e) $G(x) = \sqrt{\dfrac{x^2 - 1}{x^2 + 1}}$
    (f) $H(x) = \sqrt{\dfrac{x^2 - 1}{x^2 - 4}}$

Sketch the graph of each of the following algebraic functions.

14. $f(x) = \sqrt{x - 1}$
15. $g(x) = \sqrt{x + 2}$
16. $h(x) = \sqrt{x} + 2$
17. $F(x) = \sqrt[3]{x + 1}$
18. $G(x) = \sqrt[3]{x - 1}$
19. $H(x) = \sqrt[3]{x} - 1$
20. $f(x) = \sqrt{\dfrac{x - 2}{x + 1}}$
21. $g(x) = \sqrt{\dfrac{x + 2}{x - 1}}$
22. $h(x) = \sqrt{\dfrac{x - 1}{x - 2}}$

Piecewise Algebraic Functions

Sketch the graphs of the following piecewise algebraic functions.

23. $f(x) = |x| - 2$
24. $g(x) = |x| + 2$
25. $F(x) = |x - 2|$
26. $G(x) = |x + 2|$
27. $u(x) = [x] + 2$
28. $v(x) = [x + 2]$
29. $w(x) = [x] - x$

30. $f(x) = \begin{cases} 1 & \text{for } x \leq 0 \\ 2 & \text{for } x > 1 \end{cases}$

31. $g(x) = \begin{cases} x & \text{for } x \leq 1 \\ 1 & \text{for } x > 1 \end{cases}$

32. $h(x) = \begin{cases} -1 & \text{for } x < 0 \\ 0 & \text{for } x = 0 \\ 1 & \text{for } x > 0 \end{cases}$

33. $F(x) = \begin{cases} x & \text{for } x \leq 0 \\ x^2 & \text{for } x > 0 \end{cases}$

34. $G(x) = \begin{cases} x & \text{for } x \leq 1 \\ x^2 & \text{for } x > 1 \end{cases}$

35. $H(x) = \begin{cases} x & \text{for } x \leq 2 \\ x^2 & \text{for } x > 2 \end{cases}$

36. According to the I.R.S. in 1985, the income tax $T(x)$ for single taxpayers on x dollars of taxable income, with $\$9{,}500 \leq x < \$10{,}000$, was given by the following table. Sketch a graph of the function T.

Interval containing x	$T(x)$
[9500, 9550)	999
[9550, 9600)	1007
[9600, 9650)	1015
[9650, 9700)	1023
[9700, 9750)	1031
[9750, 9800)	1039
[9800, 9850)	1047
[9850, 9900)	1055
[9900, 9950)	1063
[9950, 10000)	1071

4.5 One-to-One and Inverse Functions

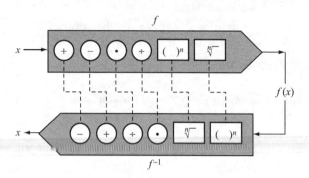

One-to-One Functions
Graphs of One-to-One Functions
Inverse Functions
Graphs of Inverse Functions

In Section R.2, we learned that every real number a has an additive inverse $-a$, and every $a \neq 0$ has a multiplicative inverse $a^{-1} = 1/a$. We now investigate the idea of an inverse for a function with respect to the operation of composition.

One-to-One Functions. If we consider the functions defined by

$$y = f(x) = 3x - 2 \tag{1}$$
$$\text{and} \quad y = F(x) = x^2 - 4, \tag{2}$$

and solve equations (1) and (2) for x, we obtain

$$x = \frac{y+2}{3} \tag{3}$$
$$\text{and} \quad x = \pm\sqrt{y+4}. \tag{4}$$

In equation (3), each real number y determines exactly one real number x, so that we may write

$$x = g(y) = \frac{y+2}{3},$$

that is, x is a function of y. But in equation (4), each $y > -4$ determines *two* real numbers x, and therefore x is *not* a function of y.

By the definition of a function $y = f(x)$, each x in the domain of f must correspond to exactly one y in the range of f. If in addition each y in the range of f corresponds to exactly one x in the domain of f, then f is called a **one-to-one function.** This means that f is one-to-one if the equation $\boldsymbol{f(x_1) = f(x_2)}$ **implies** $\boldsymbol{x_1 = x_2}$. On the other hand, f is not one-to-one if the equation $f(x_1) = f(x_2)$ can be satisfied for different values x_1 and x_2 of x.

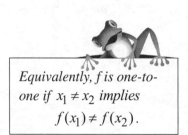

Equivalently, f is one-to-one if $x_1 \neq x_2$ implies $f(x_1) \neq f(x_2)$.

Example 1 Which of the following are one-to-one functions?

(a) $f(x) = x^3$ (b) $F(x) = \sqrt{x-4}, \quad x \geq 4$

(c) $g(x) = x^4 - 2x^2$ (d) $G(x) = x^2$, $x \leq 0$

Solution:

(a) Solving $y = x^3$ for x, we obtain $x = \sqrt[3]{y}$. Since every real number y has exactly one real cube root, f is one-to-one. To put it another way, if $x_1^3 = x_2^3$, then $x_1 = x_2$; that is, $f(x_1) = f(x_2)$ implies $x_1 = x_2$.

(b) Since $g(1) = g(-1) = -1$, g is not a one-to-one function.

(c) From $y = \sqrt{x-4}$ $(x \leq 4)$, we find that $x = y^2 + 4$. Thus, each y determines a unique x, so that F is one-to-one. Alternatively, if $\sqrt{x_1 - 4} = \sqrt{x_2 - 4}$, then $x_1 - 4 = x_2 - 4$ and $x_1 = x_2$. Therefore, $F(x_1) = F(x_2)$ implies $x_1 = x_2$.

(d) If $y = G(x) = x^2$ for $x \leq 0$, then $x = -\sqrt{y}$ for $\sqrt{y} \geq 0$. G is one-to-one because each nonnegative y has exactly one nonpositive square root. Alternately, if x_1 and x_2 are *nonpositive* and $x_1^2 = x_2^2$, then $x_1 = x_2$. Hence, $G(x_1) = G(x_2)$ implies $x_1 = x_2$. ∎

Graphs of one-to-one Functions. In Section 4.1 we used the vertical line test to distinguish between the graph of a function f and the graph of an equation that could not be put in the form $y = f(x)$. Similarly, we may use the **horizontal line test** to determine whether a function f is one-to-one.

A graph is that of a function if it passes the vertical line test, and a function is one-to-one if its graph passes the horizontal line test.

f is a one-to-one function if and only if any horizontal line $y = c$ intersects the graph of f in at most one point.

Example 2 Use the horizontal line test to determine which of the functions in Example 1 are one-to-one.

Solution: We first sketch the graphs (Figure 27-30) and then apply the horizontal line test to each.

(a) In Figure 27, any horizontal line $y = c$ intersects the graph of f in exactly one point. Therefore, f is one-to-one.

(b) In Figure 28, the line $y = c$ $(-1 < c < 0)$ intersects the graph of g in four points. Therefore, g is not one-to-one.

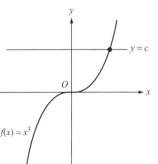

Figure 27

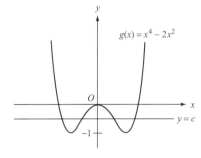

Figure 28

(c) In Figure 29, any horizontal line $y = c$ intersects the graph of F in at most one point. Therefore, F is one-to-one.

(d) In Figure 30, any horizontal line $y = c$ intersects the graph of G in at most one point. Therefore, G is one-to-one. ∎

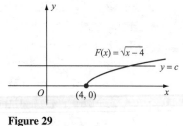

Figure 29

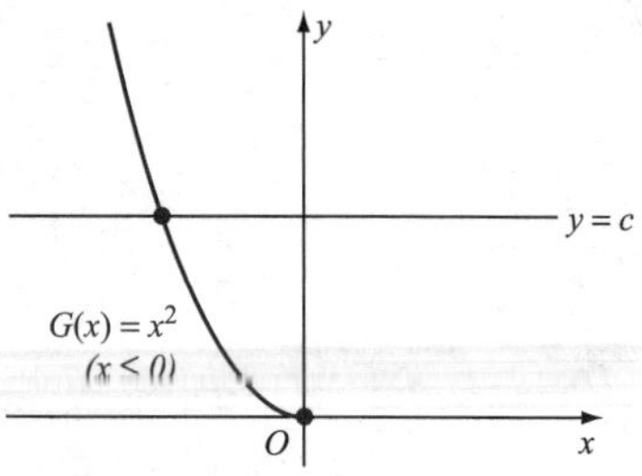

Figure 30

Note that in Figure 27 and 29 the graphs rise from left to right. Hence, the corresponding functions can be described as *increasing* functions. Similarly, the function in Figure 30 is a *decreasing* function. The algebraic definition of such functions is as follows.

> Let x_1 and x_2 be two points in D_f. Then
> f is an **increasing function** if $f(x_1) > f(x_2)$ whenever $x_1 > x_2$, and
> f is a **decreasing function** if $f(x_1) < f(x_2)$ whenever $x_1 > x_2$.

Example 3 Let $f(x) = 2x + 3$ and $g(x) = 1 - x$, where $D_f = (-\infty, \infty) = D_g$. Use the algebraic definition above the show that
(a) $f(x)$ is an increasing function and (b) $g(x)$ is a decreasing function.

Solution:

(a) Suppose x_1 and x_2 are any two real numbers satisfying $x_1 > x_2$. Then

$$x_1 > x_2 \quad \text{Given.}$$
$$2x_1 > 2x_2 \quad \text{Multiplication by a positive number 2 preserves the direction of an inequality.}$$

$$2x_1 + 3 > 2x_2 + 3. \quad \text{Addition preserves the direction of an inequality.}$$
$$f(x_1) > f(x_2)$$

Therefore, f is an increasing function.

(b) Again suppose that $x_1 > x_2$. We have

4.5. One-to-One and Inverse Functions

$$x_1 > x_2 \qquad \text{\it Given.}$$
$$-x_1 < -x_2 \qquad \text{\it Multiplication by a negative number } -1 \text{ \it reverses the direction of an inequality.}$$
$$1 - x_1 < 1 - x_2, \qquad \text{\it Addition preserves the direction of an inequality.}$$
$$g(x_1) < g(x_2).$$

Therefore, g is a decreasing function. ■

If f is an increasing or a decreasing function, then f is one-to-one (*why?*). For a continuous function (one whose graph has no breaks) with an interval domain, the converse is also true. That is, if a continuous function f is one-to-one and D_f is an interval, then f is either an increasing or a decreasing function. Note that the function g in Figure 28 is neither increasing nor decreasing, and g is also not one-to-one.

Inverse Functions. It was observed above that the function f defined in equation (1) by

$$y = f(x) = 3x - 2$$

is one-to-one, and therefore x is a function of y; that is,

$$x = g(y) = \frac{y+2}{3}.$$

Using the definition of composition of functions given in Section 4.1, we have

$$(g \circ f)(x) = g(f(x)) = g(3x - 2) = \frac{(3x - 2) + 2}{3} = x.$$

This result is shown schematically in Figure 31.

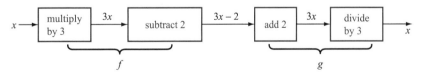

Figure 31 g undoes f

Also,

$$(f \circ g)(x) = f(g(x)) = f\left(\frac{x+2}{3}\right) = 3\left(\frac{x+2}{3}\right) - 2 = x,$$

which has the corresponding diagram (Figure 32).

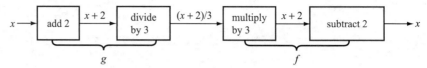

Figure 32 f undoes g

This example illustrates that when f is a one-to-one function, there is another function g for which

$$(g \circ f)(x) = x \quad \text{and} \quad (f \circ g)(x) = x. \tag{5}$$

The function g that satisfies both equations in (5) is called the **inverse** of f and is denoted by f^{-1}. Thus, in general,

$$(f^{-1} \circ f)(x) = x \quad \text{and} \quad (f \circ f^{-1})(x) = x, \tag{6}$$

as shown in Figure 33. Also, $D_{f^{-1}} = R_f$ and $R_{f^{-1}} = D_f$.

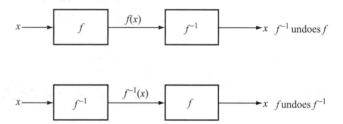

Figure 33

Using the example above, we can compute $f^{-1}(x)$ from $f(x)$ in the following three steps.

1. $y = 3x - 2 = f(x)$ *Given function.*
2. $x = \dfrac{y+2}{3} = f^{-1}(y)$ *Solve equation in step 1 for x in terms of y and set equal to $f^{-1}(y)$*
3. $y = \dfrac{x+2}{3} = f^{-1}(x)$ *Interchange x and y in step 2.*

We interchange x and y to solve for f^{-1} because we prefer x to denote the variable in the domain and y the variable in the range of f^{-1}.

Example 4   Find the inverse function for each of the one-to-one functions in Example 1, including the domain and range of each inverse function. Also verify that equations (6) are satisfied in each case.

Solution:
1. $y = x^3 = f(x)$, with $D_f = (-\infty, \infty) = R_f$
2. $x = \sqrt[3]{y} = f^{-1}(y)$

3. $y = \sqrt[3]{x} = f^{-1}(x)$, with $D_{f^{-1}} = R_f = (-\infty, \infty)$ and $R_{f^{-1}} = D_f = (-\infty, \infty)$

Here, equations (6) become

$$(f^{-1} \circ f)(x) = f^{-1}(f(x)) = f^{-1}(\sqrt{x^3}) = \sqrt[3]{x^3} = x$$
and $$(f \circ f^{-1})(x) = f(f^{-1}(x)) = f(\sqrt[3]{x}) = (\sqrt[3]{x})^3 = x.$$

(b) $g(x) = x^4 - 2x^2$ is not one-to-one and therefore has no inverse.

(c) 1. $y = \sqrt{x-4} = F(x)$, with $D_F = [4, \infty)$ and $R_F = [0, \infty)$
2. $x = y^2 + 4 = F^{-1}(y)$
3. $y = x^2 + 4 = F^{-1}(x)$, with $D_{F^{-1}} = R_F = [0, \infty)$, and $R_{F^{-1}} = D_F = [4, \infty)$

Equations (6) become

$$(F^{-1} \circ F)(x) = F^{-1}(F(x))$$
$$= F^{-1}(\sqrt{x-4}) = (\sqrt{x-4})^2 + 4$$
$$= x - 4 + 4 = x$$
and $$(F \circ F^{-1})(x) = F(F^{-1}(x)) = F(x^2 + 4)$$
$$= \sqrt{(x^2 + 4) - 4} = \sqrt{x^2} = |x| = x, \text{ since } x \geq 0$$

(d) 1. $y = G(x) = x^2$, with $D_G = (-\infty, 0]$ and $R_G = [0, \infty)$
2. $x = -\sqrt{y} = G^{-1}(y)$ (*Why the negative square root?*)
3. $y = -\sqrt{x} = G^{-1}(x)$, with $D_{G^{-1}} = R_G = [0, \infty)$ and $R_{G^{-1}} = D_G = (-\infty, 0]$

Equation (6) become

$$(G^{-1} \circ G)(x) = G^{-1}(G(x))$$
$$= G^{-1}(x^2) = -\sqrt{x^2} = -|x| = x, \text{ since } x \leq 0$$
and $$(G \circ G^{-1})(x) = G(G^{-1}(x)) = G(-\sqrt{x}) = (-\sqrt{x})^2 = x. \blacksquare$$

Question 1

x^{-1} *denotes the inverse of x with respect to multiplication, and* f^{-1} *denotes the inverse of f with respect to composition.*

An inverse element for a given operation is usually defined in terms of the identity element for that operation (see basic properties (6) through (9) in Section R.2). By definition, f^{-1} is the inverse of f with respect to the operation of composition. What is the identity element for composition?

Answer: The identity element for composition is the function $I(x) = x$ because $f \circ I = f = I \circ f$ for any function f (see Exercise 29). $\blacksquare$

Graphs of Inverse Functions. We now draw the graphs of some one-to-one functions and their inverses on the same coordinate axes. From Example 4(a), $f(x) = x^3$ and $f^{-1}(x) = \sqrt[3]{x}$ (Figure 34).

294 Chapter 4. Functions and their Graphs

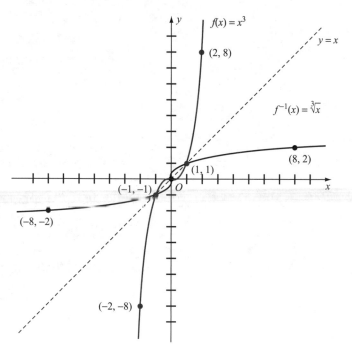

Figure 34 Graphs of $f(x) = x^3$ and $f^{-1}(x) = \sqrt[3]{x}$

The graph of f contains the points $(1,1)$, $(2,8)$, $(-1,-1)$, and $(-2,-8)$, while the graph of f^{-1} contains the corresponding points $(1,1)$, $(8,2)$, $(-1,-1)$, and $(-8,-2)$. In general, if the point (a, b) is on the graph of f, then the point (b, a) is on the graph of f^{-1}. This follows from the fact that f^{-1} is obtained from f by interchanging x and y, and it means that

the graphs of f and f^{-1} are symmetric with respect to the line $y = x$.

If we were to fold the page along the line $y = x$ in Figure 34, the graphs of f and f^{-1} would coincide. The line segment joining point (a, b) on the graph of f to point (b, a) on the graph of f^{-1} has the line $y = x$ as its perpendicular bisector. This symmetry property is true not only for the function in Example 4(a) but for *any* one-to-one function and its inverse.

Example 5 Sketch the graphs of $F(x) = \sqrt{x - 4}$ and its inverse $F^{-1}(x) = x^2 + 4$ $(x \geq 0)$ on the same axes [see Example 4(c)].

Solution: The graphs of F and F^{-1} are symmetric with respect to the line $y = x$, as shown in Figure 35. ■

Example 6 Sketch the graphs of $G(x) = x^2$ $(x \leq 0)$ and its inverse $G^{-1}(x) = -\sqrt{x}$ on the same axes [see Example 4(d)].

Solution: First the graph of $G(x) = x^2$ $(x \leq 0)$ is sketched; then the graph of G^{-1} is obtained by reflection about the line $y = x$ (Figure 36). ■

4.5. One-to-One and Inverse Functions   295

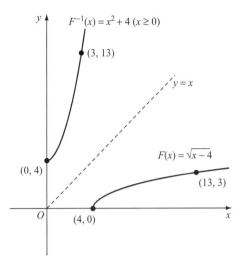

Figure 35 Graphs of $F(x) = \sqrt{x-4}$ and $F^{-1}(x) = x^2 + 4\,(x \geq 0)$

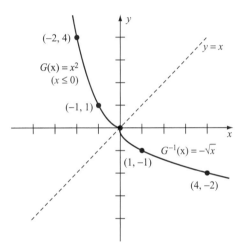

Figure 36 Graphs of $G(x) = x^2\,(x \leq 0)$ and $G^{-1}(x) = -\sqrt{x}$

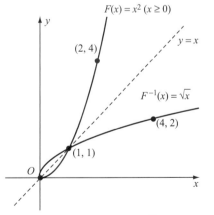

Figure 37 Graphs of $F(x) = x^2\,(x \geq 0)$ and $F^{-1}(x) = \sqrt{x}$

Some functions that are not one-to-one and therefore have no inverse can be made one-to-one by restricting their domains. This was done in Example 1(d) where the graph of $G(x) = x^2\,(x \leq 0)$ (Figure 30 and 36) is the left branch of the parabola $y = x^2$, and $G^{-1}(x) = -\sqrt{x}$. Similarly, the graph of $F(x) = x^2\,(x \geq 0)$ (Figure 37), is the right branch of the same parabola $y = x^2$, and $F^{-1}(x) = \sqrt{x}$.

Comment   In Example 4 we were able to determine the inverses of several basic one-to-one functions. However, there are many one-to-one functions whose inverse cannot be computed algebraically. For example, the function

$$f(x) = 1 + x + \frac{x^3}{27} + \frac{x^5}{256}$$

is 1 plus the sum of the increasing functions x, $x^3/27$, and $x^5/256$ and is therefore itself an increasing function. As an increasing function, f has an inverse, but to determine f^{-1} it is necessary to solve the fifth-degree equation

$$y = 1 + x + \frac{x^3}{27} + \frac{x^5}{256}$$

for x in terms of y. We cannot solve this equation, but we can obtain a graph of f^{-1} by first graphing f and then reflecting about the line $y = x$ (Figure 38).

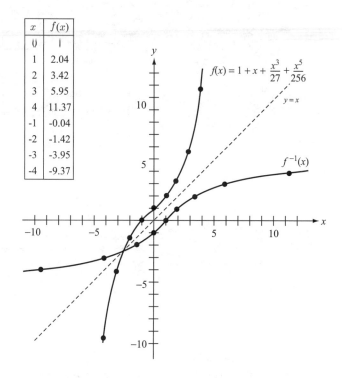

Figure 38 Graphs of $f(x) = 1 + x + x^3/27 + x^5/256$ and $f^{-1}(x)$

Exercises 4.5

Fill in the blanks to make each statement true.

1. For any function $y = f(x)$, each value of _____ in D_f corresponds to exactly one value of _____ in R_f.

2. If $y = f(x)$ has the property that each value of _____ in R_f corresponds to exactly one value of _____ in D_f, then f is called one-to-one.

3. f is one-to-one if the line $y = c$ intersects the graph of f in _____.

4. If the function g is the inverse of the function f, then _____ $= x$ for all x in D_f, and _____ $= x$ for all x in D_g.

5. The domain of f^{-1} equals _____ of f, and the range of f^{-1} equals _____ of f.

Write true or false for each statement.

6. If the equation $x_1 = x_2$ always implies that $f(x_1) = f(x_2)$, then f is a one-to-one function.

7. If the equation $f(x_1) = f(x_2)$ always implies that $x_1 = x_2$, then f is a one-to-one function.

8. If f is not a one-to-one function, then f has no inverse.

9. If f is a one-to-one function, then every horizontal line $y = c$ must intersect the graph of f in one point.

10. The graphs of f and f^{-1} are symmetric with respect to the line $y = x$.

One-to-One Functions

11. Which of the following functions are one-to-one?

(a) $f(x) = 2x + 4$
(b) $g(x) = \dfrac{1}{2x+4}$
(c) $h(x) = x^2 + 4$
(d) $F(x) = \sqrt{x^2 + 4}$
(e) $G(x) = x^3 + 4$
(f) $H(x) = \sqrt[3]{x - 4}$

12. Which of the following are *not* one-to-one functions and why?

(a) $p(x) = x^2$
(b) $q(x) = x^2 \quad (x \geq 1)$
(c) $r(x) = x^2 \quad (x \leq 1)$
(d) $s(x) = x^4$
(e) $t(x) = x^4 \quad (x \geq 0)$
(f) $u(x) = x^5$

13. Use the horizontal line test to determine which of the following are graphs of one-to-one functions.

(a)
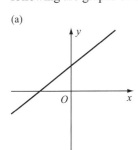

(b)
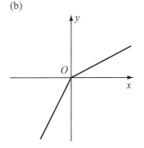

(c)
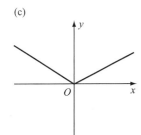

(d)
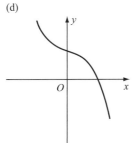

(e)
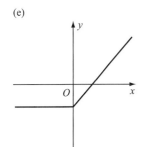

(f)
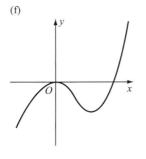

Inverse Functions

Determine $f^{-1}(x)$, including its domain and range, for each of the following.

14. $f(x) = 2x + 4$
15. $f(x) = x^3 - 1$
16. $f(x) = \dfrac{x^5}{5}$
17. $f(x) = \sqrt{x + 2}$
18. $f(x) = \sqrt[3]{x + 2}$
19. $f(x) = \sqrt[3]{x} + 2$

20. For each of the functions in Exercises 14, 16, and 18, verify that $(f^{-1} \circ f)(x) = x$ and $(f \circ f^{-1})(x) = x$.

21. For each of the functions in Exercises 15, 17, and 19, sketch the graphs of f and f^{-1} on the same axes.

Determine $g^{-1}(x)$, including its domain and range, for each of the following.

22. $g(x) = x^2 \quad (x \geq 1)$

23. $g(x) = x^2 \quad (x \leq -1)$

24. $g(x) = x^2 - 1 \quad (x \geq 1)$

25. $g(x) = x^2 - 1 \quad (x \leq -1)$

26. For each of the functions in Exercises 22 and 24, sketch the graphs of g and g^{-1} on the same axes.

27. For each of the functions in Exercises 23 and 25, verify the equations $(g^{-1} \circ g)(x) = x$ and $(g \circ g^{-1})(x) = x$.

28. Let $F(x) = \dfrac{x+1}{x} \quad (x \neq 0)$.

(a) Show that F is a one-to-one function.

(b) Solve for $F^{-1}(x)$.

(c) Verify that $(F^{-1} \circ F)(x) = x$ and $(F \circ F^{-1})(x) = x$.

(d) Sketch the graphs of F and F^{-1} on the same axes.

29. Show that if $I(x) = x$, then $(f \circ I)(x) = f(x)$ and $(I \circ f)(x) = f(x)$ for all x in D_f. Hence, conclude that $f \circ I = f$ and $I \circ f = f$.

4.6 Variation

Formulas in geometry and science give us a precise expression of the variation among related quantities.

Direct and Inverse Variation
Joint and Combined Variation

In mathematics, the term **variation** refers to a relationship of one quantity to another. For example, the equation

$$A = \pi r^2$$

says that the area of a circle varies directly as the square of its radius. Equivalently, we say that the area is *directly proportional* to the square of the radius. Here, π is called the constant of proportionality, and the word *directly* means that when r increases in magnitude, so does A. On the other hand, Newton's law of gravitation

$$F = k \left(\frac{m_1 m_2}{r^2} \right)$$

says that the force of attraction between two bodies is directly proportional to the product of their masses and is *inversely* proportional to the square of the distance between them. Here, F decreases as r increases. In either case, the variation between the quantities involved is defined by a certain kind of algebraic function.

Direct and Inverse Variation. As illustrated above, the statement that a quantity y **varies directly** as a quantity x, or y is **directly proportional** to x, means that for some nonzero number k,

$$\boldsymbol{y = kx,}$$

and k is called the **constant of variation**, or the **proportionality constant.**

Example 1 Hooke's law states that the force needed to stretch a spring d units is directly proportional to d. Translate Hooke's law into an equation. If a 6-pound weight stretches a spring 2 inches, find the constant of proportionality (spring constant).

Solution: Let F denote the force in pounds needed to stretch the spring d inches. Hooke's law says

$$F = kd,$$

where k is the proportionality constant. By substituting $F = 6$ pounds and $d = 2$ inches, we obtain

$$6 \text{ pounds} = k \cdot 2 \text{ inches}$$

or $\qquad k = 3$ pounds per inch. ∎

Example 2   If an object is dropped from above the ground, then the distance it falls before hitting the ground varies directly as the square of its falling time. A ball dropped from a hot-air balloon falls 1600 feet in 10 seconds. Find the constant of variation and the distance the ball falls in 15 seconds.

Solution: Let d denote the distance fallen and t the falling time. We have

$$d = kt^2,$$

where k is the constant of variation. By substituting $d = 1600$ feet and $t = 10$ seconds, we obtain

$$1600 \text{ feet} = k \cdot 100 \text{ seconds}^2,$$

or $\qquad k = 16$ feet per second^2.

Therefore, the distance fallen in 15 seconds is

$$d = 16 \cdot 15^2$$
$$= 3600 \text{ feet.} \quad \blacksquare$$

As in the case of Newton's inverse square law, we say that a quantity y **varies inversely** as a quantity x, or y is **inversely proportional** to x, if

$$y = \frac{k}{x},$$

and again k is the constant of variation, or the proportionality constant.

Example 3   At a constant temperature, the volume V of a gas varies inversely as the pressure P. If the volume is 900 cubic inches when the pressure is 10 pounds per square inch, what pressure will reduce the volume to 500 cubic inches?

Solution: We are told that

$$V = \frac{k}{P},$$

where k is a constant. Also,

$$900 \text{ cubic inches} = \frac{k}{10 \text{ pounds per square inch}}$$

implies that

$$k = 900 \text{ cubic inches} \cdot 10 \text{ pounds per square inch}$$
$$= 9000 \text{ inch-pounds}.$$

Therefore,

$$V = \frac{9000}{P},$$

and when $V = 500$ cubic inches, we have

$$P = \frac{9000 \text{ inch-pounds}}{500 \text{ cubic inches}}$$
$$= 18 \text{ pounds per square inch}.$$

As implied by the inverse variation, the pressure must be increased in order to decrease the volume. ∎

Joint and Combined Variation. If one quantity y is directly proportional to the *product* of two or more other quantities, we say that y **varies jointly** as each of the other quantities, Examples of **joint variation** are:

$T = kPV$ (the temperature T of a gas varies jointly as the pressure P and the volume V);

$V = RI$ [the voltage drop V across a resistor varies jointly ($k = 1$) as the resistance R and the current I];

$v = \pi r^2 h$ [the volume v of a cylinder varies jointly ($k = \pi$) as the radius r of the base squared and the height h].

If a relationship among quantities involves both direct and inverse variation, we say the relationship is one of **combined variation**. Some examples of combined variation are:

$C = k\dfrac{A}{d}$ (the capacitance C of a parallel plate capacitor varies directly as the area A of the plates and inversely as the distance d between the plates);

$E_p = k\dfrac{q_1 q_2}{r}$ (the potential energy E_p of a system of two point charges varies jointly as the values q_1 and q_2 of the charges and inversely as the distance r between them);

$F = k\dfrac{m_1 m_2}{r^2}$ (the force of attraction between two bodies varies jointly as their masses and inversely as the square of the distance between them).

Example 4 The weight of an object on the surface of the earth is defined as the force of attraction between the object and the earth. We assume that the mass of the earth is concentrated at its center. According to Newton's law of gravitation,

$$w = k\dfrac{Mm}{R^2}, \qquad \text{where} \quad \begin{cases} w = \text{weight of the object} \\ m = \text{mass of the object} \\ M = \text{mass of the earth} \\ R = \text{radius of the earth} \\ k = \text{proportionality constant} \end{cases}$$

Now k and M are constants, and for a given object, m is also constant. However, the earth is not a perfect sphere, so R and therefore w vary according to the object's position on the earth. If a person weighs 110 pounds at the north pole ($R \approx 3950$ feet), what does the person weigh at the equator ($R \approx 3963$ feet)?

Solution: Since the product kMm is the same at the north pole and the equator, we can replace it with a single constant c:

$$w = \dfrac{c}{R^2}.$$

By using the person's weight at the north pole, we have

$$110 = \dfrac{c}{3950^2}$$
$$c = 1{,}716{,}275{,}000.$$

Therefore, at the equator,

$$w = \dfrac{1{,}716{,}275{,}000}{3963^2}$$
$$= 109.28 \text{ pounds}$$
$$\approx 109 \text{ pounds}, 4 \text{ ounces}.$$

Hence, the person loses approximately 12 ounces by traveling to the equator. ■

Example 5 While at her bank, Heather notices that, on the average, she has to wait in line for 5 minutes when there are 4 people ahead of her and 2 tellers working. If she arrives at the bank with 12 people ahead of her and 3 tellers working, how long can she expect to wait?

Solution: Let

$W =$ Heather's waiting time,

$p =$ the number of people ahead of her, and

$t =$ the number of tellers working.

We assume that for some constant k,

$$W = k\left(\frac{p}{t}\right).$$

(*Why is our assumption reasonable?*) We are given $W = 5$ when $p = 4$ and $t = 2$. Therefore,

$$5 = k\left(\frac{4}{2}\right),$$

which means that $k = 2.5$. Therefore, when $p = 12$ and $t = 3$, we get

$$W = 2.5\left(\frac{12}{3}\right) = 10.$$

Hence, Heather can expect to wait about 10 minutes. ∎

Exercises 4.6

In the following exercises, whenever the constant k appears, it is assumed that $k \neq 0$.

Fill in the blanks to make each statement true.

1. If y varies directly as x, then $y/x =$ -----.
2. If $y = k/(x^2)$, then y varies-----as-----.
3. When y varies -----as x, the magnitude of y increases as that of x increases, but when y varies-----as x, the magnitude of y decreases as that of x increases.
4. The formula $A = (1/2)bh$ for the area of a triangle may be read as "the area varies-----," where $1/2$ is the-----.
5. The statement that v varies jointly as the square of x and the cube of y and inversely as z translates into the formula-----.

Write true or false for each statement.

6. Direct variation of y with x is represented by a linear equation.
7. Every linear equation in two variables x and y represents direct variation between x and y.
8. The equation $xy = k$ represents inverse variation between y and x.
9. If y varies directly as x, then x varies directly as y.
10. If y varies inversely as x, then x varies directly as y.

Direct and Inverse Variation

Translate each of the following statements into an equation, using k as the constant of variation.

11. d varies directly as t.
12. P varies inversely as V.
13. I varies directly as R.
14. F varies inversely as d^2.
15. T varies directly as the square root of L.

Translate each of the formulas in Exercises 16-20 into a statement of variation.

16. $I = kP$, where I is simple interest and P is principal invested
17. $F = kv^6$, where F is the force exerted by flowing water and v is its velocity.
18. $f = k\sqrt{T}$, where f is frequency and T is tension in a stretched string
19. $t = k/r$, where t is the time to travel between two cities and r is the average rate of speed
20. $R = k/D^2$, where R is the resistance of a conducting wire and D is its diameter
21. If y varies directly as x^2 and $y = 12$ when $x = 2$, find y when $x = 6$.
22. If y is inversely proportional to x^2 and $y = 12$ when $x = 2$, find y when $x = 6$.
23. For a given principal, simple interest varies directly as the rate of interest. If the interest earned is $80 at 5%, what is the interest at 7%?
24. For a fixed amount of simple interest, the present value of an investment varies inversely as the interest rate. Find the present value corresponding to 8% interest if the present value is $1000 at 6%.
25. The distance required for an automobile to stop varies as the square of its speed when the brakes are applied. If a car going 55 mph brakes to a stop in 151.25 ft, how fast is the car traveling if 175 ft are required to stop?
26. The cost of manufacturing a radio varies inversely as the number of radios produced. If the cost is $35 per radio to produce 1000 radios, how many must be produced to lower the cost to $25?
27. Show that if y varies directly with x, then $y_1/x_1 = y_2/x_2$ for any corresponding nonzero values x_1, y_1, and x_2, y_2.
28. Use Exercise 27 to solve Exercise 23.
29. Show that if y varies inversely as x, then $x_1 y_1 = x_2 y_2$ for any corresponding pairs x_1, y_1 and x_2, y_2.
30. Use Exercise 29 to solve Exercise 24.

Joint and Combined Variation

Translate each of the statements in Exercises 31-35 into an equation.

31. The quantity z varies jointly as x and the cube of y.
32. A force f varies directly with mass m and inversely with the square of the distance d.
33. Electrical current I varies directly with voltage V and inversely with resistance R.
34. The frequency f of vibration of a string varies directly with the square root of the tension t of the string and inversely with its length l.
35. Profit P varies directly with number n of sales and inversely with the square root of the cost c.

Translate each of the equations in Exercises 36-40 into a statement about variation.

36. $V = \pi r^2 h$
37. $h = \dfrac{V}{\pi r^2}$
38. $H = k\left(\dfrac{AT}{d}\right)$
39. $W = k\left(\dfrac{wd^2}{l}\right)$
40. $IQ = k\left(\dfrac{M}{C}\right)$

41. The volume V of a gas varies directly as the temperature T and inversely as the pressure P. If V is 30 cubic inches when T is 70° and $P = 4\,\text{lb/in}^2$, find V for $T = 65°$ and $P = 3\,\text{lb/in}^2$.
42. The weight that can be supported by a wooden beam varies jointly as its width w and the square of its depth d and inversely as its length l. A beam 4 ft long, 1 ft wide, and 2 in deep can support a weight of 90 lb. How much can a beam of the same wood support if it is 5 ft long, 2 ft wide, and 3 in deep?
43. According to Ohm's law, the resistance to the flow of current in an electric circuit varies directly with the voltage V and inversely with the current I. If $R = 11$ ohms when $V = 220$ volts and $I = 20$ amperes, find R when $V = 110$ volts and $I = 30$ amperes.
44. By using Ohm's law as defined in Exercise 43, show that

$$\frac{R_2}{R_1} = \frac{V_2 I_1}{V_1 I_2}$$

for any related triples R_1, V_1, I_1, and R_2, V_2, I_2. Also, what is the effect on the resistance when the voltage is doubled and the current is cut in half?

Chapter 4 Review Outline

4.1 Functions

A function f with domain D_f and range R_f is a rule that assigns to each element x in D_f a unique element $y = f(x)$ in R_f; x is called the independent variable, and y is called the dependent variable.

Operations with Functions:

$$(f \pm g)(x) = f(x) \pm g(x)$$
$$(fg)(x) = f(x)g(x)$$
$$\left(\frac{f}{g}\right)(x) = \frac{f(x)}{g(x)} \quad [g(x) \neq 0]$$
$$(g \circ f)(x) = g(f(x))$$

Graph of a Function:
The graph of a function f is the set of all points (x, y) with x in D_f and $y = f(x)$.
A graph is that of a function if very vertical line intersects the graph in at most one point.

4.2 Polynomial Functions

A polynomial function f of degree n is defined by

$$f(x) = a_n x^n + a_{n-1} x^{n-1} + \ldots + a_1 x + a_0,$$

where n is a nonnegative integer and $a_n \neq 0$.

Symmetry:
The graph of a function f is symmetric with respect to the y-axis $\Leftrightarrow f(-x) = f(x)$ for all x in D_f (f is an even function).
A polynomial P is an even function $\Leftrightarrow P(x)$ contains only even powers of x.
The graph of a function f is symmetric with respect to the origin $\Leftrightarrow f(-x) = -f(x)$ for all x in D_f (f is an odd function).
A polynomial P is an odd function $\Leftrightarrow P(x)$ contains only odd powers of x.

Graphing Polynomial Functions:
The three-step procedure for graphing a polynomial P includes (1) symmetry, (2) intercepts, and (3) the sign of $P(x)$.

4.3 Rational Functions

A rational function is the ratio of two polynomial functions.

Asymptotes:
The vertical line $x = a$ is a vertical asymptote for $f(x) = P(x)/Q(x)$ if $Q(a) = 0$ and $P(a) \neq 0$, which means that $|f(x)| \to \infty$ as $x \to a$.
The horizontal line $y = b$ is a horizontal asymptote for f if $f(x) \to b$ as $x \to \infty$ or as $x \to -\infty$.

Test for Horizontal Asymptotes:
Let
$$f(x) = \frac{a_m x^m + a_{m-1} x^{m-1} + \ldots + a_0}{b_n x^n + b_{n-1} x^{n-1} + \ldots + b_0} \quad (a_m \neq 0, b_n \neq 0)$$
The line $y = 0$ is a horizontal asymptote for $f \Leftrightarrow m < n$.
The line $y = a_m/b_n$ is a horizontal asymptote for $f \Leftrightarrow m = n$.
f has no horizontal asymptote $\Leftrightarrow m > n$.

Graphing Rational Functions:
The four-step procedure for graphing a rational function f includes (1) symmetry, (2) x-intercepts and vertical asymptotes, (3) the sign of $f(x)$, and (4) y-intercept and horizonal asymptote.

4.4 Algebraic Functions

An algebraic function is one generated by addition, subtraction, multiplication, division, and extraction of roots. (Polynomials and rational functions are algebraic functions.)

Graphing Algebraic Functions:
To graph an algebraic function f, consider symmetry, intercepts, the sign of $f(x)$, asymptotes, and restrictions on the domain of f.

4.5 One-to-One and Inverse Functions

Definitions
f is one-to-one $\Leftrightarrow f(x_1) = f(x_2)$ always implies $x_1 = x_2$.
f is an increasing function $\Leftrightarrow x_1 > x_2$ implies $f(x_1) > f(x_2)$ whenever x_1 and x_2 are in D_f.
f is a decreasing function $\Leftrightarrow x_1 > x_2$ implies $f(x_1) < f(x_2)$ whenever x_1 and x_2 are in D_f.
g is the inverse of f (denoted f^{-1}) $\Leftrightarrow$

$$(g \circ f)(x) = \text{ for all } x \text{ in } D_f \text{ and}$$
$$(f \circ g)(x) = x \text{ for all } x \text{ in } D_g.$$

Horizontal Line Test:
f is one-to-one $\Leftrightarrow$ every horizontal line intersects the graph of f in at most one point.

Relationships:
f has an inverse $\Leftrightarrow$ f is one-to-one.
If f is an increasing or a decreasing function, then f is one-to-one.
The graphs of f and f^{-1} are symmetric with respect to the line $y = x$.

4.6 Variation

Definitions ($k \neq 0$)

Direct Variation: $y = kx$

Inverse Variation: $y = \dfrac{k}{x}$

Joint Variation: $y = kx_1 x_2$

Combined Variation: $y = \dfrac{kx_1 x_2}{x_3}$

Chapter 4 Review Exercises

1. Given $f(x) = 2x^2 - 2$, find each of the following;
   (a) $f(4)$ (b) $f(-5)$ (c) $f(\sqrt{3})$

2. Given $g(x) = (3x+4)/(x^2+5)$, find each of the following.
   (a) $g(2x)$ (b) $g(\sqrt{x})$ (c) $g(x-1)$

3. Find the domain and range of each of the following functions.
   (a) $f(x) = 3x - 4$ (b) $g(x) = x^2 + 1$
   (c) $h(x) = |x|$

4. Find the domain for each of the following functions.
   (a) $f(x) = \dfrac{5}{3x-4}$ (b) $g(x) = \dfrac{5}{\sqrt{3x-4}}$
   (c) $h(x) = \dfrac{1}{x^2 - 4}$

5. Given $f(x) = x^2 + 2$ and $g(x) = \sqrt{x-2}$, find each of the following.
   (a) $(f+g)(x)$ (b) $(fg)(x)$ (c) $(f/g)(x)$

6. For f and g given in Exercise 5, find each of the following.
   (a) $(f+g)(5)$ (b) $(fg)(2x)$ (c) $(f/g)(x^2)$

7. For f and g given in Exercise 5, find the domain of $f+g$, fg, and f/g, respectively.

8. Given $f(x) = 1 + x$ and $g(x) = \sqrt{x}$, find each of the following.
   (a) $(g \circ f)(x)$ (b) $(f \circ g)(x)$

9. For f and g given in Exercise 8, find each of the following.
   (a) $(g \circ f)(5)$ (b) $(f \circ g)(2x)$

10. For f and g given in Exercise 8, find the respective domains of $g \circ f$ and $f \circ g$.

11. Which of the following are polynomial functions?
    (a) $f(x) = 3x^4 - 1.2x^2 - .8x + \sqrt{2}$
    (b) $g(x) = 2$
    (c) $h(x) = \dfrac{x-2}{x^2+9}$

12. Which of the following are rational functions?
    (a) $F(x) = \sqrt[3]{\dfrac{x-2}{x^2+9}}$
    (b) $G(x) = x^{-2} + 2x + 2$
    (c) $H(x) = \dfrac{x^2 + 2x}{x^3 - 1}$

13. Which of the following are even functions?
    (a) $f(x) = x^2 - 3x^4$ (b) $g(x) = x^4 + 1$
    (c) $h(x) = |x|$

14. Which of the following are odd functions?
    (a) $F(x) = x^3 - 2x - 2$
    (b) $G(x) = \sqrt[3]{x}$
    (c) $H(x) = \dfrac{x^3 + 5x}{x^2 + 1}$

15. Which of the functions in Exercise 13 have graphs that are symmetric with respect to the y-axis?

16. Which of the functions in Exercise 14 have graphs that are symmetric with respect to the origin?

17. Find all vertical asymptotes for each if the following rational functions.
    (a) $f(x) = \dfrac{x+4}{2x-5}$
    (b) $g(x) = \dfrac{x}{x^2-4}$
    (c) $h(x) = \dfrac{x^3+6}{x^2-4x+3}$

18. Find the horizontal asymptote, if any, for each of the rational functions in Exercise 17.

19. Find all vertical and horizontal asymptotes for the algebraic function $F(x) = (\sqrt{x^2+1})/(x-2)$.

Sketch the graph of each of the functions in Exercises 20-22.

20. $f(x) = x^3 - 4x^2$
21. $g(x) = \dfrac{3-x}{x+2}$
22. $h(x) = \sqrt{\dfrac{3-x}{x+2}}$

23. Use the algebraic definition of a one-to-one function to show that $f(x) = 3x - 5$ is one-to-one.

24. Use the horizonal line test to show that the function $g(x) = x^3 + 1$ is one-to-one.

25. Use the horizonal line test to show that the function in Exercise 20 is not one-to-one

26. Sketch a graph of $f(x) = 3x - 5$ and its inverse on the same axes.

27. Sketch a graph of $g(x) = x^3 + 1$ and its inverse on the same axes.

28. Which, if any, of the following functions are increasing and which are decreasing?

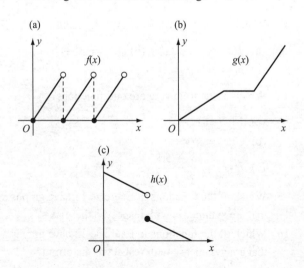

29. Which, if any, of the functions in Exercise 28 are one-to-one?

30. Let $f(x) = 3x + 5$. Use the algebraic definition to show that f is an increasing function

31. Let $g(x) = -3x + 5$. Use the algebraic definition to show that g is a decreasing function

Sketch the graph of each of the functions in Exercises 32-34.

32. $f(x) = |x - 1|$ 33. $g(x) = [x + 1]$

34. $h(x) = \begin{cases} x - 1 & \text{if } x \leq 1 \\ x^2 & \text{if } x > 1 \end{cases}$

35. What are the domain and range of the function g in Exercise 33?

36. What are the domain and range of the function h in Exercise 34?

37. Express as an equation the statement that y varies jointly as u and v and inversely as w.

38. Write the equation $L = k(wd^2)/r$ in the terminology of variation.

39. The measure of intelligence known as the I.Q. varies directly with mental age M and inversely with chronological age C. If a 10-year-old with a mental age of 12 has an I.Q. of 120, what is the I.Q. of a 15-year-old with a mental age of 12?

40. The time t that it takes the earth to complete one orbit about the sun is directly proportional to the square root of d^3, where d is its maximum distance from the sun. Using $d = 93$ million miles, and $t = 1$ year, compute the proportionality constant.

Graphing Calculator Exercises

Verify your answers to the indicated review exercises of this section by using a graphing calculator for the given functions and windows

$[x_{\min}, x_{\max}], [y_{\min}, y_{\max}]$.

41. Exercise 13:
    (a) $Y_1 = x^2 - 3x^4$, $[-.94, .94]$, $[-.5, .5]$
    (b) $Y_1 = x^4 + 1$, $[-4.4, 5]$, $[-5, 5]$
    (c) $Y_1 = \text{abs}(x)$, $[-4.4, 5]$, $[-5, 5]$

42. Exercise 14:
    (a) $Y_1 = x^3 - 2x - 2$, $[-4.4, 5]$, $[-10, 10]$
    (b) $Y_1 = \sqrt[3]{x}$, $[-4.4, 5]$, $[-3, 3]$
    (c) $Y_1 = (x^3 + 5x)/(x^2 + 1)$, $[-4.4, 5]$, $[-5, 5]$

43. Exercises 17 and 18:
    (a) $Y_1 = (x+4)/(2x-5)$, $[-2.4, 7]$, $[-15, 15]$
    (b) $Y_1 = x/(x^2 - 4)$, $[-4.4, 5]$, $[-5, 5]$
    (c) $Y_1 = (x^3 + 6)/(x^2 - 4x + 3)$, $[-2.4, 7]$, $[-30, 30]$

44. Exercise 19: $Y_1 = \sqrt{x^2 + 1}/(x - 2)$, $[-2.4, 7]$, $[-15, 15]$

45. Exercise 20: $Y_1 = x^3 - 4x^2$, $[-4.4, 5]$, $[-10, 10]$

46. Exercise 21:
    $Y_1 = (3 - x)/(x + 2)$, $[-5.4, 4]$, $[-10, 10]$

47. Exercise 22:
    $Y_1 = \sqrt{((3 - x)/(x + 2))}$, $[-4.4, 5]$, $[0, 5]$

48. Exercise 26: $Y_1 = 3x - 5$, $Y_2 = (x + 5)/3$, $[-9.4, 9.4]$, $[-7, 7]$

49. Exercise 27: $Y_1 = x^3 + 1$, $Y_2 = \sqrt[3]{x - 1}$, $[-4.4, 5]$, $[-5, 5]$

50. Exercise 32: $Y_1 = \text{abs}(x - 1)$, $[-2.4, 7]$, $[-1, 6]$

51. Exercise 33: $Y_1 = \text{Int}(x + 1)$, $[-4.4, 5]$, $[-5, 5]$ (Dot Mode)

52. Exercise 34: $Y_1 = (x - 1)(x \leq 1) + x^2(x > 1)$, $[-2.2, 2.5]$, $[-3, 6]$ (Dot Mode)

5 Exponential and Logarithmic Functions

5.1 The Exponential Function and its Graph

5.2 The Logarithmic Function and its Graph

5.3 Properties of the Logarithm

5.4 Applications of Exponential and Logarithmic Equations

Hey Students!
You will benefit more from the classroom lectures and have a better understanding of the material if you take the time to view the Gilbert Review Videos @
www.CollegeAlgebraBySchiller.com
before attending the class lecture for this chapter.

Chapter 5. Exponential and Logarithmic Functions

In this chapter we study two new functions that are quite different from the algebraic functions discussed previously. Using our knowledge of rational exponents, we define first the *exponential function* and then its inverse, the *logarithmic function*. The exponential and logarithmic functions are used extensively in calculus, as well as in many areas of applied mathematics, including growth of populations, decay of radioactive materials, and compound interest.

5.1 The Exponential Function and Its Graph

Motion in the universe is along conic sections, but growth and decay in nature are exponential.

- The Exponential Function
- Further Properties of the Exponential
- The Natural Base e

In the last chapter we graphed polynomial functions, such as $f(x) = x^3 - 3x^2 + 2x + 1$, and some other algebraic functions, such as $g(x) = \sqrt{x}$ and $h(x) = x^{2/3}$. All of these functions involve terms of the form x^a, where the base x is a variable and the exponent a is a constant. Now we reverse that situation to consider a^x with constant base a and variable exponent x. The function $f(x) = a^x$ is called an **exponential function.**

The Exponential Function. Since we learned the meaning of rational exponents in Section 1.4, most of us, if asked to draw the graph of the equation $y = 2^x$, would probably construct a table and plot the corresponding points on the graph as shown in Figure 1.

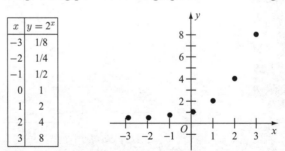

Figure 1 $y = 2^x$ for Integer Values of x

By using some nonintegral values of x, we might add additional points such as $(-3/2, .35)$ and $(1/3, 1.26)$, because we know that

$$\text{if } x = -3/2, \text{ then } 2^x = 2^{-3/2} = \sqrt{2^{-3}} = \sqrt{1/8} \approx .35,$$

$$\text{and} \quad \text{if } x = 1/3, \text{ then } 2^x = 2^{1/3} = \sqrt[3]{2} \approx 1.26.$$

5.1. The Exponential Function and its Graph

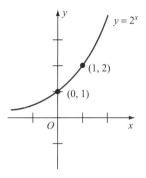

Figure 2

[1] *The rational powers of 2 were computed with a scientific calculator and rounded off to four significant digits. (See the difinition of significant digits in Appendix A.)*

[2] *By continuous we mean there are no breaks in the curve, and by smooth we mean that the curve also has no sharp turns. More precise definitions are given in calculus.*

It is then quite natural to join the plotted points by a smooth curve, obtaining the graph in Figure 2.

In drawing such a curve, we are assigning values to 2^x for every real number, although we have not yet given meaning to 2^x when x is an irrational number. We are forced to ask: *What is meant by expressions such as*

$$2^{\sqrt{2}}, \ 2^{\sqrt[3]{5}} \ \text{ or } \ 2^{\pi}?$$

It is easy to guess that $2^{\sqrt{2}} \approx 2^{1.4} \approx 2.6390$ and that successively better approximations are[1]

$$2^{\sqrt{2}} \approx 2^{1.41} \approx 2.657$$
$$2^{\sqrt{2}} \approx 2^{1.414} \approx 2.665$$
$$2^{\sqrt{2}} \approx 2^{1.4142} \approx 2.665,$$

so that the point with $x = \sqrt{2}$ and $y \approx 2.67$ is included in Figure 2.

The above reasoning can lead to a mathematically precise definition of $f(x) = 2^x$ for all *real x*, but it would require concepts from calculus. Therefore, we *assume that f* is defined by the process described above. That is, we assume that the domain of 2^x can be extended from the rational numbers to the set of all real numbers in such a way that the graph of $f(x) = 2^x$ is continuous and smooth.[2]

More generally,

---

the **exponential function with base *a*** is defined by

$$\boldsymbol{f(x) = a^x} \ (a > 0, a \neq 1)$$

with the following properties:

1. f is a function with $D_f = (-\infty, \infty)$. That is, for each real number x, a^x is a unique real number.
2. f is a continuous and smooth extension of $g(x) = a^x$, where D_g is the set of rational numbers.

The restriction $a > 0$ was already explained in Section 1.4. It is needed to avoid nonreal numbers such as $(-4)^{1/2}$, and to avoid exceptions to the power-of-a-power rule $(a^x)^y = a^{xy}$. The restriction $a \neq 1$ is made since $1^x = 1$ for all real x implies that a^x is a constant function for $a = 1$, and we do not consider constant functions to be exponential.

312 Chapter 5. Exponential and Logarithmic Functions

The graph of $y = 2^x$ (Figure 2) illustrates that $f(x) = a^x$ is an *increasing* function for base $a > 1$. That is, if $a > 1$, then $a^{x_1} > a^{x_2}$ whenever $x_1 > x_2$. The following example illustrates that $f(x) = a^x$ is *decreasing* when $0 < a < 1$. If $0 < a < 1$, then $a^{x_1} < a^{x_2}$ whenever $x_1 > x_2$.

Example 1 Sketch the graph of $y = (1/2)^x$.

Solution: We first make a table of coordinates by substituting some rational values for x and computing the corresponding values of y. We then plot the corresponding points and join them by a smooth curve, as shown in Figure 3. ∎

x	$y = (1/2)^x$
-3	8
-2	4
-1	2
0	1
1	1/2
2	1/4
3	1/8

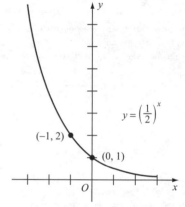

Figure 3

Figures 2 and 3 illustrate the following important **properties of the graph of an exponential function $f(x) = a^x$.**

1. $f(x) > 0$ for all real numbers x.
2. $f(0) = 1$
3. If $a > 1$, then $f(x) \to \infty$ as $x \to \infty$, and $f(x) \to 0$ as $x \to -\infty$.
3'. If $0 < a < 1$, then $f(x) \to 0$ as $x \to \infty$, and $f(x) \to \infty$ as $x \to -\infty$.

Also, if $a > 1$, then a^x is an increasing function, and if $0 < a < 1$, a^x is a decreasing function.

Property 1 says that the graph of any exponential function is always above the x-axis, and property 2 says that the curve always goes through the point $(0, 1)$. Property 3 (3') says that for any exponential function, the range R_f is $(0, \infty)$, and the x-axis is a horizontal asymptote.

We could restrict our attention to bases greater than 1 by introducing a negative sign in the exponent. That is, by definition,

$$a^{-x} = \left(\frac{1}{a}\right)^x \quad (a > 0).$$

Hence, in Example 1, we may write $y = (1/2)^x = 2^{-x}$. In general, if $f(x) = b^x$ with $0 < b < 1$, then we can write $f(x) = a^{-x}$ with $a = 1/b > 1$.

Example 2 Sketch $f(x) = 2^x$ and $g(x) = 2^{-x}$ on the same axes.

Solution: By combining Figures 2 and 3 we obtain the graph shown in Figure 4. ■

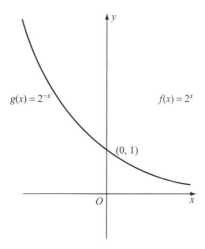

Figure 4

Figure 4 shows that the graphs of f and g, taken together, are symmetric with respect to the y-axis. More generally, to show that the graphs of $f(x) = a^x$ and $g(x) = a^{-x}$, taken together, are symmetric with respect to the y-axis, it is sufficient to observe that

$$f(-x) = a^{-x} = g(x).$$

Example 3 Sketch the graphs of $y = 3^x$ and $y = 2^x$ on the same axes.

Solution: For $y = 3^x$ we first construct a table of coordinates, and for $y = 2^x$ we use Figure 2. The graphs are shown in Figure 5. ■

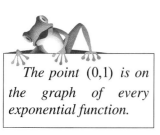

The point (0,1) is on the graph of every exponential function.

x	$y = 3^x$
-3	1/27
-2	1/9
-1	1/3
0	1
1	3
2	9
3	27

Figure 5

Figure 5 shows that the graph of 3^x starts lower but rises more rapidly than that of 2^x. The graph of 10^x would start even lower but rise more rapidly than both 2^x and 3^x. All functions $f(x) = a^x, a > 1$, are said to be **exponentially increasing,** and $b > a > 1$ implies that b^x increases more rapidly than a^x. Similarly, any function $g(x) = a^{-x}$, $a > 1$, is said to be **exponentially decreasing**, and $b > a > 1$ implies that b^{-x} decreases more rapidly than a^{-x}. As stated earlier, for both $f(x) = a^x$ and $g(x) = a^{-x}$, the domain is the set of all real numbers, the range is the set of all positive real numbers, and the x-axis is a horizontal asymptote.

Question 1 *Which function increases more rapidly, 2^x or x^2?*

Answer: The 2^x function increases more rapidly. For instance, the equation $2^{x+1} = 2 \cdot 2^x$ says that 2^x doubles in value whenever x increases by 1. On the other hand, $(\sqrt{2}x)^2 = 2 \cdot x^2$ says that x must increase by a *factor* of $\sqrt{2}$ in order for x^2 to double in value. ∎

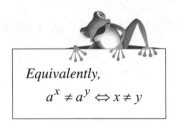

Equivalently,
$$a^x \neq a^y \Leftrightarrow x \neq y$$

Further Properties of the Exponential. Since we have seen that $f(x) = a^x$ is either increasing ($a > 1$) or decreasing ($0 < a < 1$), it follows from Section 4.5 that a^x is a *one-to-one function*. That is,

$$a^x = a^y \Leftrightarrow x = y.$$

Example 4 Solve for x in each of the following exponential equations.

(a) $2^x = 64$
(b) $2^{-x} = 4^x$
(c) $8^{x-1} = 16^{x-5/6}$

Solution:

(a) $2^x = 64 \Leftrightarrow 2^x = 2^6 \Leftrightarrow x = 6 \quad 2^x$ is one-to-one
(b) $2^{-x} = 4^x \Leftrightarrow 2^{-x} = 2^{2x} \Leftrightarrow -x = 2x \Leftrightarrow 0 = 3x \Leftrightarrow x = 0$
(c) $8^{x-1} = 16^{x-5/6} \Leftrightarrow (2^3)^{x-1} = (2^4)^{x-5/6} \Leftrightarrow 2^{3x-3} = 2^{4x-10/3}$

$$\Leftrightarrow 3x - 3 = 4x - \frac{10}{3} \Leftrightarrow x = 1/3$$

Let us check this last answer by substituting $x = 1/3$ in the given equation:

$$8^{1/3-1} = 16^{(1/3)-(5/6)}$$
$$8^{-2/3} = 16^{-1/2}$$
$$\frac{1}{(\sqrt[3]{8})^2} = \frac{1}{\sqrt{16}}$$
$$\frac{1}{2^2} = \frac{1}{4}. \quad ∎$$

In each part of Example 4, we were able to express both sides of the equation to the *same* base and then apply the one-to-one property. This is always possible, but in general we must use logarithms when several bases are involved, as we will see later in this chapter.

Since the function a^x is a continuous extension from the domain of rational numbers to the domain of all real numbers, the five **basic rules for exponents** (Section 1.4) hold for all positive bases and all real exponents. These rules are listed below.

$$a^x a^y = a^{x+y} \text{ and } \frac{a^x}{a^y} = a^{x-y} \quad \text{same-base rules}$$

$$(ab)^x = a^x b^x \text{ and } \left(\frac{a}{b}\right)^x = \frac{a^x}{b^x} \quad \text{same-exponent rules}$$

$$(a^x)^y = a^{xy} \quad \text{power-of-a-power rule}$$

Example 5 Use the rules for exponents to simplify each of the following.

(a) $a^{\sqrt{2}} a^{3\sqrt{2}}$

(b) $\dfrac{a^{\pi}}{a^{\pi-2}}$

(c) $(a^{\sqrt{2}})^{\sqrt{2}}$

(d) $10^{\sqrt{3}} 100^{\sqrt{3}}$

Solution:

(a) $a^{\sqrt{2}} a^{3\sqrt{2}} = a^{\sqrt{2}+3\sqrt{2}} = a^{4\sqrt{2}}$ Same-base rule

(b) $\dfrac{a^{\pi}}{a^{\pi-2}} = a^{\pi-(\pi-2)} = a^2$ Same-base rule

(c) $\left(a^{\sqrt{2}}\right)^{\sqrt{2}} = a^{\sqrt{2}\cdot\sqrt{2}} = a^2$ Power-of-a-power rule

(d) $10^{\sqrt{3}} 100^{\sqrt{3}} = 10^{\sqrt{3}}(10^2)^{\sqrt{3}}$
$= 10^{\sqrt{3}} 10^{2\sqrt{3}} = 10^{\sqrt{3}+2\sqrt{3}} = 10^{3\sqrt{3}}$ ∎

The Natural Base e. In Figure 5, the graphs of $y = 2^x$ and $y = 3^x$ are shown, but both in theory and in applications, the most useful base is a number between 2 and 3 denoted by the letter e [so named after the Swiss mathematician Euler (1707-1783)]. This number is irrational and hence is represented by a nonending, nonrepeating decimal, but e can be approximated as

$$e \approx 2.71828. \quad \text{natural base}$$

The number e arises most naturally in calculus, but, as we will see in Section 5.4, e also appears in the context of compound interest.

The natural base e can be defined as the value approached by the expression

$$\left(1 + \frac{1}{n}\right)^n \tag{1}$$

as n goes through the sequence of integers $1, 2, 3, \cdots$. That is, for large values of n,

$$\left(1+\frac{1}{n}\right)^n \approx e.$$

Most scientific calculators have an e^x (or exp x) key by which this function can be computed for any value of x.

Example 6 Use a calculator to compute (1) for the following values of n.

(a) $n = 10$
(b) $n = 100$
(c) $n = 10^6$

Solution: The following computations were made on a scientific calculator and rounded off to four significant digits.

(a) $\left(1+\dfrac{1}{10}\right)^{10} = (1.1)^{10} \approx 2.594$

(b) $\left(1+\dfrac{1}{100}\right)^{100} = (1.01)^{100} \approx 2.705$

(c) $\left(1+\dfrac{1}{10^6}\right)^{10^6} = (1.000001)^{1,000,000} \approx 2.718$ ∎

Exercises 5.1

Fill in the blanks to make each statement true.

1. The exponential function a^x is defined for _____ values of $a \neq 1$.

2. The exponential function a^x has domain _____ and range _____.

3. The exponential function is increasing if the base a is _____ and decreasing if a is _____.

4. If $f(x) = a^x$, then $f(x) = b^{-x}$, where $b =$ _____.

5. Because the exponential is a _____ function, $a^x = a^y \Leftrightarrow x = y$.

Write true or false for each statement.

In Exercises 6-10, it is assumed that the base a is positive and $a \neq 1$.

6. The exponential function a^x is defined for every real number x.

7. The x-axis is an asymptote for the graph of $y = a^x$.

8. $a^x \to 0$ as $x \to \infty$.

9. The graphs of $y = a^x$ and $y = a^{-x}$, taken together, are symmetric with respect to the y-axis.

10. The natural base e, approximated to one decimal place, is 2.7.

Graphs of Exponential Functions

Sketch the graph of each of the following exponential functions.

11. $f(x) = 4^x$
12. $g(x) = 10^x$
13. $h(x) = \left(\dfrac{3}{2}\right)^x$
14. $\epsilon(x) = e^x$
15. $F(x) = \left(\dfrac{1}{4}\right)^x$
16. $G(x) = \left(\dfrac{1}{10}\right)^x$
17. $H(x) = \left(\dfrac{2}{3}\right)^x$
18. $E(x) = \left(\dfrac{1}{e}\right)^x$
19. $F(x) = 4^{-x}$
20. $G(x) = 10^{-x}$
21. $H(x) = \left(\dfrac{3}{2}\right)^{-x}$
22. $E(x) = e^{-x}$

23. How do the graphs in Exercises 15-18 compare with those in Exercises 11-14, respectively?
24. How do the graphs in Exercises 19-22 compare with those in Exercises 15-18, respectively?

Sketch the graph of each function in Exercises 25-32.

25. $y = 2^x - 1$
26. $y = 2^{x-1}$
27. $y = 2^{2x}$
28. $y = 2^{(x^2)}$
29. $y = 2^{-x^2}$
30. $y = e^{-x^2}$
31. $y = 1.5e^{2t}$
32. $y = 1.5e^{-2t}$

33. Sketch the graph of $y = (e^x + e^{-x})/2$. This curve is called a **catenary** and is the shape of a cable hanging suspended between two supports.

Properties of Exponentials

Solve each of the following equations for x.

34. $2^x = 4^{-x}$
35. $2^{2x} = 4^x$
36. $3^{x+1} = 81^{x+(5/2)}$
37. $3^x 9^{x+1} = 27^{1-x}$

Solve each of the following equations for a^x. (Hint: let $a^x = y$ and find a quadratic equation in y that is equivalent to the given equation.)

38. $a^x + a^{-x} = 3$
39. $a^x - 4a^{-x} = 3$
40. $\dfrac{a^x - a^{-x}}{a^x + a^{-x}} = \dfrac{1}{3}$
41. $\dfrac{a^x + a^{-x}}{a^x - a^{-x}} = -3$

Use the rules for exponents to simplify each of the following.

42. $\left(4^{\sqrt{2}}\right)^{\sqrt{2}}$
43. $4^{\sqrt{2}} + 2^{\sqrt{2}}$
44. $\dfrac{3^{2x}}{9^x}$
45. $\dfrac{4^{-2x}}{32^{-x}}$
46. $\left(\dfrac{a^{3/\sqrt{2}}}{b^{\sqrt{2}/2}}\right)^{\sqrt{2}}$
47. $\dfrac{\left[(a^2)^x\right]^{1/2}}{(a^{-x/2})^{-4}}$
48. $\left(\dfrac{a^x + a^{-x}}{2}\right)^2 - \left(\dfrac{a^x - a^{-x}}{2}\right)^2$
49. $\dfrac{a^x + 1}{a^x + a^{-x} + 2}$

The Natural Base e

Exercises 50-60 require the use of a calculator.

50. Approximate e^2 by $(1 + 1/n)^{2n}$ for $n = 10^6$. Compare this value with $(1 + 2/n)^n$ for $n = 10^6$ and with $e^2 \approx 7.3890561$.

51. Approximate $1/e$ by $(1 + 1/n)^{-n}$ for $n = 10^6$. Compare this value with $(1 - 1/n)^n$ for $n = 10^6$ and with $1/e \approx .36787944$.

It is shown in calculus that for small $|x|$

$$e^x \approx 1 + x + \dfrac{x^2}{2} + \dfrac{x^3}{6} + \dfrac{x^4}{24} + \dfrac{x^5}{120}.$$

Use this result to approximate the following numbers and compare with the value obtained on a scientific calculator.

52. e
53. $e^{0.1}$
54. $e^{0.01}$

Replacing x by $-x$ in the above approximation for e^x gives

$$e^{-x} \approx 1 - x + \dfrac{x^2}{2} - \dfrac{x^3}{6} + \dfrac{x^4}{24} - \dfrac{x^5}{120}.$$

Use this result to approximate the following numbers and compare with the value obtained on a scientific calculator.

55. e^{-1} **56.** $e^{-0.1}$ **57.** $e^{-0.01}$

58. Approximate $3^{\sqrt{2}}$ by finding the successive values of $3^{1.4}$, $3^{1.41}$, $3^{1.414}$, and $3^{1.4142}$.

59. Approximate $3^{\sqrt[3]{2}}$ by finding the successive values of $3^{1.2}$, $3^{1.25}$, $3^{1.259}$, and $3^{1.2599}$.

60. It is shown in calculus that for small $|x|$

$$3^x \approx 1 + 1.0986x + \frac{(1.0986x)^2}{2} + \frac{(1.0986x)^3}{6} + \frac{(1.0986x)^4}{24}.$$

Use this formula to approximate $3^{\sqrt{2}}$ and compare with Exercise 58.

5.2 The Logarithmic Function and its Graph

Before calculators there were slide rules, and before slide rules there were tables of logarithms.

Definition of Logarithm
Graph of the Logarithmic Function
Change of Base

Definition of Logarithm. In the exponential function $y = a^x$, the independent variable is the exponent x, and the dependent variable is the value $y = a^x$. In the logarithmic function, the roles of independent variable and dependent variable are reversed. The independent variable is now the value $x = a^y$, and the dependent variable is now the exponent y. That is, if $a^y = x$, then y is called the **logarithm of x to the base a**, and we write

$$\log_a x = y.$$

For example, $\log_5 25 = 2$ because $5^2 = 25$,

and $\log_2 \left(\frac{1}{8}\right) = -3$ because $2^{-3} = \frac{1}{8}$.

Since the logarithm of a number to a given base is the **exponent** to which the base must be raised to yield the number, we can write

$$a^{\log_a x} = x. \tag{1}$$

Similarly, since x is the exponent to which the base a must be raised to give the number a^x,

$$\log_a(a^x) = x. \tag{2}$$

Equations (1) and (2) tell us that a^x and $\log_a x$ are **inverse functions**. That is, if $f(x) = a^x$, then equation (1) says that

$$f(\log_a x) = x, \tag{1'}$$

a^x and $\log_a x$ are inverse to each other with respect to composition of functions.

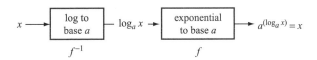

Figure 6

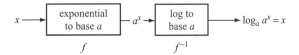

Figure 7

and equation (2) says that

$$\log_a(f(x)) = x. \qquad (2')$$

Therefore, $\log_a x = f^{-1}(x)$.

Schematically, equation (1) can be represented as shown in Figure 6. Equation (2) can be represented as shown in Figure 7. Figures 6 and 7 again show that a^x and $\log_a x$ are inverse functions.

The base a is always assumed to be positive and $a \neq 1$. If $a = e$, the natural base, then $\log_a x$ is called the **natural logarithm** of x and is denoted by **ln x**. That is, we define $\ln x = \log_e x$, so that

$$y = \ln x \Leftrightarrow x = e^y,$$

which is equivalent to saying

$$f(x) = e^x \Leftrightarrow f^{-1}(x) = \ln x.$$

The notation **log x** means $\log_{10} x$ and is called a **common logarithm**.

As illustrated below, by using the inverse property, we can convert any statement about exponents into an equivalent one about logarithms, and conversely. The most important fact to remember is that *a logarithm is an exponent*.

More precisely, $\log_a x$ is the exponent to which a must be raised in order to equal x.

Exponential Statement	Equivalent Logarithmic Statement
$2^3 = 8$	$3 = \log_2 8$
$100^{-3/2} = .001$	$-3/2 = \log_{100}(.001)$
$a^0 = 1 \ (a \neq 0)$	$0 = \log_a 1 \ (a \neq 0)$
$e^1 = e$	$1 = \ln e$

The technique of equating logarithmic statements and exponential ones is so important that we now give some additional examples.

Example 1 Write the equivalent logarithmic statement in each of the following cases.
(a) $3^2 = 9$ (b) $a^1 = a$ (c) $a^{2/3} = \sqrt[3]{a^2}$ (d) $e^{-1} \approx .3679$

Solution:

(a) $3^2 = 9 \Leftrightarrow 2 = \log_3 9$ \qquad (b) $a^1 = a \Leftrightarrow 1 = \log_a a$

(c) $a^{2/3} = \sqrt[3]{a^2} \Leftrightarrow \dfrac{2}{3} = \log_a\left(\sqrt[3]{a^2}\right)$

(d) $e^{-1} \approx .3679 \Leftrightarrow -1 \approx \ln(.3679)$ ∎

Example 2 Convert each of the following into an equivalent statement in exponential form.

(a) $\log_5 125 = 3$ \qquad (b) $\log_{10}(.0001) = -4$

(c) $\ln 10 \approx 2.3026$ \qquad (d) $\log_a(a^2) = 2$

Solution:

(a) $\log_5 125 = 3$ means $5^3 = 125$.

(b) $\log_{10}(.0001) = -4$ means $10^{-4} = .0001$.

(c) $\ln 10 \approx 2.3026$ means $e^{2.3026} \approx 10$.

(d) $\log_a(a^2) = 2$ states that the exponent of a^2 is 2, and is equivalent to saying $a^2 = a^2$. ∎

Example 3 Solve for x in each of the following.

(a) $x = \log_2 32$ \quad (b) $-2 = \log_3 x$ \quad (c) $1/2 = \log_x 100$

Solution:

(a) $x = \log_2 32 \Leftrightarrow 2^x = 32 \Leftrightarrow 2^x = 2^5 \Leftrightarrow x = 5$ (why?)

(b) $-2 = \log_3 x \Leftrightarrow 3^{-2} = x$ or $x = 1/9$

(c) $1/2 = \log_x 100 \Leftrightarrow x^{1/2} = 100 \Leftrightarrow x = 10{,}000$ ∎

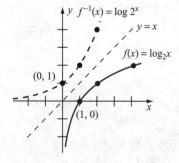

Figure 8

Graph of the Logarithmic Function. The graph of the function $f(x) = \log_a x$ can be obtained from the graph of its inverse $f^{-1}(x) = a^x$ by symmetry with respect to the line $y = x$. For example, the graph of $f(x) = \log_2 x$ (Figure 8) can be obtained from the graph of $f^{-1}(x) = 2^x$ given in Figure 2 in Section 5.1.

Figure 8 is typical of the graph of $f(x) = \log_a x$ for $a > 1$. Note that

$$D_f = (0, \infty) = R_{f^{-1}} \quad \text{and} \quad R_f = (-\infty, \infty) = D_{f^{-1}}.$$

In the following example, we consider $\log_a x$ with $0 < a < 1$. We do a direct point plot without reference to the graph of the inverse function.

Example 4 Sketch the graph of $f(x) = \log_{1/2} x$ by plotting points.

Solution: Since $y = \log_{1/2} x$ is equivalent to $x = (1/2)^y$, we can use this equation to make a coordinate table for the graph of f (Figure 9). ∎

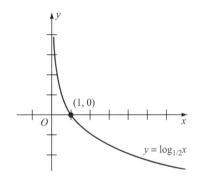

x	$y = \log_{1/2} x$
8	-3
4	-2
2	-1
1	0
1/2	1
1/4	2
1/8	3

Figure 9

In the base a, we can easily compute *exact* values of logarithms of powers of a, as we did in the table in Example 4, where $a = 1/2$. We can then use the graph to *approximate* the value of $\log_a x$ when x is not an integral power of a. However, for a more precise value of $\log_a x$, say to four significant digits, we need a calculator.

Figure 9 is typical of the graph of $\log_a x$ for $0 < a < 1$. Figures 8 and 9 illustrate the following important **properties of the logarithmic function**.

Also, $\log_a x$ is an increasing function for $a > 1$, and a decreasing function for $0 < a < 1$.

1. $\log_a x$ is defined only for $x > 0$.
2. $\log_a 1 = 0$
3. If $a > 1$, then $\log_a x \to \infty$ as $x \to \infty$ and $\log_a x \to -\infty$ as $x \to 0^+$.
3'. If $0 < a < 1$, then $\log_a x \to -\infty$ as $x \to \infty$ and $\log_a x \to \infty$ as $x \to 0^+$.

Property 1 says that the domain of the logarithmic function is $(0, \infty)$, and 2 says that for any base, the graph contains the point $(1,0)$. Property 3 (3') says that the range of the logarithmic function is $(-\infty, \infty)$ and the y-axis is a vertical asymptote.

Change of Base. As stated in Section 5.1, the natural base $e \approx 2.71828$ is the most useful base in mathematics. Also, the natural logarithm of a number x can be converted to the logarithm of x to a different positive base $b \neq 1$ by means of the formula

$$\log_b x = \frac{\ln x}{\ln b}. \tag{3}$$

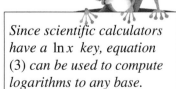

Since scientific calculators have a $\ln x$ key, equation (3) can be used to compute logarithms to any base.

For example, by using a calculator, we find

$$\log_2 17 = \frac{\ln 17}{\ln 2} = \frac{2.8332}{0.6931} \approx 4.0877.$$

Formula (3) is a special case of the more general result

$$\log_b x = \frac{\log_a x}{\log_a b}, \quad \text{change of base rule} \quad (4)$$

which enables us to convert logarithms from *any* positive base $a \neq 1$ to *any other* positive base $b \neq 1$ [see the exercises for a derivation of (4)]. For example,

$$\log_2 17 = \frac{\log_{10} 17}{\log_{10} 2} = \frac{1.2304}{0.3010} \approx 4.0877.$$

Exercises 5.2

Whenever the expression $\log_a x$ appears, it is assumed that $a > 0$ and $a \neq 1$.

Fill in the blanks to make each statement true.

1. The equation $y = \log_a x$ is equivalent to $x = $ -----.
2. The function $f(x) = \log_a x$ has the inverse $f^{-1}(x) = $ -----.
3. The domain of $f(x) = \log_a x$ is -----, and its range is -----.
4. The function $f(x) = \log_a x$ is increasing if a is ----- and is decreasing if a is -----.
5. The notation $\log x$ is used for a logarithm with base -----, and $\ln x$ is used for base -----.

Write true or false for each statement.

6. $\log_a a^x = x$ for positive x only.
7. $a^{\log_a x} = x$ for positive x only.
8. The x-axis is an asymptote for the graph of $y = \log_a x$.
9. The y-axis is an asymptote for the graph of $y = \log_a x$.
10. The graphs of $y = \log x$ and $y = 10^x$, taken together, are symmetric with respect to the line $y = x$.

Definition of Logarithm

Write a logarithmic statement equivalent to each of the following.

11. $2^4 = 16$
12. $2^{-5} = \dfrac{1}{32}$
13. $10^{2/3} = \sqrt[3]{100}$
14. $10^{-3} = .001$
15. $16^{3/2} = 64$
16. $a^b = c$

Write each of the following in exponential form.

17. $\log_{10} 100 = 2$
18. $\log_3 81 = 4$
19. $\log_3 \dfrac{1}{27} = -3$
20. $\log_{10} .00001 = -5$
21. $\log_a 1 = 0$
22. $\log_a x = 2$

Solve each of the following equations for x.

23. $\log_2 16 = x$
24. $\log_8 x = -2$
25. $\log_x 32 = 5$
26. $\log_3 (x+2) = 4$
27. $\log_3 [(x+2)^2] = -4$
28. $\log_3 [(x-2)^2] = -4$

Compute each of the following without using a calculator.

29. $10^{(\log_{10} 100)}$
30. $2^{\log_2 2}$
31. $\log_5 5^{-1000}$
32. $2 \log_a a^2$
33. $\log_{10} 100{,}000$
34. $\log_{10} 0.1$
35. $\log_{10} 1$
36. $\log_{10} \left(\sqrt[3]{10} \right)$
37. $\log_{1/2} 16$
38. $\log_3 \left(\dfrac{1}{243} \right)$

Graphs of Logarithmic Functions

Sketch each of the following graphs by first converting $y = \log_a x$ to $x = a^y$ and then plotting points. In some cases, a calculator will be helpful.

39. $y = \log_2 x$
40. $y = \log_3 x$
41. $y = \log_{1/3} x$
42. $y = \ln x$

Obtain the graph of each of the following functions by first graphing the inverse function. A calculator may be helpful.

43. $f(x) = \log_{1.5} x$

44. $g(x) = \log_{3/4} x$

45. $F(x) = \log_2(x+1)$

46. $G(x) = \log_{1/2}(x-2)$

Change of Base

47. Given two bases a and b, show that for any $x > 0$,

$$\log_b x = \frac{\log_a x}{\log_a b}.$$

(Hint: let $u = \log_a x$ and $v = \log_b x$. Then $a^u = b^v$. Now take log to base a of both sides of this exponential equation.)

Use the result in Exercise 47 to write each of the following as a ratio of common logarithms; then use a calculator to perform the division.

48. $\log_5 28$

49. $\log_2 100$

50. $\log_{.5} .15$

51. $\log_6 .25$

52. $\log_{25} 1055$

53. $\log_{17} 4981$

54. Use the result in Exercise 47 to show that $\log_b a = 1/(\log_a b)$.

55. Use the result in Exercise 47 to show that $\ln x = \log x / \log e$.

56. Use the result in Exercise 47 to show that $\log x = \ln x / \ln 10$.

Miscellaneous

57. Use Exercise 47 to show that if $b > a > 1$, then $\log_b x < \log_a x$.

58. Use the property that the exponential function $f(x) = a^x$ is increasing for $a > 1$ to show that $\log_a x$ is increasing for $a > 1$.

59. Use the property that the exponential function $f(x) = a^x$ is decreasing for $0 < a < 1$ to show that $\log_a x$ is decreasing for $0 < a < 1$.

* The Equation $a^x = x$

The equation $a^x = x$ will have a solution for x if and only if the graph of $y = a^x$ intersects the graph of $y = x$.

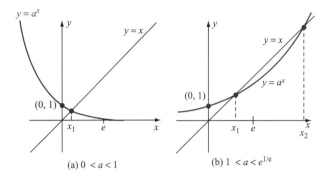

(a) $0 < a < 1$

(b) $1 < a < e^{1/e}$

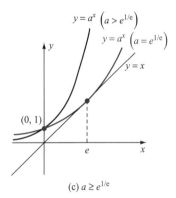

(c) $a \geq e^{1/e}$

We can see from part (a) of the figure that for $0 < a < 1$, there is exactly one solution $x_1 < 1$. It can be shown, using calculus, that for $1 < a < e^{1/e}$, there are two solutions x_1 and x_2, where $x_1 < e < x_2$. For $a = e^{1/e}$, there is one solution, namely $x = e$, and for $a > e^{1/e}$, there are no solutions.

* *Exercises 60-63 are optional.*

Use a calculator to search for approximate solutions (to three significant digits) to each of the following equations.

60. $(.5)^x = x$ (one solution)

61. $(.75)^x = x$ (one solution)

62. $(1.1)^x = x$ (two solution)

63. $(1.3)^x = x$ (two solutions)

5.3 Properties of the Logarithm

Logarithms can simplify computations by changing powers into products, products into sums, and quotients into differences.

Further Properties of Logarithms

We saw several properties of the graph of the logarithm function in Section 5.2. Here we consider some further functional properties of the logarithm.

Further Properties of Logarithms. Since logarithms are exponents, the basic rules for exponents can be translated into the following **properties of logarithms**. (We derive the first one in Example 3 and the other two in the exercises.)

$$\log_a(xy) = \log_a x + \log_a y \qquad \text{Logarithm-of-a-Product Property}$$

$$\log_a\left(\frac{x}{y}\right) = \log_a x - \log_a y \qquad \text{Logarithm-of-a-Quotient Property}$$

$$\log_a(x^y) = y \log_a x \qquad \text{Logarithm-of-a-Power Property}$$

We will apply these logarithmic properties to a variety of examples.

Since there is no algebraic process to compute exact values of logarithmic or exponential functions, we use approximate decimal values obtained from a scientific calculator. *We shall consider two numbers to be equal if they are identical to four significant digits.* (See the definition of significant digits in the Appendix.) Before calculators became readily available, tables and slide rules with logarithmic scales were used for computing logarithms. Both the construction of logarithmic and exponential tables and the internal programming of a scientific calculator involve methods covered in calculus.

Example 1 Using $\log_{10} 2 = .3010$ and $\log_{10} 3 = .4771$, compute the following.
(a) $\log_{10} 6$ (b) $\log_{10} 8$ (c) $\log_{10}(.5)$

Solution:

(a) $\log_{10} 6 = \log_{10}(2 \cdot 3)$
$\qquad = \log_{10} 2 + \log_{10} 3 \qquad$ Logarithm-of-a-Product Property
$\qquad = .3010 + .4771$
$\qquad = .7781$

(b) $\log_{10} 8 = \log_{10}(2^3)$
$\qquad = 3 \log_{10} 2 \qquad$ Logarithm-of-a-Power Property
$\qquad = 3(.3010)$
$\qquad = .9030$

(c) $\log_{10}(.5) = \log_{10}\left(\dfrac{1}{2}\right)$

$\qquad = \log_{10} 1 - \log_{10} 2 \qquad$ Logarithm-of-a-Quotient Property

$\qquad = 0 - .3010$

$\qquad = -.3010$ ■

Example 2   (a) Express $\log_a(xy/\sqrt{z})^3$ in terms of $\log_a x$, $\log_a y$ and $log_a z$.
(b) Express $5\log_a x - 1/3(\log_a y + 2\log_a z)$ as a single logarithm.

Solution:

(a) $\log_a\left(\dfrac{xy}{\sqrt{z}}\right)^3$

$\quad = \log_a\left(\dfrac{x^3 y^3}{z^{3/2}}\right) \qquad (\sqrt{z})^3 = (z^{1/2})^3 = z^{3/2}$

$\quad = \log_a(x^3 y^3) - \log_a(z^{3/2}) \qquad$ Logarithm-of-a-Quotient Property.

$\quad = \log_a x^3 + \log_a y^3 - \log_a(z^{3/2}) \qquad$ Logarithm-of-a-Product Property.

$\quad = 3\log_a x + 3\log_a y - \dfrac{3}{2}\log_a z \qquad$ Logarithm-of-a-Power Property.

(b) Give a reason for each of the following steps.

$5\log_a x - \dfrac{1}{3}(\log_a y + 2\log_a z) = \log_a(x^5) - \dfrac{1}{3}[\log_a y + \log_a(z^2)]$

$\qquad = \log_a(x^5) - \dfrac{1}{3}\log_a(yz^2)$

$\qquad = \log_a(x^5) - \log_a[(yz^2)^{1/3}]$

$\qquad = \log_a\left[\dfrac{x^5}{(yz^2)^{1/3}}\right]$

■

Example 3   Show that the property

$$\log_a(xy) = \log_a x + \log_a y$$

for the logarithm of a product follows from the corresponding rule for exponents. (Similar derivations for the logarithm of a power and the logarithm of a quotient are included in the exercises.)

Solution: Let $p = \log_a x$ and $q = \log_a y$.

Then
$$x = a^p \quad \text{and} \quad y = a^q \qquad \text{\textit{Definition of logarithm.}}$$
$$xy = a^p \cdot a^q = a^{p+q} \qquad \text{\textit{Same-base rule for exponents.}}$$
$$\log_a(xy) = p + q \qquad \text{\textit{Definition of logarithm.}}$$
$$\log_a(xy) = \log_a x + \log_a y. \qquad \text{\textit{Substitution for p and q.}} \quad \blacksquare$$

Another important property of the logarithmic function, which follows because it is the inverse of the exponential, is the fact that $f(x) = \log_a x$ is a one-to-one function. That is, for x and $y > 0$,

$$\mathbf{\log_a x = \log_a y \Leftrightarrow x = y}.$$

For instance,

$$\log_a(5^x) = \log_a(.04) \Leftrightarrow 5^x = .04 \qquad \text{\textit{The logarithmic function is one-to-one.}}$$
$$\Leftrightarrow 5^x = 5^{-2} \qquad \text{\textit{(why?).}}$$
$$\Leftrightarrow x = -2. \qquad \text{\textit{The exponential function is one-to-one.}}$$

The following example uses several of the properties of logarithms.

Example 4 Solve for x in the equation

$$\log_a x = 2\log_a 3 - \frac{2}{3}\log_a 8 - \log_a 5.$$

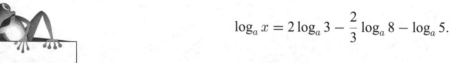
Equivalently,
$\log_a x \neq \log_a y \Leftrightarrow x \neq y$.

Solution:

$$\log_a x = \log_a(3^2) - \log_a(8^{2/3}) - \log_a 5 \qquad \text{\textit{Logarithm-of-a-Power Property.}}$$
$$= \log_a 9 - (\log_a 4 + \log_a 5)$$
$$= \log_a 9 - \log_a 20 \qquad \text{\textit{Logarithm-of-a-Product Property.}}$$
$$= \log_a\left(\frac{9}{20}\right) \qquad \text{\textit{Logarithm-of-a-Quotient Property.}}$$

Therefore, $x = \dfrac{9}{20}.$ \qquad \textit{One-to-One Property for Logarithms.} $\blacksquare$

Use of Scientific Notation As mentioned above, approximate decimal values of logarithms can be obtained by pushing a button on a scientific calculator. However, for many years, beginning with John Napier (1550-1617) who invented logarithms for computational purposes, it was necessary to express a number in scientific notation (see Exercises R.5) and

5.3. Properties of the Logarithm

to use the rules and tables to find values of logarithms. Since scientific notation uses the base 10, it became customary to omit the subscript and to write $\log x$ for $\log_{10} x$. Most scientific calculators have a button marked "log" for base 10 logarithms (**Common logarithms**) and a button labeled "ln" for base e (**natural**) logarithms.

Today the calculator has made logarithms much less important as a tool for computation, but the logarithmic function and its properties are widely used in both theoretical and applied mathematics. We include some further examples to illustrate the use of scientific notation. A calculator should be used to check the results.

Example 5 Given $\log 1.2 = 0.0792$, compute:
(a) $\log 1200$ (b) $\log .00012$ (c) $\log(-12)$

Solution: We begin by writing each number in scientific notation, and we then apply the logarithmic properties properties for products and powers.

(a) $\log 1200 = \log(1.2 \times 10^3)$
$= \log(1.2) + \log(10^3)$
$= \log(1.2) + 3 \log 10$
$= 0.0792 + 3 \cdot 1$
$= 3.0792$

(b) $\log .00012 = \log(1.2 \times 10^{-4})$
$= \log(1.2) + \log(10^{-4})$
$= \log(1.2) - 4 \log 10$
$= 0.0792 - (4.1)$
$= -3.9208$

(c) $\log(-12)$ does not exist in the real number system since $\log x$ is defined for positive numbers only. If a calculator is used to try to compute $\log(-12)$, an error signal will appear. ∎

Example 6 Given $\ln 1.2 = .1823$ and $\ln 10 = 2.3026$, find $\ln 1200$.

Solution:
$$\ln 1200 = \ln(1.2 \times 10^3)$$
$$= \ln(1.2) + 3 \ln 10$$
$$= .1823 + 3(2.3026)$$
$$= 7.0901 \quad ∎$$

Example 7 Given $\log 3.46 = .5391$, solve the following equations for x.
(a) $\log x = 1.0782$ (b) $\log x = 3.5391$ (c) $\log x = -4.4609$

Solution:

(a) $1.0782 = 2(0.5391)$. From the rule $2\log x = \log x^2$, we conclude that $x = (3.46)^2$. *Calculator check:* $(3.46)^2 = 11.9716$ and $10^{1.0782} = 11.9729$.

The difference between 11.9716 and 11.9729 is due to round-off.

(b) We write
$$\log x = .5391 + 3$$
$$= \log 3.46 + \log(10^3)$$
$$= \log(3.46 \times 10^3).$$

Therefore, $x = 3.46 \times 10^3 = 3460$.
calculator check: $10^{3.5391} = 3460.1904$

(c) We first write
$$-4.4609 = .5391 - 5. \qquad \text{\textit{Add and subtract}}$$
$$\text{\textit{5 to 4.4609 (why 5?).}}$$

Then
$$\log x = .5391 - 5$$
$$= \log 3.46 - \log(10^5)$$
$$= \log\left(\frac{3.46}{10^5}\right).$$

Therefore, $x = 3.46/10^5 = .0000346$.
Calculator check: $10^{-4.4609} = .0000346$ ∎

Comment Logarithms were the invention of John Napier, a Scottish laird who worked for 20 years on the idea before publishing his work in 1614. His purpose was to simplify trigonometric computations required in astronomy. Napier's idea was to establish a one-to-one correspondence between the numbers of two sequences so that multiplication in one sequence corresponds to addition in the other. He established the correspondence by means of the dynamic model illustrated in Figure 10.

In the model, points P and Q start at 0 on their respective scales. Point Q moves with constant speed 10^7. P starts with speed 10^7, and thereafter its speed is x, where $x = d(P, 10^7)$. Point P corresponds to point Q.

It can be shown by calculus that the correspondence between P and Q in Figure 10 satisfies

$$\frac{x}{10^7} = e^{-y/10^7} \quad \text{or} \quad \underbrace{y = -10^7 \ln\left(\frac{x}{10^7}\right)}_{y = Nap.\log x}.$$

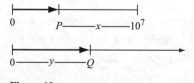

Figure 10

Napier defined his logarithms in terms of the dynamic model, not with respect to a base as we do today. Also, Napier's logarithms satisfy

$$\text{Nap. log } \frac{x_1 x_2}{10^7} = \text{Nap. log } x_1 + \text{Nap. log } x_2,$$

Which differs from our present-day relationship by the scaling factor 10^{-7}. Napier chose the scaling factor 10^{-7} because he was concerned primarily with computations in trigonometry, and trigonometric tables at the time were computed in terms of a circle of radius 10^7. Shortly after their introduction, Napier's logarithms were modified to become our common logarithms. Logarithms have been the primary tool in scientific computations from their introduction early in the seventeenth century up to the advent of the electronic computer.

Exercises 5.3

Whenever $\log_a u$ appears, it is assumed that $a > 0, a \neq 1$, and $u > 0$.

Fill in the blanks to make each statement true.

1. If $\log_a(xy) = 2\log_a x$, then $y = $ -----.
2. If $\log_a(y/x) = 2\log_a x$, then $y = $ -----.
3. If $\log_a y = x \log_a 2$, then $y = $ -----.
4. If $\log_a(x^3) = \log_a(y^2)$, then $y = $ -----.
5. If $\log_a(x^2) = (\log_a x)^2$, then $x = $ ----- or -----.

Write true or false for each statement.

6. $\log_a(x+y) = \log_a x + \log_a y$
7. $\log_a(x+y) = \log_a x \cdot \log_a y$
8. $\log_a(xyz) = \log_a x + \log_a y + \log_a z$
9. If $\log_a x = x \log_a y$, then $y = x^{1/x}$.
10. If $\log_a\left(\frac{x}{y}\right) = \log_a x + \log_a y$, then $y = 1$.

Further Properties of Logarithms

Use $\log 5 = 0.6990$ and $\log 7 = 0.8451$ to compute each of the following.

11. $\log 35$
12. $\log 3.5$
13. $\log(.00035)$
14. $\log\left(\frac{7^2}{5^3}\right)$
15. $\log(.2)$
16. $\log\left(\sqrt[3]{7}\right)$

Express each of the following as a single logarithm.

17. $3\log_a x + \frac{1}{2}\log_a(x-1) - 5\log_a(x^2)$
18. $\frac{1}{3}\log_a\left(\frac{x}{y}\right) - \log_a\left(\frac{y}{x}\right)$
19. $\log(x^2 - 4) - 2\log(x - 2)$
20. $\log(x^3) - \log(x^2) - \log x$

Express each of the following in terms of $\log b$, $\log c$, and $\log d$.

21. $\log\left(\frac{b^2 c}{d^4}\right)$
22. $\log(\sqrt[3]{b} \cdot \sqrt{c} \cdot d^4)$
23. $\log(\sqrt[3]{b} \cdot c^{1/2} \cdot d^{-1})$
24. $\log\left(\frac{b}{\frac{c}{d}}\right) - \log\left(\frac{\frac{b}{c}}{d}\right)$

Solve for x.

25. $\log_a x = \log_a 7$
26. $2\log_a x = \log_a 25$
27. $\log_a x = 3\log_a 2 - \frac{1}{3}\log_a 8 + 2\log_a 5$
28. $(\log_2 x)^2 - 4\log_2 x + 4 = 0$
29. $\log_x 20 - \log_x 5 = 2$
30. $\log_x 2 + \log_x 3 = 5$
31. $\log_x(x^2) = x$
32. $\log_x(x^4) = 2x + 1$

33. Fill in the reason for each step in the following derivation of the property $\log_a(x^y) = y \log_a x$. Let $p = \log_a x$. Then

$$x = a^p \quad \text{----------}$$
$$x^y = (a^p)^y \quad \text{----------}$$
$$x^y = a^{py} \quad \text{----------}$$
$$\log_a(x^y) = py \quad \text{----------}$$
$$\log_a(x^y) = (\log_a x)y \quad \text{----------}$$
$$\log_a(x^y) = y \log_a x.$$

34. Derive the property $\log_a(x/y) = \log_a x - \log_a y$ from the corresponding exponent rule $a^p/a^q = a^{p-q}$ (see Example 3 and the previous exercise).

Use $\log 2.40 = .3802$ *and* $\log 1.47 = .1673$ *to compute each of the following. Check your results on a calculator.*

35. $\log 240$
36. $\log .024$
37. $\log 1{,}470{,}000$
38. $\log(1.47 \times 10^{-20})$

Use $\ln 2.40 = .8755$, $\ln 1.50 = .4055$, *and* $\ln 10 = 3.3026$ *to compute each of the following. Check your results on a calculator.*

39. $\ln 240$
40. $\ln .024$
41. $\ln 1{,}500{,}000$
42. $\ln(1.5 \times 10^{-20})$

Given $\log 6.12 = .7868$ *to solve each of the following for x. Check your results on a calculator.*

43. $\log x = 3.7868$
44. $\log x = 1.7868$
45. $\log x = -0.2132$
46. $\log x = -2.2132$

Miscellaneous

47. Let $y = ca^{kx}$, where c and k are constants and $c \neq 0$. Show that if $Y = \log_a y$, then $Y = mx + b$, where m and b are constants. That is, "the logarithm of an exponential curve is a straight line."

48. With reference to Exercise 47, graph the straight line associated with $y = (1/4)10^{x/2}$.

* The Equation $\log_a x = x$

The equation $\log_a x = x$ *is equivalent to the exponential equation* $a^x = x$, *which was discussed in the exercise for Section 5.2. With reference to the figures and using a calculator, search for approximate solutions (to three significant digits) to each of the following equations.*

(a) $0 < a < 1$

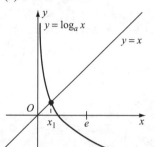

(b) $1 < a < e^{1/e}$

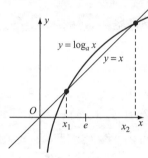

(c) $a \geq e^{1/e}$

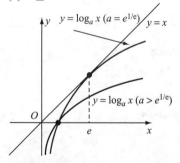

49. $\log_{.2} x = x$ (one solution)
50. $\log_{.8} x = x$ (one solution)
51. $\log_{1.2} x = x$ (two solutions)
52. $\log_{1.35} x = x$ (two solutions)

*Exercises 49-52 are optional.

5.4 Applications of Exponential and Logarithmic Equations

Compound Interest
Growth and Decay
Intensity of Sound
Richter Scale

By taking the natural logarithm of both sides, the equation $y = ce^{kx}$ becomes $\ln y = \ln c + kx$. Hence, an exponential relationship between x and y becomes a linear relationship between x and $\ln y$.

The exponential is nature's curve. The logarithm is man-made.

In earlier chapters we were able to solve problems concerning percent, motion, and work by using algebraic equations, namely linear and quadratic ones. However, many quantities in science and nature increase or decrease exponentially, and problems involving them require the solution of exponential equations. Also, because exponential functions can increase so rapidly, it is often more practical to convert exponential relationships to a logarithmic scale, such as the Richter scale for earthquakes.

In this section we will solve equations involving exponential and logarithmic expressions. Our tools for doing so include the usual algebraic operations of addition, subtraction, multiplication, and division on both sides of an equation, as well as the *functional* operations of taking logarithms or exponentiating on both sides of an equation. Because the logarithmic and exponential functions are one-to-one, any equation

$$f(x) = g(x)$$

is equivalent to

$$\log_a f(x) = \log_a g(x)$$

(assuming $f(x) > 0$ and $g(x) > 0$) and also to

$$a^{f(x)} = a^{g(x)}.$$

In the following examples, all logarithms and exponentials are evaluated by means of a scientific calculator.

Compound Interest. If $1.00 is invested at 7% compounded annually, then at the end of one year the amount of dollars accumulated is

$$A = 1.00 + .07(1.00)$$
$$= 1.07.$$

At the end of two years,

$$A = 1.07 + .07(1.07)$$
$$= 1.07(1 + .07)$$
$$= (1.07)^2,$$

and at the end of t years, the amount becomes

$$A = (1.07)^t.$$

If interest is compounded four times a year, then the interest rate per compound period is .07/4 and the number of compound periods in t years is $4t$. Therefore, the amount in t years is

$$A = \left(1 + \frac{.07}{4}\right)^{4t}.$$

More generally, for a principal of P dollars invested at an annual interest rate of R (% rate $\div$ 100) with interest compounded N times a year, the amount at the end of t years is

$$A = P\left(1 + \frac{R}{N}\right)^{Nt}. \qquad \text{compound interest} \qquad (1)$$

To see how this compound interest formula is related to the natural base e, we proceed as follows. First, we write

$$A = P\left(1 + \frac{R}{N}\right)^{Nt}$$
$$= P\left[\left(1 + \frac{R}{N}\right)^{N/R}\right]^{Rt}.$$

Next, we substitute $n = N/R$ to obtain

$$A = P\left[\left(1 + \frac{1}{n}\right)^n\right]^{Rt}.$$

In Section 5.1, we stated that the number e is approximated for large values of n by $(1 + 1/n)^n$, and the following table shows that these numbers actually increase and get close together as n increases, so that we may reasonably guess that they approach a finite value.

n	1	2	10	1000	10,000	100,000
$(1+1/n)^n$	2	2.25	2.59374	2.71692	2.71815	2.71827

5.4. Applications of Exponential and Logarithmic Equations

In fact, the number e may be defined by

$$\left(1 + \frac{1}{n}\right)^n \to e \quad \text{as} \quad n \to \infty. \quad \text{natural base}$$

Therefore, we conclude that as N, the number of compound periods per year, increases, so does $n = N/R$, and

$$A = P\left[\left(1 + \frac{1}{n}\right)^n\right]^{Rt} \to Pe^{Rt}.$$

For this reason, if P dollars are invested at an annual interest rate of R, and the amount A accumulated in t years is computed by the formula

$$A = Pe^{Rt}, \qquad (2)$$

then the interest is said to be **compounded continuously.**

Example 1 Jennifer is six years old, and her parents want to have $30,000 available when she starts college in 12 years. How much must they deposit now at 8% annual interest compounded quarterly in order to achieve this goal?

Solution: We substitute the numbers $A = 30{,}000$, $R = .08$, $N = 4$, and $t = 12$ into (1) and then solve for P.

$$30{,}000 = P\left(1 + \frac{.08}{4}\right)^{4(12)}$$
$$= P(1.02)^{48}$$
$$= P(2.5870704)$$

Therefore,

$$P = \frac{30{,}000}{2.5870704}$$
$$\approx 11596.13$$

Thus, a deposit of $11,596.13 will grow to $30,000 in 12 years. ∎

Example 2 How long will it take to double the principal if money is invested at 9% (a) compounded annually and (b) compounded continuously?

Solution:

(a) We have $R = .09$ and $N = 1$, and if P dollars are invested, we want to find the value of t for which $A = 2P$ in t years. Therefore, by substitution into (1),

$$2P = P(1.09)^t$$
$$2 = (1.09)^t \quad \text{Divide both sides by P.}$$
(key step) $\quad \log 2 = \log(1.09)^t \quad \text{Take the } \log_{10} \text{ of both sides.}$
$$\log 2 = t \log(1.09) \quad \text{Logarithm-of-a-Power Property.}$$
$$t = \frac{\log 2}{\log(1.09)} \quad \text{Solve for } t.$$
$$= \frac{.3010300}{.0374265}$$
$$\approx 8.04.$$

Hence, it takes slightly more than 8 years for money to double at 9% compounded annually. Note that the answer does not depend on the value of P (*why?*).

(b) We substitute $R = .09$ and $A = 2P$ into formula (2), and then solve for t.

$$2P = Pe^{.09t}$$
$$2 = e^{.09t} \quad \text{Divide both sides by P.}$$
(key step) $\quad \ln 2 = \ln(e^{.09t}) \quad \text{Take natural logs of both sides.}$
$$\ln 2 = .09t \quad (\text{Why?}).$$
$$t = \frac{\ln 2}{.09} \quad \text{Solve for t.}$$
$$= \frac{.69314718}{.09}$$
$$\approx 7.70 \text{ years} \quad \blacksquare$$

Comment In solving Example 2(a), we used logarithms to the base 10, but we could have base e, in which case we would have obtained

$$t = \frac{\ln 2}{\ln(1.09)}$$
$$= \frac{.69314718}{.0861777}$$
$$\approx 8.04$$

as before.

Example 3 Max inherits $50,000 and decides to use it to buy a summer home that he estimates will cost $75,000 five years from now. If he invests the money now, what rate of interest compounded daily will give him the necessary $75,000?

5.4. Applications of Exponential and Logarithmic Equations

Solution: We substitute $P = 50{,}000$, $A = 75{,}000$, $N = 365$, and $t = 5$ into (1), and then solve for R

$$75{,}000 = 50{,}000 \left(1 + \frac{R}{365}\right)^{365(5)}$$

$$\frac{75{,}000}{50{,}000} = \left(1 + \frac{R}{365}\right)^{1825}$$

$$1.5 = \left(1 + \frac{R}{365}\right)^{1825}$$

(*key step*) $\quad \log(1.5) = 1825 \log\left(1 + \frac{R}{365}\right) \quad$ (*why?*).

$$\frac{\log(1.5)}{1825} = \log\left(1 + \frac{R}{365}\right)$$

$$.00009649 = \log\left(1 + \frac{R}{365}\right)$$

(*key step*) $\quad 10^{.00009649} = 1 + \frac{R}{365} \quad use\ 10^{\log x} = x.$

$$1.0002222 = 1 + \frac{R}{365}$$

$$R \approx .0811$$

Therefore, the required interest rate is 8.11%. ■

Growth and Decay. Some quantities, such as the number of bacteria in a culture or individuals in a population, increase exponentially, while others, such as quantities of radioactive elements, decrease exponentially.

Example 4 If there are 10,000 bacteria in a culture at some initial time $t = 0$ and the number doubles every hour, then there are

$$10{,}000 \cdot 2, \quad 10{,}000 \cdot 4, \quad \text{and} \quad 10{,}000 \cdot 8$$

bacteria at the end of 1, 2 and 3 hours, respectively, and at the end of t hours, the number $N(t)$ of bacteria

$$N(t) = 10{,}000 \cdot 2^t.$$

Find the amount of time it takes for the number of bacteria to increase from 320,000 to 1,280,000.

Solution: Let t_1 be the number of hours required for the number to reach 320,000 and t_2 the number of hours to reach 1,280,000. Then

$$320{,}000 = 10{,}000 \cdot 2^{t_1} \quad \text{and} \quad 1{,}280{,}000 = 10{,}000 \cdot 2^{t_2}.$$
$$\text{Hence,} \quad 32 = 2^{t_1} \quad \text{and} \quad 128 = 2^{t_2},$$

which means $t_1 = 5$ and $t_2 = 7$. The time required is 2 hours. ∎

The base 2 is naturally involved in any situation like Example 4 in which a quantity doubles in a given unit of time. Similarly, if a quantity triples in a given unit of time, then we could write $N(t) = N_0 3^t$, where N_0 denotes the value of N at time zero: that is $N_0 = N(0)$. However, any positive base a can be written as

$$a = e^k, \qquad k = \ln a$$

and therefore,

$$a^t = e^{kt}. \qquad \text{Power-of-a-Power Rule}$$

Hence, any quantity that always changes by a fixed multiple over the same increment of time satisfies the formula

$$N(t) = N_0 e^{kt}, \qquad \text{Exponential Growth or Decay} \qquad (3)$$

where $N(t)$ denotes the amount of the quantity at time t, N_0 is the amount at time zero, and k is a constant. Such quantities are said to grow $k > 0$ or decay ($k < 0$) exponentially.

Radioactive elements disintegrate according to formula (3), and the time it takes for one half of the initial amount to disintegrate is called the **half-life** of the element.[1]

Example 5 The half-life of radium is 1600 years. If 100 grams of this element are present in a laboratory now and remain untouched for 25 years, how many grams will there be then?

Solution: First, we use the given information about the half-life to evaluate the constant k in formula (3). That is, $N_0 = 100$ and

$$N(1600) = \frac{1}{2}(100) = 50.$$

Therefore, by formula (3),

---

[1] The term *half-life* means the life of one half of the element, not one half the life of the element. Theoretically, the lifetime of the element is infinite (why?).

5.4. Applications of Exponential and Logarithmic Equations

$$50 = 100e^{k(1600)}$$

$$\frac{1}{2} = e^{1600k}$$

(*key step*) $\quad \ln\left(\frac{1}{2}\right) = 1600k \qquad$ (*why ?*).

$\qquad\qquad \ln 1 - \ln 2 = 1600k \qquad$ Lagarithm-of-a-Quotient Property.

$\qquad\qquad -\ln 2 = 1600k \qquad$ $\ln 1 = 0$.

$$k = \frac{-\ln 2}{1600} \qquad \textit{Solve for k.}$$

$$k = -.00043322.$$

Hence, in this particular case (3) becomes

$$N(t) = 100e^{-.00043322t},$$

and the amount of radium left after 25 years is

$$N(25) = 100e^{-.00043322(25)}$$
$$= 100e^{-.0108305}$$
$$= 100(.989228)$$
$$\approx 98.9 \text{ grams}.$$

Note that only 1.1 grams disappear in 25 years. Other radioactive elements disintegrate even more slowly than radium. This is why we have a problem with radioactive wastes. ∎

When a population increases exponentially according to formula (3), it is said to satisfy the **Malthusian model**.[2] The model is reasonably accurate for organisms that reproduce in a culture under controlled laboratory conditions, and also for human populations over a short time period. But if we take into account such factors as limited food supply and limited space, then a more realistic model for population growth is the **logistic law**, which is given by the formula

$$N(t) = \frac{cN_0}{N_0 + (c - N_0)e^{-kt}}, \qquad \text{Logistic Law} \qquad (4)$$

where c and k are positive constants. The graph of (4) is given in Figure 11 and shows that the growth starts out exponentially, then tapers off after N reaches $c/2$ and approaches c as a limiting value.

We can see directly from (4) that $N(t) \approx c$ for large values of t since $e^{-kt} \approx 0$ for large t. On the other hand, to investigate the graph of (4) for small values of t, we first multiply numerator and denominator by e^{kt} and rearrange the terms in the denominator to obtain

---
[2] *Thomas Malthus (1766-1834) was an English economist.*

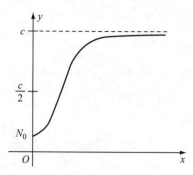

Figure 11 Logistic Growth

$$N(t) = \frac{cN_0 e^{kt}}{N_0(e^{kt}-1) + c}. \tag{4'}$$

Now, since $e^{kt} \approx 1$ for small values of t, (4') shows that initially $N(t) \approx N_0 e^{kt}$.

Example 6 Assume that the population growth of the United States satisfies the logistic law (4), where N is in millions and $k = .03$. If N was 150.7 in 1950 and 226 in 1980, find:

(a) the constant c,
(b) the population predicted by the formula for the year 2000.

Solution:

(a) Let $t = 0$ correspond to the year 1950. Then 1980 corresponds to $t = 30$, and we substitute $N_0 = 150.7$, $t = 30$, $N(30) = 226$ and $k = .03$ in (4) to obtain

$$\frac{c(150.7)}{150.7 + (c - 150.7)e^{-.03(30)}} = 226$$
$$226[150.7 + (c - 150.7)e^{-.9}] = 150.7c$$
$$34058.2 + 226ce^{-.9} - 34058.2e^{-.9} = 150.7c$$
$$c(226e^{-.9} - 150.7) = 34058.2(e^{-.9} - 1)$$
$$c = \frac{34058.2(e^{-.9} - 1)}{226e^{-.9} - 150.7}$$
$$c \approx 344.$$

(b) The year 2000 corresponds to $t = 50$. Hence, using the value $c = 344$ obtained from part (a), we substitute $t = 50$ in (4) to obtain

$$N(50) = \frac{344(150.7)}{150.7 + (344 - 150.7)e^{-.03(50)}}$$
$$\approx 267.$$

5.4. Applications of Exponential and Logarithmic Equations

Therefore, the logistic law predicts a population of approximately 267,000,000 by the year 2000. ∎

Intensity of sound. An object that is vibrating in some medium, such as air or water, produces sound waves. The intensity of the sound depends on the amount of energy transmitted by the waves, and the unit for measuring sound intensity is the decibel.[3] A sound wave that transmits energy measuring 10^{-16} watts per square centimeter is assigned a value of zero decibels. In general, if the energy transmitted by sound waves is I watts per square centimeter, then the decibel level D of the sound defined is

$$D = 10 \log(10^{16} I). \qquad \text{Decibel Level} \qquad (5)$$

[3] Named in honor of Alexander Graham Bell (1847-1922), the inventor of the telephone,

Note that 10^{-16} watts per square inch corresponds to $D = 0$ (*why?*).

Example 7 Find the decibel levels for the following:

(a) human voice for which $I = 10^{-10}$ watts per square centimeter,
(b) a jet plane for which $I = 2 \times 10^{-5}$ watts per square centimeter.

Solution:

(a) $D = 10 \log(10^{16} \, 10^{-10})$
$= 10 \log(10^6)$
$= 10 \cdot 6$
$= 60$ decibels

(b) $D = 10 \log[10^{16}(2 \times 10^{-5})]$
$= 10 \log(2 \times 10^{11})$
$= 10 [\log 2 + \log(10^{11})]$
$= 10 [.30103 + 11]$
≈ 113 decibels

The 113 decibels of the jet is close to the human tolerance level, which is about 120 decibels. ∎

Example 8 A machinist operates two machines that produce noise levels of 40 and 50 decibels, respectively. If it is unsafe for the operator to be continuously subjected to more then 85 decibels, are earplugs needed?

Solution: When both machines are running, the energy transmitted to the ear is equal to $I_1 + I_2$, where I_1 is the energy transmitted by the machine producing 40 decibels, and I_2 is that of the machine producing 50 decibels. Now

$$40 = 10 \log(10^{16} I_1) \quad \text{and} \quad 50 = 10 \log(10^{16} I_2)$$
$$4 = \log(10^{16} I_1) \quad\quad\quad\quad 5 = \log(10^{16} I_2)$$
$$10^4 = 10^{16} I_1 \quad\quad\quad\quad\quad 10^5 = 10^{16} I_2$$
$$I_1 = 10^{-12} \quad\quad\quad\quad\quad\quad I_2 = 10^{-11}.$$

Therefore,

$$I_1 + I_2 = 10^{-12} + 10^{-11}$$
$$= 10^{-12}(1 + 10)$$
$$= 11 \cdot 10^{-12},$$

and the combined decibel level is

$$D = 10 \log(10^{16} \cdot 11 \cdot 10^{-12})$$
$$= 10 \log(10^4 \cdot 11)$$
$$= 10 (4 + \log 11)$$
$$= 10 (5.0414)$$
$$\approx 50.4 \text{ decibels}.$$

The combined level of 50.4 decibels is well below 85, so no earplugs are needed. ∎

Richter Scale. The Richter scale, named for American seismologist Charles Richter (1900-1985), is mentioned in news reports whenever a serious earthquake occurs anywhere in the world. The scale measures the magnitude R of an earthquake according to the formula

$$\boldsymbol{R = \frac{2}{3} \log(10^{-4.7} E)}, \quad \text{Richter scale} \quad\quad (6)$$

where E is the energy, measured in joules, released by the earthquake.

Example 9 If the energy released by an earthquake is 10^{13} joules, what is its magnitude on the Richter scale?

Solution: We substitute $E = 10^{13}$ in (6), obtaining

$$R = \frac{2}{3} \log(10^{-4.7} 10^{13})$$
$$= \frac{2}{3} \log(10^{8.3})$$
$$= \frac{2}{3}(8.3)$$
$$\approx 5.5. \quad\blacksquare$$

Example 10 The magnitude of the San Francisco earthquake in 1906 has been estimated at 8.2 on the Richter scale. What was the amount of energy released?

Solution: By substituting $R = 8.2$ in equation (6), we obtain

$$8.2 = \frac{2}{3} \log(10^{-4.7} E)$$
$$12.3 = \log(10^{-4.7} E)$$
$$10^{12.3} = 10^{-4.7} E$$
$$E = 10^{17}.$$

Hence, the energy released was 10^{17} joules. ∎

Exercises 5.4

Compound Interest

1. Find the amount accumulated at the end of 5 years if $100 is invested at an annual rate of 7% and compounded

   (a) monthly
   (b) daily
   (c) continuously.

2. How much more will be earned in 10 years on $1000 at 8% when interest is compounded continuously than when it is compounded daily?

3. Which amounts to more in 10 years: $1000 invested at 8% compounded semiannually or $1000 invested at 7.95% compounded daily?

4. How much more interest will be earned in 1 year by transferring $1000 from a passbook savings account paying 5.5% compounded monthly to a money market account paying 7.3% compounded daily?

5. A bank advertises that its interest rate of 6.91% compounded daily is equivalent to an effective annual rate of 7.154%. Is this correct? (*Hint:* compare the amount earned on $1 for 1 year at 7.154% compounded annually with the amount earned at the given compound interest rate.)

6. What is the effective annual rate equivalent to 8.05% compounded daily? (See Exercise 5.)

7. If an interest rate of 18% compounded monthly on the unpaid balance is charged on a credit card, how much is owed on a $300 balance paid in one payment at the end of four months?

8. It is said that native Americans sold Manhattan Island for $24 in 1626. If that $24 had been put in a bank paying 7% interest compounded continuously, what would it amount to in 1988?

Growth and Decay

9. If the number of bacteria in a culture grows exponentially and doubles every hour, starting with only 10 bacteria, find the number at the end of 12 hr.

10. Solve Exercise 9 if the number of bacteria doubles every half hour.

11. The number of bacteria in a colony increases exponentially according to formula (3), and at a certain time there are 10,000 bacteria present. If the number 2 hr after that time is 80,000, what is the number 4 hr after that time?

12. Suppose the population growth of a small country follows the Malthusian law (3) and doubles every 50 yr. In how many years will the population triple?

13. Assume that the United States population, which was 226 million in 1980, grows according to the Malthusian law at 1.2% a year. Find the population predicted for the year 2000 and compare the result with that in Example 6. [*Hint:* use formula (3) with $k = 1.2\% = .012$.]

14. Show that if the half-life of a radioactive substance is H, then

    (a) the constant k in formula (3) is given by $k = -(\ln 2)/H$, and
    (b) the amount of the radioactive substance at time t is $N(t) = N_0 2^{-t/H}$.

15. The half-life of radioactive polonium is 140 days. In how many days will there be 80% of the original amount left?

16. In each living thing the ratio of the amount of radioactive carbon-14 to the amount of ordinary carbon-12 remains approximately constant. However, when a plant or animal dies, the amount of carbon-14 decreases exponentially, and its half-life is about 5700 yr.

    (a) Find k in formula (3) for carbon-14.
    (b) Determine the age of a piece of charcoal if 20% of the original amount of carbon-14 remains.

17. In 1950 Willard Libby discovered the method of carbon-14 dating described in Exercise 16, and in 1960 he received a Nobel Prize. In a famous example, paintings discovered in the Lascaux Caves in France were found to contain, in 1950, 15% of the usual amount of carbon-14 in such paint. What was the approximate age of the paintings at that time?

18. The amount of a drug in a patient's bloodstream decreases exponentially according to formula (3), where t is the number of hours after the drug is administered. If $k = -.2$ and $N_0 = .5$ ml, find the amount of the drug left after 2 hr.

19. Verify that in the logistic law (4), $N(0) = N_0$.

20. Show algebraically that equation (4) is equivalent to (4').

21. Assume that the population of deer in a certain forest follows the logistic law (4), with $k = .2$ and $N_0 = 100$, and that after 3 yr there are 170 deer.

    (a) Show that the number of deer can never exceed 1144. (*Hint:* compute c.)
    (b) Find the number of deer at the end of 6 yr.

22. For the logistic law (4), show that each of the following is true.

    (a) $t = \dfrac{1}{k} \ln\left[\dfrac{N(c - N_0)}{N_0(c - N)}\right]$
    (b) $N = \dfrac{c}{2}$ for $t = \dfrac{1}{k} \ln\left(\dfrac{c - N_0}{N_0}\right)$

Intensity of Sound

23. Find the number of decibels for the following sounds.

    (a) a whisper with $I = 10^{-14}$
    (b) rock music with $I = 10^{-6}$

24. If a sound of 130 decibels produces pain in the ear, what is the corresponding intensity I?

25. If two machines operate at noise levels of 50 and 60 decibels, respectively, what is the total number of decibels produced when both machines are operating?

26. Solve equation (5) for I in terms of D.

27. If 10^k watts produce D decibels, how many watts produce $2D$ decibels?

28. If one machine produces D decibels, how many similar machines are needed to produce $2D$ decibels?

Richter Scale

29. Find the amount of energy (in joules) released for each of the following Richter scale readings.

    (a) 0   (b) 5

30. A Himalayan earthquake in 1950 registered 8.7 on the Richter scale. What was its released energy?

31. If 10^k joules of released energy produces a Richter scale reading of R, how many joules will produce $2R$?

32. Solve equation (6) for E in terms of R.

Miscellaneous

If a cake is removed from an oven and allowed to cool, then the rate at which the cake cools is proportional to the difference between the temperature T of the cake and the temperature M of the area in which it is placed. In general, if an object of temperature T_0 is placed in a cooler medium of constant temperature M, then the temperature of the object at time t is given by the formula

$$T(t) = (T_0 - M)e^{-kt} + M, \quad \textbf{(cooling formula)}$$

where k is a positive constant.

33. Verify that $T(0)$, the temperature at time 0, is equal to T_0.

34. Suppose that the temperature of a body is 100° F and the surrounding air has temperature 50° F. If the body cools from 100° to 80° in 15 min, when will it be 60° F?

35. A metal object is heated to 250° F and then placed in a room with temperature 70°F. If the object cools to 150°F in 10 min, what will be its temperature in 10 more minutes?

36. A cup of coffee is placed on a table, and the temperature of the surrounding air is 30° C. If the coffee cools to 35° in 6 min, what was the original temperature of the coffee?

Chemists define the pH of a liquid by the equation

$$\mathbf{pH} = -\log[\mathbf{H^+}], \quad \textbf{(pH formula)}$$

*where $[\mathbf{H^+}]$ is the liquid's concentration of hydrogen ions measured in moles per liter. For pure water, $\mathbf{pH} = 7$. An **acid** has $\mathbf{pH} < 7$, and a **base** has $\mathbf{pH} > 7$.*

37. Find the **pH** for each of the following.
    (a) milk with $[\mathbf{H^+}] = 4 \times 10^{-7}$ moles per liter
    (b) vinegar with $[\mathbf{H^+}] = 6.3 \times 10^{-3}$ moles per liter

38. Find $[\mathbf{H^+}]$ for each of the following.
    (a) soda with $\mathbf{pH} = 2.6$
    (b) beer with $\mathbf{pH} = 4.8$

In the process of learning some skills, progress is rapid at the beginning and then slows down. A learning curve that describes this situation may have an equation of the form

$$y = y_0 + c(1 - e^{-kx}), \quad \textbf{(learning curve)}$$

where c and k are positive constants.

39. Show that $y = y_0$ when $x = 0$, and that $y \to y_0 + c$ as $x \to \infty$. Also sketch the graph of the learning curve for the case $y_0 = 2$, $c = 5$, $k = .04$, and $x \geq 0$.

40. Let y in the learning curve denote the number of words per minute of a typist training for x days, where $y_0 = 10$, $c = 40$, and $k = .03$.
    (a) How many words per minute can the person type after training 50 days?
    (b) What is the maximum number of words per minute the typist will ever do?

Chapter 5 Review Outline

5.1 The Exponential Function and Its Graph

Definition

The exponential function $f(x) = a^x$, where the base a is a fixed positive number ($a \neq 1$) and the exponent x is any real number, is a continuous and smooth extension of $a^{m/n}$, which was defined for rational exponents in Chapter 1.

The natural base e is the value approached by the expression

$$\left(1 + \frac{1}{n}\right)^n$$

as n goes through the sequence $1, 2, 3, 4, \ldots$. $e \approx 2.71828$

Properties of a^x

$f(x) = a^x$ is a one-to-one function that increases for $a > 1$ and decreases for $0 < a < 1$, with $D_f = (-\infty, \infty)$ and $R_f = (0, \infty)$.

$$a^x a^y = a^{x+y} \qquad \frac{a^x}{a^y} = a^{x-y}$$

$$(ab)^x = a^x b^x \qquad \left(\frac{a}{b}\right)^x = \frac{a^x}{b^x}$$

$$(a^x)^y = a^{xy}$$

5.2 The Logarithmic Function and Its Graph

Definition

$f(x) = \log_a x$ is the inverse of $g(x) = a^x$; that is,

$$\log_a(a^x) = x \text{ for all real numbers } x,$$

and $\quad a^{(\log_a x)} = x$ for all $x > 0$.

Equivalently,

$$y = \log_a x \Leftrightarrow a^y = x.$$

Properties of $\log_a x$

$f(x) = \log_a x$ is a one-to-one function that increases for $a > 1$ and decreases for $0 < a < 1$, with $D_f = (0, \infty)$ and $R_f = (-\infty, \infty)$.

$$\log_b x = \frac{\log_a x}{\log_a b}$$

5.3 Properties of the Logarithm

$$\log_a(xy) = \log_a x + \log_a y$$
$$\log_a(x/y) = \log_a x - \log_a y$$
$$\log_a(x^y) = y \log_a x$$

5.4 Applications of Exponential and Logarithmic Equations

Since the logarithmic and exponential functions are one-to-one, the equation

$$f(x) = g(x)$$

is equivalent to each of the equations

$$\log_a f(x) = \log_a g(x) \quad [f(x) > 0, g(x) > 0]$$
and
$$a^{f(x)} = a^{g(x)}.$$

Applications include compound interest:

$$A = P\left(1 + \frac{R}{N}\right)^{Nt},$$

continuous compound interest:

$$A = Pe^{Rt},$$

exponential growth and decay:

$$N(t) = N_0 e^{kt}$$

($k > 0$ for growth, $k < 0$ for decay), intensity of sound:

$$D = 10 \log(10^{16} I),$$

and the Richter scale:

$$R = \frac{2}{3} \log(10^{-4.7} E).$$

Chapter 5 Review Exercises

1. Sketch the graph of $y = (3/2)^x$ as in Section 5.1.
2. Sketch the graph of $y = (2/3)^x$ as in Section 5.1.
3. Solve $2^{3x} = 16^{x-1}$ for the exact value of x.
4. Solve $\dfrac{25^x}{5^{3x+4}} = \dfrac{5^{2x+1}}{25^{x+3}}$ for the exact value of x.
5. Simplify $\dfrac{a^x - 3 - 4a^{-x}}{1 + a^{-a}}$. (*Hint:* multiply numerator and denominator by a^x.)
6. Use a calculator to compute $\left(1 + \dfrac{1}{n}\right)^n$ for $n = 10$, 10^2, 10^3, and 10^4.
7. Sketch the graphs of $y = x$, $y = (1.1)^x$, and $y = (1.5)^x$ on the same axes as in Section 5.1.
8. Write a logarithmic statement equivalent to $8^{-2/3} = 1/4$.
9. Write $\log_2(2\sqrt[3]{4}) = 5/3$ in equivalent exponential form.
10. Find the exact value of $5^{\log_5 11}$ and of $\log_3\left(3^{\sqrt{2}}\right)$.
11. Solve for
$x : \dfrac{1}{2}\log_a x = 2\log_a 10 - 10\log_a 2 + 3\log_a 5$.
12. Solve for x : $\log_x 20 = 4 + \log_x 5$.
13. Write $\log\left[x^x \sqrt{y}/z^3\right]^{1/4}$ as an algebraic sum in $\log x$, $\log y$, and $\log z$.
14. Express $2\log_a x - 3\log_a y + \log_a(x+y)$ as a single logarithm.
15. Show that $\log_3 x = 2\log_9 x$ for any $x > 0$.
16. Write $\log_6 28$ as a ratio of common logarithms and perform the division with a calculator.
17. Express $3\log_{10} x + 4\log_5 x$ in terms of $\ln x$.
18. On the same axes, sketch the graph of $f(x) = (2.7)^x$ and the graph of $f^{-1}(x) = \log_{2.7} x$.
19. Sketch the graphs of $y = x$, $y = \log_{1.1} x$, and $y = \log_{1.5} x$ on the same axes.
20. Write the equation $\ln x = -2$ in exponential form.
21. Write the equation $x = e^3 \cdot e^{-5}$ in logarithmic form.
22. Find the exact value of x if $\log_2 x = -3$.
23. Find the exact value of x if $4^x \cdot 3^{2x} = 5 \cdot 2^{3x}$.
24. Find an approximate value of x if $\log x = 2.7404$.
25. Find an approximate value of x if $\log x = -1.2596$.

26. Find an approximate value of x if $\ln x = 2.1972$.
27. Solve $2^{3x} = 10^{x-1}$ for an approximate value of x.
28. Using $\log 5 = .69897$ and $\log 7 = .84510$, compute $\log(5^4/\sqrt{7})$.
29. Using the values given in Exercise 28, compute $\log 1.4$.
30. Use the change-of-base rule to show that $\log_{a^n}(x) = (1/n)\log_a x$.
31. Solve $(1.02)^{5t} = 10{,}000$ for t.
32. Solve $P(1.05)^{50} = 10{,}000$ for P.
33. Solve $P(1+r)^7 = 2P$ for r.
34. Find the amount accumulated at the end of 15 years if $1{,}000 is invested at an annual rate of 8.35% compounded monthly.
35. Find the amount in Exercise 34 if interest is compounded continuously.
36. To accumulate $10{,}000 in 5 years at 9% compounded daily, what principal must be invested now?
37. If money is invested at 8% compounded quarterly, how long will it take to triple the principal?
38. If a radioactive substance has a half-life of 2,000 yr, how long will it take for one fourth of it to disintegrate?
39. Using the formula for decibels, $D = 10\log(10^{16}I)$, show that if $I_2 = 10 I_1$, then $D_2 = D_1 + 10$.
40. If an earthquake on the Richter scale $R = (2/3)[\log(10^{-4.7}E)]$ measures $R = 6.2$, what is the amount of energy released?

Graphing Calculator Exercises

In each of the following, verify your answers to the indicated review exercises of this section by using a graphing calculator for the given functions and windows $[X_{\min}, X_{\max}]$, $[Y_{\min}, Y_{\max}]$.

41. Exercise 1: $Y_1 = (3/2)^x$, $[-4.4, 5]$, $[-1, 10]$
42. Exercise 2: $Y_1 = (3/2)^x$, $[-4.4, 5]$, $[-1, 10]$
43. Exercise 7: $Y_1 = x$, $Y_2 = (1.1)^x$, $Y_3 = (1.5)^x$, $[-5.8, 13]$, $[-1, 10]$
44. Exercise 18; $Y_1 = (2.7)^x$, $Y_2 = \ln x/\ln 2.7$, $[-2, 2.7]$, $[-3, 4]$
45. Exercise 19: $Y_1 = x$, $Y_2 = \ln x/\ln 1.1$, $Y_3 = \ln x/\ln 1.5$, $[0, 9.4]$, $[-5, 15]$
46. Exercise 24: $Y_1 = \log x$, $Y_2 = 2.7404$, $[0, 9.4]$, $[0.4]$
47. Exercise 25: $Y_1 = \log x$, $Y_2 = -1.2596$, $[0, 9.4]$, $[-2, 0]$
48. Exercise 26: $Y_1 = \ln x$, $Y_2 = 2.1972$, $[0, 18.8]$, $[-2, 3]$

6 Additional Topics

6.1	The Binomial Theorem
6.2	Linear Inequations in the Plane with Applications
6.3	Partial Fractions
6.4	Linear Systems: Gaussian Elimination
6.5	Matrix Algebra
6.6	Remainder Theorem, Synthetic Division, and Factor Theorem
6.7	Rational Roots, Bounds, and Descartes' Rule of Signs

Hey Students!
You will benefit more from the classroom lectures and have a better understanding of the material if you take the time to view the Gilbert Review Videos @
www.CollegeAlgebraBySchiller.com
before attending the class lecture for this chapter.

6.1 The Binomial Theorem

The binomial theorem is one of the jewels of algebra.

Powers of Binomials

- Powers of Binomials
- Pascal's Triangle
- Binomial Coefficients
- The Binomial Theorem

Powers of Binomials. In Section 1.1, we expanded the square and the cube of a binomial, that is, $(a+b)^2$ and $(a+b)^3$. Similarly, any positive integral power of a binomial $a+b$ can be obtained by successively multiplying two factors at a time. For example,

$$\begin{aligned}(a+b)^5 &= (a+b)(a+b)(a+b)(a+b)(a+b)\\&= (a^2 + 2ab + b^2)(a+b)(a+b)(a+b)\\&= (a^3 + 3a^2b + 3ab^2 + b^3)(a+b)(a+b)\\&= (a^4 + 4a^3b + 6a^2b^2 + 4ab^3 + b^4)(a+b)\\&= a^5 + 5a^4b + 10a^3b^2 + 10a^2b^3 + 5ab^4 + b^5.\end{aligned}$$

In the process of obtaining $(a+b)^5$ we first determined $(a+b)^2$, $(a+b)^3$, and $(a+b)^4$. However, there is a formula for obtaining the expansion for any positive integral power without the need to first compute all lower powers. We can acquire some insight into this formula by examining the pattern followed by the exponents and coefficients of the terms of $(a+b)^n$ for special cases of n.

The terms of $(a+b)^5$, without regard to coefficients, are

$$a^5, \ a^4b, \ a^3b^2, \ a^2b^3, \ ab^4, \ b^5.$$

We note that the exponents of a are decreasing by 1 from 5 to 0, while those of b are increasing from 0 to 5. In each term, the sum of the exponents is 5. That is, the order of each term is 5. Also, there are six terms in all.

In general, for $(a+b)^n$, where n is any positive integer, the terms without regard to coefficients are

$$a^n, \ a^{n-1}b, \ a^{n-2}b^2, \ \ldots, \ a^{n-r}b^r, \ \ldots, \ ab^{n-1}, \ b^n.$$

That is,

$$\begin{aligned}(a+b)^n = &\underline{\ \ \ } a^n + \underline{\ \ \ } a^{n-1}b + \underline{\ \ \ } a^{n-2}b^2 + \ldots \\&+ \underline{\ \ \ } a^{n-r}b^r + \ldots + \underline{\ \ \ } ab^{n-1} + \underline{\ \ \ } b^n,\end{aligned} \qquad (1)$$

where the blanks stand for the coefficients yet to be determined. Here the exponents of a decrease by 1 from n to 0, while those of b increase from

6.1. The Binomial Theorem

0 to n. The **order** (sum of the exponents) of each term is n, and there are $n + 1$ terms in all.

Example 1   What are the terms, without regard to coefficients, in the binomial expansion of $(a + b)^{20}$?

Solution: The terms are $a^{20}, a^{19}b, a^{18}b^2, a^{17}b^3, \ldots, a^3b^{17}, a^2b^{18}, ab^{19}, a^{20}$. There are 21 terms in all. ∎

We now investigate the coefficients of $(a + b)^n$ for various choices of n.

Pascal's Triangle   . The expansions of $(a + b)^n$, for n from 0 to 5, are as follows:

$$
\begin{aligned}
(a+b)^0 &= 1 \\
(a+b)^1 &= a + b \\
(a+b)^2 &= a^2 + 2ab + b^2 \\
(a+b)^3 &= a^3 + 3a^2b + 3ab^2 + b^3 \\
(a+b)^4 &= a^4 + 4a^3b + 6a^2b^2 + 4ab^3 + b^4 \\
(a+b)^5 &= a^5 + 5a^4b + 10a^3b^2 + 10a^2b^3 + 5ab^4 + b^5.
\end{aligned}
$$

The coefficients of these expansions form the following triangular array, which makes up the first six rows of what is called **Pascal's triangle**[1].

```
            1
          1   1
        1   2   1
      1   3   3   1
    1   4   6   4   1
  1   5  10  10   5   1
```

[1] Blaise Pascal (1623-62) was a precocious French mathematician who later became a religious thinker. He discovered many theorems in geometry and is one of the founders of probability theory. However, the triangle that bears his name was already in use by Chinese mathematicians early in the fourteenth century.

Pascal's triangle has the following three basic properties:

(**a**) Each row begins and ends with the number 1.
(**b**) Every interior number in a given row is the sum of the numbers immediately to the right and left in the row above.
(**c**) In each row, the pattern from right to left is the same as the pattern from left to right.

Example 2   Construct the next two rows of Pascal's triangle, corresponding to $(a + b)^6$ and $(a + b)^7$. Also, write out the binomial expansion in each case.

Solution: By Properties (a) and (b), the next two rows are

$$
\begin{array}{ccccccccc}
 & 1 & 6 & 15 & 20 & 15 & 6 & 1 & \\
1 & 7 & 21 & 35 & 35 & 21 & 7 & 1. &
\end{array}
$$

Therefore,

$$(a+b)^6 = a^6 + 6a^5b + 15a^4b^2 + 20a^3b^3 + 15a^2b^4 + 6ab^5 + b^6,$$

and

$$(a+b)^7 = a^7 + 7a^6b + 21a^5b^2 + 35a^4b^3 + 35a^3b^4 + 21a^2b^5 + 7ab^6 + b^7. \blacksquare$$

We could verify the expansions in Example 2 by direct computation. However, it can be shown that Pascal's triangle gives the correct coefficients of $(a+b)^n$ for any nonnegative integer n. Therefore, Pascal's triangle is very useful for expanding $(a+b)^n$ when n is moderately small. But, since the construction of each new row requires the entries of the previous row, the triangle becomes impractical for large values of n.

Binomial Coefficients. Let $_nC_r$ denote the coefficient of $a^{n-r}b^r$ in the expansion of $(a+b)^n$, where n is a positive integer and r is a nonnegative integer less than or equal to n. In the case $n = 5$, we have

$$(a+b)^5 = \underset{_5C_0}{1 \cdot a^5b^0} + \underset{_5C_1}{5 \cdot a^4b^1} + \underset{_5C_2}{10 \cdot a^3b^2} + \underset{_5C_3}{10 \cdot a^2b^3}$$

$$+ \underset{_5C_4}{5 \cdot a^1b^4} + \underset{_5C_5}{1 \cdot a^0b^5}.$$

The $_nC_r$ are called the **binomial coefficients**, and we want a general algebraic formula that expresses each, $_nC_r$ in terms of n and r. The aforementioned three basic properties of Pascal's triangle can now be expressed in terms of the binomial coefficients as follows.

(a) $_nC_0 = 1 = {_nC_n}$	Each row begins and ends with 1.
(b) $_nC_r = {_{n-1}C_r} + {_{n-1}C_{r-1}}$	Each interior number in row n is equal to the sum of the numbers immediately to the right and left in row $n-1$ above.
(c) $_nC_{n-r} = {_nC_r}$	The pattern from right to left in row n is the same as the pattern from left to right.

6.1. The Binomial Theorem

By means of properties (a) and (b) above, it can be proved that

$$_nC_r = \frac{n!}{(n-r)!r!} \quad (0 \leq r \leq n), \tag{2}$$

where **n!** (read "**n factorial**") is defined as

$$n! = n(n-1)(n-2) \cdot \ldots \cdot 1 \quad \text{For } n \geq 1 \text{ and } 0! = 1.$$

For example, if $n = 5$, then

$$_5C_0 = \frac{5!}{5!0!} = \frac{5 \cdot 4 \cdot 3 \cdot 2 \cdot 1}{5 \cdot 4 \cdot 3 \cdot 2 \cdot 1 \cdot 1} = 1 \quad \text{Note that } 0! = 1.$$

$$_5C_1 = \frac{5!}{4!1!} = \frac{5 \cdot 4 \cdot 3 \cdot 2 \cdot 1}{4 \cdot 3 \cdot 2 \cdot 1 \cdot 1} = 5$$

$$_5C_2 = \frac{5!}{3!2!} = \frac{5 \cdot 4 \cdot 3 \cdot 2 \cdot 1}{3 \cdot 2 \cdot 1 \cdot 2 \cdot 1} = 10$$

$$_5C_3 = \frac{5!}{2!3!} = \frac{5 \cdot 4 \cdot 3 \cdot 2 \cdot 1}{2 \cdot 1 \cdot 3 \cdot 2 \cdot 1} = 10$$

$$_5C_4 = \frac{5!}{1!4!} = \frac{5 \cdot 4 \cdot 3 \cdot 2 \cdot 1}{1 \cdot 4 \cdot 3 \cdot 2 \cdot 1} = 5$$

$$_5C_5 = \frac{5!}{0!5!} = \frac{5 \cdot 4 \cdot 3 \cdot 2 \cdot 1}{1 \cdot 5 \cdot 4 \cdot 3 \cdot 2 \cdot 1} = 1.$$

These values agree with the coefficients obtained for $(a + b)^5$ at the beginning of this section.

Example 3 Find the coefficient of $a^7 b^3$ in $(a + b)^{10}$.

Solution: The coefficient is $_{10}C_3$, and by equation (2),

$$_{10}C_3 = \frac{10!}{7!3!} = \frac{10 \cdot 9 \cdot 8 \cdot 7 \cdot 6 \cdot 5 \cdot 4 \cdot 3 \cdot 2 \cdot 1}{7 \cdot 6 \cdot 5 \cdot 4 \cdot 3 \cdot 2 \cdot 1 \cdot 3 \cdot 2 \cdot 1} = 120. \quad \blacksquare$$

In Example 3, note that $10! = 10 \cdot 9 \cdot 8 \cdot 7!$ so that we can cancel 7! from the numeration and denominator when computing $_{10}C_3$. That is,

$$_{10}C_3 = \frac{10 \cdot 9 \cdot 8 \cdot 7!}{7! \cdot 3!} = \frac{10 \cdot 9 \cdot 8}{3!}.$$

In general, $n! = n(n-1)(n-2)\ldots(n-r+1)\cdot(n-r)!$, and we can cancel $(n-r)!$ from the numerator and denominator on the right side of equation (2) when computing ${}_nC_r$. That is,

$$_nC_r = \frac{n(n-1)(n-2)\ldots(n-r+1)}{r!}. \tag{3}$$

Equation (3) is usually more convenient than (2) when computing the binomial coefficients. (*Why?*)

The Binomial Theorem. By filling in the blanks of equation (1) with the binomial coefficients of equation (3), we arrive at the **binomial theorem**.

Equivalently,
$(a+b)^n = a^n + {}_nC_1 a^{n-1}b + {}_nC_2 a^{n-2}b^2 + \ldots + {}_nC_r a^{n-r}b^r + \ldots + b^n.$

For any positive integer n,

$$(a+b)^n = a^n + na^{n-1}b + \frac{n(n-1)}{2}a^{n-2}b^2 + \ldots$$
$$+ \frac{n(n-1)(n-2)\ldots(n-r+1)}{r!}a^{n-r}b^r + \ldots$$
$$+ nab^{n-1} + b^n.$$

Example 4 Use the binomial theorem to expand $(a+b)^{10}$.

Solution: We use equation (3) and property (c) of the binomial coefficients.

$${}_{10}C_1 = \frac{10}{1} = 10 = {}_{10}C_9 \qquad {}_{10}C_4 = \frac{10\cdot 9\cdot 8\cdot 7}{4\cdot 3\cdot 2\cdot 1} = 210 = {}_{10}C_6$$

$${}_{10}C_2 = \frac{10\cdot 9}{2\cdot 1} = 45 = {}_{10}C_8 \qquad {}_{10}C_5 = \frac{10\cdot 9\cdot 8\cdot 7\cdot 6}{5\cdot 4\cdot 3\cdot 2\cdot 1} = 252.$$

$${}_{10}C_3 = \frac{10\cdot 9\cdot 8}{3\cdot 2\cdot 1} = 120 = {}_{10}C_7$$

Therefore,

$$(a+b)^{10} = a^{10} + 10a^9b + 45a^8b^2 + 120a^7b^3 + 210a^6b^4 + 252a^5b^5$$
$$+ 210a^4b^6 + 120a^3b^7 + 45a^2b^8 + 10ab^9 + b^{10}. \blacksquare$$

6.1. The Binomial Theorem

Example 5 Derive an expansion for $(a-b)^5$.

Solution: Replace b by $-b$ in the binomial theorem with $n=5$ to obtain

$$(a-b)^5 = a^5 + 5a^4(-b) + 10a^3(-b)^2 + 10a^2(-b)^3 + 5a(-b)^4 + (-b)^5$$
$$= a^5 - 5a^4b + 10a^3b^2 - 10a^2b^3 + 5ab^4 - b^5. \blacksquare$$

Example 6 Expand $(2x+y)^6$.

Solution: We substitute $a=2x$, $b=y$ and $n=6$ in the binomial theorem to obtain

$$(2x+y)^6 = (2x)^6 + 6(2x)^5 y + 15(2x)^4 y^2 + 20(2x)^3 y^3 + 15(2x)^2 y^4$$
$$+ 6(2x)y^5 + y^6$$
$$= 64x^6 + 192x^5 y + 240x^4 y^2 + 160x^3 y^3 + 60x^2 y^4$$
$$+ 12xy^5 + y^6. \blacksquare$$

Comment Isaac Newton (1642-1727) extended the binomial theorem to the case $(a+x)^{p/q}$, where p/q is a rational number. Although $n!$ is defined only when n is a nonnegative integer, formula (3) can be used for the coefficient of $a^{n-1}x^r$ when n is a rational number p/q. For example, the expansion of $(1+x)^{1/2}$ becomes

$$1 + \frac{1}{2}x - \frac{1 \cdot 1}{2 \cdot 4}x^2 + \frac{1 \cdot 1 \cdot 3}{2 \cdot 4 \cdot 6}x^3 - \frac{1 \cdot 1 \cdot 3 \cdot 5}{2 \cdot 4 \cdot 6 \cdot 8}x^4 + \cdots.$$

When n is a positive integer, the binomial expansion has $n+1$ terms, corresponding to $r=0,1,2,\ldots,n$, respectively. However, when n is a rational number p/q other than a positive integer, the expansion contains a term corresponding to every nonnegative integer r. For example, the above expansion contains an *infinite* number of terms. Sums with an infinite number of terms were not completely understood by mathematicians in Newton's day. It was not until the nineteenth century that a precise treatment for such sums was developed. Today the subject is called *infinite series* and is studied in calculus. Note that if we substitute $x=1$ in $(1+x)^{1/2}$, we get $2^{1/2}$, or $\sqrt{2}$. Use a calculator to add the first 7 terms of the above expansion with $x=1$ and see how close the sum comes to the value of $\sqrt{2}$ obtained with the square root key on the calculator.

Exercises 6.1

Whenever n, r, or $_nC_r$ appear in the following exercises, it is assumed that n is a positive integer and r is a nonnegative integer less than or equal to n.

Fill in the blanks to make each statement true.

1. In the expansion of $(a+b)^n$, there are ----- terms.
2. In the expansion of $(a+b)^n$, the sum of the exponents of each term is -----.
3. In the expansion of $(a+b)^n$, the coefficient of $a^{n-r}b^r$ is given by the formula -----.
4. The coefficient of $a^8 b^4$ in the expansion of $(a+b)^{12}$ is same as the coefficient of -----.
5. The largest coefficient in the expansion of $(a+b)^{12}$ is -----.

Write true or false to make each statement true.

6. One of the terms in the expansion of $(a-b)^{10}$ is $120a^7 b^3$.
7. According to Pascal's triangle, $_{12}C_7 = {}_{11}C_7 + {}_{11}C_6$.
8. $_nC_r = {}_nC_{r+1}$
9. $_nC_r = {}_{n+r}C_{n-r}$
10. $0! = 0$

Powers of Binomials

Expand each of the following by successively multiplying two factors at a time.

11. $(x-1)^3$
12. $(2x+3)^3$
13. $(x+2)^4$
14. $(3x-1)^4$
15. $(x-1)^5$
16. $(2x-3)^5$

Give all the terms, without regard to coefficients, in the expansions of each of the following.

17. $(x+1)^5$
18. $(1+x^2)^5$
19. $(ax+y)^4$
20. $(a^2 b + xy^2)^4$

Pascal's Triangle

Use Pascal's triangle to expand each of the following in Exercises 21-26

21. $(x+2)^4$
22. $(x-2)^5$
23. $(2x+3)^4$
24. $(2x-1)^5$
25. $(1+x)^6$
26. $(1-x)^7$
27. Use Pascal's triangle to find the coefficient of x^4 in the expansion of $(2x+1)^8$.
28. Use Pascal's triangle to find the coefficient of x^5 in the expansion of $(1-3x)^8$.
29. First add the numbers in the nth row of Pascal's triangle for $n = 1, 2, 3, 4, 5$, then use these results to predict the sum of the coefficients for the following.
    (a) $(a+b)^6$
    (b) $(a+b)^n$, where n is any positive integer
30. First add *all* of the numbers in the first n rows of Pascal's triangle for $n = 1, 2, 3, 4, 5$, then use these results to predict the corresponding sum for the following.
    (a) $n = 6$
    (b) any positive integer n

Binomial Coefficients

31. Evaluate each factorial.
    (a) $4!$ (b) $5!$ (c) $7!$
32. Evaluate each binomial coefficient.
    (a) $_7C_0$ (b) $_7C_3$ (c) $_7C_5$
33. Simplify each expression.
    (a) $\dfrac{(n+2)!}{n!}$ (b) $\dfrac{(n+1)!}{(n-1)!}$ (c) $\dfrac{[2(n+1)]!}{(2n)!}$
34. Simplify each binomial coefficient.
    (a) $_{n+2}C_n$ (b) $_{n+1}C_{n-1}$ (c) $_{2(n+1)}C_{2n}$

Verify each statement in Exercises 35 and 36.

35. (a) $_5C_3 = {_4C_3} + {_4C_2}$ (b) $_6C_1 = {_5C_1} + {_5C_0}$
36. (a) $_{10}C_5 = {_9C_5} + {_9C_4}$ (b) $_{11}C_7 = {_{10}C_7} + {_{10}C_6}$

Binomial Theorem

Use the binomial theorem to expand each of the following.

37. $(x+y)^4$ 38. $(x-y)^4$
39. $(2x-3y)^4$ 40. $(3x+5)^4$
41. $(x+2)^6$ 42. $\left(y - \frac{1}{2}\right)^6$
43. $\left(2x + \frac{3}{4}\right)^5$ 44. $\left(3x - \frac{1}{3}\right)^5$

Find the coefficient of x^6 in each of the following.

45. $(x+1)^8$ 46. $(2x-1)^8$
47. $(x^2+1)^8$ 48. $\left(x + \frac{1}{x}\right)^8$
49. $\left(x^2 + \frac{2}{x}\right)^6$ 50. $(x^3 - 2x)^6$

51. Evaluate $(1.1)^5$ by means of the binomial theorem. (*Hint*: $1.1 = 1 + .1$.)
52. Use the binomial theorem to evaluate $(.9)^5$. (*Hint*: see Exercises 51.)

As in Exercises 51 and 52, use the binomial theorem to evaluate each of the following.

53. $(1.01)^4$ 54. $(.99)^4$
55. $(9.8)^3$ 56. $(10.2)^3$
57. Show that $_nC_0 + {_nC_1} + {_nC_2} + \ldots + {_nC_n} = 2^n$. (*Hint*: let $a = b = 1$ in the binomial theorem.)
58. Show that the sum of the coefficients in the expansion of $(a-b)^n$ is zero. (*Hint*: see Exercise 57.)
59. Expand $(x + y + z)^3$. (*Hint*: let $x + y = a$ and expand $(a + z)^3$. Then replace a by $x + y$ and complete the expansion.)
60. Use the method from Exercise 59 to expand $(1 + x + x^2)^4$.

Miscellaneous

*The formula for **compound interest** (see exercise 106 in Section R.5) is $A = P\left(1 + \frac{R}{N}\right)^{Nt}$, where A is the amount accumulated in t years when P dollars are invested, interest is compounded N times per year, and R is the annual percentage interest rate divided by 100.*

In each of the following cases, use the binomial theorem to find A.

61. $P = \$1$, $N = 4$, $t = 1$, and the annual interest rate is 4%. (*Hint*: see Exercise 53.)
62. $P = \$1$, $N = 2$, $t = 2$, and the annual interest rate is 12%.
63. $P = \$100$, $N = 1$, $t = 3$, and the annual interest rate is 8%.
64. $P = \$100$, $N = 6$, $t = .5$, and the annual interest rate is 6%.

*It can be shown that $_nC_r$ is the number of ways of selecting r objects from a collection of n distinct objects, called the number of **combinations** of n distinct objects taken r at a time. For example, from the set $\{a, b, c, d\}$, 2 letters can be selected in 6 ways, namely ab, ac, ad, bc, bd, cd, and $_4C_2 = 4!/(2!2!) = 6$. Use this interpretation of $_nC_r$ to do each of the following exercises.*

65. From a group of 30 students, in how many ways can a committee of 3 be selected?
66. In how many ways can a jury of 12 be selected from a panel of 25 persons?
67. If a student must answer 8 out of 10 questions on a test, how many choices are possible?
68. In how many ways can a student choose 4 required courses out of a set of 9 possible courses?
69. If 2 out of every 100 items on a production line are inspected, in how many ways can the 2 items be selected?
70. A fair coin is tossed 5 times. In how many ways can exactly 2 heads result? (*Hint*: number the tosses 1, 2, 3, 4, 5.)

6.2 Linear Inequalities in the Plane with Applications

A thin line separates the "greater thans" from the "less thans."

Linear Inequalities and Half-Planes
Systems of Linear Inequalities
Applications
Linear Programming

Linear Inequalities and Half-Planes. We have seen that the solution of a linear inequality in one variable x is an infinite interval along the real line. We now show that the solution of a linear inequality in two variables x and y is a *half-plane*. To see why this is so, let L be a nonhorizontal line. Then L partitions the plane into a right **half-plane** $\mathfrak{R}$ and a left **half-plane** $\mathcal{L}$. $\mathfrak{R}$ consists of all points $(x_0 + h, y_0)$, where (x_0, y_0) is on L and $h > 0$; all points $(x_0 - h, y_0)$ make up $\mathcal{L}$ (Figure 1). The line L itself is the boundary of both $\mathfrak{R}$ and $\mathcal{L}$.

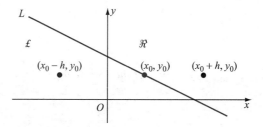

Figure 1 Right Half-Plane $\mathfrak{R}$ and Left Half-plane $\mathcal{L}$

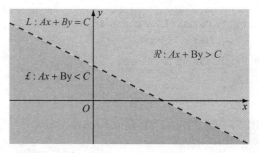

Figure 2 Half-Plane Rule ($A > 0$)

The equation of line L can be written in the form

$$Ax + By = C,$$

where $A > 0$ (*why may we assume that $A > 0$?*). Suppose that (x_0, y_0) lies on L and that $h > 0$. Then

$$A(x_0 + h) + By_0 = Ax_0 + By_0 + Ah = C + Ah > C$$

and
$$A(x_0 - h) + By_0 = Ax_0 + By_0 - Ah = C - Ah < C.$$

6.2. Linear Inequations in the Plane with Applications 357

Therefore, as indicated in Figure 2, we obtain the following **half-plane rule**.

> Given line $L : Ax + By = C$, where $A > 0$, then $Ax + By > C$ for all points (x, y) in the right half-plane $\Re$, and
> $Ax + By < C$ for all points (x, y) in the left half-plane $\pounds$.

The solution to $Ax + By \geq C$ consists of all points in $\Re$ and on L, which we denote by $\Re \cup L$. Similarly, the solution to $Ax + By \leq C$ is $\pounds \cup L$.

Example 1   Graph the solution to each of the following inequalities.

(a) $2x + 3y > 6$
(b) $2x + 3y < 6$
(c) $2x + 3y \geq 6$
(d) $2x + 3y \leq 6$

Solution: In Figure 3, a solid line for L indicates that L is included in the solution; a broken line indicates that L is not in the solution.   ∎

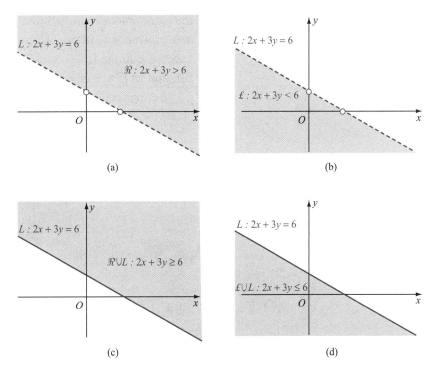

Figure 3

Example 2   Graph the solution to each of the following inequalities.

(a) $-3x + 4y > 12$
(b) $-3x + 4y \leq 12$

Solution: The half-plane rule requires that the coefficient of x be positive. We therefore multiply both sides of each inequality by -1, which changes all the signs and reverses the direction of each inequality. That is, we replace (a) and (b) by the equivalent inequations (a') $3x - 4y < -12$ and (b') $3x - 4y \geq -12$. The graphs of (a) and (b) are equal to the graphs of (a') and (b'), respectively. See Figure 4. ∎

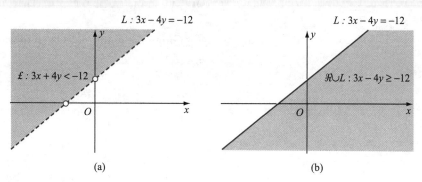

Figure 4

A horizontal line L given by an equation $y = c$ does not partition the plane into a right half and a left half, but into an **upper half-plane** and a **lower half-plane.** All points in the upper half-plane satisfy $y > c$, and those in the lower half-plane satisfy $y < c$ (Figure 5). Finally, the solution to $y \geq c$ consists of all points on or above the line $y = c$, and $y \leq c$ consists of all points on or below the line.

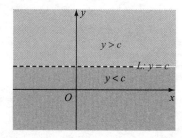

Figure 5

Example 3 Graph the solution to each of the following inequalities.

(a) $y \geq -2$
(b) $y < 4$

6.2. Linear Inequations in the Plane with Applications

Solution: The half-plane solutions are indicated in Figure 6. ∎

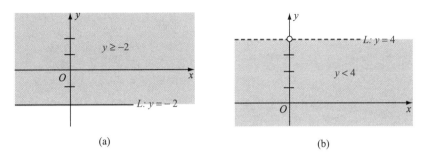

Figure 6

Systems of Linear Inequalities. We have seen that inequalities involving linear expressions in x and y have for their solutions half-planes with or without their boundaries. For a system of two or more such inequalities, the solution is the region of intersection of the individual solutions.

Example 4 Graph the solution of the system

$$2x - y > 4$$
$$x + y < 6.$$

Solution: In Figure 7, the area shaded in light gray denotes the solution of $2x - y > 4$, and the dark gray shaded area indicates the solution of $x + y < 6$. The region of intersection, the double-shaded portion of the figure, is the solution of the system. ∎

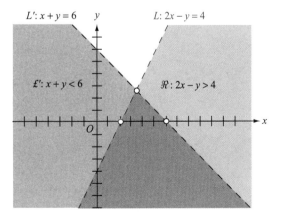

Figure 7

Example 5 Graph the solution to the system

$$2x + 3y \leq 12$$
$$3x + y \leq 9$$
$$x \geq 0$$
$$y \geq 0.$$

Solution: The last two inequalities tell us that the solution is confined to the first quadrant and its boundary. The shaded region in Figure 8, including its boundary, indicates the solution of the first two inequalities in the first quadrant. ■

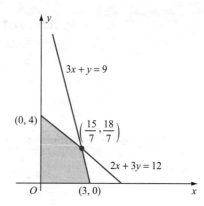

Figure 8

Applications. The following two examples illustrate some applications of systems of linear inequalities in the plane.

Example 6 A clock company manufactures two types of clocks, A and B. Each type A clock costs the company $25 in parts and requires 5 hours of labor. Each type B clock costs $40 in parts but requires only 4 hours of labor. If $2000 is available for parts, what combinations of A and B clocks can be produced with 320 hours of labor?

Solution: Let x = the number of type A clocks, and y = the number of type B clocks.
Then

$$25x + 40y \leq 2000 \quad \textit{Parts}$$
$$5x + 4y \leq 320 \quad \textit{Labor}$$
$$x \geq 0$$
$$y \geq 0.$$

Any pair of integers (x, y) in the shaded region of Figure 9 represents a combination of the number of A and B clocks that can be manufactured within the given constraints. ∎

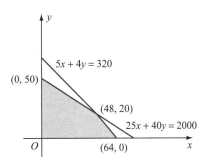

Figure 9

Example 7   A construction company buys lumber in 8-foot and 14-foot board lengths. If at least 75 5-foot lengths and 125 3-foot lengths are needed to complete a project, what combinations of 8-foot and 14-foot boards will be sufficient? Assume that from each 8-foot board, one 5-foot and one 3-foot length will be cut. Also, from each 14-foot board, one 5-foot and three 3-foot lengths will be cut.

Solution: Let x = the number of 8-foot boards, and y = the number of 14-foot boards.
Then

$$x + y \geq 75 \quad \text{\textit{5-foot lengths}}$$
$$x + 3y \geq 125 \quad \text{\textit{3-foot lengths}}$$
$$x \geq 0$$
$$y \geq 0.$$

Each pair of integers (x, y) in the shaded region of Figure 10 represents a combination of 8-foot and 14-foot boards that will be sufficient for the job. ∎

Linear Programming.   Linear programming is a method for making the optimum choice in a situation where several choices are possible. We can apply linear programming to the previous two examples.

Suppose, in Example 6, that the clock company makes $10 profit on each type A clock and $12 profit on each type B clock. Then the company would want to know which combinations of type A and B clocks would provide the *most profit*. One way to proceed would be to compute the profit for each production point in the shaded region. A more efficient

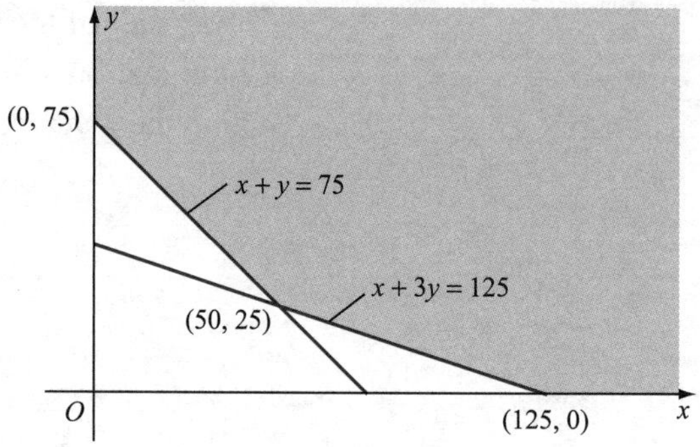

362 Chapter 6. Additional Topics

Figure 10

way is to begin by setting up an equation for the profit P for x type A clocks and y type B clocks:

$$P = 10x + 12y. \tag{1}$$

Now for each value of P, the graph of equation (1) is a straight line L_P with slope $m = -5/6$; different values of P correspond to parallel lines. Furthermore, by Example 6 of Section 2.6, the distance from the origin to L_P is $P/\sqrt{244}$. Hence, *the maximum profit will occur on the line with slope $-5/6$ that is farthest from the origin but still touches the shaded region of Figure 9*. As indicated in Figure 11, the line of maximum profit is L_{720}, which passes through the vertex (48, 20). Hence, the maximum profit of \$720 is achieved by producing 48 type A clocks and 20 type B clocks.

If the respective profit for type A or type B clocks changes, then the slope of L_P may change. However, because of the shape of the shaded region, *the maximum P will always occur among the vertices* (0, 50), (48, 20), *or* (64, 0). Hence, only these three points need be tested.

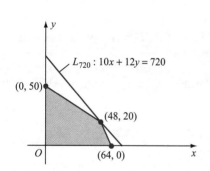

Figure 11 The Real Number Line

Example 6 (continued) Let P_A and P_B stand for the profit that the company makes on each A and B clock, respectively. Find the maximum profit in each of the following.

(a) $P_A = 10$, $P_B = 18$
(b) $P_A = 10$, $P_B = 16$
(c) $P_A = 14$, $P_B = 12$
(d) $P_A = 10$, $P_B = 8$
(e) $P_A = 14$, $P_B = 10$

Solution: The profit P at each vertex (x, y) for each case is listed in the following table. The maximum in each case is indicated by a star.

The line of maximum profit in each case is shown in Figure 12 below the table.

Vertex	(a) $P = 10x + 18y$	(b) $P = 10x + 16y$	(c) $P = 14x + 12y$	(d) $P = 10x + 8y$	(e) $P = 14x + 10y$
(0, 50)	900*	800*	600	400	500
(48, 20)	840	800*	912*	640*	872
(64, 0)	640	640	896	640*	896*

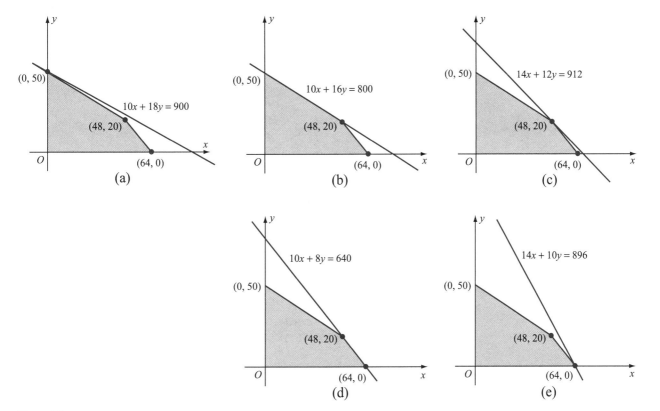

Figure 12

We note that in case (b) the maximum profit of $800 occurs at any production point on the line $10x + 16y = 800$ between (0, 50) and (48, 20), namely at the points (0, 50), (8, 45), (16, 40), (24, 35), (32, 30), (40, 25), and (48, 20). Similarly, in case (d) the maximum profit of $640 occurs at any production point on the line $10x + 8y = 640$ between (48, 20) and (64, 0), namely at (48, 20), (52, 15), (56, 10), (60, 5) and (64, 0). In all the other cases, the maximum profit occurs at a unique point. ∎

In Example 7, the objective is not to maximize profits but but to *minimize costs*. For instance, if the company pays 50 cents per foot for each board, then the cost C (in dollars) for x 8-foot and y 14-foot boards is

$$C = 4x + 7y. \qquad (2)$$

For each value of C, equation (2) is a straight line L_C with slope $m = -4/7$. *The minimum cost will correspond to the line of slope $-4/7$ that is closest to the origin but still touches the shaded region of Figure 10 (why?)*. These properties are satisfied by the line

$$L_{375} : 4x + 7y = 375,$$

which passes through the vertex (50, 25), as shown in Figure 13. Hence, the minimum cost is $375, which is the price of 50 8-foot and 25 14-foot boards.

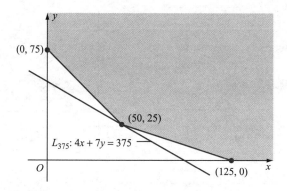

Figure 13

If either of the respective prices of the 8-foot and 14-foot boards changes, then the slope of the cost line L_C may change. However, *the minimum cost will always occur among the three vertices (0, 75), (50, 25), and (125, 0) (why?)*.

Example 7 (continued) In order to reduce inventory, the lumber supplier occasionally lowers the prices on either the 8-foot or the 14-foot boards. In each of the following cases, determine the minimum cost and graph the line L_C of minimum cost.

(a) 8 - foot board : $4
    14 - foot board : $4

(b) 8 - foot board : $2
    14 - foot board : $6

(c) 8 - foot board : $2
    14 - foot board : $7

Solution: The cost at each vertex for each case is listed in the following table. The minimum in each case is indicated by a star. The line of minimum cost in each case is shown under Figure 14 below the table.

6.2. Linear Inequations in the Plane with Applications 365

Vertex	(a) $C = 4x + 4y$	(b) $C = 2x + 6y$	(c) $C = 2x + 7y$
(0, 75)	300*	450	525
(50, 25)	300*	250*	275
(125, 0)	500	250*	250*

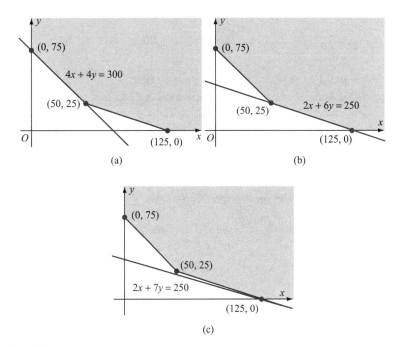

Figure 14

In case (a) the minimum cost of \$300 occurs at any purchase point on the line $4x + 4y = 300$ from (0, 75) to (50, 25), namely at the points (0, 75), (5, 70), (10, 65), (15, 60), ..., (45, 30), (50, 25). Similarly, in case (b) the minimum cost of \$250 occurs at any purchase point on the line $2x + 6y = 250$ from (50, 25) to (125, 0), namely at (50, 25), (53, 24), (56, 23), (59, 22), ..., (122, 1), (125, 0). In case (c) the minimum cost of \$250 occurs only at (125, 0). ■

Comment The basic idea in linear programming is that only the corners (vertices) of the region need to be tested. The number of corners depends on the number of inequalities, and the dimension of the region is equal to the number of variables involved. We have considered problems with two variables, hence, two dimensions. Large corporations, such as airlines and telephone companies, routinely consider problems with thousands of variables. In these cases the corner points are tested by techniques carried out on high-speed computers. ■

Exercises 6.2

Linear Inequalities and Half-Planes

In the plane, graph the solution to each of the following inequalities.

1. $x + y > 1$
2. $x - y < 1$
3. $-x + y \geq 1$
4. $-x - y \leq 1$
5. $x > 2$
6. $y \leq 5$
7. $2x + 3y \geq 6$
8. $4x - 5y \leq 20$
9. $3x + 4y < 12$
10. $2x + 5y > 10$
11. $-x + 4y > 3$
12. $-3x - 5y < 4$

Systems of Linear Inequalities

In the plane, graph the solution to each of the following system of inequalities.

13. $x + y > 2$
    $x - 2y < 2$

14. $x - y \leq 2$
    $2x - y \geq 4$

15. $-x + y < 3$
    $2x - y < 5$

16. $x + y \geq 1$
    $x + 2y \geq 2$

17. $2x - y > 0$
    $x - 2y < 0$

18. $2x - 3y < 0$
    $-3x + 4y > 0$

19. $x \geq 3$
    $y \leq 5$
    $y \geq 0$
    $x + y \leq 8$

20. $x < 2$
    $y < 4$
    $x > 0$
    $y > 0$

21. $2x + 3y \leq 6$
    $4x + 5y \leq 20$
    $x \geq 0$
    $y \geq 0$

22. $3x + 4y > 12$
    $x + y < 5$
    $x > 0$
    $y > 0$

23. $3x + 5y \leq 30$
    $x + y \geq 4$
    $x \geq 2$
    $y \geq 2$

24. $4x + 7y \geq 28$
    $x + y \leq 10$
    $x \leq 7$
    $y \geq 3$

In each of the following cases, find a system of inequalities whose solution is the given graph.

25.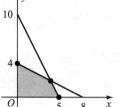

26.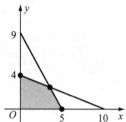

27.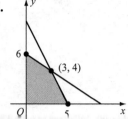

28.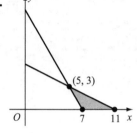

29.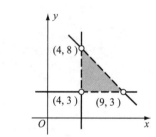

30.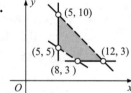

Applications

Graph the solution for each problem.

31. A company packages and sells 16-oz containers of mixed nuts. Brand A contains 12 oz of peanuts and 4 oz of cashews. Brand B contains 8 oz of peanuts and 8 oz of cashews. If 120 lb of peanuts and 96 lb of cashews are available, how many containers of each brand can be packaged?

32. A company packages and sells 32-oz cartons of mixed fruit juice. Type 1 contains 24 oz of pineapple juice and 8 oz of grapefruit juice. Type 2 contains 12 oz of pineapple juice and 20 oz grapefruit juice. If 150 gal of pineapple juice and 120 gal of grapefruit juice are available, how many cartons of each type can be packaged?

33. A gardener has a 2304-sq-in flower garden in which she wants to plant begonias and marigolds. Each begonia requires 64 sq in of space, and each marigold requires 144 sq in. If the gardener wants to put in at least 26 plants, what combinations of begonias and marigolds are possible?

34. A 60-sq-ft vegetable garden is to be planted in tomatoes and peppers. Each tomato plant requires 4 sq ft of space, and each pepper plant requires 1 sq ft. If there must be at least 24 plants, what combinations of tomatoes and peppers are possible?

35. A stadium has 10,000 seats for $5 each and 4,000 seats for $8. To make a profit on a given event, the stadium must take in at least $50,000. What seating combinations will result in a profit?

36. A theater has 50 seats at $20 each and 150 seats at $12 each. To make a profit on a given performance, the theater must take in at least $1,500. What seating combinations will result in a profit?

37. A small toy company makes two kinds of stuffed animals. A stuffed puppy costs $2 in materials and requires 4 hr of labor. A teddy bear costs $2.50 and takes 3 hr. The company has $40 available for materials and has orders for 5 animals of each type. How many stuffed animals can the company make in 64 hr and still meet its orders?

38. A retired carpenter makes toy airplanes and trains out of wood. An airplane costs $3 for materials and requires 2 hr to cut, shape, and finish. A train costs $8 and requires 4 hr of work. The carpenter has orders for 4 airplanes and 3 trains. With a budget of $72, how many of each type can be made in 40 hr and still meet the orders?

39. Marie plans to have a spring garden with at least 60 daffodils and 48 tulips but no more than 128 bulbs in all. A type A package contains 5 daffodil and 3 tulip bulbs. A type B package contains 2 daffodil and 2 tulip bulbs. What combinations of type A and B packages will be sufficient?

40. Sarah's fall garden is to have at least 15 yellow chrysanthemums and 12 white ones but no more than 36 plants in all. A type A variety pack contains 3 yellow and 2 white plants, and a type B pack contains 3 yellow and 3 white. What combinations of type A and B variety packs will be sufficient?

Linear Programming

41. In Exercise 31, let P_A equal the profit in selling a 16–oz container of brand A, and let P_B be the profit for brand B. Find the maximum profit in each of the following cases.

    (a) $P_A = \$.75$, $P_B = \$1.00$
    (b) $P_A = \$1.00$, $P_B = \$.25$
    (c) $P_A = \$1.00$, $P_B = \$.75$

42. In Exercise 32, let P_1 equal the profit in selling a 32–oz carton of type 1 juice, and let P_2 be the profit for type 2. Find the maximum profit in each of the following cases.

    (a) $P_1 = \$.50$, $P_2 = \$.25$
    (b) $P_1 = \$.50$, $P_2 = \$.10$
    (c) $P_1 = \$.50$, $P_2 = \$.50$

43. In Exercise 33, let C_B equal the cost of each begonia, and let C_M be the cost of each marigold. Find the minimum cost to plant the garden in each of the following cases.

    (a) $C_B = \$3$, $C_M = \$2$
    (b) $C_B = \$2$, $C_M = \$3$
    (c) $C_B = \$2$, $C_M = \$2$

44. In Exercise 34. let C_T equal the cost of each tomato plant, and let C_P be the cost of each pepper plant. Find the minimum cost to plant the garden in each of the following cases.

    (a) $C_T = \$1.00$, $C_P = \$.50$
    (b) $C_T = \$.25$, $C_P = \$.50$
    (c) $C_T = \$.75$, $C_P = \$.75$

45. In Exercise 35, what is the least number of tickets that must be sold in order to make a profit?

46. In Exercise 36, what is the least number of tickets that must be sold in order to make a profit?

47. In Exercise 37, let P_P equal the profit made on each stuffed puppy, and let P_T be the profit on each teddy bear. Find the maximum profit in each of the following cases.

(a) $P_P = \$12$, $P_T = \$16$
(b) $P_P = \$15$, $P_T = \$10$
(c) $P_P = \$15$, $P_T = \$12$

48. In Exercise 38, let P_A equal the profit made on each toy airplane, and let P_T be the profit on each train. Find the maximum profit in each of the following cases.

(a) $P_A = \$10$, $P_T = \$30$
(b) $P_A = \$15$, $P_T = \$28$
(c) $P_A = \$12$, $P_T = \$30$

49. In Exercise 39, let C_A be the cost of each type A package of bulbs, and let C_B equal the cost of a type B package. Find the minimum cost in each of the following cases.

(a) $C_A = \$8$, $C_B = \$3$
(b) $C_A = \$6$, $C_B = \$5$
(c) $C_A = \$7$, $C_B = \$4$

50. In Exercise 40, let C_A be the cost of a type A variety pack, and let C_B equal the cost of a type B pack. Find the minimum cost in each of the following cases.

(a) $C_A = \$12$, $C_B = \$10$
(b) $C_A = \$9$, $C_B = \$15$
(c) $C_A = \$10$, $C_B = \$12$

6.3 Partial Fractions

In partial fraction problems, we are given the answer, and the object is to find the question.

Partial Fraction Decomposition

Decomposition of Proper Rational Expressions

Decomposition of Improper Rational Expressions

Partial Fraction Decomposition. In Section 1.3 we added two or more rational expressions by means of a common denominator in order to obtain a single rational expression. Here we reverse the process. That is, we decompose a given rational expression into a sum of fractional expressions whose denominators are factors of the given denominator. The fractional expressions in the sum are called **partial fractions.** For example, each of the decompositions

$$\frac{x-2}{x(x-1)^2} = \frac{-2}{x} + \frac{2}{x-1} + \frac{-1}{(x-1)^2}$$

and

$$\frac{x^2 - x + 13}{(x+1)(x^2+4)} = \frac{3}{x+1} + \frac{-2x+1}{x^2+4}$$

can be verified by adding the fractions on the right to obtain the rational expressions on the left. Partial fraction decomposition is an important operation in calculus, and it is made possible by the property that every polynomial with real coefficients has a real factored form as defined in Section 6 of this chapter.

6.3. Partial Fractions

Decomposition of Proper Rational Expressions. Let $P(x)/Q(x)$ be a given rational expression, where $P(x)$ and $Q(x)$ are polynomial in x. We consider first the case where $P(x)/Q(x)$ is a **proper rational expression**, which means that the degree of the numerator $P(x)$ is *less than* the degree of the denominator $Q(x)$. Then $P(x)/Q(x)$ can be decomposed into partial fractions in the following three steps.

A quadratic factor $x^2 + bx + c$ is irreducible if $b^2 - 4c < 0$.

1. Factor $Q(x)$ into its real factored form of linear and irreducible quadratic factors:

$$Q(x) = a \cdot \ldots \cdot (x-r)^m \cdot \ldots \cdot (x^2+bx+c)^n \cdot \ldots$$

If $m > 1$, the factor $(x-r)^m$ is called a **repeated linear factor,** and if $n > 1$, $(x^2+bx+c)^n$ is called a **repeated quadratic factor.**

2. Write

$$\frac{P(x)}{Q(x)} = \cdots + \frac{A_1}{x-r} + \frac{A_2}{(x-r)^2} + \cdots + \frac{A_m}{(x-r)^m} + \cdots$$
$$\cdots + \frac{B_1 x + C_1}{x^2+bx+c} + \frac{B_2 x + C_2}{(x^2+bx+c)^2} + \cdots$$
$$\cdots + \frac{B_n x + C_n}{(x^2+bx+c)^n} + \cdots,$$

where the A's, B's, and C's are constant real numbers to be determined. Note that there are m terms corresponding to each factor $(x-r)^m$ and n terms corresponding to each factor $(x^2+bx+c)^n$.

3. Solve for the constant A's, B's, and C's.

It can be shown by advanced algebra that every proper rational expression has a decomposition into partial fractions as indicated in step 2. The most general method for finding the constants is illustrated in the following examples. Another method, which requires less computation in some cases, is described in Exercises 35 and 36. Also, we note in step 1 that for a general polynomial $Q(x)$ of degree ≥ 5, its real factored form must be given (*why?*).

Example 1 Decompose the proper rational function $1/(x^2+x-2)$ into partial fractions.

Solution: First factor the denominator into $(x-1)(x+2)$. These factors are linear and nonrepeating. Now let

$$\frac{1}{x^2+x-2} = \frac{A}{x-1} + \frac{B}{x+2}$$

in accordance with step 2. We now solve for A and B. By adding on the right side with the LCD, we obtain

Chapter 6. Additional Topics

$$\frac{1}{x^2+x-2} = \frac{A(x+2)}{(x-1)(x+2)} + \frac{B(x-1)}{(x+2)(x-1)} \quad (1)$$

or

$$\frac{1}{x^2+x-2} = \frac{(A+B)x + (2A-B)}{(x-1)(x+2)}.$$

Since the denominators in equation (1) are equal, the numerators must also be equal. Therefore,

$$1 = (A+B)x + (2A-B), \quad (2)$$

which implies the system of equations

$$\begin{array}{l} A+B=0 \\ 2A-B=1, \end{array} \quad \left\{ \begin{array}{l} \textit{In order for equation (2) to hold for all values} \\ \textit{of x, the coefficient } A+B \textit{ of x must equal 0,} \\ \textit{and the constant term } 2A-B \textit{ must equal 1.} \end{array} \right.$$

whose solution is $A = 1/3$ and $B = -1/3$. Therefore,

$$\frac{1}{x^2+x-2} = \frac{\frac{1}{3}}{x-1} + \frac{-\frac{1}{3}}{x+2}$$

$$= \frac{1}{3}\left(\frac{1}{x-1}\right) - \frac{1}{3}\left(\frac{1}{x+2}\right). \blacksquare$$

Example 2 Decompose $(2-x)/[x(x-1)^2]$ into partial fractions.

Solution: The numerator has degree 1 and the denominator has degree 3, so the given rational expression is proper. The factors of the denominator are linear, but $(x-1)^2$ is a repeated linear factor. Therefore, let

$$\frac{2-x}{x(x-1)^2} = \frac{A}{x} + \frac{B}{x-1} + \frac{C}{(x-1)^2}$$

in accordance with step 2. We now find A, B and C. The partial fractions on the right have LCD $= x(x-1)^2$. Therefore,

$$\frac{2-x}{x(x-1)^2} = \frac{A(x-1)^2 + Bx(x-1) + Cx}{x(x-1)^2}$$

or

$$\frac{2-x}{x(x-1)^2} = \frac{(A+B)x^2 + (-2A-B+C)x + A}{x(x-1)^2}. \quad (3)$$

By equating the numerators of equation (3), we obtain

$$2 - x = (A+B)x^2 + (-2A-B+C)x + A, \quad (4)$$

which is equivalent to the system of equations

$$A + B = 0$$
$$-2A - B + C = -1$$
$$A = 2,$$

In order for equation (4) to hold for all values of x, the coefficient of each power of x on the right must equal the coefficient of the same power of x on the left.

whose solution is $A = 2$, $B = -2$, $C = 1$. Therefore,

$$\frac{2-x}{x(x-1)^2} = \frac{2}{x} - \frac{2}{x-1} + \frac{1}{(x-1)^2}. \quad \blacksquare$$

Example 3   Decompose the proper rational expression $x/(x+1)^2$ into partial fractions.

Solution: As in step 2,

$$\frac{x}{(x+1)^2} = \frac{A}{x+1} + \frac{B}{(x+1)^2}$$
$$= \frac{A(x+1) + B}{(x+1)^2}$$
$$= \frac{Ax + (A+B)}{(x+1)^2}.$$

By equating numerators, we obtain the equations

$$A = 1$$
$$A + B = 0.$$

Therefore, $A = 1$, $B = -1$, and

$$\frac{x}{(x+1)^2} = \frac{1}{x+1} - \frac{1}{(x+1)^2}. \quad \blacksquare$$

Example 4   Decompose the proper rational expression $(x-5)/[(x+1)(x^2+1)]$ into partial fractions.

Solution: The denominator is factored into one linear and one irreducible quadratic factor, both nonrepeating. Hence, by step 2,

$$\frac{x-5}{(x+1)(x^2+1)} = \frac{A}{x+1} + \frac{Bx+C}{x^2+1}.$$

As in the preceding examples, we add the partial fractions with their **LCD**, thereby obtaining

$$\frac{x-5}{(x+1)(x^2+1)} = \frac{A}{x+1} + \frac{Bx+C}{x^2+1}$$
$$= \frac{A(x^2+1) + (Bx+C)(x+1)}{(x+1)(x^2+1)}$$
$$= \frac{(A+B)x^2 + (B+C)x + (A+C)}{(x+1)(x^2+1)}.$$

Therefore,

$$x - 5 = (A+B)x^2 + (B+C)x + (A+C),$$

which implies

$$A + B = 0$$
$$B + C = 1$$
$$A + C = -5.$$

The solution of this system is $A = -3$, $B = 3$, and $C = -2$. Hence,

$$\frac{x-5}{(x+1)(x^2+1)} = \frac{-3}{x+1} + \frac{3x-2}{x^2+1}. \quad \blacksquare$$

Example 5 Decompose the proper rational expression

$$\frac{x^4 + 3x^3 + 3x^2 + 4x + 1}{x(x^2+1)^2}$$

into partial fractions.

Solution: The denominator is factored into one nonrepeating linear factor and one repeating, irreducible quadratic factor. Therefore, by step 2,

$$\frac{x^4 + 3x^3 + 3x^2 + 4x + 1}{x(x^2+1)^2}$$
$$= \frac{A}{x} + \frac{Bx+C}{x^2+1} + \frac{Dx+E}{(x^2+1)^2}$$
$$= \frac{A(x^2+1)^2 + (Bx+C)x(x^2+1) + (Dx+E)x}{x(x^2+1)^2}$$
$$= \frac{(A+B)x^4 + Cx^3 + (2A+B+D)x^2 + (C+E)x + A}{x(x^2+1)^2}$$

By equating numerators as in the previous examples, we arrive at the system of equations

$$A + B = 1$$
$$C = 3$$
$$2A + B + D = 3$$
$$C + E = 4$$
$$A = 1,$$

whose solution is $A = 1$, $B = 0$, $C = 3$, $D = 1$, and $E = 1$. Thus, the partial fraction decomposition is

$$\frac{x^4 + 3x^3 + 3x^2 + 4x + 1}{x(x^2 + 1)^2} = \frac{1}{x} + \frac{3}{x^2 + 1} + \frac{x + 1}{(x^2 + 1)^2}. \quad \blacksquare$$

Comment We note that in all cases the number of constants to be determined is equal to the degree of the given denominator. In example 1, the degree of the denominator $x^2 + x - 2$ is 2, and we found two constants A and B. Similarly, in Example 5, in which the denominator $x(x^2 + 1)^2 = x^5 + 2x^3 + x$ has degree 5, we solved for five constants.

Decomposition of Improper Rational Expressions. Suppose that $P(x)/Q(x)$ is an improper rational expression, which means that degree $P(x) \geq$ degree $Q(x)$. We then divide $Q(x)$ into $P(x)$ to obtain

$$\frac{P(x)}{Q(x)} = q(x) + \frac{R(x)}{Q(x)},$$

where the quotient $q(x)$ is a polynomial, and the remainder $R(x)$ is a polynomial whose degree is *less than* the degree of $Q(x)$. The partial fraction decomposition may now be applied to the *proper* rational expression $R(x)/Q(x)$.

Example 6 Decompose $(3x^2 + 3x - 5)/(x^2 + x - 2)$ into partial fractions.

Solution:

$$\frac{3x^2 + 3x - 5}{x^2 + x - 2} = 3 + \frac{1}{x^2 + x - 2} \qquad \text{By division.}$$
$$= 3 + \frac{1}{3}\left(\frac{1}{x - 1}\right) - \frac{1}{3}\left(\frac{1}{x + 2}\right) \qquad \text{By Example 1.} \quad \blacksquare$$

Example 7 Decompose $(3x^3 + 7x^2 + 3x + 1)/(x^2 + x)$ into partial fractions.

Solution:

$$\frac{3x^3 + 7x^2 + 3x + 1}{x^2 + x} = 3x + 4 + \frac{-x + 1}{x^2 + x} \qquad \text{By division.}$$

Now

$$\frac{-x+1}{x^2+x} = \frac{-x+1}{x(x+1)} = \frac{A}{x} + \frac{B}{x+1}, \qquad \text{By step 2.}$$

where we obtain, by the methods of the previous examples, $A = 1$ and $B = -2$. Therefore

$$\frac{3x^3 + 7x^2 + 3x + 1}{x^2 + x} = 3x + 4 + \frac{1}{x} - \frac{2}{x+1}. \qquad \blacksquare$$

Caution If the denominator of a proper rational expression is a *nonrepeating* linear or irreducible quadratic polynomial, then no further partial fraction decomposition is possible. For example,

$$\frac{x^4 + 5x^2 + 6x + 11}{x^2 + 4} = x^2 + 1 + \frac{6x + 7}{x^2 + 4},$$

and the proper rational expression $(6x+7)/(x^2+4)$ is already in partial fraction from.

Exercises 6.3

Fill in the blanks to make each statement true.

1. Partial fraction decomposition is applied to _____ rational expressions.
2. In a proper rational expression, the numerator has _____ less than that of the denominator.
3. If the degree of the numerator is _____ or _____ the degree of the denominator, the first step needed before a partial fractional decomposition is _____.
4. The type of partial fraction decomposition of a proper rational expression depends upon the _____ of the denominator.
5. For each factor $(x-r)^n$ of the denominator, there are _____ constants to be determined; for each irreducible factor $(x^2 + bx + c)^m$, there are _____ constants to be determined.

Write true or false for each statement.

6. For some constants A and B,

$$\frac{3x^2 + 3x - 5}{x^2 + x - 2} = \frac{A}{x-1} + \frac{B}{x+2}.$$

7. If a proper rational expression has denominator $(x+3)(x-3)^2$, its decomposition is of the form

$$\frac{A}{x+3} + \frac{B}{(x-3)^2}.$$

8. If a proper rational expression has denominator $x^2 + 3x + 2$, its decomposition is of the form

$$\frac{A}{x+1} + \frac{B}{x+2}.$$

9. If a proper rational expression has denominator $(x+2)(x^2+x+2)$, its decomposition is of the form

$$\frac{A}{x+2} + \frac{Bx+C}{x^2+x+2}.$$

10. If a proper rational expression has denominator $x^2(x+1)^3$, its decomposition is of the form

$$\frac{A}{x} + \frac{B}{x^2} + \frac{C}{x+1} + \frac{D}{(x+1)^2} + \frac{E}{(x+1)^3}.$$

Partial Fraction Decomposition of Rational Expressions

Decompose, if possible, each of the following rational expressions into partial fraction. If necessary, divide first.

11. $\dfrac{x-13}{x^2-x-6}$

12. $\dfrac{2x+1}{x^2+5x+6}$

13. $\dfrac{5x-4}{x^2-4x}$

14. $\dfrac{2x^2}{x^2-3x+2}$

15. $\dfrac{2x^2+3}{x(x-1)^2}$

16. $\dfrac{2x+4}{x^2(x-2)}$

17. $\dfrac{x^2-2x-8}{x(x-2)^3}$

18. $\dfrac{x-1}{(x+1)^2}$

19. $\dfrac{1}{(x-1)^3}$

20. $\dfrac{x}{(x-1)^3}$

21. $\dfrac{x^2}{(x-1)^3}$

22. $\dfrac{x^3}{(x-1)^3}$

23. $\dfrac{x^3+2x^2-2x+2}{x^2-3x+2}$

24. $\dfrac{x^2}{x^2-3x+2}$

25. $\dfrac{x}{x^2+1}$

26. $\dfrac{x}{x^2+x+1}$

27. $\dfrac{1}{x(x^2+x+1)}$

28. $\dfrac{3x^2+x+13}{(x-1)(x^2+16)}$

29. $\dfrac{4}{(x^2-4)(x^2+4)}$

30. $\dfrac{3x^3+3x^2+3x+6}{(x^2+1)(x^2+4)}$

31. $\dfrac{1}{(x^2+1)^2}$

32. $\dfrac{x^2}{(x^2+1)^2}$

33. $\dfrac{x^2}{(x^2-1)^2}$

34. $\dfrac{x^3}{(x^2+1)^2}$

35. The general method used in this section for finding the constants in a partial fraction decomposition requires the solution of a system of equations. An alternate (and sometimes easier) method is now illustrated. The decomposition in Example 1 of the text is

$$\dfrac{1}{x^2+x-2} = \dfrac{A}{x-1} + \dfrac{B}{x+2} = \dfrac{A(x+2)+B(x-1)}{(x-1)(x+2)},$$

and by setting numerators equal, we have

$$1 = A(x+2) + B(x-1). \quad \text{(i)}$$

Substitution of $x=1$ in (i) gives $1 = A(3)$ or $A = 1/3$.
Substitution of $x=-2$ in (i) gives $1 = B(-3)$ or $B = -1/3$. Thus,

$$\dfrac{1}{x^2+x-2} = \dfrac{\tfrac{1}{3}}{x-1} - \dfrac{\tfrac{1}{3}}{x+1}.$$

Apply this substitution method to solving Exercises 11-14 on the preceding page. Note that these examples involve only nonrepeated linear factors.

36. The substitution method in Exercise 35 may be applied with some modification to cases in which the factors are not all linear and nonrepeated. We solve Example 2 by this method as follows. We have

$$\dfrac{2-x}{x(x-1)^2} = \dfrac{A}{x} + \dfrac{B}{x-1} + \dfrac{C}{(x-1)^2}$$
$$= \dfrac{A(x-1)^2+Bx(x-1)+Cx}{x(x-1)^2}$$

and by setting numerators equal,

$$2-x = A(x-1)^2 + Bx(x-1) + Cx. \quad \text{(ii)}$$

Substituting $x=0$ in (ii) gives $2 = A$, and substituting $x=1$ in (ii) gives $1 = C$, but this method will not work for B *(why?)*. However, if we substitute $A=2$, $C=1$, and any value of x ($\neq 0$ or 1) in (ii), we will get an equation for B. For instance, if $x=2$, then (ii) becomes

$$0 = 2(1) + B(2)(1) + 1(2)$$

so that $-4 = 2B$ or $B = -2$. Therefore,

$$\dfrac{2-x}{x(x-1)^2} = \dfrac{2}{x} - \dfrac{2}{x-1} + \dfrac{1}{(x-1)^2}.$$

Solve Exercises 15, 16, 28, and 29 by this method of substitution.

6.4 Linear Systems: Gaussian Elimination

First we eliminate variables, and then we eliminate equations.

Reduction Method

Augmented Matrix of a Linear System

Gaussian Elimination in Matrix Form

Gauss-Jordan Elimination

Reduction Method. As shown in Section 2.7, a linear system in three variables can be solved by reducing it to one with two variables. At each step, one variable is eliminated from two equations by adding a multiple of one of the equations to the other.

Example 1   Solve the following system by reduction

$$2x + 5y + 4z = 4 \quad (1)$$
$$x + 4y + 3z = 1 \quad (2)$$
$$x - 3y - 2z = 5 \quad (3)$$

Solution: By multiplying equation (2) by -2 and adding the result to (1), we eliminate x from the first two equations and obtain equation (4) below. Similarly, if (3) is multiplied by -1 and added to (2), the result is (5).

$$\begin{array}{ll} 2x + 5y + 4z = 4 \\ \underline{-2x - 8y - 6z = -2} \\ -3y - 2z = 2 \quad (4) \end{array} \qquad \begin{array}{ll} x + 4y + 3z = 1 \\ \underline{-x + 3y + 2z = -5} \\ 7y + 5z = -4 \quad (5) \end{array}$$

The reduced system

$$-3y - 2z = 2 \quad (4)$$
$$7y + 5z = -4 \quad (5)$$

has the solution $y = -2$ and $z = 2$. Substituting these values in one of the original equations, say (2), we have $x + 4(-2) + 3 \cdot 2 = 1$ or $x = 3$. Hence, the solution of the given system is $x = 3$, $y = -2$ and $z = 2$. ∎

The reduction procedure used in Example 1 is satisfactory for a linear system in three variables. Although reduction can be applied to more than three variables, it becomes increasingly inefficient and impractical as the number of variables increases. Here we develop a method that can be used conveniently on any linear system, no matter how many equations and variables are involved and regardless of the number of solutions. The method, called **Gaussian elimination**, can be easily im-

6.4. Linear Systems: Gaussian

plemented both by hand computation and on a computer.
to systematically eliminate one variable at a time until we re
system to a triangular form, as illustrated below with the sys.
Example 1.

Solution of Example 1 by Gaussian Elimination:

Step 1: Make the coefficient of x equal to 1 in the first equation. This is easily done by interchanging the first two equations, giving the equivalent system

$$x + 4y + 3z = 1$$
$$2x + 5y + 4z = 4$$
$$x - 3y - 2z = 5.$$

Step 2: Eliminate x in all equations following the first one.

$$x + 4y + 3z = 1$$
$$-3y - 2z = 2 \quad -2 \text{ times the first equation added to the second.}$$
$$-7y - 5z = 4 \quad -1 \text{ times the first equation added to the third.}$$

Step 3: Make the coefficient of y equal to 1 in the second equation. In this example, we can do this by multiplying the second equation above by $-1/3$. Our system now is

$$x + 4y + 3z = 1 \qquad (1')$$

$$y + \frac{2}{3}z = -\frac{2}{3} \qquad (2')$$

$$-7y - 5z = 4. \qquad (3')$$

Step 4: Eliminate y in all equations following the second one.

$$x + 4y + 3z = 1$$
$$y + \frac{2}{3}z = -\frac{2}{3}$$
$$-\frac{1}{3}z = -\frac{2}{3} \quad \text{7 times } (2') \text{ was added to } (3')$$

Step 5: Make the coefficient of z equal to 1 in the third equation by multiplying both sides by -3.

$$x + 4y + 3z = 1 \qquad (1'')$$

$$y + \frac{2}{3}z = -\frac{2}{3} \qquad (2'')$$

$$z = 2. \qquad (3'')$$

Since there are no equations following the third one in this system, the elimination process stops here. We now proceed to find the solution by **back-substitution.**

Step 6: We substitute $z = 2$ from (3″) into (2″) to get $y = -2$. Then both $z = 2$ and $y = -2$ are substituted in (1″) to give $x = 3$. Thus, we have $x = 3$, $y = -2$, and $z = 2$, as in our first method for solving Example 1. ∎

We note that each step in the general elimination procedure used above involves one of the following operations:

(1) interchanging two equations,
(2) multiplying an equation by a constant $k \neq 0$,
(3) adding a constant multiple of one equation to another equation.

Before solving any more examples by this method, we introduce a more efficient way to carry out these operations.

Performing these elementary row operations results in an equivalent system of equations, that is, one that has the same solution as the original system.

Augmented Matrix of a Linear System. The general elimination procedure used above to solve a linear system can be simplified if we observe that only the *positions* of the variables x, y, and z are important, not the symbols used to denote them. That is, if we keep the variables in the same order in each equation, then a given linear system is completely determined by the array of coefficients of the variables and the constant terms. For instance, the system in Example 1,

$$2x + 5y + 4z = 4$$
$$x + 4y + 3z = 1$$
$$x - 3y - 2z = 5,$$

can be represented by the rectangular array of numbers

$$\begin{bmatrix} 2 & 5 & 4 & | & 4 \\ 1 & 4 & 3 & | & 1 \\ 1 & -3 & -2 & | & 5 \end{bmatrix},$$

which is called the **augmented matrix** of the linear system. The adjective "augmented" is used here to distinguish that matrix from the **coefficient matrix** of the system, namely

6.4. Linear Systems: Gaussian Elimination

$$\begin{bmatrix} 2 & 5 & 4 \\ 1 & 4 & 3 \\ 1 & -3 & -2 \end{bmatrix}.$$

The vertical line in the augmented matrix is optional, but it is a convenient device to separate the coefficient matrix from the column of constant terms.

Each horizontal row of the augmented matrix represents one equation, and to solve the linear system, we simply perform on the *rows* of the augmented matrix the operations that correspond exactly to the operations performed on the *equations* in the general elimination procedure. These three operations, which are called **elementary row operations**, are:

(1) interchanging two rows,
(2) multiplying each entry of a row by a constant $k \neq 0$,
(3) adding a constant times each entry in a row to the corresponding entry in another row.

We repeat the solution to Example 1 below with the reduction of the equations to triangular form on the left and the corresponding row operations applied to the augmented matrix on the right.

Solution:

Linear System	**Augmented Matrix**	**Operations**
$2x + 5y + 4z = 4$	$\begin{bmatrix} 2 & 5 & 4 & \mid & 4 \\ 1 & 4 & 3 & \mid & 1 \\ 1 & -3 & -2 & \mid & 5 \end{bmatrix}$	
$x + 4y + 3z = 1$		
$x - 3y - 2z = 5$		

Step 1: Make the coefficient of the first variable in the first equation (first entry in first row) equal to 1.

$$\begin{aligned} x + 4y + 3z &= 1 \\ 2x + 5y + 4z &= 4 \\ x - 3y - 2z &= 5 \end{aligned} \qquad \begin{bmatrix} 1 & 4 & 3 & \mid & 1 \\ 2 & 5 & 4 & \mid & 4 \\ 1 & -3 & -2 & \mid & 5 \end{bmatrix}$$

Equations (rows) 1 and 2 were interchanged.

Step 2: Make the coefficient of the first variable (entry in the first column) equal to 0 in all equations (rows) below the first.

$$\begin{aligned} x + 4y + 3z &= 1 \\ -3y - 2z &= 2 \\ -7y - 5z &= 4 \end{aligned} \qquad \begin{bmatrix} 1 & 4 & 3 & \mid & 1 \\ 0 & -3 & -2 & \mid & 2 \\ 0 & -7 & -5 & \mid & 4 \end{bmatrix}$$

-2 times equation (row) 1 was added to equation (row) 2, and -1 times equation (row) 1 was added to equation (row) 3.

Step 3: Make the coefficient of the second variable in the second equation (second entry in second row) equal to 1.

$$\begin{aligned} x + 4y + 3z &= 1 \\ y + \frac{2}{3}z &= -\frac{2}{3} \\ -7y - 5z &= 4 \end{aligned} \qquad \left[\begin{array}{ccc|c} 1 & 4 & 3 & 1 \\ 0 & 1 & 2/3 & -2/3 \\ 0 & -7 & -5 & 4 \end{array}\right] \qquad \begin{array}{l} \textit{Equation (row)} \\ \textit{2 was multiplied} \\ \textit{by } -1/3. \end{array}$$

Step 4: Make the coefficient of the second variable (entry in second column) equal to 0 in all equations (rows) below the second.

$$\begin{aligned} x + 4y + 3z &= 1 \\ y + \frac{2}{3}z &= -\frac{2}{3} \\ -\frac{1}{3}z &= -\frac{2}{3} \end{aligned} \qquad \left[\begin{array}{ccc|c} 1 & 4 & 3 & 1 \\ 0 & 1 & 2/3 & -2/3 \\ 0 & 0 & -1/3 & -2/3 \end{array}\right] \qquad \begin{array}{l} \textit{7 times equation} \\ \textit{(row) 2 was added} \\ \textit{to equation (row) 3.} \end{array}$$

Step 5: Make the coefficient of the third variable in the third equation (third entry in third row) equal to 1.

$$\begin{aligned} x + 4y + 3z &= 1 \\ y + \frac{2}{3}z &= -\frac{2}{3} \\ z &= 2 \end{aligned} \qquad \left[\begin{array}{ccc|c} 1 & 4 & 3 & 1 \\ 0 & 1 & 2/3 & -2/3 \\ 0 & 0 & 1 & 2 \end{array}\right] \qquad \begin{array}{l} \textit{Equation (row) 3 was} \\ \textit{multiplied by } -3. \end{array}$$

Step 6: Use back-substitution of $z = 2$ to obtain $y = -2$ from equation (row) 2 and then use $z = 2$ and $y = -2$ to obtain $x = 3$ from equation (row) 1. ■

Gaussian Elimination in Matrix Form. The matrix in step 5 above corresponding to the triangular form of the linear system is said to be in **row-echelon form**, and the general elimination method, when applied to the augmented matrix of the system, is called *Gaussian elimination*.[1] As mentioned previously, Gaussian elimination is our most general and efficient way to solve linear systems. We now apply this method to the augmented matrix of several linear systems to illustrate that it works equally well if the system has one solution, none, or an infinite number of solutions and regardless of the number of variables or equations.

---

[1] Carl Friedrich Gauss (1777 – 1855) is considered to be one of the greatest mathematicians of all time. See his fundamental theorem in the chapter on roots of polynomials.

Example 2 Solve the following system by Gaussian elimination in matrix for

$$\begin{aligned} x + y + z &= 4 \\ 2x + y + z &= 3 \\ 3x + 2y - z &= 1 \end{aligned}$$

Solution: The augmented matrix of the system is

$$\begin{bmatrix} 1 & 1 & 1 & 4 \\ 2 & 1 & 1 & 3 \\ 3 & 2 & -1 & 1 \end{bmatrix},$$

which can be reduced to row-echelon form as follows.

We begin by noting that the entry in the first row column is 1. Therefore, we proceed to introduce 0's in the rest of column 1. To do this, row 1 is multiplied by -2 and the result is added to row 2. Similarly, row 1 is multiplied by -3 and added to row 3, with the following result:

$$\begin{bmatrix} 1 & 1 & 1 & 4 \\ 0 & -1 & -1 & -5 \\ 0 & -1 & -4 & -11 \end{bmatrix}.$$

Next, the entry in row 2, column 2 is made equal to 1 by multiplying the second row by -1:

$$\begin{bmatrix} 1 & 1 & 1 & 4 \\ 0 & 1 & 1 & 5 \\ 0 & -1 & -4 & -11 \end{bmatrix}.$$

To introduce 0 as the entry in row 3 and column 2, we now add row 2 to row 3:

$$\begin{bmatrix} 1 & 1 & 1 & 4 \\ 0 & 1 & 1 & 5 \\ 0 & 0 & -3 & -6 \end{bmatrix}.$$

Finally, row 3 is multiplied by $-1/3$ to make its first nonzero entry equal to 1:

$$\begin{bmatrix} 1 & 1 & 1 & 4 \\ 0 & 1 & 1 & 5 \\ 0 & 0 & 1 & 2 \end{bmatrix}.$$

The last matrix represents the system

$$x + y + z = 4$$
$$y + z = 5$$
$$z = 2.$$

Back-substitution yields $y = 3$ and then $x = -1$. Therefore, the solution is $x = -1$, $y = 3$ and $z = 2$. ∎

In the following examples, we will simplify our explanations and use $R_1, R_2, \ldots$ to denote row 1, row 2, and so on.

Example 3 Use Gaussian elimination in matrix form to solve

$$x - 2y - 3z = 2$$
$$x - 4y - 13z = 14$$
$$-3x + 5y + 4z = 2.$$

Solution: We first write the augmented matrix and then perform the indicated elementary row operations. (Read the $+$ sign in the steps as "added to.")

$$\begin{bmatrix} 1 & -2 & -3 & | & 2 \\ 1 & -4 & -13 & | & 14 \\ -3 & 5 & 4 & | & 2 \end{bmatrix}$$

1. $(-1)R_1 + R_2,$
   $3R_1 + R_3:$
   $$\begin{bmatrix} 1 & -2 & -3 & | & 2 \\ 0 & -2 & -10 & | & 12 \\ 0 & -1 & -5 & | & 8 \end{bmatrix}.$$

2. $(-1/2)R_2:$
   $$\begin{bmatrix} 1 & -2 & -3 & | & 2 \\ 0 & 1 & 5 & | & -6 \\ 0 & -1 & -5 & | & 8 \end{bmatrix}.$$

3. $R_2 + R_3:$
   $$\begin{bmatrix} 1 & -2 & -3 & | & 2 \\ 0 & 1 & 5 & | & -6 \\ 0 & 0 & 0 & | & 2 \end{bmatrix}.$$

The last row of the reduced matrix represents the equation

$$0x + 0y + 0z = 2,$$

which has no solution. Therefore, the given system has no solution. ∎

When a system of equations has no solution, as in Example 3, we say that the system is **inconsistent.**

Example 4 By reducing the augmented matrix to row-echelon form, solve

$$x + y + 2z + 3t = 13$$
$$3x + y + z - t = 1$$
$$x - 2y + z + t = 8.$$

6.4. Linear Systems: Gaussian Elimination

Solution: The augmented matrix is

$$\begin{bmatrix} 1 & 1 & 2 & 3 & | & 13 \\ 3 & 1 & 1 & -1 & | & 1 \\ 1 & -2 & 1 & 1 & | & 8 \end{bmatrix},$$

which is reduced as follows. (Read the + sign in the steps as "added to.")

1. $(-3)R_1 + R_2$,
   $-R_1 + R_3$:
$$\begin{bmatrix} 1 & 1 & 2 & 3 & | & 13 \\ 0 & -2 & -5 & -10 & | & -38 \\ 0 & -3 & -1 & -2 & | & -5 \end{bmatrix}.$$

2. $(-1/2)R_2$:
$$\begin{bmatrix} 1 & 1 & 2 & 3 & | & 13 \\ 0 & 1 & 5/2 & 5 & | & 19 \\ 0 & -3 & -1 & -2 & | & -5 \end{bmatrix}.$$

3. $3R_2 + R_3$:
$$\begin{bmatrix} 1 & 1 & 2 & 3 & | & 13 \\ 0 & 1 & 5/2 & 5 & | & 19 \\ 0 & 0 & 13/2 & 13 & | & 52 \end{bmatrix}.$$

4. $(2/13)R_3$:
$$\begin{bmatrix} 1 & 1 & 2 & 3 & | & 13 \\ 0 & 1 & 5/2 & 5 & | & 19 \\ 0 & 0 & 1 & 2 & | & 8 \end{bmatrix}.$$

The last matrix is in row-echelon form and corresponds to the system of equations

$$x + y + 2z + 3t = 13$$
$$y + \frac{5}{2}z + 5t = 19$$
$$z + 2t = 8.$$

By back-substitution, we first get

$$z = 8 - 2t,$$

where t can be *any real number*. If this expression for z is substituted into the second equation, we obtain

$$y = -1,$$

and if $y = -1$ and the expression for z are substituted in the first equation, we get

$$x = -2 + t.$$

Hence, all solutions can be written as $x = -2 + t$, $y = -1$, $z = 8 - 2t$, and t, where t is any real number. In other words, the system has *infinitely many* solutions. ∎

When reducing the augmented matrix to row-echelon form, it is possible to obtain an entire row of 0's. If this occurs, we simply move the row of 0's to the bottom row and continue with the reduction process.

Example 5 Solve

$$y + z = 2$$
$$x + y + z = 5$$
$$x + 2y + 2z = 7$$
$$2x + y - z = 4.$$

Solution: The augmented matrix is

$$\begin{bmatrix} 0 & 1 & 1 & | & 2 \\ 1 & 1 & 1 & | & 5 \\ 1 & 2 & 2 & | & 7 \\ 2 & 1 & -1 & | & 4 \end{bmatrix},$$

which we reduce to row-echelon form in the following steps:

1. $R_1 \leftrightarrow R_2$: $\begin{bmatrix} 1 & 1 & 1 & | & 5 \\ 0 & 1 & 1 & | & 2 \\ 1 & 2 & 2 & | & 7 \\ 2 & 1 & -1 & | & 4 \end{bmatrix}$. *Row 1 interchanged with Row 2.*

2. $-R_1 + R_3$, $-2R_1 + R_4$: $\begin{bmatrix} 1 & 1 & 1 & | & 5 \\ 0 & 1 & 1 & | & 2 \\ 0 & 1 & 1 & | & 2 \\ 0 & -1 & -3 & | & -6 \end{bmatrix}$.

3. $-R_2 + R_3$, $R_2 + R_4$: $\begin{bmatrix} 1 & 1 & 1 & | & 5 \\ 0 & 1 & 1 & | & 2 \\ 0 & 0 & 0 & | & 0 \\ 0 & 0 & -2 & | & -4 \end{bmatrix}$.

4. $R_3 \leftrightarrow R_4$: $\begin{bmatrix} 1 & 1 & 1 & | & 5 \\ 0 & 1 & 1 & | & 2 \\ 0 & 0 & -2 & | & -4 \\ 0 & 0 & 0 & | & 0 \end{bmatrix}$.

5. $(-1/2)R_3$:
$$\begin{bmatrix} 1 & 1 & 1 & | & 5 \\ 0 & 1 & 1 & | & 2 \\ 0 & 0 & 1 & | & 2 \\ 0 & 0 & 0 & | & 0 \end{bmatrix}.$$

The last matrix is in row-echelon form and represents the system

$$x + y + z = 5$$
$$y + z = 2$$
$$z = 2,$$

whose solution is $x = 3$, $y = 0$, and $z = 2$. ■

Gauss-Jordan Elimination Method. As an alternative to the back-substitution used in the preceding examples, we can continue row operations on the row-echelon form of the augmented matrix in order to introduce 0's above the leading 1's in each row. For instance, in Example 2 above, the row-echelon form of the augmented matrix was determined to be

$$\begin{bmatrix} 1 & 1 & 1 & | & 4 \\ 0 & 1 & 1 & | & 5 \\ 0 & 0 & 1 & | & 2 \end{bmatrix}.$$

We now work from the last row upward to introduce 0's *above* the leading 1's as follows:

1. $-R_3 + R_2$,
$-R_3 + R_1$:
$$\begin{bmatrix} 1 & 1 & 0 & | & 2 \\ 0 & 1 & 0 & | & 3 \\ 0 & 0 & 1 & | & 2 \end{bmatrix}.$$

2. $-R_2 + R_1$:
$$\begin{bmatrix} 1 & 0 & 0 & | & -1 \\ 0 & 1 & 0 & | & 3 \\ 0 & 0 & 1 & | & 2 \end{bmatrix}$$

This matrix represents the solution

$$x = -1$$
$$y = 3$$
$$z = 2.$$

The matrix in step 2 above is said to be in **reduced row-echelon form**, and the procedure used to obtain it is called *Gauss-Jordan[2] elimination*. We now summarize the previous two matrix methods.

Wilhelm Jordan (1842-1899) used the method to solve systems of equations that occur in the subject of geodesy.

To solve a linear system by the method of **Gaussian elimination**, we first perform the following steps on the augmented matrix by means of elementary row operations.

---

1. Make the element in the first row and first column equal to 1, if possible. This may require interchanging rows.
2. Make all elements below the leading 1 of step 1 equal to 0.
3. Place any row consisting entirely of 0's under all nonzero rows.
4. Repeat steps 1, 2, and 3 on the smaller matrix remaining after the first row and column are deleted.
5. Repeat step 4 until the process automatically stops. The augmented matrix is now in *row-echelon form*.

We then solve for the variable represented by the leading 1 in the last nonzero row and back-substitute to obtain the complete solution.

To solve a linear system by the method of **Gauss-Jordan elimination**, we perform steps 1 through 5 above as well as the following step.

---

6. Make all elements above the leading 1's in each nonzero row of the row-echelon form equal to 0. The matrix is now in *reduced row-echelon form* and represents the simplest form of the original system of equations.

Exercises 6.4

Fill in the blanks to make each statement true.

1. A linear system of three equations in three unknowns may be solved by reducing the system to _____ equations in _____ unknowns.
2. When solving a linear system by the Gaussian elimination method, the aim is to reduce the system to _____ form.
3. In order to represent a linear system by its augmented matrix, we write the system with the _____ in the same order in each equation and with the _____ on the right side of each equation.
4. If an augmented matrix represents a system of three equations in three variables, then its first three columns constitute its _____ matrix, and its last column contains the _____ terms.

5. In the Gaussian elimination method, the three types of row operations that may be performed on the augmented matrix are _____, _____ and _____.

Write true or false each statement.

6. The general elimination method can be used for solving linear systems in any number of variables.
7. The reduction method is satisfactory for solving linear systems in three variables.
8. In the Gaussian elimination method, a row of the augmented matrix may be multiplied by any constant.
9. Interchanging two rows of the augmented matrix of a linear system corresponds to interchanging two equations in the system.
10. If the last column of an augmented matrix contains all 0's, then that column is not changed in the Gaussian elimination method.

Reduction Method

In Exercises 11−14, solve the system by reducing it to two equations in two unknowns, as in Section 2.7.

11. $x + y - z = 4$
    $2x + 2y - 3z = 3$
    $3x + y - z = 2$

12. $2x - y + 2z = 9$
    $x + y + z = 13$
    $x - 5y + 2z = 12$

13. $x + y - 3z = 1$
    $2x - y - 3z = -1$
    $x + 3y - 5z = 3$

14. $x + y + z = 3$
    $2x + y - z = 0$
    $3x + 2y + z = 7$

15-18. Solve Exercise 11-14 by the Gaussian elimination procedure.

Use the Gaussian elimination procedure to solve the system in Exercises 19-20

19. $x + y - 3z = 1$
    $x + 3y - 5z = 2$

20. $x + y + 2z + 2t = 7$
    $x - y - z - t = 0$
    $2x + 3y + z + t = 5$
    $2x + y - z - 2t = -5$

Augmented Matrix of a Linear System

In Exercises 21-25, write a linear system represented by the given augmented matrix.

21. $\begin{bmatrix} 2 & 1 & | & -3 \\ 4 & -5 & | & 6 \end{bmatrix}$

22. $\begin{bmatrix} 1 & -3 & 2 & | & 5 \\ 6 & 1 & -2 & | & 7 \\ 2 & 5 & -1 & | & 3 \end{bmatrix}$

23. $\begin{bmatrix} 1 & 2 & -3 & 4 & | & 5 \\ 2 & -1 & 0 & 7 & | & 6 \end{bmatrix}$

24. $\begin{bmatrix} 1 & -2 & 0 & | & 3 \\ 2 & 0 & 1 & | & 5 \\ 3 & 1 & 1 & | & 1 \\ 4 & -5 & 2 & | & 3 \end{bmatrix}$

25. $\begin{bmatrix} 1 & 0 & 0 & | & 2 \\ 0 & 1 & 0 & | & -3 \\ 0 & 0 & 1 & | & 4 \end{bmatrix}$

In Exercises 26-30, write the augmented matrix that represents the given linear system.

26. $x - 2y = 5$
    $2x + 3y = 1$

27. $x - y - 6 = 0$
    $y + 2x - 3 = 0$

28. $x - y + z = 5$
    $2x + y - 3z = 7$
    $4x + y - 2z = 6$

29. $2x - y + 3z - 8 = 0$
    $y + 7x - z = 4$
    $z - x + 2y = 3$

30. $x - y + z = 8$
    $x + 2y + 3t = 5$
    $x - z + 2t = 6$
    $x + y + 3z = 0$

Gaussian Elimination Method

In Exercises 31-35, the matrix given is the reduced augmented matrix obtained by Gaussian elimination. Write the solution of the corresponding system of equations

31. $\begin{bmatrix} 1 & 0 & | & 2 \\ 0 & 1 & | & 3 \end{bmatrix}$

32. $\begin{bmatrix} 1 & 1 & | & 2 \\ 0 & 0 & | & 0 \end{bmatrix}$

33. $\begin{bmatrix} 1 & 0 & 0 & | & 3 \\ 0 & 1 & 0 & | & -2 \\ 0 & 0 & 1 & | & 1 \end{bmatrix}$

34. $\begin{bmatrix} 1 & 0 & 1 & | & 3 \\ 0 & 1 & -2 & | & 1 \\ 0 & 0 & 1 & | & 2 \end{bmatrix}$

35. $\begin{bmatrix} 1 & 0 & 1 & | & 3 \\ 0 & 1 & -2 & | & 1 \\ 0 & 0 & 0 & | & 0 \end{bmatrix}$

In Exercises 36-40, solve by Gaussian elimination.

36. $\begin{aligned} x - 2y &= 5 \\ 2x + 3y &= -4 \end{aligned}$

37. $\begin{aligned} x - y + z &= 6 \\ 2x + y + z &= 3 \\ x + y + z &= 2 \end{aligned}$

38. $\begin{aligned} 2x + 5y - 4z &= -3 \\ 4x + y - z &= 6 \\ 3x - 2y + z &= 7 \end{aligned}$

39. $\begin{aligned} x - 2y - 3z &= 2 \\ x - 4y - 13z &= 14 \\ -3x + 5y + 4z &= 2 \end{aligned}$

40. $\begin{aligned} x - 2y - 3z &= 2 \\ x - 4y - 13z &= 14 \\ -3x + 5y + 4z &= 0 \\ y + 5z &= -6 \end{aligned}$

Gauss-Jordan Elimination Method

41-45. *Solve Exercises 36-40 by Gauss-Jordan elimination.*

Use the Gauss-Jordan elimination method to solve the systems in Exercises 46-50.

46. $\begin{aligned} x + y &= 9 \\ 2x - \tfrac{1}{2}y &= 8 \end{aligned}$

47. $\begin{aligned} x - 3y &= 5 \\ 4x - 12y &= 20 \end{aligned}$

48. $\begin{aligned} x + y + z &= 0 \\ 2x - y + 4z &= 1 \\ x + 2y - z &= -3 \end{aligned}$

49. $\begin{aligned} x + y + z &= 0 \\ 2x - y + 4z &= 1 \\ x + 2y - z &= -3 \\ -x - y + 2z &= 6 \end{aligned}$

50. $\begin{aligned} x + y + z + 2t &= 0 \\ 3x + y + 3z + 5t &= 1 \\ x - y - z + 3t &= -1 \\ 2x - 3y + z + t &= 0 \end{aligned}$

6.5 Matrix Algebra

In matrix algebra, $(A + B)^2 \neq A^2 + 2AB + B^2$.

Matrices: Size and Notation

Matrix Addition and Multiplication by a Scalar

Matrix Multiplication

Inverse of a Square Matrix

Matrices were used in the previous section to simplify the solution of linear systems. They have other uses as well, and their study has been intensified because operations with matrices can be programmed easily on a computer. With certain restrictions, matrices can be added, subtracted, and multiplied, and although we do not call it division for matrices, the equivalent operation of multiplication by an inverse is possible for one class of matrices. This section gives an introduction to the algebra of matrices.

Matrices: Size and Notation. A matrix has already been described as a **rectangular array** of real numbers. It may have any number of rows and any number of columns, and these two numbers specify its size. For example, in the following matrices,

$$\begin{bmatrix} 1 \\ 2 \\ 3 \end{bmatrix}, \quad \begin{bmatrix} 1 & -2 \\ 4 & 1 \\ 0 & 3 \end{bmatrix}, \quad \begin{bmatrix} 1 & 4 & 0 \\ -2 & 1 & 3 \end{bmatrix},$$

the first has 3 rows and 1 column (size 3×1), the second has 3 rows and 2 columns (3×2), and the third has 2 rows and 3 columns (2×3). A 1×1 matrix contains only one real number, and in this case we sometimes drop the bracket notation. It is customary to denote matrices by capital letters and to represent each real number in the array by a lowercase letter. In this context, a real number is called a **scalar** and the symbol a_{ij} is used to denote the entry in row i and columns j of matrix A.

As stated earlier, a matrix with n rows and n columns is called **a square matrix of order n**. The entries a_{ii} ($i = 1, 2, \ldots, n$) in an $n \times n$ square matrix A are said to form the **main diagonal** of A.

Two matrices are **equal** if and only if they are identical; that is, if they have the same size and all corresponding entries are the same. In symbols, for two $m \times n$ matrices A and B,

$A = B$ if and only if $a_{ij} = b_{ij}$ for $i = 1, 2, \ldots, m$
and $j = 1, 2, \ldots, n$.

For instance, $\begin{bmatrix} 1 & 2 \\ 3 & 4 \end{bmatrix} = \begin{bmatrix} 1 & 2 \\ 3 & 4 \end{bmatrix}$,

but $\begin{bmatrix} 1 & 2 \\ 3 & 4 \end{bmatrix} \neq \begin{bmatrix} 2 & 3 \\ 1 & 4 \end{bmatrix}$, and $\begin{bmatrix} 1 & 2 \\ 3 & 4 \end{bmatrix} \neq \begin{bmatrix} 1 & 2 & 0 \\ 3 & 4 & 0 \end{bmatrix}$.

Matrix Addition and Multiplication by a scalar. If two matrices have the same size, their sum is obtained by adding the corresponding entries.

Thus, if A and B are both of size $m \times n$, then

$A + B = [a_{ij}] + [b_{ij}] = [a_{ij} + b_{ij}]$
for $i = 1, 2, \ldots, m$ and $j = 1, 2, \ldots, n$.   Addition of Matrices.

Similarly,

$A - B = [a_{ij}] - [b_{ij}] = [a_{ij} - b_{ij}]$
for $i = 1, 2, \ldots, m$ and $j = 1, 2, \ldots, n$.   Subtraction of Matrices.

Caution If A and B are not the same size, then $A \pm B$ are not defined.

Example 1   Suppose an appliance dealer sells 500 radios, 450 TVs and 225 VCRs in 2015, and 575 radios, 380 TVs and 500 VCRs in 2016. Use matrices to represents the sales each year and the sales for the two years together.

Solution: If we let

$$A = \begin{bmatrix} 500 \\ 450 \\ 225 \end{bmatrix} \quad \text{and} \quad B = \begin{bmatrix} 575 \\ 380 \\ 500 \end{bmatrix}$$

represent the sales of the three appliances for 2015 and 2016, respectively, then the total sales of these items for the two years are represented by the matrix sum

$$A + B = \begin{bmatrix} 500 + 575 \\ 450 + 380 \\ 225 + 500 \end{bmatrix} = \begin{bmatrix} 1075 \\ 830 \\ 725 \end{bmatrix} \quad \begin{matrix} Radios. \\ TVs. \\ VCRs. \end{matrix}$$

Thus, the dealer sold 1075 radios, 830 TVs, and 725 VCRs in the two years. ∎

To multiply a matrix by a real number (scalar), c, means to multiply every entry by c. That is,

If $A = [a_{ij}]$, **then** $cA = [ca_{ij}]$.   Multiplication of a Matrix by a Scalar

Example 2   If in 2016 the appliance dealer in Example 1 had doubled the 2015 sales of the three items, what matrix would represent that result?

Solution: The sales for 2015 were given by

$$A = \begin{bmatrix} 500 \\ 450 \\ 225 \end{bmatrix},$$

so doubling those sales in 2016 would be indicated by the matrix

$$2A = \begin{bmatrix} 2 \cdot 500 \\ 2 \cdot 450 \\ 2 \cdot 225 \end{bmatrix} = \begin{bmatrix} 1000 \\ 900 \\ 450 \end{bmatrix}. \quad ∎$$

Since matrix addition and multiplication by a scalar involve ordinary addition and multiplication of real numbers, it is not surprising that these operations on matrices have many of the properties discussed earlier for the real number system. We list these properties here and leave their verification to the exercises. It is understood that matrices to be added must be of the same size.

Matrix Addition

$A + B = B + A$ commutative property of addition
$(A + B) + C = A + (B + C)$ associative property of addition

If O is a matrix whose entries are all zeros, then

$$A + O = O + A = A. \qquad \text{Identity Matrix for Addition}$$

If $-A$ denotes $(-1)A$, then

$$A + (-A) = (-A) + A = O. \qquad \text{Additive Inverse of a Matrix}$$

Multiplication by a Scalar (c and k are scalars)

$$\begin{aligned} c(kA) &= (ck)A & &\text{Associative Property} \\ (c+k)A &= cA + kA & &\text{Distributive Property} \\ c(A+B) &= cA + cB & &\text{Distributive Property} \end{aligned}$$

Matrix Multiplication. We might expect that multiplication of matrices would involve multiplying corresponding entries, but this definition turns out not to be useful. Instead, we define multiplication of matrices only if *the number of columns of the first matrix equals the number of rows of the second matrix*, and we give the definition in two steps. Examples will illustrate the usefulness of our definition. (See also Exercise 37 and 44.)

We first consider the case in which A has size $1 \times n$ and B has size $n \times 1$. Here, A is called a **row matrix** and B is called a **column matrix**, and we define AB as follows.

$$[a_{11} \ a_{12} \ \ldots \ a_{1n}] \begin{bmatrix} b_{11} \\ b_{21} \\ \vdots \\ b_{n1} \end{bmatrix} = [a_{11}b_{11} + a_{12}b_{21} + \ldots + a_{1n}b_{n1}].$$

That is, if A is a $1 \times n$ matrix and B is an $n \times 1$ matrix, then AB is a 1×1 matrix.[1]

---

[1] *The n-tuples $(a_{11}, a_{12} \cdots, a_{1n})$ and $(b_{11}, b_{21}, \cdots, b_{n1})$ are n-dimensional vectors, and the real number $a_{11}b_{11} + a_{12}b_{21} + \cdots + a_{1n}b_{n1}$ is called their dot product.*

Example 3 In 2015 the appliance dealer in Example 1 made profits of \$5, \$30, and \$50 on each radio, TV, and VCR, respectively. Indicate the total profits as a matrix multiplication.

Solution: We let the profits on the three items be represented by the matrix

$$P = [5 \quad 30 \quad 50].$$

To compute the total profit, we multiply the profit on each item by the number of that item sold and then add these three products. Hence, the total profit is equal to the matrix product

$$PA = \begin{bmatrix} 5 & 30 & 50 \end{bmatrix} \begin{bmatrix} 500 \\ 450 \\ 225 \end{bmatrix} = [5 \cdot 500 + 30 \cdot 450 + 50 \cdot 225]$$

$$= [27{,}250]. \blacksquare$$

We are now ready to define the product AB of matrices A and B in which the number of columns of A equals the number of rows of B. The entry in row i and column j of AB is the product of row i of A and column j of B, as defined above. That is, if A is an $m \times n$ matrix and B is an $n \times p$ matrix, then, by definition,

$$\begin{bmatrix} a_{11} & a_{12} & \ldots & a_{1n} \\ \vdots & \vdots & & \vdots \\ a_{i1} & a_{i2} & \ldots & a_{in} \\ \vdots & \vdots & & \vdots \\ a_{m1} & a_{m2} & \ldots & a_{mn} \end{bmatrix} \begin{bmatrix} b_{11} & \ldots & b_{1j} & \ldots & b_{1p} \\ b_{21} & \ldots & b_{2j} & \ldots & b_{2p} \\ \vdots & & \vdots & & \vdots \\ b_{n1} & \ldots & b_{nj} & \ldots & b_{np} \end{bmatrix} = \begin{bmatrix} c_{11} & \ldots & c_{1j} & \ldots & c_{1p} \\ \vdots & & \vdots & & \vdots \\ c_{i1} & \ldots & c_{ij} & \ldots & c_{ip} \\ \vdots & & \vdots & & \vdots \\ c_{m1} & \ldots & c_{mj} & \ldots & c_{mp} \end{bmatrix},$$

where

$$c_{ij} = a_{i1}b_{1j} + a_{i2}b_{2j} + \cdots + a_{in}b_{nj} \qquad \text{Matrix}$$
$$\text{for } i = 1, 2, \ldots, m \quad \text{and} \quad j = 1, 2, \ldots, p. \qquad \text{Multiplication}$$

c_{ij} is the dot product of the ith row of A with the jth column of B.

Hence, the product AB of an $m \times n$ matrix A and an $n \times p$ matrix B is an $m \times p$ matrix C. *If the number of columns of A is not equal to the number of rows of B, then AB is not defined.*

Example 4 Find AB for $A = \begin{bmatrix} 1 & 2 & 3 \\ 4 & 5 & 6 \end{bmatrix}$ and $B = \begin{bmatrix} 2 & -1 & 3 \\ 1 & 2 & -1 \\ 4 & 0 & -4 \end{bmatrix}$

Solution: Since A is 2×3 and B is 3×3, we know AB is 2×3. We start by multiplying row 1 by column 1:

$$\begin{bmatrix} 1 & 2 & 3 \\ -- & -- & -- \end{bmatrix} \begin{bmatrix} 2 & -- & -- \\ 1 & -- & -- \\ 4 & -- & -- \end{bmatrix}$$

$$= \begin{bmatrix} 1 \cdot 2 + 2 \cdot 1 + 3 \cdot 4 & -- & -- \\ -- & -- & -- \end{bmatrix} = \begin{bmatrix} 16 & -- & -- \\ -- & -- & -- \end{bmatrix}.$$

Similarly, we obtain the remaining entries in the first row of AB by multiplying row 1 by column 2 and then row 1 by column 3, giving

$$\begin{bmatrix} 1 & 2 & 3 \\ \text{---} & \text{---} & \text{---} \end{bmatrix} \begin{bmatrix} \text{---} & -1 & 3 \\ \text{---} & 2 & -1 \\ \text{---} & 0 & -4 \end{bmatrix}$$

$$= \begin{bmatrix} 16 & 1(-1)+2\cdot 2+3\cdot 0 & 1\cdot 3+2(-1)+3(-4) \\ \text{---} & \text{---} & \text{---} \end{bmatrix}$$

$$= \begin{bmatrix} 16 & 3 & -11 \\ \text{---} & \text{---} & \text{---} \end{bmatrix}.$$

Finally, multiplying row 2 of A by each of the columns of B, we get

$$\begin{bmatrix} \text{---} & \text{---} & \text{---} \\ 4 & 5 & 6 \end{bmatrix} \begin{bmatrix} 2 & -1 & 3 \\ 1 & 2 & -1 \\ 4 & 0 & -4 \end{bmatrix}$$

$$= \begin{bmatrix} \text{---} & \text{---} & \text{---} \\ 4\cdot 2+5\cdot 1+6\cdot 4 & 4(-1)+5\cdot 2+6\cdot 0 & 4\cdot 3+5(-1)+6(-4) \end{bmatrix}$$

$$= \begin{bmatrix} \text{---} & \text{---} & \text{---} \\ 37 & 6 & -17 \end{bmatrix}.$$

Therefore, $AB = \begin{bmatrix} 16 & 3 & -11 \\ 37 & 6 & -17 \end{bmatrix}$. ∎

Example 5 Suppose the appliance dealer in Example 1 purchased from each of two suppliers 550 radios, 500 TVs, and 300 VCRs in 2015 and 600 radios, 450 TVs, and 500 VCRs in 2015. If the prices were the same in both years and supplier 1 charges $25 per radio, $200 per TV, and $395 per VCR, while supplier 2 charges $20 per radio, $185 per TV, and $425 per VCR, use matrix multiplication to display total costs for the two years from the two suppliers.

Solution: The matrix product

	Numbers of items			Cost per item from supplier			Total cost from supplier	
	Radio	TV	VCR	1	2		1	2
(2015)	550	500	300	25	20	=	232,250	231,000	(2015)
(2016)	600	450	500	200	185		302,500	307,750	(2016)
				395	425			

shows that in 2015 the cost from supplier 1 is $232,250 and from supplier 2 is $231,000, while in 2016, the cost from supplier 1 is $302,500 and from 2 is $307,750. ∎

Example 6 Show how a linear system of m equations in n unknowns can be represented in terms of matrices.

Solution: By the definitions of matrix multiplication and equality of matrices, the system

$$\begin{aligned} a_{11}x_1 + a_{12}x_2 + \ldots + a_{1n}x_n &= b_1 \\ a_{21}x_1 + a_{22}x_2 + \ldots + a_{2n}x_n &= b_2 \\ &\vdots \\ a_{m1}x_1 + a_{m2}x_2 + \ldots + a_{mn}x_n &= b_m \end{aligned}$$

can be written

$$\begin{bmatrix} a_{11} & a_{12} & \ldots & a_{1n} \\ a_{21} & a_{22} & \ldots & a_{2n} \\ \vdots & \vdots & & \vdots \\ a_{m1} & a_{m2} & \ldots & a_{mn} \end{bmatrix} \begin{bmatrix} x_1 \\ x_2 \\ \vdots \\ x_n \end{bmatrix} = \begin{bmatrix} b_1 \\ b_2 \\ \vdots \\ b_m \end{bmatrix}$$

or more briefly $AX = B$, where A is the coefficient matrix and X and B are both column matrices. ∎

In the algebra of matrices, we have some additional properties similar to those for real numbers, namely

$(AB)C = A(BC)$ Associative Property of Multiplication
$A(B + C) = AB + AC$ Left Distributive Property
$(A + B)C = AC + BC.$ Right Distributive Property

Caution One major difference between real numbers and matrices is that *matrix multiplication (even when both AB and BA are defined) is not a commutative operation.* For example, if $A = \begin{bmatrix} 1 & -2 \\ 4 & 3 \end{bmatrix}$ and $B = \begin{bmatrix} 2 & 5 \\ -1 & 3 \end{bmatrix}$, then $AB = \begin{bmatrix} 4 & -1 \\ 5 & 29 \end{bmatrix}$ but $BA = \begin{bmatrix} 22 & 11 \\ 11 & 11 \end{bmatrix}$. Also, for matrices *we may have $AB = 0$ with $A \neq 0$ and $B \neq 0$*, and $AB = AC$ with $A \neq 0$ and $B \neq C$ (see Exercise 36).

Inverse of a Square Matrix. For square matrices of order n, there exists an **identity matrix I** for multiplication. I has 1's on its main diagonal and 0's elsewhere. For example, if $n = 3$, then

$$I = \begin{bmatrix} 1 & 0 & 0 \\ 0 & 1 & 0 \\ 0 & 0 & 1 \end{bmatrix}.$$

It is easy to verify by matrix multiplication that

$$\begin{bmatrix} a_{11} & a_{12} & a_{13} \\ a_{21} & a_{22} & a_{23} \\ a_{31} & a_{32} & a_{33} \end{bmatrix} \begin{bmatrix} 1 & 0 & 0 \\ 0 & 1 & 0 \\ 0 & 0 & 1 \end{bmatrix} = \begin{bmatrix} a_{11} & a_{12} & a_{13} \\ a_{21} & a_{22} & a_{23} \\ a_{31} & a_{32} & a_{33} \end{bmatrix};$$

that is, $AI = A$, and similarly, $IA = A$.

An $n \times n$ matrix B is called the **multiplicative inverse** of an $n \times n$ matrix A if

$$AB = BA = I. \qquad \text{inverse of a square matrix}$$

If such an inverse matrix B exists for A, then B is denoted by A^{-1}, and A is called an **invertible matrix**. It can be shown that if there is a matrix B satisfying $AB = I$, then B also satisfies $BA = I$, and B is unique. Hence, to determine whether a square matrix A is invertible, we determine whether the single equation $AB = I$ has a solution for B. If a solution exists, then $B = A^{-1}$.

Example 7 Find A^{-1}, if possible, for

$$A = \begin{bmatrix} 2 & 3 \\ 4 & 5 \end{bmatrix}.$$

Solution: The matrix equation $AB = I$, or

$$\begin{bmatrix} 2 & 3 \\ 4 & 5 \end{bmatrix} \begin{bmatrix} x & u \\ y & v \end{bmatrix} = \begin{bmatrix} 1 & 0 \\ 0 & 1 \end{bmatrix}$$

requires $\begin{cases} 2x + 3y = 1 \\ 4x + 5y = 0 \end{cases}$ and $\begin{cases} 2u + 3v = 0 \\ 4u + 5v = 1 \end{cases}$, for which the solutions are

$$\begin{aligned} x &= -5/2 \\ y &= 2 \end{aligned} \quad \text{and} \quad \begin{aligned} u &= 3/2 \\ v &= -1. \end{aligned}$$

Hence,

$$A^{-1} = \begin{bmatrix} -5/2 & 3/2 \\ 2 & -1 \end{bmatrix}.$$

This result is verified by showing that

$$AA^{-1} = \begin{bmatrix} 2 & 3 \\ 4 & 5 \end{bmatrix} \begin{bmatrix} -5/2 & 3/2 \\ 2 & -1 \end{bmatrix} = \begin{bmatrix} 1 & 0 \\ 0 & 1 \end{bmatrix},$$

and

$$A^{-1}A = \begin{bmatrix} -5/2 & 3/2 \\ 2 & -1 \end{bmatrix} \begin{bmatrix} 2 & 3 \\ 4 & 5 \end{bmatrix} = \begin{bmatrix} 1 & 0 \\ 0 & 1 \end{bmatrix}. \qquad \blacksquare$$

Example 8 Find A^{-1}, if possible, for $A = \begin{bmatrix} 1 & 0 \\ 0 & 0 \end{bmatrix}$.

Solution: Consider the matrix equation $AB = I$, or

$$\begin{bmatrix} 1 & 0 \\ 0 & 0 \end{bmatrix} \begin{bmatrix} x & u \\ y & v \end{bmatrix} = \begin{bmatrix} 1 & 0 \\ 0 & 1 \end{bmatrix}$$

$$\begin{bmatrix} x & u \\ 0 & 0 \end{bmatrix} = \begin{bmatrix} 1 & 0 \\ 0 & 1 \end{bmatrix},$$

which implies

$$\begin{matrix} x = 1 \\ 0 = 0 \end{matrix} \quad \text{and} \quad \begin{matrix} u = 0 \\ 0 = 1 \end{matrix}.$$

Since $0 \neq 1$, we have shown that there is *no* solution for the matrix B. Hence, A is *not* invertible, which means A^{-1} does not exist. ∎

Although solving the equation $AB = I$ for B is easy enough when A is of order 2, as in Example 7, for matrices of order 3, 4, 5,..., or n, the above procedure involves the solution of 9, 16, 25,..., or n^2 equations, respectively. Hence, we want a more efficient method for finding the inverse of a matrix. We now describe such a method for finding A^{-1}, if it exists, when A is or order 2, and this method can be generalized to matrices of higher order. Given matrix $A = \begin{bmatrix} a & b \\ c & d \end{bmatrix}$, we seek a matrix $B = \begin{bmatrix} x & u \\ y & v \end{bmatrix}$ such that $AB = I$:

$$\begin{bmatrix} a & b \\ c & d \end{bmatrix} \begin{bmatrix} x & u \\ y & v \end{bmatrix} = \begin{bmatrix} 1 & 0 \\ 0 & 1 \end{bmatrix}$$

Hence, we must try to solve the systems

$$\begin{cases} ax + by = 1 \\ cx + dy = 0 \end{cases} \quad \text{and} \quad \begin{cases} au + bv = 0 \\ cu + dv = 1. \end{cases} \quad (1)$$

The corresponding augmented matrices are

$$\left[\begin{array}{cc|c} a & b & 1 \\ c & d & 0 \end{array} \right] \quad \text{and} \quad \left[\begin{array}{cc|c} a & b & 0 \\ c & d & 1 \end{array} \right],$$

and if there is a unique solution by Gauss-Jordan elimination, these matrices must row-reduce to

$$\begin{bmatrix} 1 & 0 & | & x \\ 0 & 1 & | & y \end{bmatrix} = \begin{bmatrix} I & | & x \\ & & y \end{bmatrix} \quad \text{and} \quad \begin{bmatrix} 1 & 0 & | & u \\ 0 & 1 & | & v \end{bmatrix} = \begin{bmatrix} I & | & u \\ & & v \end{bmatrix},$$

respectively. In both cases, this means that the matrix A is reduced to I, and when this is possible, the other two columns $\begin{bmatrix} 1 \\ 0 \end{bmatrix}$ and $\begin{bmatrix} 0 \\ 1 \end{bmatrix}$, which are the columns of I, will be replaced by $\begin{bmatrix} x \\ y \end{bmatrix}$ and $\begin{bmatrix} u \\ v \end{bmatrix}$, which are the columns of A^{-1}. To organize this row-reduction method in compact form, we represent both systems (1) by one matrix, namely,

$$\begin{bmatrix} a & b & | & 1 & 0 \\ c & d & | & 0 & 1 \end{bmatrix} = [A \mid I] \tag{2}$$

If it is possible to row-reduce this matrix in such a way that A is replaced by I, then at the same time I will be replaced by

$$\begin{bmatrix} x & u \\ y & v \end{bmatrix} = A^{-1}.$$

That is, matrix (2) will be replaced by

$$\begin{bmatrix} 1 & 0 & | & x & u \\ 0 & 1 & | & y & v \end{bmatrix} = [I \mid A^{-1}].$$

If we find that A cannot be row-reduced to I, then A^{-1} does not exist, and A is not invertible.

Example 9 Repeat Example 7 by the row-reduction method.

Solution:

Since $A = \begin{bmatrix} 2 & 3 \\ 4 & 5 \end{bmatrix}$, we row-reduce the matrix $\begin{bmatrix} 2 & 3 & | & 1 & 0 \\ 4 & 5 & | & 0 & 1 \end{bmatrix}$ as follows.

1. $\frac{1}{2}R_1$: $\begin{bmatrix} 1 & 3/2 & | & 1/2 & 0 \\ 4 & 5 & | & 0 & 1 \end{bmatrix}$.

2. $-4R_1 + R_2$: $\begin{bmatrix} 1 & 3/2 & | & 1/2 & 0 \\ 0 & -1 & | & -2 & 1 \end{bmatrix}$.

3. $-1R_2$: $\begin{bmatrix} 1 & 3/2 & | & 1/2 & 0 \\ 0 & 1 & | & 2 & -1 \end{bmatrix}$.

4. $-\frac{3}{2}R_2 + R_1$: $\begin{bmatrix} 1 & 0 & | & -5/2 & 3/2 \\ 0 & 1 & | & 2 & -1 \end{bmatrix}$.

We conclude that A is invertible and

$$A^{-1} = \begin{bmatrix} -5/2 & 3/2 \\ 2 & -1 \end{bmatrix},$$

as in Example 7. ∎

The procedure explained above for finding A^{-1} is applicable if A is of order $n > 2$, as is illustrated in the following example.

Example 10   Find A^{-1}, if possible, for

$$A = \begin{bmatrix} 1 & 2 & 3 \\ 3 & 2 & 1 \\ 1 & 2 & 1 \end{bmatrix}.$$

Solution: Use elementary row operations on $[A \mid I]$ as follows.

1. $-3R_1 + R_2,$
   $-R_1 + R_3:$
   $$\begin{bmatrix} 1 & 2 & 3 & | & 1 & 0 & 0 \\ 0 & -4 & -8 & | & -3 & 1 & 0 \\ 0 & 0 & -2 & | & -1 & 0 & 1 \end{bmatrix}.$$

2. $-\dfrac{1}{4}R_2:$
   $$\begin{bmatrix} 1 & 2 & 3 & | & 1 & 0 & 0 \\ 0 & 1 & 2 & | & 3/4 & -1/4 & 0 \\ 0 & 0 & -2 & | & -1 & 0 & 1 \end{bmatrix}.$$

3. $-\dfrac{1}{2}R_3:$
   $$\begin{bmatrix} 1 & 2 & 3 & | & 1 & 0 & 0 \\ 0 & 1 & 2 & | & 3/4 & -1/4 & 0 \\ 0 & 0 & 1 & | & 1/2 & 0 & -1/2 \end{bmatrix}.$$

4. $-2R_3 + R_2,$
   $-3R_3 + R_1:$
   $$\begin{bmatrix} 1 & 2 & 0 & | & -1/2 & 0 & 3/2 \\ 0 & 1 & 0 & | & -1/4 & -1/4 & 1 \\ 0 & 0 & 1 & | & 1/2 & 0 & -1/2 \end{bmatrix}.$$

5. $-2R_2 + R_1:$
   $$\begin{bmatrix} 1 & 0 & 0 & | & 0 & 1/2 & -1/2 \\ 0 & 1 & 0 & | & -1/4 & -1/4 & 1 \\ 0 & 0 & 1 & | & 1/2 & 0 & -1/2 \end{bmatrix}.$$

Therefore,

$$A^{-1} = \begin{bmatrix} 0 & 1/2 & -1/2 \\ -1/4 & -1/4 & 1 \\ 1/2 & 0 & -1/2 \end{bmatrix}.$$

It is a good idea to check that $AA^{-1} = I$. ∎

6.5. Matrix Algebra

A system of n equations in n unknowns may be written in matrix form (Example 6) as

$$AX = B,$$

where A is $n \times n$, and X and B are $n \times 1$. Now if A is invertible, then

$$A^{-1}AX = A^{-1}B$$
$$IX = A^{-1}B$$
$$X = A^{-1}B.$$

Hence, *a linear system $AX = B$, in which A is invertible, has the matrix solution $X = A^{-1}B$.*

Example 11 Solve the system $\begin{cases} 2x + 3y = 1 \\ 4x + 5y = 6 \end{cases}$ by inverting the coefficient matrix.

Solution: The matrix representation of the given system is

$$\begin{bmatrix} 2 & 3 \\ 4 & 5 \end{bmatrix} \begin{bmatrix} x \\ y \end{bmatrix} = \begin{bmatrix} 1 \\ 6 \end{bmatrix}.$$

The coefficient matrix is $A = \begin{bmatrix} 2 & 3 \\ 4 & 5 \end{bmatrix}$, and by Example 9 we have

$$A^{-1} = \begin{bmatrix} -5/2 & 3/2 \\ 2 & -1 \end{bmatrix}.$$

Therefore, the solution of the system can be given by

$$X = A^{-1}B$$

or

$$\begin{bmatrix} x \\ y \end{bmatrix} = \begin{bmatrix} -5/2 & 3/2 \\ 2 & -1 \end{bmatrix} \begin{bmatrix} 1 \\ 6 \end{bmatrix} = \begin{bmatrix} 13/2 \\ -4 \end{bmatrix}.$$

That is, $x = 13/2$ and $y = -4$ is the solution. ∎

Exercises 6.5

Fill in the blanks to make each statement true.

1. A matrix has size 2×5 if it has 2 _____ and 5 _____.
2. Matrices A and B can be added only if they have the _____.
3. If every entry in matrix A is multiplied by a constant c, then the resulting matrix is denoted by _____.
4. The product AB of matrices A and B is defined if and only if the number of _____ of A equals the number of _____ of B.
5. A square matrix A has a multiplicative inverse B if and only if $AB = BA =$ _____, and then A is said to be _____.

Write true or false for each statement.

6. If A and B are matrices of different size, then $A + B = 0$.
7. To add two matrices A and B of the same size, we add corresponding entries a_{ij} and b_{ij}.
8. We can multiply two matrices if they have the same size.
9. Multiplication of matrices is not an associative operation.
10. A square matrix has a multiplicative inverse if and only if it can be row reduced to the identity matrix for multiplication.

Matrices: Size and Notation

11. State the size of each of the following matrices.

    (a) $\begin{bmatrix} 1 \\ 3 \\ 2 \end{bmatrix}$

    (b) $\begin{bmatrix} 1 & 3 & 2 \end{bmatrix}$

    (c) $\begin{bmatrix} 1 & -2 & 3 \\ 6 & 5 & 7 \end{bmatrix}$

    (d) $\begin{bmatrix} 2 & 1 & 3 \\ 4 & 6 & 8 \\ 1 & 5 & 7 \end{bmatrix}$

    (e) $\begin{bmatrix} 2 & 1 & 3 & 9 & 7 \\ 4 & 6 & 8 & 1 & 1 \\ 1 & 5 & 7 & 2 & 3 \end{bmatrix}$

12. For the matrix $A = \begin{bmatrix} 2 & 1 & 3 \\ 4 & 7 & -1 \\ 3 & -5 & 6 \end{bmatrix}$, find each of the following.

    (a) a_{32} (b) a_{13}
    (c) a_{22}
    (d) the entries on the main diagonal

13. Find x and y for each equation.

    (a) $\begin{bmatrix} 2 & 1 \\ -3 & 4 \end{bmatrix} = \begin{bmatrix} x & 1 \\ -3 & y \end{bmatrix}$

    (b) $\begin{bmatrix} 1 & 0 & 0 \\ 0 & 2x & 0 \\ 0 & 0 & 2+y \end{bmatrix} = \begin{bmatrix} 1 & 0 & 0 \\ 0 & x-3 & 0 \\ 0 & 0 & 3y \end{bmatrix}$

Matrix Addition and Multiplication by a Scalar

In Exercises 14-19, add the matrices A and B or explain why the addition is not possible.

14. $A = \begin{bmatrix} 1 & 0 \\ 3 & -4 \end{bmatrix}; B = \begin{bmatrix} 2 & 1 \\ 2 & 1 \end{bmatrix}$

15. $A = \begin{bmatrix} 1 & 2 & 3 \\ 4 & 5 & 6 \\ 7 & 8 & 9 \end{bmatrix}; B = \begin{bmatrix} 1 & -1 & 1 \\ 2 & -5 & -2 \\ 3 & 2 & 1 \end{bmatrix}$

16. $A = \begin{bmatrix} 1 & 0 \\ 3 & -4 \end{bmatrix}; B = \begin{bmatrix} 2 & 1 & 3 \\ 2 & 1 & 4 \end{bmatrix}$

17. $A = \begin{bmatrix} 1 \\ 2 \\ 3 \end{bmatrix}; B = \begin{bmatrix} -1 \\ -2 \\ -3 \end{bmatrix}$

18. $A = \begin{bmatrix} 1 \\ 2 \\ 3 \end{bmatrix}; B = \begin{bmatrix} 3 & -1 & 0 \end{bmatrix}$

19. $A = \begin{bmatrix} 1 & 2 & 3 \\ 4 & 5 & 6 \end{bmatrix}; B = \begin{bmatrix} 1 & 4 \\ 2 & 5 \\ 3 & 6 \end{bmatrix}$

In Exercises 20-23, find the indicated matrix.

20. $2 \begin{bmatrix} 1 & 2 \\ 4 & 3 \end{bmatrix}$

21. $-3\begin{bmatrix} 1 & 0 & -1 \\ 2 & -2 & 1 \\ 3 & 4 & 0 \end{bmatrix}$

22. $0\begin{bmatrix} 1 & 2 \\ 4 & -3 \end{bmatrix}$

23. $\dfrac{1}{2}\begin{bmatrix} 2 & -4 & 6 \\ 3 & 5 & 7 \end{bmatrix}$

24. For the matrices

$$A = \begin{bmatrix} 1 & 3 & -2 \\ 0 & -4 & 3 \end{bmatrix}, B = \begin{bmatrix} 2 & -1 & 1 \\ 3 & 0 & 5 \end{bmatrix}, C = \begin{bmatrix} 0 & -3 & 1 \\ 2 & 3 & -1 \end{bmatrix},$$

verify each of the following equations.

(a) $A + B = B + A$
(b) $(A + B) + C = A + (B + C)$
(c) $A + O = A$
(d) $B + (-B) = O$

25. For the matrices given in Exercise 24, verify each of the following.

(a) $2(3A) = 6A$
(b) $(2 + 3)B = 2B + 3B$
(c) $-2(A + C) = -2A - 2C$

26. For the matrices

$$A = \begin{bmatrix} a_{11} & a_{12} \\ a_{21} & a_{22} \end{bmatrix}, B = \begin{bmatrix} b_{11} & b_{12} \\ b_{21} & b_{22} \end{bmatrix}, C = \begin{bmatrix} c_{11} & c_{12} \\ c_{21} & c_{22} \end{bmatrix},$$

verify each of the following

(a) $A + B = B + A$
(b) $(A + B) + C = A + (B + C)$
(c) $c(kA) = (ck)A$
(e) $k(B + C) = kB + kC$

Matrix Multiplication

In Exercises 27-34, find the matrix AB or explain why AB is not defined.

27. $A = \begin{bmatrix} 3 & 4 \end{bmatrix}; B = \begin{bmatrix} 1 \\ 2 \end{bmatrix}.$

28. $A = \begin{bmatrix} 1 \\ 2 \end{bmatrix}; B = \begin{bmatrix} 3 & 4 \end{bmatrix}$

29. $A = \begin{bmatrix} 3 & -4 \\ 1 & 2 \end{bmatrix}; B = \begin{bmatrix} 5 & -1 \\ 7 & 3 \end{bmatrix}$

30. $A = \begin{bmatrix} -1 & 2 & 0 \\ 3 & 1 & 4 \end{bmatrix}; B = \begin{bmatrix} 2 & -1 & 5 \\ 3 & 0 & 4 \\ 1 & 6 & 2 \end{bmatrix}$

31. $A = \begin{bmatrix} 2 & -3 & 5 \\ 1 & -1 & 0 \\ 0 & 4 & 6 \end{bmatrix}; B = \begin{bmatrix} 1 & -2 & 4 \\ 1 & 3 & -5 \\ 1 & 2 & 4 \end{bmatrix}$

32. $A = \begin{bmatrix} 1 & 0 & -1 \\ 0 & 2 & 3 \\ -4 & 5 & 2 \end{bmatrix}; B = \begin{bmatrix} 1 \\ 2 \\ -3 \end{bmatrix}$

33. $A = \begin{bmatrix} 1 & 2 \\ 2 & 0 \\ 3 & 5 \\ -4 & 6 \end{bmatrix}; B = \begin{bmatrix} 2 & 1 & 0 \\ -1 & 1 & 3 \end{bmatrix}$

34. $A = \begin{bmatrix} 1 & 2 \\ 2 & 0 \\ 3 & 5 \\ 4 & 6 \end{bmatrix}; B = \begin{bmatrix} 1 & 2 \\ 1 & 3 \\ 0 & 4 \\ 1 & -1 \end{bmatrix}$

35. For the matrices

$$A = \begin{bmatrix} 1 & -1 \\ 2 & 0 \\ 3 & 1 \end{bmatrix}, B = \begin{bmatrix} 3 & 1 \\ 2 & -4 \end{bmatrix}, C = \begin{bmatrix} -5 & -2 \\ 1 & 6 \end{bmatrix},$$

verify each of the following.

(a) $(AB)C = A(BC)$
(b) $A(B + C) = AB + AC$

36. Let $A = \begin{bmatrix} 2 & -1 \\ 6 & -3 \end{bmatrix}, B = \begin{bmatrix} 1 & 4 \\ 2 & 8 \end{bmatrix}$, and $C = \begin{bmatrix} 2 & 3 \\ 4 & 6 \end{bmatrix}.$

(a) Show that $A \neq 0$ and $B \neq 0$ but $AB = 0$.
(b) Show that $AB = AC$, but $B \neq C$.

37. Let $f(x) = ax + b$ and $g(x) = cx + d$. Also, let $\begin{bmatrix} a & b \\ 0 & 1 \end{bmatrix}$ correspond to f and $\begin{bmatrix} c & b \\ 0 & 1 \end{bmatrix}$ correspond to g.

Show that $\begin{bmatrix} a & b \\ 0 & 1 \end{bmatrix}\begin{bmatrix} c & d \\ 0 & 1 \end{bmatrix}$ corresponds to $f \circ g$.

Inverse of Square Matrix

In Exercises 38-43, find A^{-1} or show that A is not invertible.

38. $A = \begin{bmatrix} 2/5 & -3/5 \\ -4 & 6 \end{bmatrix}$

39. $A = \begin{bmatrix} 1 & 2 \\ -1 & 3 \end{bmatrix}$

40. $A = \begin{bmatrix} 1 & 0 & -1 \\ 0 & 2 & 3 \\ 1 & 2 & 0 \end{bmatrix}$

41. $A = \begin{bmatrix} 1 & 0 & -1 \\ 2 & 2 & 1 \\ 0 & 2 & 3 \end{bmatrix}$

42. $A = \begin{bmatrix} 1 & -1 & 1 \\ 3 & 4 & 5 \end{bmatrix}$

43. $A = \begin{bmatrix} 2 & 1 & 0 \\ -1 & 1 & 3 \\ 3 & -1 & 5 \end{bmatrix}$

44. Let $f(x) = ax + b$ correspond to $\begin{bmatrix} a & b \\ 0 & 1 \end{bmatrix}$ as in Exercise 37. Assume that $a \neq 0$ and show that $f^{-1}(x)$ corresponds to $\begin{bmatrix} a & b \\ 0 & 1 \end{bmatrix}^{-1}$.

In Exercises 45-47, write the given system in the matrix form $AX = B$, and then solve, if possible, by finding $X = A^{-1}B$.

45. $2x - y = 4$
    $8x + y = 1$

46. $x + 2y = 0$
    $y + 3z = 1$
    $x - z = 3$

47. $3x + 6y + z = 0$
    $3x - 3y + 2z = 2$
    $6x + 9y + 2z = 1$

In Exercises 48-49, show that all parts can be put in the matrix form $AX = B$ with the same A but different B's. Find A^{-1} and then solve for $X = A^{-1}B$.

48. Henry Rich invests $10,000, some in a money market fund paying 3% per year and the rest in a mutual fund paying 5% per year. How much is each investment if his yearly income from both amounts to the following?
    (a) $400
    (b) $440
    (c) $460

49. A manufacturer produces two clocks, type I and type II. The cost of parts is $25 for type I and $40 for type II. The type I clock requires 5 hr of labor and type II requires 4 hr of labor. If 320 hr of labor are used, how many clocks of each type are produced if the amount the manufacturer spends on parts is as follows?
    (a) $1700
    (b) $2000
    (c) $2500

6.6 Remainder Theorem, Synthetic Division, and Factor Theorem

Introduction
Remainder Theorem
Synthetic Division
Factor Theorem
Complex and Real Factored Forms

Introduction The traditional goal of algebra has been to solve equations, and we are concerned here with polynomial equations of degree n. That is, we wish to solve equations of the form

$$a_n x^n + a_{n-1} x^{n-1} + \cdots + a_1 x + a_0 = 0,$$

where the coefficients a_j ($j = 0, 1, 2, \ldots, n$) are real numbers and $a_n \neq 0$. So for we have learned how to solve linear equations

$$ax + b = 0$$

and quadratic equations

$$ax^2 + bx + c = 0.$$

It is also possible to solve cubic equations

$$ax^3 + bx^2 + cx + d = 0$$

and quartic equations

$$ax^4 + bx^3 + cx^2 + dx + e = 0$$

by algebraic methods. However, the procedures for solving a given cubic or quartic equation involve lengthy computations with complex numbers and therefore are not very practical. For polynomial equations of degree ≥ 5, the problem is more than one of practicality because of the following **insolvability theorem.**[1]

[1] *Niels Henrik Abel (1802-1829) proved the insolvability theorem for $n = 5$ around 1828. The result for $n > 5$ follows from the work of Evariste Galois (1811-1832).*

The general polynomial equation of degree ≥ 5 cannot be solved by algebraic methods.

More specifically, the insolvability theorem says that for a general polynomial $P(x)$ of degree ≥ 5, the roots of $P(x) = 0$ cannot be expressed in terms of sums, differences, products, quotients, and radicals involving the coefficients of $P(x)$. This "can't do" theorem is in sharp contrast to the following result, which is called the **fundamental theorem of algebra.**[2]

[2] *Carl Friedrich Gauss (1777-1855) proved the fundamental theorem for his Ph.D. dissertation in 1799.*

Every polynomial equation of degree n has n roots in the complex number system.

Hence, the fundamental theorem tells us that the roots are "out there," but the insolvability theorem says that we can't find them by algebra! As a sort of compromise to these two great opposing theorems, we are able to *approximate* roots of polynomial equations by numerical methods. In place of formulas, computers can be programmed to search for roots by a systematic trial-and-error procedure. Although answers obtained in this manner are not exact, they can be made to approximate exact roots to

Remainder Theorem. We saw in Section 1.1 that if a polynomial $P(x)$ of degree $n \geq 1$ is divided by a linear expression $x - a$, then

$$P(x) = (x - a)Q(x) + R, \quad \text{Division Formula} \quad (1)$$

where the quotient $Q(x)$ is a polynomial of degree $n - 1$, and the remainder R is a constant. For instance, if $2x^3 + 7x^2 + 8x + 9$ is divided by $x + 2$, then

$$2x^3 + 7x^2 + 8x + 9 = (x + 2)(2x^2 + 3x + 2) + 5. \quad (2)$$

Here $a = -2$ and $R = 5$. The division formula (1) holds for any real or complex value of a. If we let $x = a$ in (1), then the term $(x - a)Q(x)$ drops out, and the equation becomes

$$P(a) = R. \quad (3)$$

For instance, when $x = -2$, equation (2) becomes

$$2(-2)^3 + 7(-2)^2 + 8(-2) + 9 = 5. \quad (4)$$

Hence, by combining equations (1) and (4), we obtain the following result, which is called the **remainder theorem**.

If a polynomial $P(x)$ of degree ≥ 1 is divided by $x - a$, then the remainder R is equal to $P(a)$.

Example 1 Find the remainder when $3x^5 - 2x^4 + x^3 + 7x^2 - 5x + 4$ is divided by $x + 1$.

Solution: Here $a = -1$, and by the remainder theorem,

$$\begin{aligned} R &= P(-1) \\ &= 3(-1)^5 - 2(-1)^4 + (-1)^3 + 7(-1)^2 - 5(-1) + 4 \\ &= 10. \quad \blacksquare \end{aligned}$$

Example 2 Verify the remainder theorem for the polynomial $P(x) = x^3 - 2x^2 - 5x + 6$ by actually dividing $P(x)$ by $x - a$.

Solution:

$$
\begin{array}{r}
x^2 + (a-2)x + (a^2 - 2a - 5) \\
x-a \overline{\smash{\big)}\, x^3 - 2x^2 - 5x + 6 } \\
\underline{x^3 - ax^2 } \\
(a-2)x^2 - 5x + 6 \\
\underline{(a-2)x^2 - (a-2)ax } \\
\underbrace{-5x + 6 + (a-2)ax}_{} \\
(a^2 - 2a - 5)x + 6 \\
\underline{(a^2 - 2a - 5)x - a(a^2 - 2a - 5)} \\
\underbrace{6 + a(a^2 - 2a - 5)}_{} \\
a^3 - 2a^2 - 5a + 6 \\
\textit{Remainder}
\end{array}
$$

Hence, the quotient is $x^2 + (a-2)x + (a^2 - 2a - 5)$ and, as stated in the remainder theorem, $R = a^3 - 2a^2 - 5a + 6 = P(a)$. ∎

Synthetic Division. Synthetic division is a shortcut method to divide a polynomial $P(x)$ by a linear expression $x - a$. However, its importance from a computational point of view is that, whether we are computing by hand or by a calculator or computer, synthetic division enables us to rapidly compute values of $P(x)$ for given values of x. To explain, suppose we wish to determine $P(a)$. The remainder theorem tells us that $P(a)$ is the remainder when $P(x)$ is divided by $x - a$. Therefore, if we have quick way to divide $P(x)$ by a $x - a$, then we have a quick way to determine $P(a)$.

We illustrate the method of synthetic division by means of the following example, in which $P(x) = 3x^4 - 2x^3 - 10x^2 + 15$ is divided by $x - 2$. In the usual division process, we have the following.

$$
\begin{array}{r}
3x^3 + 4x^2 - 2x - 4 \text{Quotient}\\
\textit{Divisor} \quad x-2 \overline{\smash{\big)}\, 3x^4 - 2x^3 - 10x^2 + 0x + 15} \quad \text{Dividend}\\
\underline{3x^4 - 6x^3 } \\
4x^3 - 10x^2 \\
\underline{4x^3 - 8x^2 } \\
-2x^2 + 0x \\
\underline{-2x^2 + 4x } \\
-4x + 15 \\
\underline{-4x + 8} \\
7 \quad \text{Remainder}
\end{array}
$$

Only the second term above and the first term below each division line are essential to the process; the others are repetitions. We delete the repeated terms and move the essential ones upward as follows.

406   Chapter 6. Additional Topics

$$\begin{array}{r} 3x^3 + 4x^2 - 2x - 4 \\ x-2 \overline{\smash{\big)}\, 3x^4 - 2x^3 - 10x^2 + 0x + 15} \\ -6x^3 - 8x^2 + 4x + 8 \\ \hline 4x^3 - 2x^2 - 4x + 7 \end{array}$$

Next we drop all plus signs and all $x's$, which results in the following.

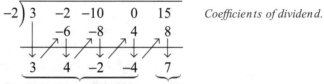

The coefficients of x in the divisor is always 1 and may be omitted. Also, the coefficients in the quotient line may be deleted since they are repeated in the remainder line, provided we include the first coefficient 3 as follows:

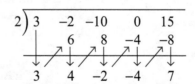

In the display above, the vertical arrows ↓ indicate subtraction, and the diagonal arrows ↗ indicate multiplication by -2. Finally, it is customary to replace the subtraction by addition by changing -2 in the divisor to 2 and letting the vertical arrows represent addition. Hence, for the **Synthetic division** of $3x^4 - 2x^3 - 10x^2 + 15$ by $x - 2$, we obtain

$$\begin{array}{r|rrrrr} 2 & 3 & -2 & -10 & 0 & 15 \\ & & 6 & 8 & -4 & -8 \\ \hline & 3 & 4 & -2 & -4 & 7 \end{array}$$

where ↓ means *addition* and ↗ means *multiplication* by 2 (in general, by a). Since $P(x) = 3x^4 - 2x^3 - 10x^2 + 15$, the result can be stated as

$$P(x) = (x - 2)(3x^3 + 4x^2 - 2x - 4) + 7,$$

and therefore $P(2) = 7$.

Example 3   Find $P(-3)$ for $P(x) = 3x^4 - 2x^3 - 10x^2 + 15$.

Solution: We use synthetic division to divide $P(x)$ by $x - (-3)$. We get

6.6. Remainder Theorem, Synthetic Division, and Factor Theorem

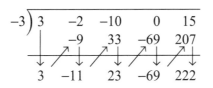

Each ↓ means addition, and each ↗ means multiplication by −3.

and conclude that $P(-3) = 222$. ∎

To compute $P(a)$ directly, we must compute *powers* of a, whereas in synthetic division, we compute simple *multiples* of a. Powers, especially high powers of large numbers, take a longer time to compute. Another method for computing $P(a)$ quickly, called **Horner's Method**, is covered in the exercises.

Factor Theorem. From equation (1), we see that if $R = 0$, then $x - a$ is a *factor* of $P(x)$. But $R = P(a)$ by the remainder theorem, and therefore the condition $R = 0$ is equivalent to $P(a) = 0$, or a is a *root* of $P(x) = 0$. Hence, we obtain the following result, which is called the **factor theorem**.[1]

[1] *The factor theorem appeared in the work* La Géométrie *by René Descartes in 1637.*

> If a is a root of $P(x) = 0$,
>
> then $x - a$ is a factor of $P(x)$.

Of course, the converse is also true, that is, if $x-a$ is a factor of $P(x)$, then a is a root of $P(x) = 0$.

For instance, we saw in Chapter 3 that if a_1 and a_2 are the roots of a quadratic, so equation $ax^2 + bx + c = 0$. then

$$ax^2 + bx + c = a(x - a_1)(x - a_2). \qquad (4)$$

Here, a_1 and a_2 can be fractions, irrational numbers, or even complex conjugates, so equation (4) goes beyond the method of factoring covered in Chapter 1.

Example 4 Factor $x^2 - 2x + 5$.

Solution: By the quadratic formula, the roots of $x^2 - 2x + 5 = 0$ are $1 + 2i$ and $1 - 2i$. Therefore, $[x - (1 + 2i)]$ and $[x - (1 - 2i)]$ are factors of $x^2 - 2x + 5$. That is,

$$x^2 - 2x + 5 = [x - (1 + 2i)][x - (1 - 2i)]. \quad \blacksquare$$

Example 5 Factor $P(x) = 8x^3 - 12x^2 - 10x + 15$, given that $3/2$ is a root of $P(x) = 0$.

Solution: Since $3/2$ is a root of $P(x) = 0$, $x - 3/2$ is a factor of $P(x)$. Now if $8x^3 - 12x^2 - 10x + 15$ is divided by $x - 3/2$, we obtain the quotient $8x^2 - 10$. That is,

$$8x^3 - 12x^2 - 10x + 15 = \left(x - \frac{3}{2}\right)(8x^2 - 10).$$

The roots of $8x^2 - 10 = 0$ are $\sqrt{5}/2$ and $-\sqrt{5}/2$. Therefore,

$$8x^2 - 10 = 8\left(x - \frac{\sqrt{5}}{2}\right)\left(x + \frac{\sqrt{5}}{2}\right),$$

and

$$8x^3 - 12x^2 - 10x + 15 = \left(x - \frac{3}{2}\right) 8\left(x - \frac{\sqrt{5}}{2}\right)\left(x + \frac{\sqrt{5}}{2}\right)$$

$$= 8\left(x - \frac{3}{2}\right)\left(x - \frac{\sqrt{5}}{2}\right)\left(x + \frac{\sqrt{5}}{2}\right). \blacksquare$$

Complex and Real Factored Forms. Let

$$P(x) = a_n x^n + a_{n-1} x^{n-1} + \cdots + a_1 x + a_0 \tag{5}$$

be a polynomial of degree n, where the coefficients $a_0, a_1, \ldots, a_{n-1}, a_n$ are real numbers and $a_n \neq 0$. By the fundamental theorem of algebra, $P(x)$ has n zeros $R_1, R_2, \ldots, R_n$ (not necessarily distinct), and by the factor theorem we can deduce (see Exercise 51) that

$$P(x) = a_n(x - R_1)(x - R_2) \ldots (x - R_n). \tag{6}$$

The roots of $P(x) = 0$ are considered in the complex number system, and it is understood that the **order** of each root is taken into account. For instance, the polynomial

$$P(x) = (x - 7)(x + 5)^2(x - 4)^3(x - i)(x + i)$$
$$= (x - 7)(x + 5)(x + 5)(x - 4)(x - 4)(x - 4)(x - i)(x + i)$$

is of degree 8, and the eight roots of $P(x) = 0$ are

$$7, -5, -5, 4, 4, 4, i, -i.$$

That is, 7 is a root of order 1, as are i and $-i$, -5 is a root of order 2, and 4 is a root of order 3.

The factored form of $P(x)$ given by (6) tells us that the coefficients of $P(x)$ are determined, up to a constant multiple, by its zeros. That is, if

$$Q(x) = b_n x^n + b_{n-1} x^{n-1} + \cdots + b_1 x + b_0$$

is another polynomial that also has $R_1, R_2, \ldots R_n$ for its zeros, then

$$Q(x) = b_n(x - R_1)(x - R_2) \ldots (x - R_n),$$

which means that
$$Q(x) = \frac{b_n}{a_n} P(x).$$

Hence, the coefficients b_k of $Q(x)$ are related to the a_k of $P(x)$ by
$$b_k = ca_k, \quad k = 0, 1, 2, \ldots, n,$$

where c is the constant b_n/a_n.

Example 6 Find all polynomials whose zeros are $1, 2, 5, 5$.

Solution: Any polynomial $P(x)$ with the given zeros has the factored form
$$P(x) = c(x-1)(x-2)(x-5)^2,$$

where c is a nonzero constant. Therefore,
$$\begin{aligned} P(x) &= c(x^2 - 3x + 2)(x^2 - 10x + 25) \\ &= c(x^4 - 13x^3 + 57x^2 - 95x + 50). \end{aligned}$$ ∎

We now derive two more specific versions of (6). First, we know from Chapter 3 that if a complex number $z = \alpha + \beta i$ is a root of a quadratic equation $ax^2 + bx + c = 0$, where a, b and c are real numbers, then so is the **complex conjugate** $\bar{z} = \alpha - \beta i$. We obtained this result from the quadratic formula, but we could also obtain it as follows.

Let $P(z) = az^2 + bz + c$, where a, b and c are real numbers. Then

$$\begin{aligned} P(\bar{z}) &= a\bar{z}^2 + b\bar{z} + c \\ &= a\overline{z^2} + b\bar{z} + c &&\overline{z}^2 = \overline{z^2} \text{ because the product of complex} \\ &&&\text{conjugates equals the conjugate of the} \\ &&&\text{product (see Exercises 62).} \\ &= \bar{a}\overline{z^2} + \bar{b}\bar{z} + \bar{c} &&\text{The conjugate of a real number is equal to itself.} \\ &= \overline{az^2} + \overline{bz} + \bar{c} &&\text{The conjugate of a product equals the product} \\ &&&\text{of the conjugates.} \\ &= \overline{az^2 + bz + c} &&\text{The conjugate of a sum equals the sum of the} \\ &&&\text{conjugates (see Exercise 61).} \\ &= \overline{P(z)}. \end{aligned}$$

Therefore, if z is a root of $P(x) = 0$, then
$$P(\bar{z}) = \overline{P(z)} = \bar{0} = 0,$$

which means that $\bar{z}$ is also a root.

The above method can be applied to a polynomial $P(x)$ of any degree, provided $P(x)$ has real coefficients. That is, if $z = \alpha + \beta i$ is a root of $P(x) = 0$, where $P(x)$ is any polynomial with real coefficients, then $\bar{z} = \alpha - \beta i$ is also a root.[2]

[2] This result was proved by Sir Isaac Newton in his work Arithmetica Universalis, which was published in 1707.

Now suppose that the n roots of $P(x) = 0$ separate into k distinct real roots $r_1, r_2, \ldots, r_k$ of order $m_1, m_2, \ldots, m_k$, respectively, and j distinct pairs of complex conjugate roots $c_1, \bar{c}_1, c_2, \bar{c}_2, \ldots, c_j, \bar{c}_j$ of orders $n_1, n_2, \ldots, n_j$, respectively. Then the factoring (6) of $P(x)$ becomes

$$P(x) = a_n(x - r_1)^{m_1}(x - r_2)^{m_2} \ldots (x - r_k)^{m_k}(x - c_1)^{n_1}(x - \bar{c}_1)^{n_1} \cdot \\ (x - c_2)^{n_2}(x - \bar{c}_2)^{n_2} \ldots (x - c_j)^{n_j}(x - \bar{c}_j)^{n_j}, \qquad (7)$$

which is called the **complex factored form** of $P(x)$. The expression in (7) looks complicated, but it is just equation (6) with a little more detail. Now for any complex number $c = \alpha + \beta i$, we have

$$[x - (\alpha + \beta i)]^n [x - (\alpha - \beta i)]^n = \{[x - (\alpha + \beta i)][x - (\alpha - \beta i)]\}^n \\ = [x^2 - 2\alpha x + (\alpha^2 + \beta^2)]^n \\ = (x^2 + Ax + B)^n,$$

where $A = -2\alpha$ and $B = \alpha^2 + \beta^2$ are real numbers. Therefore, (7) can be written as

$$P(x) = a_n(x - r_1)^{m_1}(x - r_2)^{m_2} \ldots (x - r_k)^{m_k}(x^2 + A_1 x + B_1)^{n_1} \cdot \\ (x^2 + A_2 x + B_2)^{n_2} \ldots (x^2 + A_j x + B_j)^{n_j}, \qquad (8)$$

which is called the **real factored form** of $P(x)$. In this form, $P(x)$ factors into powers of linear and quadratic factors with real coefficients. The quadratic factors are **irreducible** because they cannot be factored in the real number system. Note that the coefficient a_n appears as a constant factor in both the complex and the real factored forms.

Example 7 Find the complex and real factored forms of a polynomial $P(x)$ whose real zeros are 2 and -3 of orders 1 and 2, respectively, and whose complex zeros are $3 \pm 4i$ and $1 \pm 2i$ of order 1 and 3, respectively. What is the degree of $P(x)$?

Solution: The complex factored form of $P(x)$ is

$$P(x) = c(x - 2)(x + 3)^2[x - (3 + 4i)][x - (3 - 4i)] \cdot \\ [x - (1 + 2i)]^3[x - (1 - 2i)]^3,$$

where c can be any nonzero real number. Now

$$\text{and} \quad \begin{aligned} [x - (3 + 4i)][x - (3 - 4i)] &= x^2 - 6x + 25, \\ [x - (1 + 2i)][x - (1 - 2i)] &= x^2 - 2x + 5. \end{aligned}$$

Therefore, the real factored form of $P(x)$ is

$$P(x) = c(x-2)(x+3)^2(x^2 - 6x + 25)(x^2 - 2x + 5)^3.$$

The degree of $P(x)$ is obtained by adding the degrees of its factors in either form. From the real factored form we get

$$\text{degree } P(x) = 1 + 2 + 2 + 3 \cdot 2 = 11. \quad \blacksquare$$

Question *Can we find the real or complex form for any polynomial P(x)?*

Answer: Here we again see the opposition between the fundamental theorem of algebra and the insolvability theorem mentioned at the beginning of this section. The existence of the real and complex factored forms of $P(x)$ follows directly from the fundamental theorem of algebra, but to actually perform the factorization, we must know the roots of $P(x) = 0$, and the insolvability theorem says that we cannot find the roots by algebra for general polynomials of degree ≥ 5. Hence, we cannot write down the factored forms for a general polynomial $P(x)$ of degree ≥ 5 even though we know they exist!

We can state the situation in other words as follows: the fundamental theorem, by way of the factored forms, tells us that the n roots of a polynomial equation $P(x) = 0$ of degree n determine its coefficients, up to a constant multiple. Conversely, we also know that the coefficients of $P(x) = 0$ somehow determine its roots. But the insolvability theorem tells us that, for $n \geq 5$, the "somehow" goes beyond addition, subtraction, multiplication, division, and radicals. $\blacksquare$

Exercises 6.6

Fill in the blanks to make each statement true.

1. If 5 is a root of the polynomial equation $P(x) = 0$, then _____ is a factor of $P(x)$.
2. If $x + 3$ is a factor of $P(x)$, then _____ is a root of $P(x) = 0$.
3. The roots of $64x^3 +$ _____ $x^2 +$ _____ $x +$ _____ $= 0$ are $1/2$, $1/4$, and $1/8$.
4. The polynomial $x^2 - 2x + 2$ factors into _____.
5. The notation $4\overline{)6 \; -5 \; 1 \; 0 \; 3}$ in synthetic division indicates that the polynomial _____ is to be divided by _____.

Write true or false for each statement.

6. If a is a root of a polynomial equation $P(x) = 0$, then $x + a$ is a factor of $P(x)$.
7. If $x + a$ is a factor of a polynomial $P(x)$, then $-a$ is a root of $P(x) = 0$.
8. The roots of $x^7 - 4x^6 - 14x^5 + 56x^4 + 49x^3 - 196x^2 - 36x + 144 = 0$ are $1, -1, 2, -2, 3, -3, 4$ and -4.
9. We can find the factors of any polynomial with real coefficients by algebraic methods.
10. Synthetic division provides us with a rapid method for evaluating polynomials.

Remainder Theorem

In Exercises 11-16, determine the remainder when the given polynomial is divided by
(a) $x - 1$ and (b) $x + 2$.

11. $2x^3 - 4x^2 + 3x + 9$
12. $5x^3 - 7x^2 - 10x + 8$
13. $2x^4 + x^3 - 7x + 4$
14. $x^4 - 2x^3 + x^2 - 4$
15. $x^5 - 3x^4 - 5x^3 + 15x^2 + 4x - 12$
16. $2x^5 - x^4 + 3x^2 - x + 5$

As in Example 2, verify the remainder theorem in Exercises 17-20 by actually dividing $P(x)$ by $x - a$.

17. $P(x) = x^2 + 2x - 3$
18. $P(x) = 2x^2 - 3x + 4$
19. $P(x) = 2x^3 + x^2 - 4x + 5$
20. $P(x) = x^3 - 6x^2 + 3x - 2$

Synthetic Division

In Exercises 21-24, Perform the indicated division by the method of synthetic division. State the quotient and the remainder.

21. $x-2 \overline{)x^3 - 3x^2 + 2x - 5}$
22. $x-1 \overline{)x^3 + x^2 + x + 1}$
23. $x-1 \overline{)2x^5 + 5x^3 - 3x + 4}$
24. $x+2 \overline{)3x^5 - 4x^4 + 2x^2 - 1}$

Use synthetic division to compute $P(5)$ and $P(-5)$ in each of the following.

25. $P(x) = x^4 + 2x^3 - x + 4$
26. $P(x) = 4x^5 + 2x^4 - 3x^3 + 1$
27. $P(x) = 6x^6 - 2x^5 + 4x^3 - x^2 + 3$
28. $P(x) = 5x^6 + 6x^5 - 3x^4 - x^3 + 7$

Use synthetic division to compute $P(3/2)$ and $P(-2.5)$ for the following.

29. $P(x) = x^3 + 3x^2 - 2x + 1$
30. $P(x) = x^3 - x^2 + x - 1$

Factor Theorem

Use the factor theorem to verify each of the following.

31. $x + 1$ is a factor of $P(x) = 4x^2 + 12x + 8$.
32. $x - 1$ is a factor of $P(x) = 4x^3 - 8x^2 + 3x + 1$.
33. $x - 2$ is a factor of
    $P(x) = 3x^5 - 8x^4 + 5x^3 - 7x^2 + 11x - 2$.
34. $x + 2$ is a factor of $P(x) = x^4 - x^3 - 2x^2 + 3x - 10$.
35. $x - \dfrac{1}{2}$ is factor of $P(x) = 2x^3 - x^2 - 6x + 3$.
36. $x - \sqrt{2}$ is a factor of $P(x) = x^4 + x^3 - x^2 - 2x - 2$.

Use the factor theorem in each of the following to find all factors $x - a$, where a is an integer from -3 to 3. State the order of each factor.

37. $x^3 - 7x + 6$
38. $x^3 - 3x + 2$
39. $x^3 + x^2 - 8x - 12$
40. $x^3 + 2x^2 - 5x - 6$

Complex and Real Factored Forms

In each of the following, give the complex and real factored forms of a polynomial $P(x)$ of lowest degree with real coefficients and the given zeros.

41. $a_1 = 2$, $a_2 = -2$, $a_3 = 1$
42. $a_1 = \dfrac{1}{2}$, $a_2 = 2$, $a_3 = \dfrac{1}{3}$
43. $a_1 = -1$, $a_2 = 1$, $a_3 = -2$, $a_4 = 2$
44. $a_1 = 2$, $a_2 = -1$, $a_3 = i$, $a_4 = -i$
45. $a_1 = -2$, $a_2 = 1+i$, $a_3 = 1-i$
46. $a_1 = 1$, $a_2 = -2$, $a_3 = 3$, $a_4 = -4$
47. $a_1 = 1$, $a_2 = -\dfrac{1}{2}$, $a_3 = \dfrac{1}{3}$, $a_4 = -\dfrac{1}{4}$
48. $a_1 = \dfrac{1}{\sqrt{2}}$, $a_2 = -\dfrac{1}{\sqrt{2}}$, $a_3 = \dfrac{1}{\sqrt{3}}$, $a_4 = -\dfrac{1}{\sqrt{3}}$

49. $a_1 = \dfrac{1}{3}$, $a_2 = \dfrac{1}{3}$, $a_3 = 3 - 2i$

50. $a_1 = \sqrt{2}$, $a_2 = -\sqrt{2}$, $a_3 = 2i$, $a_4 = 2 - i$

Miscellaneous

51. Let $P(x)$ be a polynomial of degree n with *distinct* zeros $a_1, a_2, \ldots, a_n$. Fill in the blanks for each of the following steps, which lead to the conclusion that $P(x) = c(x - a_1)(x - a_2)\ldots(x - a_n)$.

$P(x) = (x - a_1)Q_1(x)$ by the _____ theorem, where $Q_1(x)$ is a polynomial of degree _____.

$Q_1(x) = (x - a_2)Q_2(x)$, where $Q_2(x)$ is a polynomial of degree _____.

$\vdots$

$Q_{n-1}(x) = (x - a_n)Q_n(x)$, where $Q_n(x)$ is a polynomial of degree _____.

Therefore $Q_n(x)$ is a _____, and
$P(x) = c(x - a_1)(x - a_2)\ldots(x - a_n)$.

52. Given that n is a positive integer, show that:

(a) $(x - a)$ is a factor of $x^n - a^n$;

(b) $(x + a)$ is a factor of $x^n - a^n$, if n is even;

(c) $(x + a)$ is a factor of $x^n + a^n$, if n is odd.

53. Why is it true that every polynomial with real coefficients of degree n, where n is *odd*, has at least one real zero?

54. Let $P(x)$ and $Q(x)$ be polynomials.

(a) Show that if $P(x) = 0$ for all real x, then all the coefficients of $P(x)$ are zero.

(b) Show that if $P(x) = Q(x)$ for all real x, then $P(x)$ and $Q(x)$ have the same coefficients.

Horner's Method

The standard computation of
$P(x) = 2x^3 + 4x^2 + 5x + 3$ *for a given value of x requires six multiplications and three additions. However, by successively factoring out x, we can obtain*

$$P(x) = (2x^2 + 4x + 5)x + 3$$
$$= [(2x + 4)x + 5]x + 3.$$

We call the last expression **Horner's form** *of $P(x)$. In this form only three multiplications and three additions are needed to evaluate $P(x)$. For example,*

$P(11) = [(2 \times 11 + 4)11 + 5]11 + 3$	Horner's Form
$= [(22 + 4)11 + 5]11 + 3$	Multiplication
$= [26 \times 11 + 5]11 + 3$	Addition
$= [286 + 5]11 + 3$	Multiplication
$= 291 \times 11 + 3$	Addition
$= 3201 + 3$	Multiplication
$= 3204.$	Addition

This method of successively factoring out x in order to obtain Horner's form for rapid computation can be applied to any polynomial and is known as **Horner's method**.

Apply Horner's method in Exercises 55-58.

55. $P(x) = 2x^3 - 4x^2 + 5x + 7$, $x = 6$

56. $P(x) = 4x^3 + 3x^2 - 8x + 10$, $x = 5$

57. $P(x) = x^4 + x^3 + 12x^2 - 4x - 15$, $x = 7$

58. $P(x) = 2x^4 - 5x^3 - 15x^2 + 12x + 3$, $x = 8$

59. Use synthetic division to evaluate the polynomials in Exercises 55 and 57.

60. Use synthetic division to evaluate the polynomials in Exercises 56 and 58.

Complex Conjugates

61. Show that $\overline{(3 + 2i) + (4 + 5i)} = \overline{(3 + 2i)} + \overline{(4 + 5i)}$, and in general,
$\overline{(a + bi) + (c + di)} = \overline{(a + bi)} + \overline{(c + di)}$ (the conjugate of a sum equals the sum of the conjugates).

62. Show that $\overline{(3 + 2i)(4 + 5i)} = \overline{(3 + 2i)}\,\overline{(4 + 5i)}$, and in general, $\overline{(a + bi)(c + di)} = \overline{(a + bi)}\,\overline{(c + di)}$ (the conjugate of a product equals the product of the conjugates).

6.7 Rational Roots, Bounds, and Descartes' Rule of Signs

Rational Roots
Upper and Lower Bounds
Descartes's Rule of Signs

Rational Roots. In Example 5 of the previous section we were given that $3/2$ is a root of $8x^3 - 12x^2 - 10x + 15 = 0$. Note that 3 divides the constant term 15 and 2 divides the coefficient 8 of x^3. More generally, suppose that a rational number p/q is a root of the polynomial equation

$$a_3 x^3 + a_2 x^2 + a_1 x + a_0 = 0, \qquad (1)$$

where the coefficients a_3, a_2, a_1, and a_0 are *integers*, $a_3 \neq 0$, and $a_0 \neq 0$. As usual, we assume that p and q are integers with no common factors except ± 1, and of course, $q \neq 0$. Then (1) becomes

$$a_3 \left(\frac{p}{q}\right)^3 + a_2 \left(\frac{p}{q}\right)^2 + a_1 \left(\frac{p}{q}\right) + a_0 = 0. \qquad (2)$$

If we multiply both sides of (2) by q^3, we get

$$a_3 p^3 + a_2 p^2 q + a_1 p q^2 + a_0 q^3 = 0. \qquad (3)$$

Equation (3) is equivalent to

$$a_3 p^3 + a_2 p^2 q + a_1 p q^2 = -a_0 q^3. \qquad (4)$$

The left side of (4) is divisible by p, and therefore the right side must also be divisible by p. But, by assumption, p and q have no common factors. It can then be shown that p and q^3 have no common factors, and we may conclude that a_0 *is divisible by* p.

Similarly, equation (3) is equivalent to

$$a_3 p^3 = -a_2 p^2 q - a_1 p q^2 - a_0 q^3. \qquad (5)$$

The right side of (5) is divisible by q and therefore the left side must also be divisible by q. Since p and q have no common factors, it can be shown that p^3 and q have no common factors. Hence, we may conclude that a_3 *is divisible by* q.

The above procedure can be applied to any polynomial equation with integer coefficients and leads to the following general result, which is called the **rational root theorem**.[1]

---

[1] Descartes utilized the idea of the rational root theorem in his work LA Géométrie (1637).

6.7. Rational Roots, Bounds, and Descartes' Rule of Signs

> If the rational number p/q is a root of the equation
>
> $$a_n x^n + a_{n-1} x^{n-1} + \cdots + a_1 x + a_0 = 0,$$
>
> where the coefficients a_k are integers, $a_n \neq 0$, and $a_0 \neq 0$, then a_0 is divisible by p, and a_n is divisible by q.

Question 1 *In the rational root theorem, why do we assume that $a_0 \neq 0$?*

Answer: $a_0 = 0$ if and only if 0 is a root of $P(x) = 0$, but a_0 is not divisible by 0. However, if 0 is a root of order k of a polynomial equation $P(x) = 0$ then by the factor theorem, $P(x) = x^k Q(x)$, where the constant term of $Q(x)$ is not zero. We can then apply the rational root theorem to the polynomial $Q(x)$. ■

As illustrated in the following examples, we can use the rational root theorem to determine all the rational roots of a given polynomial equation with integral *or* rational coefficients. However, even small values of a_0 and a_n can generate many possible rational numbers p/q to be tested. Therefore, a rapid method for computing $P(p/q)$ is needed. We can use synthetic division or a calculator to help with the computations. For large values of a_0 and a_n that may have many factors, it is helpful to use a programmable calculator or a computer to test all the possibilities.

Example 1 Find all rational roots of the equation

$$15x^4 + 8x^3 + 6x^2 - 9x - 2 = 0.$$

Solution: Here, $a_0 = -2$ and $a_4 = 15$. The factors of -2 are ± 1 and ± 2, and those of 15 are ± 1, ± 3, ± 5, and ± 15. Therefore, by the rational root theorem, the *possible* rational roots are

$$\pm 1, \pm \frac{1}{3}, \pm \frac{1}{5}, \pm \frac{1}{15}, \pm 2, \pm \frac{2}{3}, \pm \frac{2}{5}, \pm \frac{2}{15}.$$

If each of these numbers is tested in the given equation, we find that

$$\frac{2}{3} \quad \text{and} \quad -\frac{1}{5}$$

are the only rational roots. ■

Example 2 Find all rational roots, including order, of the equation

$$x^4 - \frac{3}{2}x^3 + \frac{41}{16}x^2 - 3x + \frac{9}{8} = 0.$$

Solution: We first clear the fractions by multiplying both sides of the equation by 16. We get the equivalent equation

$$16x^4 - 24x^3 + 41x^2 - 48x + 18 = 0.$$

Here, $a_0 = 18$ has the positive factors 1, 2, 3, 6, 9, and 18, and $a_4 = 16$ has the positive factors 1, 2, 4, 8, and 16. The possible rational roots are $\pm$ the following.

$$1, 2, 3, 6, 9, 18, \frac{1}{2}, \frac{3}{2}, \frac{9}{2}, \frac{1}{4}, \frac{3}{4}, \frac{9}{4}, \frac{1}{8}, \frac{3}{8}, \frac{9}{8}, \frac{1}{16}, \frac{3}{16}, \frac{9}{16}.$$

Since there are 36 numbers to be tested, a programmable calculator would be helpful. We discover that 3/4 is the only rational root. Division by $x - 3/4$ gives

$$16x^4 - 24x^3 + 41x^2 - 48x + 18$$
$$= \left(x - \frac{3}{4}\right)(16x^3 - 12x^2 + 32x - 24).$$

We find that 3/4 is also a zero of the quotient $16x^3 - 12x^2 + 32x - 24$, and when the quotient is divided by $x - 3/4$, we get

$$16x^3 - 12x^2 + 32x - 24 = \left(x - \frac{3}{4}\right)(16x^2 + 32)$$
$$= 16\left(x - \frac{3}{4}\right)(x^2 + 2).$$

Therefore,

$$16x^4 - 24x^3 + 41x^2 - 48x + 18 = 16\left(x - \frac{3}{4}\right)^2 (x^2 + 2).$$

Hence, 3/4 is a root of order 2 of the original equation. Also, the quadratic $x^2 + 2$ is irreducible over the real number system (*why?*), and therefore the right side of the above equation is the real factored from of the polynomial on the left side. ∎

Caution The rational root theorem does not say that a polynomial with integer coefficients *must* have a rational zero. For example, $x^2 - 2 = 0$ has no rational roots.

6.7. Rational Roots, Bounds, and Descartes' Rule of Signs   417

Question 2 *What can be said about the rational zeros of a polynomial with integer coefficients if the leading coefficient a_n is 1?*

Answer: Since any rational root p/q must have the property that q divides a_n, it follows that p/q must be an integer if $a_n = 1$. ∎

Upper and Lower Bounds. We can sometimes reduce the number of rational numbers that must be tested in a given polynomial equation by obtaining upper and lower bounds for the real roots of the equation. First, recall that if a polynomial $P(x)$ of degree $n \geq 1$ is divided by $x - a$, then the division formula says that

$$P(x) = (x - a)Q(x) + R, \tag{6}$$

where $Q(x)$ is a polynomial of degree $n - 1$ and R is a real number. Suppose R and the coefficients of $Q(x)$ are ≥ 0. Then, if $b > a$, we must have $P(b) > 0$ (*why?*). Hence, all real roots of $P(x) = 0$ are less than or equal to a, in which case we call a an **upper bound** for the real roots of $P(x) = 0$. We now have the following **upper bound test**.

> If a polynomial $P(x)$ of degree ≥ 1 is divided by $x - a$ and the remainder R as well as the coefficients of the quotient $Q(x)$ are ≥ 0, then a is an upper bound for the real roots of $P(x) = 0$.

Caution We apply the upper bound test to polynomials $P(x)$ whose leading coefficient a_n is *positive*. Otherwise, the leading coefficient of $Q(x)$ in equation (6) would always be negative. Hence, if a_n is negative, first replace $P(x)$ by $-P(x)$ and then apply the upper bound test to $-P(x)$. The roots of $-P(x) = 0$ are the same as those of $P(x) = 0$.

We can use synthetic division to divide $P(x)$ by $x - a$ for $a = 1, 2, 3, \ldots$ until we reach a positive integer a that gives a nonnegative R and nonnegative coefficients for $Q(x)$. Such an a is an upper bound for the real roots of $P(x) = 0$. If $R = 0$, then a also a root of $P(x) = 0$. If $R > 0$, then all real roots of $P(x) = 0$ are strictly less than a.

Example 3 Use the upper bound test to find an upper bound for the real roots of the polynomial equation $16x^4 - 24x^3 + 41x^2 - 48x + 18 = 0$ in Example 2.

Solution: We first use synthetic division to divide the given polynomial by $x - 1$.

$$\begin{array}{r|rrrrr} 1 & 16 & -24 & 41 & -48 & 18 \\ & & 16 & & & \\ \hline & 16 & -8 & & & \end{array}$$

Because the negative term -8 appears in the quotient, we stop dividing and proceed to $x - 2$.

$$\begin{array}{r|rrrrr} 2 & 16 & -24 & 41 & -48 & 18 \\ & & 32 & 16 & 114 & 132 \\ \hline & 16 & 8 & 57 & 66 & 150 \end{array}$$

Since the quotient's coefficients and the remainder are nonnegative, we conclude that 2 is an upper bound. ■

If we had performed the upper bound test before doing Example 2, we could have eliminated the rational numbers 2, 3, 6, 9, 18, 9/2, and 9/4 from consideration (*why is 2 eliminated?*)

A number $-a$ is called a **lower bound** for the real roots of $P(x) = 0$ if all the real roots are greater than or equal to $-a$. The upper bound test can also be used to find a lower bound. We first replace $P(x)$ by $P(-x)$. Now r is a positive root of $P(-x) = 0$ if and only if $-r$ is a negative root of $P(x) = 0$. We therefore obtain the following **lower bound test**.

If a is a positive upper bound for the real roots of $P(-x) = 0$, then $-a$ is a negative lower bound for the real roots of $P(x) = 0$.

Example 4 Use the lower bound test to find a lower bound for the real roots of the polynomial equation $16x^4 - 24x^3 + 41x^2 - 48x + 18 = 0$ in Example 2.

Solution: We first replace $P(x) = 16x^4 - 24x^3 + 41x^2 - 48x + 18$ by $P(-x) = 16x^4 + 24x^3 + 41x^2 + 48x + 18$. We then use synthetic division to divide $P(-x)$ by $x - 1$.

$$\begin{array}{r|rrrrr} 1 & 16 & 24 & 41 & 48 & 18 \\ & & 16 & 40 & 81 & 129 \\ \hline & 16 & 40 & 81 & 129 & 147 \end{array}$$

Since the quotient's coefficients and the remainder are nonnegative, we can conclude that 1 is an upper bound for the real roots of $P(-x) = 0$. Hence, -1 is lower bound for the real roots of $P(x) = 0$. Also, since $R = 147 \neq 0$, -1 is not of $P(x) = 0$ and all of the real roots are strictly greater than -1. ■

The upper bound test enabled us to eliminate the rational numbers 2, 3, 9, 18, 9/2, and 9/4 from consideration as roots of $P(x) = 0$ in Example 2. The lower bound test now allows us to eliminate also the rational numbers $-1, -2, -3, -6, -9, -18, -3/2, -9/2, -9/4$, and $-9/8$. Hence, of the original 36 possible rational roots, we can eliminate 17 of them by our upper and lower bound tests. Furthermore, all of the real roots are contained in the interval $(-1, 2)$.

Descartes' Rule of Signs. If all the coefficients of a polynomial equation are positive (negative), then the equation cannot have a positive root (*why?*). Hence, if such an equation does have a positive root, then there must be at least one sign change among the coefficients. The following rule, which is known as **Descartes' rule of signs**,[2] gives us more precise information concerning the positive roots. We omit the proof.

[2] *The rule of signs appeared in Descartes' work La Géométrie (1637).*

> The number of positive roots of the polynomial equation
>
> $$a_n x^n + a_{n-1} x^{n-1} + \cdots + a_1 x + a_0 = 0 \quad (a_n \neq 0)$$
>
> is at most equal to the number of sign changes in the sequence of coefficients
>
> $$a_n, a_{n-1}, \ldots, a_1, a_0.$$

For instance, in the polynomial equation

$$3x^4 - 2x^3 - x^2 + 5x + 4 = 0,$$

the coefficients

$$3, -2, -1, 5, 4$$

have two sign changes, namely from 3 to -2 and from -1 to 5. Hence, according to Descartes' rule of signs, the equation can have at most two positive roots.

In counting the number of sign changes for a polynomial, we adopt the convention that a zero coefficient has the same sign as the coefficient of its preceding higher-power term. For example, the polynomial

$$P(x) = x^5 - 3x^2 + 2$$

has coefficients

$$1, 0, 0, -3, 0, 2$$

whose signs are

$$+, +, +, -, -, +,$$

and there are two sign changes.

Question 3 *Does Descartes' rule of signs take into account the order of a positive root?*

Answer: Yes. A positive root of order k counts as k roots in Descartes' rule of signs. For example, $P(x) = (x-1)^3 = x^3 - 3x^2 + 3x - 1$ has three sign changes, and the root 1 of order 3 counts as three positive roots. Hence, Descartes' rule says that the number of positive roots, *including order*, is at most equal to the number of sign changes in the coefficients. ■

We can also use Descartes' rule of signs to determine the maximum number of negative roots of a polynomial. We use the property that $-r$ is a negative root of $P(x) = 0$ if and only if r is a positive root of $P(-x) = 0$. Hence, the number of negative roots of $P(x) = 0$ is equal to the number of positive roots $P(-x) = 0$. Therefore, we can state the following version of Descartes' rule of signs.

The number of negative roots of a polynomial equation $P(x) = 0$ is at most equal to the number of sign changes in the coefficients of $P(-x)$.

Finally, by using descartes' rule of signs for both the positive and the negative roots, and by testing 0 separately, we can determine the maximum number of *real roots*, including order, of a polynomial equation.

If $P(x)$ has degree n, the number of real roots (counting multiplicity) + the number of complex roots (always an even number because they come in conjugate pairs) = n. Hence, if m = the maximum number of real roots, then $n - m$ = the minimum number of complex roots.

Example 5 Given $P(x) = x^5 - 3x^2 + 2$, find each of the following.

(a) the maximum number of real roots of $P(x) = 0$
(b) upper and lower bounds for the real roots of $P(x) = 0$
(c) all rational roots of $P(x) = 0$

Solution:

(a) Since $P(x)$ has two sign changes, there are at most two positive roots. Now

$$P(-x) = (-x)^5 - 3(-x)^2 + 2$$
$$= -x^5 - 3x^2 + 2.$$

Hence, $P(-x)$ has one sign change, so there is at most one negative root. Finally, $P(0) = 2$, so 0 is not a root. Therefore, $P(x) = 0$ has

6.7. Rational Roots, Bounds, and Descartes' Rule of Signs

at most three real roots, including order. (Since $P(x)$ is of degree 5, it follows that $P(x) = 0$ has at least two complex roots.)

(b) To find an upper bound for the real roots, we use synthetic division to divide $P(x)$ by $x - a$, for $a = 1, 2, 3, \ldots$. When we divide by $x - 1$, there are negative coefficients in the quotient, but when the divisor is $x - 2$, we get the following.

$$
\begin{array}{r|rrrrrr}
2) & 1 & 0 & 0 & -3 & 0 & 2 \\
 &   & 2 & 4 & 8 & 10 & 20 \\
\hline
 & 1 & 2 & 4 & 5 & 10 & 22
\end{array}
$$

Since the remainder and the coefficients of the quotient are all non-negative, 2 is an upper bound for the real roots of $P(x) = 0$. Also, since $R = 22 \neq 0$, 2 is not a root and all real roots are strictly less than 2.

To find a lower bound, we apply the upper bound test to $P(-x)$. However, $P(-x) = -x^5 - 3x^2 + 2$ has a negative leading coefficient, so we must replace $P(-x)$ by $-P(-x) = x^5 + 3x^2 - 2$ (see Caution following the statement of the upper bound test). When $x^5 + 3x^2 - 2$ is divided by $x - 1$, we get

$$
\begin{array}{r|rrrrrr}
1) & 1 & 0 & 0 & 3 & 0 & -2 \\
 &   & 1 & 1 & 1 & 4 & 4 \\
\hline
 & 1 & 1 & 1 & 4 & 4 & 2.
\end{array}
$$

Since the remainder and the coefficients of the quotient are all non-negative, -1 is a lower bound for the real roots of $P(x) = 0$. Also, since $R = 2 \neq 0$, -1 not a root and all real roots are strictly greater than -1. We have found that the real roots must be in the interval $(-1, 2)$.

(c) By the rational root theorem, the only possible rational roots are ± 1 and ± 2. However, -1 and ± 2 were eliminated in part (b). This leaves 1, which is in fact a root. ∎

Example 5 illustrates the general approach to polynomial equations. We use Descartes' rule of signs to determine the maximum number of real roots, the bound tests to find upper and lower bounds for the real roots, and the rational root theorem to find all rational roots. To implement the bound tests, we use synthetic division. We can also use synthetic division to implement the rational root theorem if there are a large number of rational numbers to be tested.

Exercises 6.7

Fill in the blanks to make each statement true.

1. According to the rational root theorem, the possible rational roots of $2x^2 + bx + 3 = 0$, where b is an integer, are _____.
2. If $x^3 + bx + c = 0$ has 5 as a root, where b and c are integers and $c \neq 0$, then the possible values of c are _____.
3. If $ax^4 + bx^3 + cx + 4 = 0$, where a, b, and c are integers, has 2/3 as a root, then the possible nonzero values of a are _____.
4. The polynomial $ax^4 + bx^3 - cx + d = 0$, where $a, b, c,$ and d are positive integers, has at most _____ positive root(s).
5. The polynomial $ax^3 - bx^2 + cx + d = 0$, where $a, b, c,$ and d are positive integers, has at most _____ negative root(s).

Write true or false for each statement.

6. If the coefficients of $P(x) = a_n x^n + a_{n-1} x^{n-1} + \cdots + a_1 x + a_0$ are integers and p/q is a rational number for which a_0 is divisible by p and a_n is divisible by q, then $P(p/q) = 0$.
7. The equation $7x^3 + bx^2 + cx - 11 = 0$, where b and c are integers, could have 7 as a root.
8. The equation $6x^3 + bx^2 + cx + 6 = 0$, where b and c are integers, has only ± 1 as its possible rational roots.
9. The equation $x^4 + x^3 + x^2 + x + 1 = 0$ has no rational roots.
10. The equation $x^6 + 2x^4 + 5x^2 + 3 = 0$ has no real roots.

Rational Roots

Find all rational roots of the following equations.

11. $x^3 - 2x^2 - x + 2 = 0$
12. $x^3 - 6x^2 + 11x - 6 = 0$
13. $x^4 - 5x^3 + 20x - 16 = 0$
14. $2x^3 - 7x^2 - 2x + 12 = 0$
15. $4x^4 + 7x^3 + 2x^2 + 7x - 2 = 0$
16. $x^4 + 2x^3 - x - 2 = 0$

For the polynomials in Exercises 17-24, perform each of the following steps.

(a) Find all rational roots of $P(x) = 0$.
(b) Use the factor theorem to find any remaining real roots and the real factored form of $P(x)$.
(c) Find all complex roots of $P(x) = 0$.

17. $P(x) = x^3 - x^2 - 4x + 4$
18. $P(x) = x^3 - x^2 - 4x - 2$
19. $P(x) = 2x^3 - x^2 + 2x - 1$
20. $P(x) = 2x^3 + x^2 + 2x + 1$
21. $P(x) = x^4 + x^3 - 5x^2 - 3x + 6$
22. $P(x) = x^4 + 2x^3 - 25x^2 - 50x$
23. $P(x) = x^4 + 2x^3 + x^2 - 2x - 2$
24. $P(x) = x^4 + 2x^3 - 3x^2 - 2x + 2$

Upper and Lower Bounds

Find integers a and b that are lower and upper bounds, respectively, for the real roots of each of the following polynomial equations.

25. $2x^3 - 4x^2 + 5x + 7 = 0$
26. $4x^3 + 3x^2 - 8x + 10 = 0$
27. $x^4 + 7x^3 + 12x^2 - 4x - 16 = 0$
28. $2x^4 - 5x^3 - 15x^2 + 12x + 3 = 0$

Descartes' Rule of Signs

Use Descartes' rule of signs to determine the maximum number of real roots of each of the following equations.

29. $2x^3 - x^2 - 2x - 1 = 0$
30. $x^4 - x^3 + x^2 - 4x - 5 = 0$
31. $5x^5 + 4x^3 - x + 2 = 0$
32. $x^4 - 2x^2 + 7 = 0$

33. $3x^6 + x^5 + 2x^3 + x = 0$
34. $2x^6 - 4x^4 + x^2 + x = 0$

For the polynomials in Exercises 35-40, perform each of the following steps.

(a) *Find the maximum number of positive and negative roots of* $P(x) = 0$.
(b) *Find integers a and b for which $[a, b]$ contains the real roots of* $P(x) = 0$.

35. $P(x) = x^3 + 3x^2 + 3x - 7$
36. $P(x) = x^3 - 3x^2 + 3x + 7$
37. $P(x) = x^3 - 3x^2 - 21x + 55$
38. $P(x) = x^3 + 9x^2 + 21x + 13$
39. $P(x) = x^4 + 8x^3 + 6x^2 - 4x - 2$
40. $P(x) = x^4 - 8x^3 + 30x^2 - 56x + 43$

Miscellaneous

41. Use the equation $x^2 - 2 = 0$ to conclude that $\sqrt{2}$ is an irrational number.
42. Use the equation $x^2 + 1 = 0$ to conclude that $\sqrt{-1}$ is not a real number.

Chapter 6 Review Outline

6.1 The Binomial Theorem

$$(a+b)^n = a^n + na^{n-1}b + \ldots + {}_nC_r a^{n-r} b^r + \ldots + b^n,$$

where

$$_nC_r = \frac{n!}{(n-r)!r!} = \frac{n(n-1)\ldots(n-r+1)}{r!}$$

(n a positive integer, $r = 0, 1, \ldots, n$)

Also, Pascal's triangle can be used to determine ${}_nC_r$.
Properties of Binomial Coefficients:

$$_nC_0 = 1 = {}_nC_n$$
$$_nC_r = {}_{n-1}C_r + {}_{n-1}C_{r-1}$$
$$_nC_{n-r} = {}_nC_r$$

6.2 Linear Inequalities in the Plane with Applications

A line L with equation $Ax + By = C$ $(A > 0)$ divides the plane into a right half-plane $\Re$ and a left half-plane $\mathcal{L}$, and

$Ax + By > C$ for all (x, y) in $\Re$
$Ax + By < C$ for all (x, y) in $\mathcal{L}$.

The solution of a system of linear inequalities is the intersection of their respective half-plane solutions. In linear programming, a point of maximum profit (minimum cost) occurs at a vertex of the region of solution.

6.3 Partial Fractions

Definition
A rational expression $P(x)/Q(x)$ is called a proper rational expression if the degree of $P(x)$ is less than the degree of $Q(x)$.

Partial Fraction Decomposition:
A proper rational expression $P(x)/Q(x)$ can be decomposed into partial fractions by using the following steps.

(1) Factor $Q(x)$ into its real factored form.
(2) For each linear factor of order $m, (x-r)^m$, there are m terms

$$\frac{A_j}{(x-r)^j} \quad (j = 1, 2, \ldots, m),$$

and for each irreducible quadratic factor of order n, $(x^2 + bx + c)^n$, there are n terms

$$\frac{B_j x + C_j}{(x^2 + bx + c)^j} \quad (j = 1, 2, \ldots, n).$$

(3) By setting $P(x)/Q(x)$ equal to its partial fraction decomposition and equating coefficients of like powers of x, a system of linear equations in the unknowns A_j, B_j, and C_j is determined. The system can be solved by the methods in the chapter on linear systems. If degree $P(x) \geq$ degree $Q(x)$, divide $Q(x)$ into $P(x)$ to obtain

$$\frac{P(x)}{Q(x)} = q(x) + \frac{R(x)}{Q(x)},$$

where $q(x)$ is a polynomial and $R(x)/Q(x)$ is a proper rational expression. Partial fraction decomposition can be applied to $R(x)/Q(x)$.

6.4 Linear Systems: Gaussian Elimination

Solving Linear Systems:

Three general elimination procedures for solving a linear system are as follows.

(1) Reduce the given system to a triangular system of equations by means of elementary (row) operations; then back-substitute to obtain the solution.
(2) (Gaussian elimination) Reduce the augmented matrix of the system to row-echelon form by means of elementary row operations. The corresponding linear system is now in triangular form and can be solved by back-substitution.
(3) (Gauss-Jordan elimination) Further reduce the row-echelon form of the augmented matrix to reduced row-echelon form, which represents the simplest form of the original system.

6.5 Matrix Algebra

Definitions

A matrix A with m rows and n columns is called an $m \times n$ matrix.
Two matrices are equal if their entries are identical.
If A is $m \times n$ and B is $m \times n$, then

$$A \pm B = C,$$

where

$$c_{ij} = a_{ij} \pm b_{ij} \quad (i = 1, \ldots, m; j = 1, \ldots, n).$$

If A is $m \times n$ and B is $n \times p$, then

$$AB = C,$$

where

$$c_{ij} = a_{i1}b_{1j} + a_{i2}b_{2j} + \cdots + a_{in}b_{nj}$$
$$(i = 1, 2, \ldots, m; j = 1, 2, \ldots, p).$$

The $n \times n$ multiplicative identity matrix I has 1's on its main diagonal and 0's elsewhere.
A square matrix A of order n is invertible if there is a matrix B of order n satisfying $AB = BA = I$. If such a B exists, it is denoted by A^{-1}.

Properties

A square matrix A is invertible $\Leftrightarrow$ the matrix $[A|I]$ can be row-reduced to $[I|B]$, and then $B = A^{-1}$.
A linear system of n equations in n unknowns can be written in matrix form as

$$AX = B,$$

where A is the $n \times n$ coefficient matrix, X is the $n \times 1$ matrix of unknowns, and B is the $n \times 1$ matrix of constants. If A is invertible, then the solution is

$$X = A^{-1}B.$$

6.6 Remainder Theorem, Synthetic Division, and Factor Theorem

Insolvability Theorem:

The general polynomial equation of degree ≥ 5 cannot be solved by algebraic methods (Abel, Galois).

Fundamental Theorem of Algebra:

Every polynomial equation of degree n has n roots in the complex number system (Gauss).

Remainder Theorem:
If a polynomial $P(x)$ of degree ≥ 1 is divided by $x - a$, the remainder is $P(a)$.
Synthetic division provides a rapid method to evaluate polynomials.

Factor Theorem:
If a is a root of the polynomial equation $P(x) = 0$, then $x - a$ is a factor of $P(x)$.

Complex Conjugate Roots:
If $P(x)$ has real coefficients and the complex number $z = \alpha + \beta i$ is also a root of $P(x) = 0$, then the complex conjugate $\bar{z} = \alpha - \beta i$ is also a root.

Complex and Real Factored Forms:
If the real roots of $P(x) = 0$ are $r_1, \ldots, r_k$, of orders $m_1, \ldots, m_k$, respectively, and the complex roots are $c_1, \bar{c}_1, \ldots, c_j, \bar{c}_j$ of orders $n_1, \ldots, n_j$, respectively, then

$$P(x) = a_n(x - r_1)^{m_1} \ldots (x - r_k)^{m_k}(x - c_1)^{n_1} \cdot (x - \bar{c}_1)^{n_1} \ldots (x - c_j)^{n_j}(x - \bar{c}_j)^{n_j}.$$

By combining the factors of complex conjugate roots into real irreducible quadratics $x^2 + Ax + B$,
$$P(x) = a_n(x - r_1)^{m_1} \ldots (x - r_k)^{m_k} \cdot (x^2 + A_1 x + B_1)^{n_1} \ldots (x^2 + A_j x + B_j)^{n_j}.$$

6.7 Rational Roots, Bounds, and Descartes' Rule of Signs

Rational Root Theorem:
If the rational number p/q is a root of
$a_n x^n + a_{n-1} x^{n-1} + \cdots + a_1 x + a_0 = 0$, where the a_k are integers, $a_n \neq 0$, and $a_0 \neq 0$, then p divides a_0 and q divides a_n.

Upper and Lower Bounds:
If the remainder and the coefficients of the quotient are nonnegative when $P(x)$ is divided by $x - a$, then a is an upper bound for the real roots of $P(x) = 0$.
If a is a positive upper bound for the real roots of $P(-x) = 0$, then $-a$ is a lower bound for the real roots of $P(x) = 0$.

Descartes' Rule of Signs:
The number of positive roots of the polynomial equation $P(x) = 0$ is at most equal to the number of sign changes in the coefficients of $P(x)$.
The number of negative roots of $P(x) = 0$ is at most equal to the number of sign changes in the coefficients of $P(-x)$.

Chapter 6 Review Exercises

1. Construct the first 6 rows of Pascal's triangle, and use the triangle to expand $(a + b)^n$ for $n = 2, 3, 4, 5$.

2. Evaluate:
   (a) $_7C_2$
   (b) $_5C_0$
   (c) $_{10}C_3$.

3. Verify that $_8C_4 = {_7C_4} + {_7C_3}$.

4. Expand the binomial theorem:
   (a) $(x + 2)^5$
   (b) $(x - 3)^4$.

5. Use the binomial theorem to expand $(x - y + 1)^4$. (*Hint*: let $a = x - y$.)

6. In the plane, graph the solution of the system $x \geq 3$, $y \geq 2$, $x + y \leq 10$.

7. In the plane, graph the solution of the system
$$y + 2x \leq 4$$
$$x + y < 3$$

8. For a rock concert, 20,000 tickets were sold for $560,000. If tickets cost either $25 or $40, how many were sold at each price?

9. The price of a car is $10,269.50 after a 5% discount on the sticker price. If the sticker price includes the dealer's cost plus a 15% markup on that cost, what is the dealer's cost?

10. Steve drove his car 165 mi. He drove that first 1.5 hr at one speed and the next 2.5 hr at 10 mph faster. Find his speeds for each part of the trip.

11. A college bookstore must order two books A and B, with more of A than B and a total of 60 or less. The order can include at most 40 of A. If A nets a $5 profit and B a $2 profit, how many of each should be ordered to obtain the maximum profit?

Decompose the rational expressions in Exercises 12-15 into partial fractions, and solve for the constants.

12. $\dfrac{6x^2 + x - 6}{x^3 + x^2 - 6x}$

13. $\dfrac{1}{x^4 - x^2}$

14. $\dfrac{2x^3 - x^2 + 2x - 7}{x^2 - x - 2}$

15. $\dfrac{x^5 + x^2 + 2x - 1}{x^3 + x}$

Apply Gaussian elimination to the systems in Exercises 16-19.

16. $\begin{aligned} x + y + z &= 3 \\ x - 2y + 2z &= 1 \\ 2x + 3y + z &= 4 \end{aligned}$

17. $\begin{aligned} x + y - 3z &= 1 \\ x - y + z &= 1 \\ 3x + y - 5z &= 3 \end{aligned}$

18. $\begin{aligned} x - y + z &= 5 \\ 2x + 3y + z &= 7 \\ 3x + 2y + 2z &= 4 \end{aligned}$

19. $\begin{aligned} x + 2y - z &= 0 \\ 2x - y + z &= 0 \\ 5x + 5y - 2z &= 0 \end{aligned}$

20. Solve the following system by Gauss-Jordan elimination.

$$\begin{aligned} x + y + z + t &= 4 \\ x - y - z &= 0 \\ y + 2z - t &= 0 \\ 2y + 3t &= 2 \end{aligned}$$

Explain why the indicated operations cannot be performed.

21. $\begin{bmatrix} 2 & 1 & 0 \\ 3 & 4 & 0 \end{bmatrix} + \begin{bmatrix} 3 & -2 \\ 5 & 4 \end{bmatrix}$

22. $\begin{bmatrix} 2 & 1 & 0 \\ 3 & 4 & 0 \end{bmatrix} \begin{bmatrix} 3 & -2 \\ 5 & 4 \end{bmatrix}$

23. $\begin{bmatrix} 3 & -2 \\ -6 & 4 \end{bmatrix}^{-1} \begin{bmatrix} 2 & 1 \\ 3 & 4 \end{bmatrix}$

24. $\begin{bmatrix} 2 & 1 & 3 \\ 4 & 6 & 5 \end{bmatrix}^{-1}$

Perform the indicated operations in Exercises 25-30

25. $\begin{bmatrix} 2 & -3 \\ 1 & 4 \end{bmatrix} + 3 \begin{bmatrix} 0 & 5 \\ 4 & 3 \end{bmatrix}$

26. $\begin{bmatrix} 2 & -3 \\ 1 & 4 \end{bmatrix} \begin{bmatrix} 0 \\ 4 \end{bmatrix}$

27. $\begin{bmatrix} 1 & 2 & -3 \\ 3 & -1 & 2 \\ 4 & 5 & 1 \end{bmatrix} \begin{bmatrix} 1 & 3 \\ 1 & -2 \\ 2 & 1 \end{bmatrix}$

28. $\begin{bmatrix} 2 & 8 \\ 1 & 4 \end{bmatrix} \begin{bmatrix} 4 & 8 \\ -1 & -2 \end{bmatrix}$

29. $\begin{bmatrix} 2 & -3 \\ 1 & 4 \end{bmatrix} \begin{bmatrix} 1 & -3 \\ -2 & 5 \end{bmatrix}$

30. $\begin{bmatrix} 2 & -3 \\ 1 & 4 \end{bmatrix} \begin{bmatrix} 2 & -3 \\ 1 & 4 \end{bmatrix}^{-1}$

31. Find the multiplicative inverse of

$$A = \begin{bmatrix} 1 & 1/2 \\ -2 & -3 \end{bmatrix}$$

by solving $AB = I$ for B.

32. Find A^{-1} for matrix A in Exercise 31 by row reduction of the matrix $[A|I]$.

33. Find the inverse of $\begin{bmatrix} 1 & -1 & 1 \\ 0 & 2 & -2 \\ 1 & 3 & -1 \end{bmatrix}$.

34. The equation of any plane in 3-space has the form $ax + by + cz = d$. Find an equation for the plane that passes through the points $(1, -1, 1), (2, 0, 2)$, and $(0, -2, 1)$.

35. Find the point of intersection of the planes

$$2x - 5y - 2z = 9$$
$$x - y + z = 2$$
$$x - 3y - 2z = 6.$$

36. Write the system

$$2x - y = 7$$
$$x + 3y = 0$$

in the matrix form $AX = B$, find A^{-1}, and use it to solve for $\begin{bmatrix} x \\ y \end{bmatrix}$.

37. Write the following system in the form $AX = B$. Then find A^{-1} and use it to solve for $\begin{bmatrix} x_1 \\ x_2 \\ x_3 \end{bmatrix}$.

$$x_1 + x_2 + 3x_3 = 2$$
$$-x_2 + x_3 = -4$$
$$x_1 + x_3 = 1$$

38. For any matrix A, its transpose A^t is obtained by inter-changing the rows and columns of A. For example,

if $A = \begin{bmatrix} a & b \\ c & d \end{bmatrix}$,

then $A^t = \begin{bmatrix} a & c \\ b & d \end{bmatrix}$.

Show that

if $B = \begin{bmatrix} a' & b' \\ c' & d' \end{bmatrix}$,

then $(AB)^t = B^t A^t$.

39. A company makes three kinds of dresses, and each type requires a certain amount of time for cutting, sewing, and finishing, as indicated in the following matrix.

	Type A	Type B	Type C
Cutting	15 min	10 min	20 min
Sewing	20 min	15 min	30 min
Finishing	30 min	20 min	20 min

If the company uses a total of 8 1/3, 11 2/3, and 15 hr labor for cutting, sewing, and finishing, respectively, how many dresses of each type are produced?

Follow the directions or fill in the blanks in Exercises 40-67, required.

40. Sate the fundamental theorem of algebra.
41. State the insolvability theorem.
42. Without dividing, find the remainder when $P(x) = x^4 - 3x^3 + 2x^2 - 5x + 1$ is divided by $x - 1$.
43. Without dividing, find the remainder when $P(x) = 2x^3 + 3x^2 - 4x + 4$ is divided by $x + 2$.
44. Use synthetic division to evaluate the polynomial in Exercise 42 at $x = 5$.
45. Use synthetic division to evaluate the polynomial in Exercise 43 at $x = -5$.
46. Without dividing, verify that $x - \sqrt{2}$ is a factor of $P(x) = x^5 + 3x^4 - 4x^3 - x^2 + 4x - 10$.
47. Without dividing, verify that $x - i$ is a factor of $P(x) = 3x^4 - 2x^3 + 4x^2 - 2x + 1$.
48. Find all factors $x - a$, where a is an integer, of the polynomial $P(x) = x^5 + x^4 - x^3 + x^2 - 2x$.
49. Find a polynomial $P(x)$ with real coefficients of lowest degree that has 1/2 and i among the roots of $P(x) = 0$.
50. Find the complex factored form of the polynomial $P(x)$, if $-2, 3, 3, \sqrt{2}, 1 + 2i$, and $1 - 2i$ are the roots of $P(x) = 0$.
51. Find the real factored form of the polynomial in Exercise 50.
52. The numbers $4, 5/2, -\sqrt{3}, 2 + i, 2 - i$, and $3i$ cannot be the roots of $P(x) = 0$ where $P(x)$ is a polynomial of degree 6 with real coefficients, because _____.
53. Find all rational roots of $2x^4 - x^3 - 5x^2 + 2x + 2 = 0$.
54. Show that the polynomial equation $x^5 - 5x^4 + 7x^3 - 2x^2 + 3x + 2 = 0$ has no rational roots.

55. Find upper and lower bounds for the real roots in Exercise 53.
56. Find upper and lower bounds for the real roots in Exercise 54.
57. Find all rational roots, including order, for $x^4 - 2x^3 + 2x^2 - 2x + 1 = 0$.
58. Find the real factored form of the polynomial in Exercise 57, (*Hint:* use the factor theorem.)
59. If $2x^3 + bx^2 + cx + d = 0$, where $b, c,$ and d are integers, has $5/2$ for a root, then the possible nonzero values for d are _____.
60. The possible rational roots of $x^3 + bx^2 + x = 0$, where b is a nonzero integer, are _____.
61. According to Descartes' rule of signs, the polynomial equation $3x^4 - 2x^3 + 6x^2 - 3x - 5 = 0$ can have at most _____ positive root(s).
62. By Descartes' rule of signs, the polynomial equation in Exercise 61 can have at most _____ negative roots(s).
63. Use Descartes' rule of signs to determine the maximum number of real roots of $x^6 + 2x^5 - 4x^4 - 7x^2 - 5x + 5 = 0$.
64. The equation $ax^5 + bx^3 + cx^2 - dx + e = 0$ where $a, b, c, d,$ and e are positive integers, can have at most _____ real root(s).
65. Use Descartes' rule of signs to show that the equation $3x^6 + 5x^4 + 2x^2 + 8 = 0$ has no real roots.
66. Find all *positive* rational roots of the polynomial equation $10x^6 + 3x^5 + 2x^4 + 3x^3 + x^2 + 2x + 2 = 0$.
67. Find all real roots of $x^4 - 3x^3 + 6x - 4 = 0$. (*Hint:* first find all rational roots and then apply the factor theorem.)

Appendices

Appendix A: Significant Digits

When performing computations, we often work with approximate values of numbers. This may be due to the numbers themselves or to the way in which the values are obtained. For example, to express $1/3$ or $\sqrt{2}$ as a decimal, we can use only a finite number of decimal places. Also, if values are obtained by measurements, then the numbers can be only as accurate as the measuring devices used. As a general rule,

> any number computed by means of approximate values is only as accurate as the least accurate number used.

For example, suppose we measure the diameter of a circle as 11.5 centimeters using a tape measure graded in tenths of a centimeter. If we compute the circumference C of the circle on an 8-place calculator that has a key for π, we get

$$C = \pi d = 36.128316 \text{ centimeters.}$$

However, the measured value 11.5 of d has only three significant digits. Therefore, according to the above rule, our answer should be expressed with three significant digits. That is, we should take for our answer,

$$C = 36.1 \text{ centimeters.}$$

We now define precisely what is meant by "significant digit" as well as the rules for performing computations with approximate values. First note that numbers in a table are usually rounded off according to the following convention. If, for example, 12.73 is an approximation of the number x to two decimal places, then it is understood that the difference between 12.73 and x is at most .005 in magnitude, That is,

$$12.725 \le x \le 12.735.$$

Similarly, an approximation 12.730 for x means that the difference is at most .0005 in magnitude, or

$$12.7295 \le x \le 12.7305.$$

If the whole number 1273 is an approximation to x, then we assume

$$1272.5 \le x \le 1273.5.$$

We call 3 the round-off digit in 12.73 and 1273, and 0 the round-off digit in 12.730. In general, the last digit of any decimal whatsoever or of any whole number not ending in 0 is by definition the **round-off digit**. If the last digit of a whole number is 0, then the round-off digit must be indicated in some manner. Here are several examples in which the round-off digit is indicated by underlining in the first column and by scientific notation in the third column.

Approximate Value of x	True Value of x	Scientific Notation
2<u>0</u>	$15 \leq x \leq 25$	2×10
2<u>0</u>	$19.5 \leq x \leq 20.5$	2.0×10
<u>2</u>00	$150 \leq x \leq 250$	2×10^2
2<u>0</u>0	$195 \leq x \leq 205$	2.0×10^2
20<u>0</u>	$199.5 \leq x \leq 200.5$	2.00×10^2

The number of **significant digits** in a decimal or whole number is by definition the number of digits, counting from the first nonzero digit on the left to the round-off digit. We give several examples below.

Number	Number of Significant Digits
0.0215	3
4.0215	5
0.003	1
0.0030	2
405	3
40<u>5</u>0	3
40.50	4
4.2×10^6	2
4.20×10^{-6}	3

We note that every nonzero digit in a number is significant, and 0 is significant unless its only purpose is to fix the location of the decimal point.

If we wish to round off a given number x to fewer significant digits, we proceed as illustrated in the following examples.

x	Rounded–Off Approximation	Significant Digits
2.7148674	2.715	4
2.7148674	2.71	3
2.7148674	2.7	2
2.7148674	3	1

That is, the approximation differs from x by at most 1/2 unit in the round-off digit. When the last significant digit of x is 5 and we wish to round off to one less significant digit, then our convention is to round off to an *even* digit. Here are some examples.

x	**Rounded-Off Value**
.0315	.014
.0125	.012
2715	2720
2.715	2.72

Note that if we first round off 2.7148674 to four digits, we get 2.715, and if 2.715 is then rounded off to three digits, we get 2.72. However, if 2.7148674 is rounded off in one step to three digits, we get 2.71. Hence, successive round-offs are not necessarily equivalent to a single one.

Our convention for **addition** and **substraction** with approximate values depends more on the number of decimal places (digits after the decimal point) than on the number of significant digits of the numbers. The convention is as follows.

> Add or subtract the numbers as given, and round off
> the result to the decimal place of the number(s) with
> the fewest decimal places.

For examples, when adding approximate values 23.714, .0002, and 5.0129, we first obtain 28.7271, which is then rounded off to 28.727.

The number of significant digits is the important factor in our convention for **multiplication** and **division,** which is given below.

> Multiply or divide the numbers as given, and round
> off the result to have as many significant digits
> as the number(s) with the fewest significant digits.

For example, as indicated earlier, the product $\pi \cdot 11.5$, when performed on an 8-place calculator, gives $3.1415927 \cdot 11.5 = 36.128316$, which is rounded off to 36.1

The above conventions for performing operations with approximate values are merely guidelines, any they do not guarantee the accuracy of the final result. A precise analysis of error buildup when computing with approximate values requires methods of advanced mathematics. Error analysis is especially important these days when problems requiring thousands or even millions of calculations are performed on computers. The computational examples and exercises in this text do not require a large number of calculations, and our conventions will give accurate results.

When working with a calculator, we should keep in mind that all numbers entered and computed have the same number of significant digits. Rounding off occurs in the last place only. If our numbers actually have fewer significant digits than those of the calculator, we must do our own rounding off, according to the above conventions.

Appendix B: Base-10 Logarithms

The material in this appendix is independent of the treatment of exponentials and logarithms in Chapter 5. It is intended for students using the text for an intermediate algebra course that would not necessarily cover Chapter 5. The emphasis is on practical applications of logarithms (mostly pH applications), and the treatment is therefore restricted to base-10.

Introduction Base-10 logarithms, also called common logarithms, are used to measure the intensity of sound (decibel level), the magnitude of earthquakes (Richter scale), and the level of acidity or alkalinity of an aqueous solution (pH scale). We denote the common logarithm of a number x by $\log x$. By definition,

$$y = \log x \Leftrightarrow 10^y = x.$$

That is, log of x is the *exponent* to which 10 must be raised in order to equal x. For example,

$$\log 100 = 2 \text{ because } 10^2 = 100 \quad \text{and} \quad \log 0.01 = -2 \text{ because } 10^{-2} = \frac{1}{10^2} = \frac{1}{100} = 0.01.$$

Some further examples.

$$\log 10^6 = 6, \quad \log 10^{-6} = -6, \quad \log 10 = \log 10^1 = 1,$$
$$\text{and } \log 1 = \log 10^0 = 0$$

Since every power of 10 is positive, whether the exponent is positive, negative, or zero, we can take logs only of *positive numbers*. So expressions such as $\log 0$ or $\log(-20)$ are not defined. When x is 10 raised to an integer exponent, then $\log x$ is simply that integer, but for all other values of x, we need a scientific calculator, computer, or a log table to determine $\log x$. For example, using a calculator, we find that (to five decimal places),

$$\log 25 = 1.39794, \quad \log 750 = 2.87506, \quad \log 76{,}458 = 4.88342.$$

Although we must rely on the accuracy of the calculator for these results, we know they are in the "right ballpark." That is, 25 is between 10 and 100 so its log must be between 1 and 2; 750 is between 100 and 1000 so its log must be between 2 and 3, and 76,458 is between 10,000 and 100,000 so its log must be between 4 and 5. It is a good idea when computing logs with a calculator to check the magnitude of the result for accuracy, not as a check on the calculator but as a check that we keyed in the right number, at least up to order of magnitude.

Example 1   Without using a calculator, determine two consecutive integers that $\log x$ lies between for each of the following value of x.
(a) $x = 5351$ (b) $x = 5.351$ (c) $x = 0.5351$

Solution:
(a) $3 < \log 5351 < 4$ (b) $0 < \log 5.351 < 1$
(c) $-2 < \log 0.5351 < -1$ ■

Computations Involving Logarithms   Logarithms were the idea of John Napier (1550-1617) who wanted to simplify the tedious computations made by scientists of his day. His idea was to transform multiplication into addition, division into subtraction, and extraction of roots into multiplication by a fraction by means of the following properties:

L1. $\log(xy) = \log x + \log y$
L2. $\log\left(\dfrac{x}{y}\right) = \log x - \log y$
L3. $\log(x^r) = r \log x$

To implement these properties, he constructed a table of logarithms of numbers, consisting of some ninety pages, over a period of 20 years. Computations involving tables of logarithms are illustrated in the following two examples.

Example 2   Compute the product $75{,}438 \times 2{,}947$ using logarithms.

Solution: Using an 8-place log table, we would find that $\log(75{,}438) = 4.8775902$ and $\log(2{,}947) = 3.4693801$. Then, by property L1,

$$\begin{aligned}\log(75{,}438 \times 2{,}947) &= \log(75{,}438) + \log(2{,}947) \\ &= 4.8775902 + 3.4693801 \\ &= 8.3469703\end{aligned}$$

Therefore, $75{,}438 \times 2{,}947$ is the number whose log is 8.3469703. Using the same table, we find that 222,315,785 is the number (called the antilog of 8.3469703). The actual value is 222,315,786, so we lost one unit due to the limitations of an 8–place table. Early log tables had 14-place accuracy, so with such a table, we would have gotten the exact value. ∎

Example 3 Compute $\sqrt{3{,}256}$ using logarithms.

Solution: Again using an 8-place log table, we find that $\log(3{,}256) = 3.5126844$. Then, by property L3, $\log(\sqrt{3{,}256}) = \log(3{,}256^{1/2}) = \frac{1}{2}(3.5126844) = 1.7563422$. Therefore, $\sqrt{3{,}256}$ is the number whose logarithm is 1.7563422. From the table, we find that the antilog of 1.7563422 is 57.061370, correct to eight places. ∎

Applications of Logarithms With the advent of computers, and now calculators, computations are no longer performed using logarithms. However, logarithms are still important both for theoretical purposes and for practical applications. For example, as stated earlier, common logarithms are used in measuring the intensity of sound, the magnitude of an earthquake, and the level of acidity or alkalinity of an aqueous solution. These measures are given by the following formulas.

Decibel Level: $D = 10 \cdot \log(10^{16} I)$, where $I = $ energy of the sound wave emitted, measured in watts per square centimeter. An energy level of 10^{-16} watts per square centimeter, which is the threshold of hearing, has decibel level

$$D = 10 \cdot \log(10^{16} 10^{-16}) = 10 \cdot \log 10^0 = 10 \cdot \log 1 = 0.$$

Richter Scale: $R = \frac{2}{3} \log(10^{-4.7} E)$, where $E = $ energy released by the earthquake, measured in joules. Movement of the earth caused by an earthquake releasing $10^{4.7}$ joules is barely detectable, and is assigned a Richter value $R = \frac{2}{3} \log(10^{-4.7} 10^{4.7}) = \frac{2}{3} \log 1 = 0$.

pH Formula: $\text{pH} = -\log[\text{H}^+]$, where $[\text{H}^+]$ is the solution's concentration of hydrogen ions measured in moles per liter. Pure water has a pH of 7; solutions with a pH of less than 7 are acidic, and those with a pH greater than 7 are alkaline.

The pH values of some common liquids are listed in the following table in order of decreasing acidity or increasing alkalinity.

Battery Acid	0.3	Milk	6.4
Lime & Lemon juice	2.0	Pure Water	7.0
Coca-Cola	3.0	Sodium Bicarbonate solution	8.4
Wine	2.8 − 3.8	Liquid Ammonia	11.6
Orange juice	3.0 − 4.0	Bleach	12.6
Beer	4.0 − 5.0	Sodium Hydroxide solution	14.0

Since the logarithm of a number is the exponent to which 10 must be raised in order to equal the number, we can convert the pH formula to one in exponential form as follows.

$$pH = -\log[H^+]$$
$$-pH = \log[H^+]$$
$$10^{-pH} = [H^+]$$

we therefore have the following relationship between pH and hydrogen ion concentration $[H^+]$.

$$(*) \qquad pH = -\log[H^+] \Leftrightarrow [H^+] = 10^{-pH} \text{ moles per liter}$$

For example, if the pH of a solution is 8, then the hydrogen ion concentration of the solution is 10^{-8} moles per liter. Here are some examples involving pH.

Example 4 What is the pH for grapefruit juice with $[H^+] = 0.63 \times 10^{-3}$?

Solution: Using property L1 and a calculator, we get

$$\begin{aligned} pH &= -\log[0.63 \times 10^{-3}] \\ &= -\left(\log 0.63 + \log(10^{-3})\right) \\ &= -(\log 0.63 + (-3)) \\ &= 0.2 + 3 \\ &= 3.2 \end{aligned}$$

More simply, we could first compute $0.63 \times 10^{-3} = 0.00063$, and then use a calculator to find that $-\log(0.00063) = 3.2$. ∎

Example 5 What is the hydrogen ion concentration for a solution with pH = 5?

Solution: $5 = -\log[H^+] \Rightarrow -5 = \log[H^+] = 10^{-5}$ moles per liter. Alternatively, we could use (*) above to conclude that $5 = -\log[H^+] \Leftrightarrow [H^+] = 10^{-5}$ moles per liter. ∎

Example 6 If the pH of a solution changes from 5 to 4, by what factor does $[H^+]$ change?

Solution: Computing as in Example 5, we would find that $[H^+]$ changes from 10^{-5} to 10^{-4}. Since $10^{-4}/10^{-5} = 10^{-4+5} = 10$, $[H^+]$ increases by a factor of 10. Hence, the solution becomes 10 times more acidic. ∎

Example 7 Generalize Example 11 to show that if the pH of a solution decreases by 1 unit, then $[H^+]$ increases by a factor of 10, and if the pH increases by 1 unit, $[H^+]$ decreases by a factor of 10.

Solution: Suppose the pH decreases from x to $x - 1$. Then $[H^+]$ changes from 10^{-x} to $10^{-(x-1)} = 10 \cdot 10^{-x}$. Hence, $[H^+]$ increases by a factor of 10. Similarly, if pH increases from x to $x + 1$, then $[H^+]$ changes from 10^{-x} to $10^{-(x+1)} = 10^{-x} \cdot 10^{-1} = 10^{-x}/10$. Hence, $[H^+]$ decreases by a factor of 10. ∎

Example 7 shows that as the pH decreases, the level of acidity increases, and the pH increases, the level of alkalinity increases. In the following examples involving mixtures, we assume that the hydrogen ions mix freely and no chemical reactions between the solutions disturb the relative concentrations of hydrogen ions.

Example 8 Suppose solution A with pH $= 6$ is mixed with an equal amount of solution B with pH $= 4.5$. What is the pH of the mixture?

Solution: For simplicity, assume that there is one liter of each solution so two liters of the mixture. Solution A contributes 10^{-6} moles of hydrogen ions and B contributes $10^{-4.5}$, so the concentration of the mixture is $(10^{-6} + 10^{-4.5})/2$ moles per liter. Therefore, the pH of the mixture is $-\log[(10^{-6} + 10^{-4.5})/2] = 4.7875$. ∎

We note that the pH of the mixture in Example 8 is not equal to the average of the two pH values, which 5.25. This is because $-\log[(x + y)/2] \neq -[\log x + \log y]/2$. Rather, by property L2, $-\log[(x+y)/2] = -[\log(x + y) - \log 2] = -\log(x + y) + \log 2$. We can check that $-\log(10^{-6} + 10^{-4.5}) + \log 2 = 4.7875$.

Example 9 Suppose 15 liters of solution A having pH $= 8$ are mixed with 10 liters of solution B having pH $= 6$ forming solution C. What is the pH of solution C?

Solution: Solution C consists of 25 liters to which A contributes $15 \cdot 10^{-8}$ moles of hydrogen ions, and B contributes $10 \cdot 10^{-6} = 10^{-5}$ moles, for a total of $15 \cdot 10^{-8} + 10^{-5}$ moles. So the concentration of C in moles per liter is $(15 \cdot 10^{-8} + 10^{-5})/25$ and the pH of C is $-\log\left[(15 \cdot 10^{-8} + 10^{-5})/25\right] = 6.3915$. ∎

Example 10 How many liters of solution A with pH $= 6$ must be mixed with 10 liters of solution B having pH $= 8$ to make solution C with pH $= 7$?

Solution: Let x equal the number of liters of Solution A needed, so that the volume of C is $x + 10$ liters. A contributes $10^{-6}x$ moles of hydrogen ions, and B contributes $10 \cdot 10^{-8} = 10^{-7}$ moles for a total of $10^{-6}x + 10^{-7}$ moles. We want the concentration of C to be 10^{-7} moles per liter, so we must solve the equation

$$\frac{10^{-6}x + 10^{-7}}{x + 10} = 10^{-7}$$

for x. Multiplying both sides by $10^7(x + 10)$ gives

$$10x + 1 = x + 10$$
$$9x = 9$$
$$x = 1$$

so one liter of solution A is needed. ∎

Exercises

The following exercises are derived from the above table of pH values. In mixture problems, we assume, as above, that the hydrogen ions mix freely and no chemical reactions between the solutions disturb the relative concentrations on hydrogen ions.

1. What is the hydrogen ion concentration of Coca-Cola?
2. How much more acidic is lemon juice than orange juice ?
3. How much more alkaline is sodium bicarbonate than milk?
4. How much less alkaline is ammonia than bleach?
5. If one liter of orange juice (pH $= 3.5$) is diluted with one liter of water, what is the pH of the solution?
6. If a half liter of ammonia is mixed with a half liter of bleach, what is the pH of the solution?
7. If wine having pH $= 2.8$ is mixed with an equal amount of wine having pH $= 3.8$, what is the pH of the wine formed?
8. How many liters of water must be added to one liter of milk to increase the pH to 6.5?
9. How many liters of orange juice having a pH of 3.0 must be mixed with 2 liters of orange juice having a pH of 4.0 to produce an orange juice with pH 3.5?
10. How many liters of orange juice having a pH of 4.0 must be mixed with 2 liters of orange juice having a pH of 3.0 to produce an orange juice with pH 3.5?

Answers:

1. 10^{-3} moles per liter
2. Lemon juice to 10 times more acidic than an orange juice having pH $= 3.0$, and is 100 times more acidic than an orange juice having pH $= 4.0$
3. Sodium bicarbonate is 100 times more alkaline (less acidic) than milk.
4. Ammonia is 10 times less alkaline (more acidic) than bleach.
5. 3.8
6. 11.86
7. 3.06
8. 0.379 liters
9. 0.632 liters
10. 6.32 liters

Answers to Selected Exercises

Chapter R

Section R.1

1. rational, irrational **2.** rational **3.** neither terminates nor has an indefinitely repeating pattern **4.** 0
5. addition, subtraction, multiplication **6.** true **7.** false **8.** false **9.** true **10.** true **11.** 7/10, 3/10
13. .34999... **15.** .666999... **17.** .333... **19.** 123/1000 **21.** 5/9 **23.** One possible answer is .50005.
25. $-3.5 < -\sqrt{9.1} < -3 < 1.41 < \sqrt{2} < 3.14 < \pi < 22/7$ **27.** [number line with points at -4.25, -5, -2, 0, $.75$, 1.5, 2.75, 3.5]
29. $100 < 110 < 135 < 140 < 150$; median $= 135$ Ib **31.** $16 < 17 < 18 < 19 < 20 < 21 < 22$; median
$= 19$ yr **33.** $125 < 250 < 325 < 350 < 475 < 500 < 625 < 750$; median $= \$412.50$
35. $5.25 < 5.5 < 8.5 < 9.75 < 10 < 10.5$; median $= 9.125\%$ **37.** $\cdot, \div$ **39.** $+, -, \cdot$ **41.** One possible
answer is the set of odd integers. **43.** $0

Section R.2

1. commutative property, associative property **2.** $a \cdot (b+c) = a \cdot b + a \cdot c$, $(a+b) \cdot c = a \cdot c + b \cdot c$ **3.** 3, 3
4. 0, 0, 1, 1 **5.** 1, 0, 1/0 is not a real number **6.** false **7.** true **8.** false **9.** true **10.** false
11. (**a**) commutative property of multiplication (**b**) associative property of addition (**c**) left distributive
property **13.** associative property of addition **15.** right distributive property **17.** all three **19.** 2, -2
21. $\pm 1, \pm 2$ **23.** $\$25(1.06) = \$26.50$ **25.** $\$2662.50/1.065 = \$2,500$ **27.** The solution of the equation is
$x = -2$, and therefore $x + 2 = 0$. Hence, multiplying by $(x+2)^{-1}$ is equivalent to dividing by 0. **33.** additive
inverse property, 16, 32 **35.** (**a**) 24 (**b**) 152 (**c**) 216 (**d**) 153 **37.** (**a**) 34.607477 (**b**) 16.080997
(**c**) 34.607477 (**d**) 24.700685 **39.** extended associative property of addition **41.** extended commutative and
associative properties of addition, right distributive property **43.** $-E = E$; $-O = O$ **45.** (**a**) 4 (**b**) 8 (**c**) 6.5
(**d**) 6 (**e**) 12 (**f**) 12 **47.** (**a**) yes (**b**) no (**c**) yes (**d**) yes **49.** 6M(2A12) $= 7$, but (6M2)A(6M12) $= 9$

Section R.3

1. Subtract the smaller absolute value from the larger and precede the result with the sign of the term with the
larger absolute value. **2.** changing the sign **3.** $a + (-b)$ **4.** $a \cdot b^{-1}$ or $a \cdot (1/b)$ **5.** positive, negative **6.** true
7. false **8.** true **9.** true **10.** false **11.** -9 **13.** 13 **15.** 15 **17.** -2 **19.** -64 **21.** 6 **23.** -28 **25.** $-1/3$
27. $x - 7$ **29.** $-6 + y$ **31.** $-x - y$ **33.** $-4b$ **35.** $-5y$ **37.** 1 **39.** -5 **41.** $\pm 2, -3$ **42.** positive **45.** 6
47. < 3 **49.** algebraic definition of greater than, algebraic property of equality, associative and commutative
properties of addition, algebraic definition of greater than **51.** algebraic definition of greater than, algebraic
definition of greater than, substitution for b, associative property of addition, positive numbers are closed under
addition, algebraic definition of greater than **63.** preservation property for addition, transitive property of
greater than

Section R.4

1. 0, 0 **2.** b, 0 **3.** $(a+c)/b$, $(ac+ab)/(bc)$ **4.** numerators, denominators **5.** invert, multiplication **6.** false
7. false **8.** false **9.** true **10.** true **11.** 3 **13.** 9 **15.** 2 **17.** $b/3$; $a \neq 0$ **19.** $(1+2b)/(2b)$; $a \neq 0, b \neq 0$

21. $(b-c)/(b+c)$; $a \neq 0$, $b+c \neq 0$ **23.** $1/(2a)$; $a \neq 0$, $2x - 3y + 4z \neq 0$ **25.** $a + b$; $x \neq y$
27. $4/(x+1)$; $x \neq -1$ **29.** $2x/(2x-1)$; $x \neq 1/2$ **31.** 1; $x \neq 1/3$ **33.** $(2x+6)/(3x+1)$; $x \neq -1/3$
35. $3/x$; $x \neq 0$, $x \neq -1$ **37.** $8x/15$ **39.** $(x+3)/6$ **41.** $(x+1)/x$; $x \neq 0$
43. $(7x-2)/[(x+2)(x-2)]$; $x \neq \pm 2$ **45.** $(7x-3)/5$ **47.** 0; $x \neq a$
49. $(2ax + 2bc)/[(2x+a)(2x-b)]$; $x \neq -a/2$, $x \neq b/2$ **51.** $1/4$; $a \neq 0$ **53.** $3a/2$ **55.** $1/4$; $x \neq -3/2$
57. $5/4$; $x \neq 1/2$ **59.** $1/4$; $x \neq 1/2$, $x \neq 3$ **61.** $1/4$; $a \neq 0$ **63.** $27ad/(8bc)$; $b \neq 0$, $c \neq 0$, $d \neq 0$
65. 1; $a \neq 0$ **67.** $2/(2x+3)$; $x \neq -3/2$ **69.** $1/(2x+2)$; $x \neq -1$ **71.** $(x-1)/(x+2)$; $x \neq \pm 2$
73. $8/3$; $x \neq -1$, $x \neq 0$ **75.** $8/3$; $x \neq -1/2$, $x \neq 2$ **77.** 1; $a \neq \pm b$ **79.** a/b; $a \neq 0$, $b \neq 0$, $ab + 1 \neq 0$
81. $2x + 7$; $x \neq -3$, $x \neq -4$ **83.** $(x+3)(x+1)/[(x+2)(2x+3)]$; $x \neq -1$, $x \neq -2$, $x \neq -3/2$
85. 1; $x \neq 1$, $x \neq -1$ **87.** $bc/[a(b+c)]$; $a \neq 0$, $b \neq 0$, $c \neq 0$, $(b+c) \neq 0$ **89.** $c(ab+1)/b$; $b \neq 0$, $c \neq 0$
91. $(a+b)/(a-b)$; $a \neq 0$, $b \neq 0$, $a \neq b$ **93.** a/c; $a \neq 0$, $b \neq 0$, $c \neq 0$, $abc \neq -1$
95. $(2x+3)(x+3)/[(x+1)(x+2)]$; $x \neq -1$, $x \neq -2$, $x \neq -3$ **97.** $3 - 2x$; $x \neq 1$, $x \neq 2$

Section R.5

1. (a) $4^3 \cdot 4^5 = 4^8$ **(b)** $(3^2)^4 = 3^8$ **(c)** $(5 \cdot y)^5 = 5^5 \cdot y^5$ **2.** x^n **3.** $2^6 = 64$, $y^n = x$ **4.** x, $|x|$ **5.** two, no;
one, one **6.** true **7.** false **8.** false **9.** true **10.** true **11. (a)** $(-2)^3$ **(b)** $(-2)^4$ **(c)** $(-2)^5$ **(d)** $(-2)^6$
(e) $(-2)^{11}$ **13. (a)** 2^5 **(b)** 2^8 **(c)** 2^{18} **15. (a)** $2^4 \cdot 3^3$ **(b)** $2^9 \cdot 3^2$ **(c)** $2^{11} \cdot 3^7$ **17.** x^{15}
19. $(-3)^{17}$ or $(-3)^{17}$ **21.** a^{12} **23.** $(-2)^{15}$ or -2^{15} **25.** x^{10n} **27.** 12^9 **29.** $a^{12}b^{12} = (ab)^{12}$ **31.** $3^3 \cdot 2^3 = 6^3$
33. $x^4 y^5$; $x \neq 0$, $y \neq 0$ **35.** 3^4 **37.** 2^3 **39.** 8^2 **41.** a; $a \neq 0$ **43.** xy^4; $x \neq 0$, $y \neq 0$
45. $a^2 b^{16}$; $a \neq 0$, $b \neq 0$ **47.** a^5/b^{10}; $a \neq 0$, $b \neq 0$ **49.** $1{,}680$ **51.** $-6ab^2$ **53.** $2a^a/(5b)$; $b \neq 0$ **55.** -15
57. -15 **59.** $8\sqrt{2}$ **61.** $2abc^2 \sqrt[3]{2ab^2}$ **63.** $|ab|c^2 d^2 \sqrt{bd}$ **65.** $(a^2/b)\sqrt[3]{a/b^2}$; $b \neq 0$ **67.** 5.28×10^3
69. 1.76×10^2 **71.** 2.5×10^{-4} **73.** 1.76×10^5 **75.** 2.5×10^7 **77.** 1.0×10^{-7} **79.** 1.32×10^{12}
81. 6.25×10^6 **83.** 1.0×10^{-7} **85.** 2.25×10^{-4} **87.** 3.0×10^{-11} **89.** 2.99776×10^{10} cm/sec
91. $36{,}000{,}000$ mi **93.** $92{,}956{,}000$ mi **95.** $483{,}400{,}000$ mi **97.** $1{,}782{,}000{,}000$ mi **99.** $3{,}664{,}000{,}000$ mi
101. 6.624×10^{-27} **103.** $.000\,000\,000\,000\,000\,000\,000\,01328$ grams **105.** $.000\,000\,000\,000\,000\,000\,000$
327 grams **107. (a)** $\$260.96$ **(b)** $\$205.79$ **(c)** $\$617.17$ **(d)** $\$482.77$

Chapter R Review Exercises

1. All except $\sqrt{-8}$ and $4/0$ are real numbers. Of the real numbers, $\sqrt{5}$ is irrational, and the rest are rational.
2. $75/99 = 25/33$ **3.** $0.571428571428\ldots$ **4.** Let $x = .9999\ldots$. Then $10x = 9.9999\ldots$, $9x = 9$, $x = 1$.
5. $3/8 = (3 \cdot 11)/(8 \cdot 11) = 33/88$; $3/8 = (3 \cdot 111)/(8 \cdot 111) = 333/888$ **6. (a)** commutative property of
addition **(b)** associative property of multiplication **(c)** left distributive property **(d)** additive identity

7. (a)
$$7 + a = 7 + b$$
$$-7 + (7 + a) = -7 + (7 + b)$$
$$(-7 + 7) + a = (-7 + 7) + b$$
$$0 + a = 0 + b$$
$$a = b$$
(b)
$$5 \cdot a = 5 \cdot b$$
$$5^{-1} \cdot (5 \cdot a) = 5^{-1} \cdot (5 \cdot b)$$
$$(5^{-1} \cdot 5) \cdot a = (5^{-1} \cdot 5) \cdot b$$
$$1 \cdot a = 1 \cdot b$$
$$a = b$$

8. (a) $1/2 = 1/3 + 1/6$, and $1/6$ is positive
(b) $-1/3 = -1/2 + 1/6$, and $1/6$ is positive

9.
$$2x - 18 > 5x - 3$$
$$2x - 18 + 18 > 5x - 3 + 18$$
$$2x > 5x + 15$$
$$2x - 5x > 5x + 15 - 5x$$
$$-3x > 15$$
$$x < -5$$

10. (a) true (b) false (unless $c \neq 0$) (c) true (d) true (e) true (f) true (g) false (unless $a \geq 0$) (h) false (i) true (j) true **11.** (a) 23/20 (b) 5/9 (c) 36 (d) 1 (e) 12/5 (f) 5 **12.** (a) $10a + 10b + 12c$ (c) $12a + 12b + 66$ **13.** If $a \neq b$ ($b - a \neq 0$), then $(a-b)/(b-a) = -(b-a)/(b-a) = -1$. **14.** If $x \neq -3$, then $(x+3)/(x^2 - 9) = (x+3)/[(x+3)(x-3)] = 1/(x-3) \neq 0$. If $x = -3$, then $(x+3)/(x^2-9) = 0/0$ is not defined. **15.** (a) 1, if $x \neq 3$ (b) $(x-3)/6$ (c) $(x+4)/(2x)$, if $x \neq 0$ **16.** $2^{13} \cdot 3^9$ **17.** (a) $2/x$ (b) $5x^2$ (c) $4x^{10}$ (d) a^2/b^2 (e) a/b **18.** (a) 2 (b) -2 (c) 1 **19.** (a) $5|a|b\sqrt{2b}$ (b) $2ab\sqrt[3]{2a^2}$ (c) $|x|$ (d) $-x$ **20.** (a) 5/2 (b) 20 (c) $7\sqrt{2}$ **21.** 1.414213562; 4 **22.** 3, 6, 7, or 9; 3, 6, or 9; 6 **23.** (a) 7962624 (b) .03125 (c) .0022675737 (d) 3.16227766 (e) 2.924017738 **24.** (a) =(b) = (e) = 0.50505051; (c) = (d) 10.22727273 **25.** (a) 9.722222222 (b) 12.9 (c) 68.25 **26.** (a) 2.99776E10 (b) 6.023E23 (c) 1.6E-19 (d) 6.624E-27 **27.** (a) 92,956,000 (b) 141,600,000 (c) 1.673E-24 (d) 1.328E-23 **28.** (a) 1.805550848E34 (b) 24,154,589.37 (c) 1.31625696E16 (d) .1259789157 **29.** (a) $12,833.59 (b) $16,470.09 (c) $27,126.04 **30.** (a) $313.36 (b) $244. (c) $202.76

Chapter 1

Section 1.1

1. addition, subtraction, multiplication **2.** $-3x^2 + 5x + 2$ **3.** multiplying each term of P by each term of Q **4.** quotient, remainder **5.** root, zero **6.** true **7.** true **8.** false **9.** false **10.** false **11.** $2x^2 + 5xy + 4y$ **13.** $-7x + 5$ **15.** $3x^3 - 2x^2 + x - 8$ **17.** $3ax^2 + 4bx + 5c$ **19.** $(1+a)x^3 + (1+b)x^2 + (1+c)x + (1+d)$ **21.** $6x^2 + 8x$ **23.** $2x^3y + 2xy$ **25.** $x^2 + 3x + 2$ **27.** $x^2 - x - 6$ **29.** $x^3y^3 + 2x^2y^2 + xy$ **31.** $25x^2 + 40x + 16$ **33.** $4x^4 + 12x^2 + 9$ **35.** $4x^4 - 25$ **37.** $x^3 + 6x^2 + 12x + 8$ **39.** $x^4 + 8x^3 + 24x^2 + 32x + 16$ **41.** $6x^3 - 12x^2 - 90x$ **43.** $15x^5 + 20x^4 + 15x^3 + 20x^2$ **45.** $24x^3 - 98x^2 + 133x - 60$ **47.** $x^4 - 5x^2 + 4$ **49.** $x^5 + 10x^4 + 35x^3 + 50x^2 + 24x$ **51.** $x^2 + x - 1$ **53.** $x^3 + 2x^2 + 3x + 4 + 5/(x-1)$ **55.** $x^4 + 4x^3 + 11x^2 + 44x + 178 + 716/(x-4)$ **57.** $2 + 12/(2x-3)$ **59.** $3x^2 + 4x - 2$ **61.** $3x^2 - 4x + 11 + (-36x + 16)/(x^2 + 2x - 1)$ **63.** $x^3 - x^2 + x + 1/(x^2 + x + 1)$ **65.** $x^6 - 5x^4 + 3x^2 - 6 + 5/(x^2 + 1)$ **67.** $5 + (10x + 12)/(x^2 - 2x - 1)$ **69.** $x - 1 + 1/(x^3 + x^2 + x + 1)$ **71.** $5x - 2$, 4 **73.** $x^6 + x^5 + x^4 + x^3 + x^2 + x + 1$, 0 **75.** 2, $7x + 3$ **77.** $x + 1$, $-3x + 2$ **79.** $x^4 - x^3 + x^2 - x + 1$, 0 **81.** (a) -1 (b) 1 (c) 4 **83.** (a) 15 (b) 39 (c) 0 **85.** (a) $-2 - 2\sqrt{2}$ (b) -2 (c) -2 **87.** $-2/3$, $5/3$, -2, 7 **89.** $\sqrt{5}$, $-\sqrt{5}$ **91.** 2, 3, -2, -3 **93.** (a) $(20-1)^2 = 400 - 40 + 1 = 361$ (b) $(20-2)^2 = 400 - 80 + 4 = 324$ (c) $(30-1)^2 = 900 - 60 + 1 = 841$ (d) $(30-2)^2 = 900 - 120 + 4 = 784$

Section 1.2

1. $x^2 + 4x - 5$ **2.** $(x+6)(x-a)$ **3.** $-$, $a - b$, $x - y$ **4.** $x^k + m$, $x^k + n$ **5.** B, AC, grouping **6.** true **7.** true **8.** false **9.** false **10.** false **11.** $(x+2)^2$ **13.** not a perfect square **15.** not a perfect square **17.** $(2x - 5y)^2$ **19.** $(x^2 + 3)^2$ **21.** $(x-8)(x+8)$ **23.** $(x-y)(x+y)(x^2+y^2)$ **25.** $4ax$ or $(2a)(2x)$ **27.** $3(x-2)(x+2)$ **29.** $2(x+2)^2$ **31.** $(x+4)(x+1)$ **33.** $3(x+3)(x-4)$ **35.** $x(x-9)(x+1)$ **37.** $(x-1)(x-2)$ **39.** $(x+1)(x-2)$ **41.** not factorable by integers **43.** $3(x+1)(x-2)$ **45.** $(x-1)(2x-3)$ **47.** $(x-8)(x-10)$ **49.** $(x+4)(x+25)$ **51.** $(x+11)(x+12)$ **53.** $3(x+1)(x+4)$ **55.** $2(x+1)(x-9)$ **57.** $(2x+1)(x+4)$ **59.** $(5x+1)(x+4)$ **61.** $(5x+7)(x+1)$ **63.** $(2x-1)(2x-5)$ **65.** $(2x-5)(5x+2)$ **67.** $(x+2)(x^2 - 2x + 4)$ **69.** $(2x-3y)(4x^2 + 6xy + 9y^2)$ **71.** $2(2a^2 - 3b)(4a^4 + 6a^b + 9b^2)$ **73.** $(x-b)(x-4b)$ **75.** $(x-a)(x-b)$ **77.** $(ax+2)(ax+3)$ **79.** $(a-b)(x-1)(x-2)$ **81.** $x(x+2a)$ **83.** $(x-a+2)(x+a+4)$ **85.** $(a+2)(x-y)(x+y)$

87. $2a(x+y)(x-y)$ **89.** $a(2x+3)(3x+2)$ **91.** $x(x+4)(5x-2)$ **93.** $(3x^2+5)(x^2+1)$
95. $(2x^2+1)(x^2+a^2)$ **97.** $(x-\sqrt{2})(x+\sqrt{2})$ **99.** $4(x-\sqrt{3})(x+\sqrt{3})$ **101.** $(x+1/2)(x+3/2)$

Section 1.3

1. least, factor **2.** LCM **3.** multiple **4.** $2(x-6)(x+6)^2$ **5.** rational numbers **6.** true **7.** false **8.** true
9. true **10.** false **11.** 126 **13.** $(x^2-1)(x^2-9)$ **15.** $3x^3y^2$ **17.** $2(x+1)(x^2-4)$ **19.** $-2/63$
21. $-h/[x(x+h)]$ **23.** $(1-2x)/x^3$ **25.** $(x^2+5x-10)/[2(x-1)(x-3)]$
27. $(-6x^2+5x-6)/[2(2x-1)(2x+1)]$ **29.** $(x^2+5x+8)/[(x+1)(x+2)(x+4)]$
31. $(-x^2+4x+2)/[x(x-3)(2x+1)]$ **33.** $3/(x^2-4)$ **35.** $(2x^2-4x+1)/[(x-2)^2(x-3)]$
37. $(4x^2-15x+18)/[x(x-2)^2]$ **39.** $(x^4+3x^3-6x^2+9)/[x^2(x+3)^2]$
41. $(-3x^2-11x-8)/[(x+1)(x+2)(x+3)]$ **43.** $(x+a-b)/[(x+a)(x+b)]$ **45.** $(x+a^2)/(x^3-a^3)$
47. $4a^2y^2/(3bx)$ **49.** $(5x-2)(2x-1)/[(3x-2)(2x-5)]$ **51.** $5bc^2yz/(4x)$ **53.** $5(x+3)$ **55.** $3/(x^2-1)$
57. $x/(x^2-1)$ **59.** $3x+2$

Section 1.4

1. $\sqrt[n]{x}$, even **2.** positive nth **3.** $(x^{1/n})^m$, $1/x^{m/n}$, 0 **4.** $(x^{1/n})^m$, $(x^m)^{1/n}$ **5.** positive **6.** true **7.** false
8. false **9.** false **10.** true **11. (a)** 2 **(b)** not a real number **(c)** -4 **13. (a)** 729 **(b)** 1/729 **(c)** not a real number **15.** $(9^{1/2})^3 = 3^3 = 27$, and $(9^3)^{1/2} = \sqrt{729} = 27$ **17.** $[(-32)^{1/5}]^2 = [-2]^2 = 4$, and $[(-32)^2]^{1/5} = \sqrt[5]{1024} = 4$ **19.** $(1250^{1/5})^2 \approx (4.162766)^2 \approx 17.3286$, and $(1250^2)^{1/5} = \sqrt[5]{1,562,500} \approx 17.3286$ **21. (a)** $64^{1/6} = 2$ **(b)** $64^{1/2} = 8$ **(c)** -1 **23. (a)** $8^4 = 4096$ **(b)** $1/27^3 = 1/19,683$ **(c)** $2^3 = 8$ **25.** $2^{1/3} \cdot 3^{3/4}$ **27. (a)** x^4 **(b)** x^9 **(c)** $1/x^{1/4}$ **29.** 1/2
31. $5^{2/3}/(7^4 x^{4/3} y^{12})$ **33.** any negative x **35.** $y^{2/3}(x^{2/3}+y)$ **37.** $x^2(y^3+1)/y^2$ **39.** 0

Section 1.5

1. $\sqrt{x}/x$ **2.** $x^{2/3}/x$ or $\sqrt[3]{x^2}/x$ **3.** $x^{3/4}/x^2$ or $\sqrt[4]{x^3}/x^2$ **4.** $\sqrt{a}+\sqrt{b}$ **5.** $(x+2\sqrt{a}\sqrt{x}+a)/(x-a)$ **6.** true
7. true **8.** false **9.** false **10.** true **11.** $(x^2+1)^{1/2}$ **13.** $(x^2+3)^{3/4}$ **15.** $(x+x^{1/2})^{1/2}$ **17.** $(x+2)^{3/4}$
19. $\sqrt[6]{(x^2+1)^5}$ **21.** $(x+1)\sqrt[6]{x+1}$ **23.** $(x^2-1)\sqrt{x^2-1}$ **25.** $x-2$ **27.** 1 **29.** $6\sqrt[3]{x^2-1}$
31. $(2a+3b)\sqrt{x^2+x+1}$ **33.** $(x+2)\sqrt[3]{x-4}$ **35.** $\sqrt{x-3}/(x-3)$ **37.** $x\sqrt{x}$ **39.** $\sqrt[3]{x+1}/(x+1)$
41. $(\sqrt{x+1}-\sqrt{x-1})/2$ **43.** $(1+x)\sqrt{x}/x$ **45.** $(a\sqrt{a}-b\sqrt{ab})/(ab)$ **47.** $(a+\sqrt{a}+a\sqrt{a})/a^2$
49. $\sqrt{x}+x\sqrt[4]{x^3}+x^3\sqrt[6]{x^5}$ **51.** $(1+2a)\sqrt{a^2-b^2}/(a^2-b^2)$ **53.** $2/(x-1)$ **55.** $(2x^2+1)/\sqrt{x^2+1}$ or $(2x^2+1)\sqrt{x^2+1}/(x^2+1)$ **57.** $-(x+10)/(2x^2\sqrt{x+5})$ or $-(x+10)\sqrt{x+5}/[2x^2(x+5)]$
59. $(5x^2-3)/(3\sqrt[3]{(x^2-1)^2})$ or $(5x^2-3)\sqrt[3]{x^2-1}/[3(x^2-1)]$ **61.** $-2/(x^2\sqrt{x^2+2})$ or $-2\sqrt{x^2+2}/[x^2(x^2+2)]$

Chapter 1 Review Exercises

1. $8ab-4a+1$ **2.** $3x^9-48x$ **3.** $(x^3-3x^2+5)/(x+3) = x^2-6x+18-49/(x+3)$ or $x^3-3x^2+5 = (x^2-6x+18)(x+3)-49$ **4.** quotient $= x^3+3x$; remainder $= 4x-3$
5. $(2x+1)^2 = 4x^2+4x+1$; $(2x-1)^3 = 8x^3-12x^2+6x-1$ **6.** $P(-2) = -11$; $P(\sqrt{2}) = 1+2\sqrt{2}$
7. $(\sqrt{2}-2)^2+4(\sqrt{2}-2)+2 = 2-4\sqrt{2}+4+4\sqrt{2}-8+2 = 0$ **8.** $P(x) \geq 1$ for all real x.
9. $(4x-7y)(4x+7y)$ **10.** $5(x-3)^2$ **11.** $(3x-2)(2x+5)$ **12.** $2(x-2y)(x^2+2xy+4y^2)$ **13.** $(5x+2)(x-6)$

14. $(a^2 + b^2 + 1)(a^2 - b^2 + 1)$ **15.** $(x^5 - 32)/(x - 2) = x^4 + 2x^3 + 4x^2 + 8x + 16$; therefore, $x^5 - 32 = (x^4 + 2x^3 + 4x^2 + 8x + 16)(x - 2)$ **16.** $2(x - \sqrt{3})(x + \sqrt{3})$ **17.** $(x + 2)(x + 1)$
18. $24(x - 1)(x + 1)^2$ **19.** $(2x + 3)/[(x - 1)(x + 2)(x + 3)]$ **20.** $(2x^2 + 11x - 4)/(x^2 - 4)$ **21.** $(2x - 1)/2$
22. $(2 - x)/[3(x - 1)]$ **23.** (a) 64   (b) 1/64   (c) undefined   (d) 1   (e) undefined   **24.** $2 \cdot 5 = 10$
25. $64^{-2/3} = 1/64^{2/3} = 1/16$; $64^{-2/3} = (64^{1/3})^{-2} = 4^{-2} = 1/16$; $64^{-2/3} = (64^{-2})^{1/3} = \sqrt[3]{1/4096} = 1/16$
26. The base must be positive in the power-of-a-rule   **27.** $4x^{1/2}y$ **28.** $x^{2/3}$ **29.** $9a^2$ **30.** $(x - y)^2/(x + y)$
31. $(x + 2 + \sqrt{2x + 4})/x$ **32.** $3/(x - 1)$

Chapter 2

Section 2.1

1. equation   **2.** equivalent   **3.** linear   **4.** $-b/a$ **5.** the set of all real numbers, the empty set   **6.** true   **7.** false
8. true   **9.** false   **10.** false   **11.** 2   **13.** 10   **15.** $-57/2$ **17.** 2.6   **19.** $-122/43$ **21.** all real numbers   **23.** no solution   **25.** $(\sqrt{3} + 8)/(\sqrt{3} - 3)$ **27.** $(3 - \sqrt{5})/3$ **29.** $-51/13$ **31.** $12/(2 - c)$ **33.** If $c = 5$, $x =$ any real number; if $c \neq \pm 5$, $x = 1/(c + 5)$.   **35.** $-(by + c)/a$ **37.** $-2ab/(a + b)$ **39.** $-a^2 y/b^2$ **41.** -1 **43.** 7/2
45. $-1/5$ **47.** 5   **49.** no solution   **51.** -4 **53.** 1   **55.** all real numbers except ± 2 **57.** $-9/2$ **59.** no solution   **61.** 3   **63.** 5/4   **65.** -3 **67.** all real numbers except -1 and -2 **69.** $-8/3$ **71.** $y/(1 - y); y \neq 1$
73. $y/(y - 1); y \neq 0, y \neq 1$ **75.** $(a - b)/c$ **77.** $a/2$ **79.** 1/2 if $y \neq 0$, any nonzero real number if $y = 0$
81. 15/2   **83.** no solution   **85.** -4 **87.** 2/3   **89.** 3/2   **91.** -2 **93.** no solution   **95.** 0   **97.** $-3/4$ **99.** $-3/4$
101. all real numbers except 0   **103.** $-1, 5$ **105.** 4, 2/3   **107.** no solution   **109.** $-37/4, -47/4$ **111.** 21/13, 19/17   **113.** 1/2, 3/2   **115.** 5, 1/7   **117.** $5, -3/5$ **119.** $16/7, -4/7$ **121.** $8/3, -2/7$

Section 2.2

1. inequality   **2.** $>, \geq, <, \leq$ **3.** $<, >$ **4.** $M > c, M < -c$ **5.** $M < c, M > -c$ **6.** true   **7.** false   **8.** true
9. false   **10.** true

11. $x \geq 2, (2, \infty)$ **13.** $x \leq 7/2, (-\infty, 7/2]$ **15.** $x < 7/2, (-\infty, 7/2)$ **17.** $x \geq -1/3, [-1/3, \infty)$

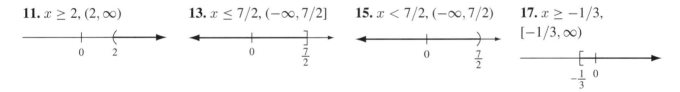

19. $x > 3/2, (3/2, \infty)$ **21.** $x \leq 14, (-\infty, 14]$ **23.** $x > -5, (-5, \infty)$ **25.** all $x, (-\infty, \infty)$

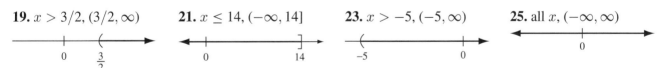

27. $x < -b/a$ **29.** $x < a(b - y)/b$ **31.** $x > (d - b)/(a - c)$ **33.** $(-1/4, \infty)$

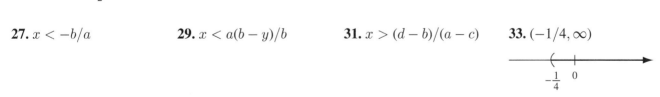

35. $[0, 1)$ **37.** $(-3, 1/2)$ **39.** $(1, 3/2]$ **41.** $(-\infty, -3) \cup (-1, \infty)$

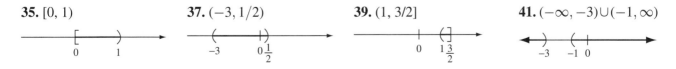

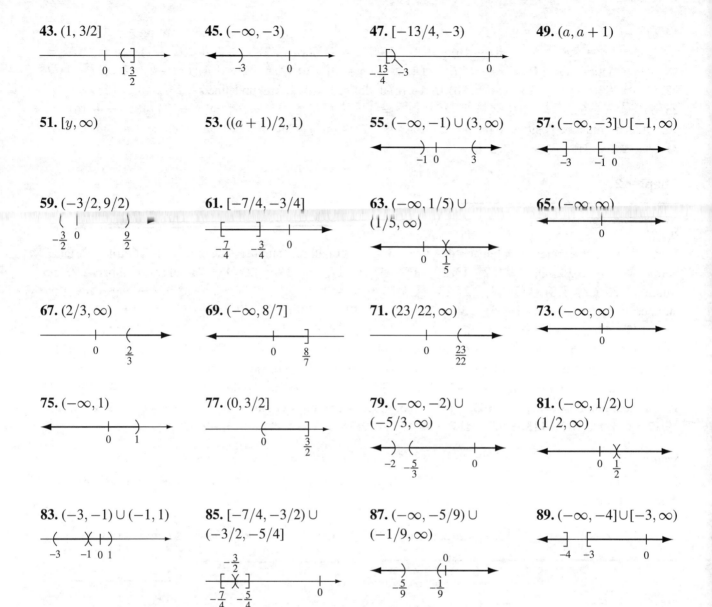

Section 2.3

1. $19.96 **3.** 25% **5.** $33\frac{1}{3}$% **7.** decrease of 1.6% **9.** $236 **11.** about 38.46 mph **13.** They meet 48 mi from the starting point of the car traveling at 40 mph, at 11:12 A.M. **15.** 3.33 P.M. **17.** $33\frac{1}{3}$ meters **19.** Patrick rows at 12.5 mph, and the current is 2.5 mph. **21.** 38% **23.** 2.5 liters **25.** 1 lb **27.** 34% **29.** 90.8 **31.** 1/3 hr, or 20 min **33.** 2.4 hr **35.** The two machines are 1.5 times as fast as A and 3 times as fast as B. **37.** 1.5 min **41.** 92 **45.** .5 and 5.5 **47.** about 60 micrograms **49.** 6

Section 2.4

1. abscissa, ordinate **2.** 0 **3.** III, I **4.** IV, IV **5.** Pythagorean **6.** false **7.** false **8.** false **9.** true **10.** false

11.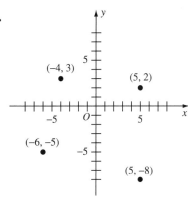

13.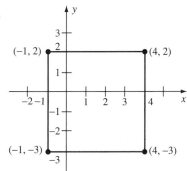

15.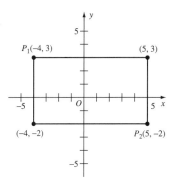

17.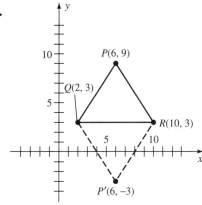

19. One possibility is shown below.
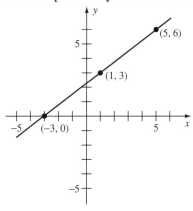

21. (a) $d = \sqrt{34}$ (b) $d = 4\sqrt{5}$ (c) $d = 2\sqrt{10}$ (d) $d = 13\sqrt{2}$

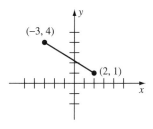 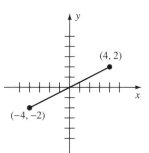 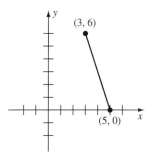 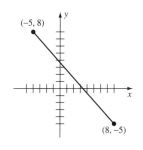

23. $\sqrt{61}$ and $\sqrt{41}$

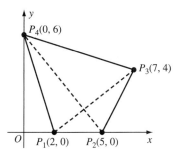

25. $y = 7$

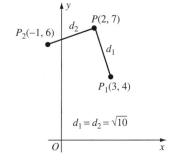

27. $d(P_1, P_2) = d(P_2, P_3) = d(P_1, P_3)$

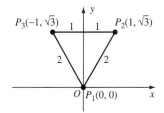

29. no; $d(P_1, P_3) \neq d(P_2, P_4)$

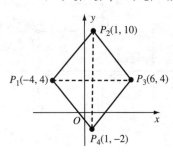

31. yes; center of circle at $(9/2, 5/2)$ and radius $= \sqrt{130}/2$

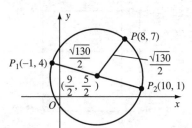

33. $d(P_1, P_2) + d(P_2, P_3) = d(P_1, P_3)$

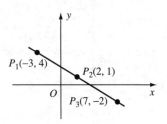

35. $4 + 2\sqrt{2}$ mi

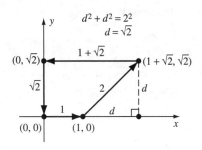

37.

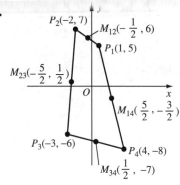

39. $x = -5/2$; $y_2 = 3$

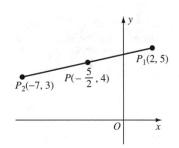

41.

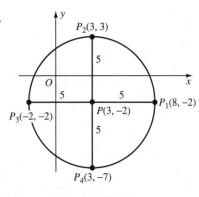

43. $P_3(5, 6)$, $P_4(-1, 4)$ (shown) and $P_3'(17, 6)$, $P_4'(11, 4)$

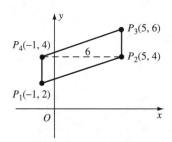

Section 2.5

1. graph **2.** slope **3.** $(3, 5)$, $(1, -3)$ **4.** point-slope, slope-intercept, general linear **5.** $y = -4, x = 3$ **6.** true
7. true **8.** false **9.** true **10.** true **11.** -1 **13.** $-1/4$ **15.** undefined slope **17.** $y = 7$ **19.** $x = -5/2$
21. $y - 4 = 3(x + 2)$ **23.** $y = x - 3$ **25.** $y - 4 = 2x$ **27.** $y = 2x - 1$

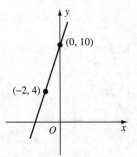 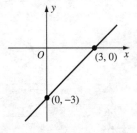 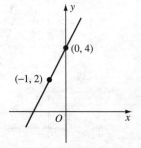 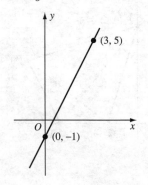

29. $y = 3x - 8$ **31.** $y = -x + 6$ **33.** $m = 4/3; b = -7/3$ **35.** undefined slope; no y-intercept

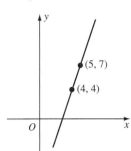 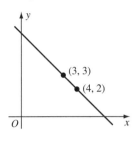 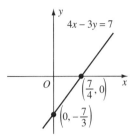 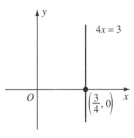

37. $m = -1; b = 1$ **39.** $x/4 + y/3 = 1$ **41.** $x/5 + y/(20/3) = 1$ **43.** $x/4 + y/(-4) = 1$

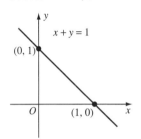 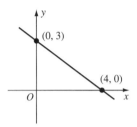 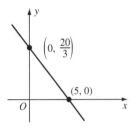 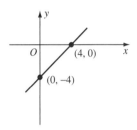

45. $x = 2 + 3r$, $y = -5 + 3r$ **47.** $x = 6, y = 2 - 4r$ **49.** Using $P_2(8, 5)$, $x = 7 + r, y = 2 + 3r$. **51.** $x_1 x + y_1 y = 0$

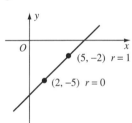 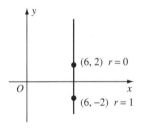 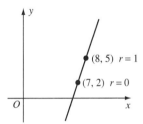

53. (a) 5 mph (b) .5 mi (c) $y = 5x + .5$ **55.** (a) $151 (b) 1% (c) $P = 1.51t + 151; t \geq 0$

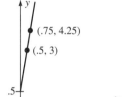 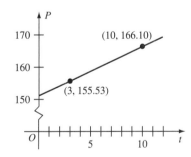

Section 2.6

1. $m_1 = m_2$ **2.** $A_1 B_2 - A_2 B_1 = 0$ **3.** $m_1 m_2 = -1$ **4.** $A_1 A_2 + B_1 B_2 = 0$ **5.** $A_1 B_2 - A_2 B_1 \neq 0$ **6.** true
7. false **8.** true **9.** true **10.** true **11.** parallel **13.** perpendicular **15.** perpendicular **17.** coincident
19. parallel **21.** perpendicular **23.** $4x - 7y = -23$ **27.** $m_1 = 4/3$ and $m_2 = -3/4$ **33.** $x = 28/13$, $y = 4/13$ **35.** $x = 5, y = 2$ **37.** $x = 25, y = 32$ **39.** $x = 1/2, y = 4$ **41.** $x = 38, y = -29$

43. $x = 9, y = 13$ **45.** $x = -23/6, y = 11/3$ **47.** $x = 9/16, y = -37/32$ **49.** $x = 7/2, y = 13/3$ **51.** 12 ml of 20% acid and 6ml of 50% acid **53.** $27.90 fixed charge and $1.00 for each word after the fifteenth **55.** 2000 of the first style and 1500 of the second **57.** 6

Section 2.7

1. one, none, infinitely many **2.** planes, point **3.** infinite **4.** two **5.** constant **6.** true **7.** false **8.** false **9.** true **10.** false **11.** $x = 3, y = 2, z = -1$ **13.** $x = 5/2, y = 3/2, z = 0$ **15.** $x = -2, y = 1, z = 3$ **17.** The reduced system is of the form $x + y = c_1, x + y = c_2$, where $c_1 \neq c_2$. **19.** The reduced system is of the form $x + y = c_1, x + z = c_2$, where $c_1 \neq c_2$. **21.** $x = -2z, y = 2z, z =$ any real number **23.** $x = -20, y =$ any real number, $z = 4y - 32$ **25.** $x = -4 - t, y = -t, z = 6 + t, t =$ any real number **27.** $x = 2, y = 3, z = 1, t = 2$ **29.** Each pair of planes intersects in a line, resulting in three parallel lines. **31. (a)** The reduced system is of the form $y - z = c_1, y - z = c_2$, where $c_1 \neq c_2$. **(b)** $x - y - z = D$, where D is any real number **33.** $1000 in certificates, $1500 in bounds, and $2500 in mutual funds **35.** 20 valued at $45, 30 at $30, and 50 at $25 **37.** 500 of type 1, 400 of type 2, and 100 of type 3 **39.** $7x^2 + 7y^2 + 31x + 29y = 126$

Chapter 2 Review Exercises

1. $x = -7/2$ **2.** $x = 725/300$ **3.** $x > -7/2$ **4.** $x \geq 1$ **5.** $x = 7$ **6.** no solution **7.** $x = 0$ **8.** $x = 9$ **9.** no solution **10.** $-6 < x < 2$ **11.** $x \leq 1/3$ or $x \geq 3$ **12. (a)** all x **(b)** no x **13.** $x = 3$ or $x = 5$ **14.** $x = \pm 2$ **15.** If $a = -4$, $x =$ any real number; if $a \neq \pm 4$, $x = 1/(a-4)$. **16.** $x = (5 - 3a)/5; a \neq 5/3, a \neq -5/2$ **17.** $-3 < x < -1/2$ **18.** $(-\infty, -3] \cup (-1/2, \infty)$ **19.** $-a < x < 0$ **20.** $M = -1/2$ and $d = 13$ **21.** $M = (-1/2, 1/2)$ and $d = \sqrt{194}$ **22. (a)** $d^2(P_1, P_3) + d^2(P_2, P_3) = d^2(P_1, P_2)$; that is, $41 + 41 = 82$ **(b)** $m(P_2, P_3) = -5/4$, and $m(P_1, P_3) = 4/5$ **23.** $-3/10$ **24.** $x = 7, y = 3$ **25.** $6x - 8y = -11$ **26.** $y = 4, x = 6$ **27.** $y = (-3/5)x - 1/5$ **28.** $y = (3/2)x - 2$ **29.** $6x - 8y = -11$ **30.** $2x - 3y = -17$ **31.** $x = 4, y = -3$ **32.** $x = 4, y = -3$ **33.** $x = 4, y = -3$ **34.** $x = 4, y = 2$ **35.**

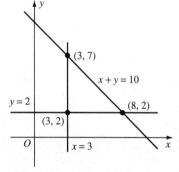

36.

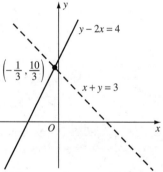

37. Unique solution: $x = -3, y = 2, z = 4$
38. infinite number of solutions: $x = 1 + z, y = 2z, z =$ any value. **39.** no solution **40.** infinite number of solutions; $x = -z/5, y = 3z/5, z =$ and value

Chapter 3

Section 3.1

1. $ax^2 + bx + c = 0$, constant real numbers, a **2.** two, one **3.** factoring, completing the square, quadratic formula **4.** $(-b \pm \sqrt{b^2 - 4ac})/(2a)$ **5.** $b^2 - 4ac > 0, b^2 - 4ac = 0, b^2 - 4ac < 0$ **6.** true **7.** true **8.** true **9.** true **10.** true **11.** 3,4 **13.** $-2,8$ **15.** 4 **17.** $\pm 3/2$ **19.** $-1, -2/3$ **21.** $-a, -b$ **23.** $x^2 - 9 = 0$

25. $6x^2 - 17x + 12 = 0$ **27.** $4x^2 - 28x + 49 = 0$ **29.** $x^2 - 6x + 7 = 0$ **31.** $(1 \pm \sqrt{5})/2$ **33.** $-2 \pm \sqrt{2}$
35. $(3 \pm \sqrt{3})/2$ **37.** $(5 \pm \sqrt{13})/3$ **39.** $(5 \pm \sqrt{41})/4$ **41.** a, b **43.** $0, 6$ **45.** $-1, -1/2$ **47.** $(1 + \sqrt{3})/2$
49. $1/2, -3/2$ **51.** 2 **53.** 1 **55.** 0 **57.** 2 **59.** 1 **61.** $2(x - 5/2)(x - 4)$
63. $\left[x - (5 + \sqrt{3})\right]\left[x - (5 - \sqrt{3})\right]$ **65.** $4(x - 3/2)^2$ **67.** irreducible
69. $4\left[x - (1 + \sqrt{2})/2\right]\left[x - (1 - \sqrt{2})/2\right]$ **71.** $(5 \pm \sqrt{5})/2$ **73.** $(2 \pm \sqrt{10})/2$ **75.** $(-1 \pm \sqrt{2})/2$
77. $1, -1/3$ **79.** $2(x + 3)^2 - 15$ **81.** $4(x + 1/4)^2 - 21/4$ **83.** $1(x + 1/2)^2 + 3/4$ **87.** Exercise 51: $a_1 + a_2 = 100, a_1 a_2 = 275$; Exercise 53: $a_1 + a_2 = -24, a_1 a_2 = 144$; Exercise 57; $a_1 + a_2 = -75/107, a_1 a_2 = -240/107$; Exercise 59; $a_1 + a_2 = 17/3, a_1 a_2 = 289/36$

Section 3.2

1. real numbers, $\sqrt{-1}$ **2.** $a - bi$ **3.** 0 **4.** pure imaginary **5.** diagonal **6.** false **7.** true **8.** true **9.** true
10. false **11.** $7 - 4i$ **13.** $3 + 6i$ **15.** $14 - 5i$ **17.** 25 **19.** $3 + 3i$ **21.** $(5/2) - (1/2)i$
23. $(-2/29) + (5/29)i$ **25.** $4 - 7i$ **27.** (a) i (b) -1 (c) $-i$ (d) 1 **29.** $\pm 3i$ **31.** $2 \pm i$ **33.** $-4 \pm 3i$
35. $(1 \pm i)/2$ **37.** $(2 \pm \sqrt{2}i)/2$ **39.** $(x - 3i)(x + 3i)$ **41.** $[x - (-4 + 3i)][x - (-4 - 3i)]$
43. $[x - (-5 + i)][x - (-5 - i)]$ **45.** $2\left[x - (-1 + \sqrt{2}i/2)\right]\left[x - (-1 - \sqrt{2}i/2)\right]$

47. $z_1 + z_2 = 5 + 5i$ **49.** $z_1 + z_2 = -2 + 6i$ **51.** $z_1 + z_2 = 4i$

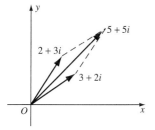
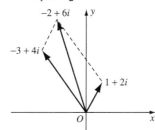
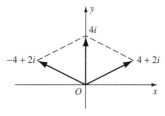

53. $z_1 + z_2 = 2 + 2i$ **57.** (a) $-1 + 7i$ (b) $7 + i$ (c) $-7 - i$

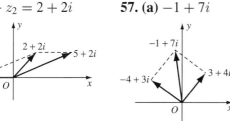
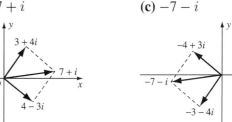

Section 3.3

1. substitution **2.** 0 **3.** the original equation **4.** nonnegative **5.** $(x + 1)^{3/2}$ **6.** false **7.** false **8.** false
9. true **10.** false **11.** $\pm 1, \pm \sqrt{2}$ **13.** $81/16$ **15.** $(3 \pm \sqrt{5})/2$ **17.** $\pm \sqrt{5}$ **19.** $-8/5, -13/8$ **21.** $2, 3$
23. $1, 4$ **25.** -3 **27.** $2, -1$ **29.** $(11 \pm \sqrt{73})/4$ **31.** $4, 5$ **33.** $1/3$ **35.** 12 **37.** 5 **39.** $-1, -15/2$

Section 3.4

1. a, roots **2.** less than **3.** $(-\infty, \infty), \varnothing$ **4.** $a < 0$ **5.** $(-\infty, a_1] \cup [a_2, \infty)$ **6.** false **7.** false **8.** true
9. false **10.** false
11. $(-\infty, 1] \cup [3, \infty)$ **13.** $(-4, 2)$ **15.** $(-\infty, 1/2] \cup [3, \infty)$ **17.** $[-2/3, 4]$

19. $(-\infty, -1/2] \cup [8/3, \infty)$ **21.** all real x **23.** ∅ **25.** ∅ **27.** $1/4$ **29.** $(-\infty, 3/4) \cup (3/4, \infty)$ **31.** all real x **33.** all real x **35.** ∅ **37.** $(-\infty, -3/2] \cup [3/2, \infty)$ **39.** $(-1/2, 3)$ **41.** $(-2, 4/5)$ **43.** $(-\infty, 1/2) \cup (5, \infty)$ **45.** $(-\sqrt{2}, \sqrt{2})$ **47.** $[1, 3]$ **49.** $(x, 0), -4 \leq x \leq 4$ **51.** $1/4 \leq t \leq 7/4$

Section 3.5

1. 5 **3.** 40 **5.** 25% the first year and 50% the second **7.** 7 **9.** Jean: 10 mph; Margo: 15 mph **11.** 300 mph **13.** 1 **15.** (a) 4.5 sec after firing (b) 324 ft (c) 9 sec after firing **17.** yes **19.** $16\sqrt{6}$ ft/sec **21.** one carrier: 2 hr; other carrier: 3 hr **23.** A: 100 min; B: 15 min **25.** A: 4.8 min; B: 8 min; C: 12 min **27.** 2 and 7 **29.** 75 yd by 100 yd **31.** 1 in

Section 3.6

1. A, B **2.** completing the square **3.** A, B **4.** axis of symmetry **5.** $a > 0, a < 0$ **6.** true **7.** false **8.** true **9.** true **10.** true **11.** $(x - 2)^2 + (y - 1)^2 = 9$ **13.** $(x - 2)^2 + (y + 2)^2 = 8$ **15.** $(x + 1)^2 + y^2 = 1$ **17.** $x^2 + y^2 = 1/2$ **19.** $(x - 2)^2 + (y - 1)^2 = 20$

21. circle

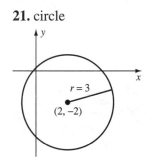

23. circle

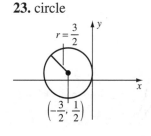

25. point

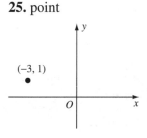

27. ∅ **29.** ∅ **31.** $x^2 + 2y - 3 = 0$ **33.** $x^2 - 4x + 4y - 4 = 0$

35. vertex: (0,0); axis of symmetry: y–axis

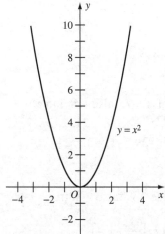

37. vertex: (0,2); axis of symmetry: y–axis

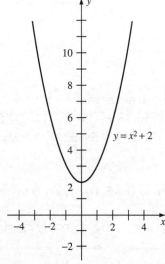

39. vertex: $(-3, 1)$; axis of symmetry: $x = -3$

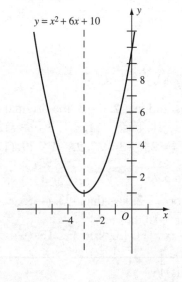

41. vertex: $(2, -4)$; axis of symmetry: $x = 2$

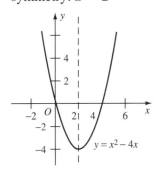

43. vertex: $(-2, -1)$; axis of symmetry: $x = -2$

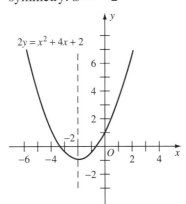

45. vertex: $(0,0)$; axis of symmetry: x-axis

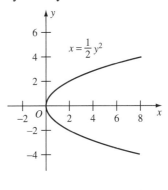

47. vertex: $(-9, 3)$; axis of symmetry: $y = 3$

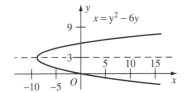

49. vertex: $(-25/8, 3/2)$; axis of symmetry: $y = 3/2$

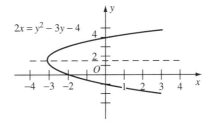

51. length: 500 ft; width: 250 ft   **53.** 2 weeks   **55.** (a) 8 sec   (b) 512 ft   (c) 256 ft   (d) $y = -x^2/256 + 2x$

Section 3.7

1. $x' = x - x_0$ and $y' = y - y_0$ **2.** $(x')^2 + (y')^2 = r^2$ **3.** $x^2 + y^2 = r^2$ **4.** completing the square   **5.** four
6. true   **7.** true   **8.** true   **9.** false   **10.** true   **11.** (a) $(1, -7)$ (b) $(-2, -5)$ (c) $(0, 0)$
13. $x' = x + 2$; $y' = y - 4$
15. $x' = x + 1$; $y' = y - 2$ **17.** $x' = x + 1$; $y' = y - 2$ **19.** $x' = x$; $y' = y + 2$

21. $2x' + 3y' = 8$

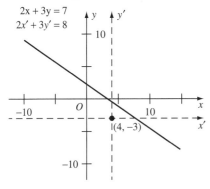

23. $(x' + 2)^2 + (y' - 4)^2 = 5$

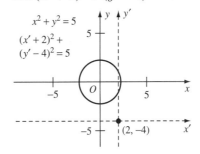

25. $y' = (x')^2 - 5$

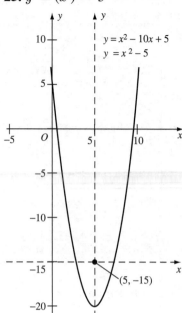

27. $x^2 + y^2 = 25$

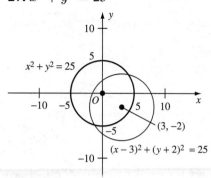

29. $y = x^2 + 1$

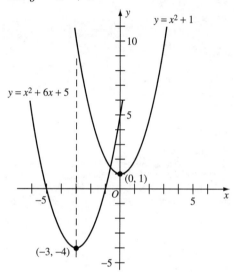

31. $y' = (x')^2$, where $x' = x + 3/2$ and $y' = y + 5/4$
33. $x' = 2(y')^2$, where $x' = x + 9/2$ and $y' = y + 1/2$ **35.** $(x')^2 + (y')^2 = 35/4$, where $x' = x + 3/2$ and $y' = y + 2$
37. $(x_0, y_0) = (-3/4, -49/8)$; $y' = 2(x')^2$
39. $(x_0, y_0) = (1/4, -1/2)$; $(x')^2 + (y')^2 = 25/16$

41. $(-1, 3)$ and $(3, -5)$

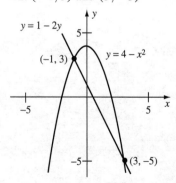

43. $(3, 5)$ and $(-3, -5)$

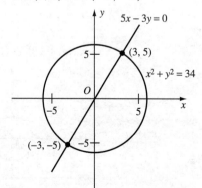

45. $(\sqrt{2}, 1)$ and $(-\sqrt{2}, 1)$

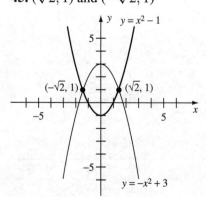

47. (3,1) and (3,3)   **48.** ($\sqrt{2}, \sqrt{2}$) and ($-\sqrt{2}, \sqrt{2}$)   **51.**

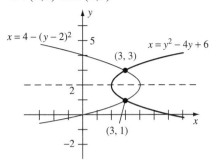 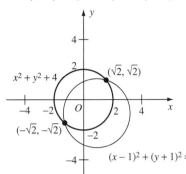 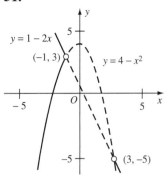

53.   **55.**

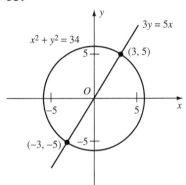 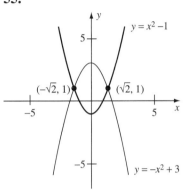

Section 3.8

1. $d(P_1, P) + d(P_2, P) = c$, $d(P_2, P) + d(P_1, P) = \pm c$ **2.** focus   **3.** hyperbola, $(3, -5)$ **4.** $2\sqrt{10}, 4$
5. $0, >, <$
6. false   **7.** false   **8.** true   **9.** false   **10.** false   **11.** $3x^2 + 4y^2 = 12$ **13.** $9x^2 + 5y^2 = 45$

15.   **17.**   **19.**

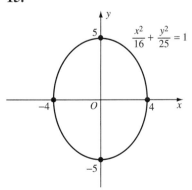 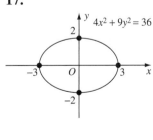 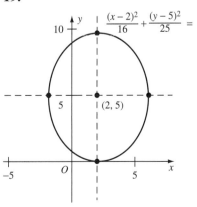

21.

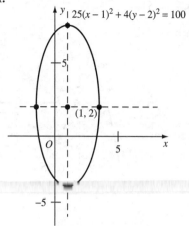

23. ellipse $(x+2)^2 + (y-2)^2 = 100$
25. point $(-1, -2)$
27. ellipse $4(x+2)^2 + 9(y-1)^2 = 36$
31. $12x^2 - 4y^2 = 3$
33. $3y^2 - x^2 = 3$
35. two lines $y = 2x, y = -2x - 1$
37. hyperbola $5(y+1)^2 - 2(x-1)^2 = 5$
39. hyperbola $8(y-1/2)^2 - x^2 = 1$

43.

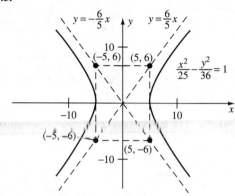

45.

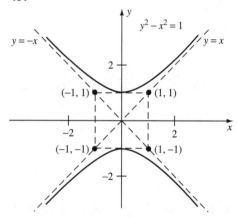

47.

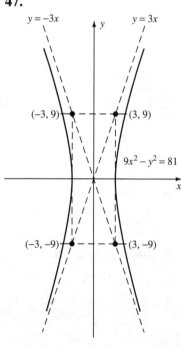

49.

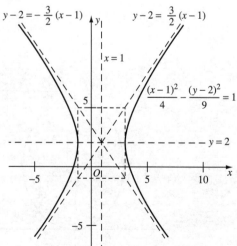

Chapter 3 Review Exercises

1. $-5, 10$ 2. $-2/3, 4$ 3. -3 4. $(-1 \pm \sqrt{5i})/3$ 5. $(-1 \pm \sqrt{7})/3$ 6. $(5 \pm \sqrt{23i})/12$
7. $8(x + 2/4)(x - 1/2)$ 8. $x^2 - 6x + 11$ has no real factors because its discriminant is equal to -8, which is less then 0. 9. $12x^2 + x - 6 = 0$ 10. $\pm\sqrt{3}$ 11. 4 12. $4, -4/5$ 13. $8 - i$

14.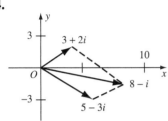

15. $11 + 2i, 1 + 2i$ 16. $(-\infty, 1] \cup [3, \infty)$ 17. $(-\infty, -4] \cup [2/3, \infty)$
18. $5/2$ 19. $\oslash$
20.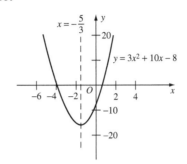
21.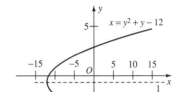

22. $(x - 4)^2 + (y - 2)^2 = 37$

23.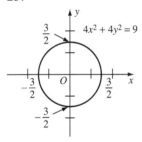

24.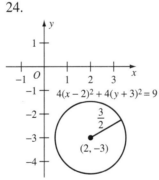

25.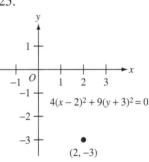

26.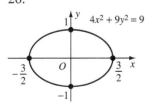

27.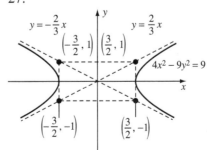

28.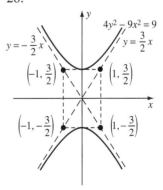

29. center:(1, 6); circle   **30.** center:(-5, 4); hyperbola   **31.** $(x')+(y')^2=5$ **32.** $4(x')^2-(y')^2=35$
33. $y'=2(x')^2$ **34.** $(5,0),(-4,-3)$ **35.** $(3,0),(-1/2,-7/4)$ **36.**

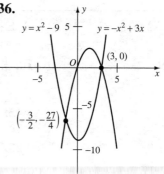

37. 3(3 % or .03)   **38.** Dick: 8 mph; Frank: 7.5 mph
39. Janet: 15 min; Dan: 10 min   **40.** .5 (.5% or .005)

Chapter 4

Section 4.1

1. exactly one element   **2.** domain   **3.** 8, −8, range   **4.** in both D_f and D_g **5.** $f, f(x)$ **6.** false   **7.** false
8. true   **9.** false   **10.** false   **11. (a)** 1   **(b)** 9   **(c)** 9   **(d)** 5   **(e)** 5   **(f)** $2\pi^2+1$ **13. (a)** 1   **(b)** 1   **(c)** 1
(d) −1
(e) −1   **(f)** 1   **15.** $(-\infty,\infty)$ **17.** $[5/2,\infty)$ **19.** $(-\infty,5/2)\cup(5/2,\infty)$ **21.** $(-\infty,-1]\cup[1,\infty)$
23. $(-\infty,0)\cup(0,\infty)$
25. $(-\infty,1]\cup[2,\infty)$ **27.** ± 2 **29.** 2   **31.** ∅   **33. (a)** $\dfrac{2x-1}{2x+1}$ **(b)** $\dfrac{x^2-1}{x^2+1}$ **(c)** $\dfrac{x}{x+2}$
(d) $\dfrac{1-x}{1+x}$ if $x\ne 0$, undefined if $x=0$ **(e)** $\dfrac{x+h-1}{x+h+1}$ **(f)** $\dfrac{x-h-1}{x-h+1}$ **35.** not a function of x
37. $f(x)=\sqrt[3]{x}$ **39.** $f(x)=x^2/(x-1)$
41. the graph of a function of x **43.** not the graph of a function of x

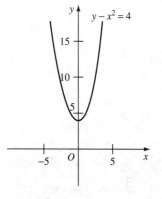

45. not the graph of a function of x **47.** the graph of a function of x
49. $(f+g)(x)=5x-7, (f-g)(x)=-x+7, (fg)(x)=6x^2-14x, (f/g)(x)=2x/(3x-7); D_{f+g}=D_{f-g}=D_{fg}=(-\infty,\infty),$
$D_{f/g}=(-\infty,7/3)\cup(7/3,\infty)$ **51.** $(f+g)(x)=x^2+\sqrt{x},(f-g)(x)=x^2-\sqrt{x},(fg)(x)=x^2\sqrt{x},(f/g)(x)=x^2/\sqrt{x}; D_{f+g}=D_{f-g}=D_{fg}=[0,\infty) D_{f/g}=(0,\infty)$
53. $(f+g)(x)=x^3+1/x^3, (f-g)(x)=x^3-1/x^3, (fg)(x)=1$ if $x\ne 0$, undefined if $x=0, (f/g)(x)=x^6$ if $x\ne 0$, undefined if $x=0; D_{f+g}=D_{f-g}=D_{fg}=D_{f/g}=(-\infty,0)\cup(0\infty)$
55. $(f+g)(x)=x/x^2+1+x^2+1, (f-g)(x)=x/(x^2+1)-(x^2+1), (fg)(x)=x, (f/g)(x)=x/(x^2+1)^2; D_{f+g}=D_{f-g} D_{fg}=D_{f/g}=(-\infty,\infty)$

Answers to Selected Exercises 457

57. $(g \circ f)(x) = 4x^2, D_{g \circ f} = (-\infty, \infty)$ **59.** $(g \circ f)(x) = x, D_{g \circ f} = (-\infty, \infty)$
61. $(g \circ f)(x) = |x|, D_{g \circ f} = (-\infty, \infty)$ **63.** $(g \circ f)(x) = \sqrt{x+1}, D_{g \circ f} = [-\infty, \infty)$
65. $(g \circ f)(x) = 1/(x^2 - 4), D_{g \circ f} = (-\infty, -2) \cup (-2, 2) \cup (2, \infty); (f \circ g)(x) = 1/x^2 - 4, D_{f \circ g} = (-\infty, 0) \cup (0, \infty)$
69. $A(r) = \pi r^2$ **71.** $V(s) = s^3$ **73.** $C(x) = 5000 + 1500x$ **75.** $S(x) = 10x^2$ **77.** $L(x) = 2x + 10{,}000/x$
79. (a) 80 ft/sec **(b)** $(32 + 16\Delta t)$ ft/sec

Section 4.2

1. $a_4 \neq 0$ **2.** the set of all real numbers **3.** even, symmetric with respect to the y-axis **4.** $0, -2, 3$ **5.** 4
6. true **7.** false **8.** true **9.** true **10.** false **11. (a)** and **(b)** are polynomial functions. **13.** The graphs of **(a)**,
(c), **(d)**, and **(e)** are symmetric with respect to the y-axis. **15. (a)**, **(c)** and **(d)** are odd functions.
17. (a) $x = -1, y = 1$ **(b)** $x = 1, y = 1$ **(c)** no x-intercept, $y = 7$ **(d)** no x-intercept, $y = -1$
(e) $e = \pm 2, y = -4$ **(f)** no x-intercept, $y = 4$

19.

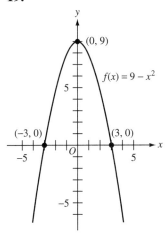

21.

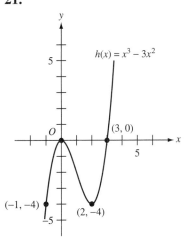

23.

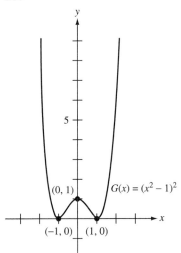

25.

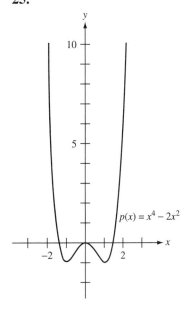

27.

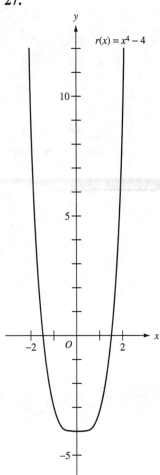

29.

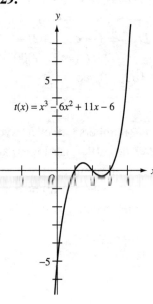

Section 4.3

1. two polynomial functions **2.** $Q(a) = 0$ **3.** 3/2 **4.** the degree of the numerator is less than that of the denominator **5.** the degree of the numerator is greater than that of the denominator **6.** true **7.** true **8.** false **9.** false **10.** true **11.** (a), (b), (c), (d), and (f) are rational functions. **13.** $x = -1$ **15.** $x = 1, x = -1$ **17.** $x = -1, x = -3$ **19.** $x = 1$ **21.** $y = 2$ **23.** $y = -3/2$ **25.** no horizontal asymptote **27.** no horizontal asymptote

29.

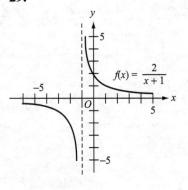

31.

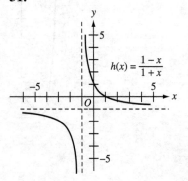

33.

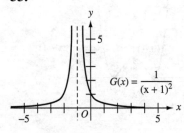

Answers to Selected Exercises   459

35.
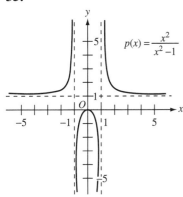

37.
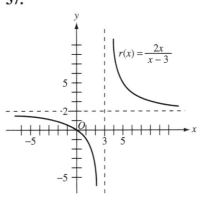

39.
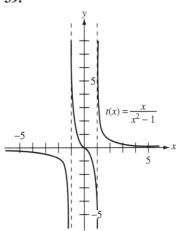

Section 4.4

1. addition, subtraction, multiplication, division, extraction of roots   **2.** the radicand is negative   **3.** $(-\infty, \infty)$
4. $\geq 1, \leq 1$ **5.** algebraic, polynomial   **6.** false   **7.** true   **8.** true   **9.** true   **10.** false   **11.** (b), (c), (e), and (f) are not rational functions.   **13.** (a) $(-\infty, 1) \cup (1, \infty)$ (b) $(-\infty, 1) \cup (1, \infty)$ (c) $(-\infty, -2] \cup (1, \infty)$ (d) $(-\infty, 1) \cup (1, \infty)$ (e) $(-\infty, -1] \cup [1, \infty)$ (f) $(-\infty, -2) \cup [-1, 1] \cup (2, \infty)$

15.
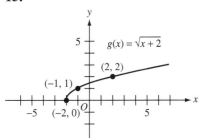

17.
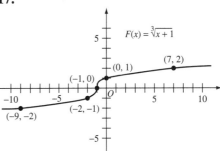

19.
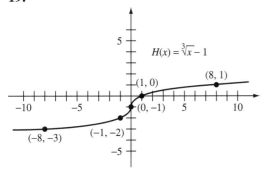

21.
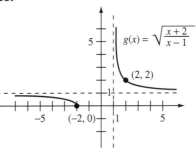

23.
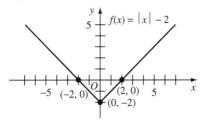

25.
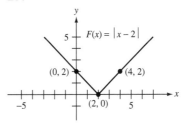

27.
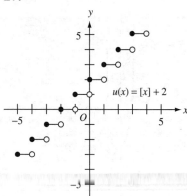

29.
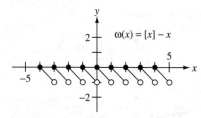

31.
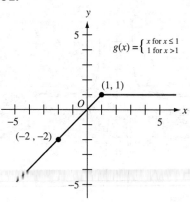

33.
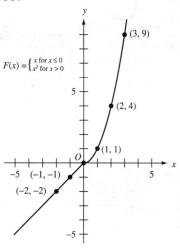

35.
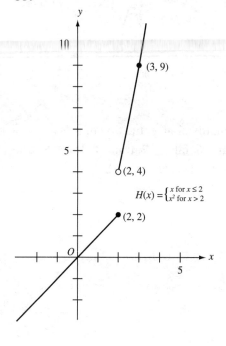

Section 4.5

1. x, y **2.** y, x **3.** at most one point **4.** $(g \circ f), (f \circ g)(x)$ **5.** the range, the domain **6.** false **7.** true **8.** true **9.** false **10.** true **11.** (a), (b), (e), and (f) are one-to-one. **13.** (a), (b), and (d) are graphs of one-to-one functions. **15.** $f^{-1}(x) = \sqrt[3]{x+1}$; $D_{f^{-1}} = R_{f^{-1}} = (-\infty, \infty)$ **17.** $f^{-1}(x) = x^2 - 2$; $D_{f^{-1}} = [0, \infty)$; $R_{f^{-1}} = [-2, \infty)$ **19.** $f^{-1}(x) = (x-2)^3$; $D_{f^{-1}} = R_{f^{-1}} = (-\infty, \infty)$

21.
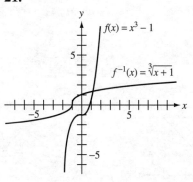

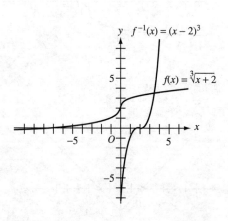

23. $g^{-1}(x) = -\sqrt{x}$; $D_{g^{-1}} = [1, \infty)$; $R_{g^{-1}} = (-\infty, -1]$ **25.** $g^{-1}(x) = -\sqrt{x+1}$; $D_{g^{-1}} = [0, \infty)$; $R_{g^{-1}} = (-\infty, -1]$ **27.** For Exercise 23, $(g^{-1} \circ g)(x) = g^{-1}(g(x)) = g^{-1}(x^2) = -\sqrt{x^2} = -|x| = -(-x) = x$ (since x is in D_g, x is negative and $|x| = -x$); $(g \circ g^{-1})(x) = g(g^{-1}(x)) = g(-\sqrt{x}) = (-\sqrt{x})^2 = x$ (since x is in $D_{g^{-1}}$, x is positive).
For Exercise 25, $(g^{-1} \circ g)(x) = g^{-1}(g(x)) = g^{-1}(x^2 - 1) = -\sqrt{x^2 - 1 + 1} = -\sqrt{x^2} = -|x| = -(-x) = x$ (since x is in D_g, x is negative and $|x| = -x$); $(g \circ g^{-1})(x) = g(g^{-1}(x)) = g(-\sqrt{x+1}) = (-\sqrt{x+1})^2 - 1 = x + 1 - 1 = x$ (since x is in $D_{g^{-1}}$, $x + 1$ is positive).

Section 4.6

1. a nonzero constant   **2.** inversely, x^2 **3.** directly, inversely   **4.** jointly as the base and the height, constant of variation   **5.** $v = kx^2y^3/z$ $(k \neq 0)$ **6.** true   **7.** false   **8.** true   **9.** true   **10.** false   **11.** $d = kt$ **13.** $I = kR$ **15.** $T = k\sqrt{L}$ **17.** The force exerted by flowing water varies directly as the sixth power of its velocity.   **19.** The time to travel between two cities varies inversely as the average rate of speed.   **21.** 108   **23.** \$112   **25.** 59.65 mph   **31.** $z = kxy^3$ **33.** $I = kV/R$ **35.** $P = kn/\sqrt{c}$ **37.** h varies directly as V and inversely as the square of r.   **39.** W varies jointly with w and the square of d and inversely with l **41.** 37 1/7 cubic inches   **43.** 3 2/3 ohms

Chapter 4 Review Exercises

1. (a) 30   **(b)** 48   **(c)** 4   **2. (a)** $(6x + 4)/(4x^2 + 5)$ **(b)** $(3\sqrt{x} + 4)/(x + 5)$ **(c)** $(3x + 1)/(x^2 - 2x + 6)$ **3. (a)** $D_f = R_f = (-\infty, \infty)$ **(b)** $D_g = (-\infty, \infty)$, $R_g = [1, \infty)$ **(c)** $D_h = (-\infty, \infty)$, $R_h = [0, \infty)$ **4. (a)** $(-\infty, 4/3) \cup (4/3, \infty)$ **(b)** $(4/3, \infty)$ **(c)** $(-\infty, -2) \cup (-2, 2) \cup (2, \infty)$ **5. (a)** $x^2 + 2 + \sqrt{x - 2}$ **(b)** $(x^2 + 2)\sqrt{x - 2}$ **(c)** $(x^2 + 2)/\sqrt{x - 2}$ **6. (a)** $27 + \sqrt{3}$ **(b)** $(4x^2 + 2)\sqrt{2x - 2}$ **(c)** $(x^4 + 2)/\sqrt{x^2 - 2}$ **7.** $D_{f+g} = [2, \infty)$, $D_{fg} = [2, \infty)$, $D_{f/g} = (2, \infty)$ **8. (a)** $\sqrt{1 + x}$ **(b)** $1 + \sqrt{x}$ **9. (a)** $\sqrt{6}$ **(b)** $1 + \sqrt{2x}$ **10.** $D_{g \circ f} = [-1, \infty)$, $D_{f \circ g} = [0, \infty)$ **11. (a)** and **(b)** and are polynomial functions.   **12. (b)** and **(b)** are rational functions.   **13. (a), (b),** and **(c)** are even functions.   **14. (b)** and **(c)** are odd functions.   **15. (a), (b),** and **(c)** have graphs that are symmetric with respect to the y-axis.   **16. (b),** and **(c)** have graphs that are symmetric with respect to the origin.   **17. (a)** $x = 5/2$ **(b)** $x = 2, x = -2$ **(c)** $x = 1, x = 3$ **18. (a)** $y = 1/2$ **(b)** $y = 0$ **(c)** no horizontal asymptote   **19.** vertical asymptote $x = 2$, horizontal asymptotes $y = 1$ and $y = -1$

20.

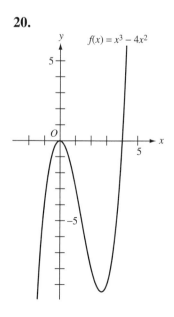

21.

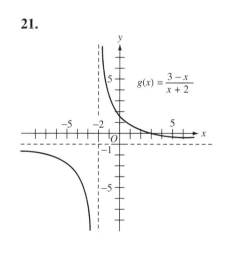

22.

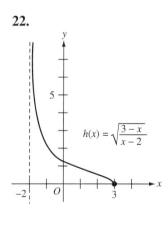

24. The graph of $g(x) = x^3 + 1$ is intersected in exactly one point by any horizontal line $y = c$.

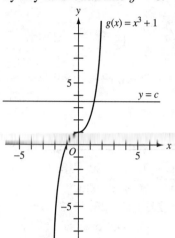

25. The x-axis intersects the graph in the two points $(0, 0)$ and $(4, 0)$

32.

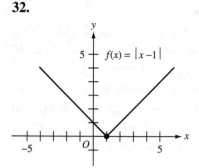

26.

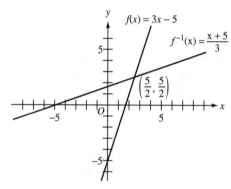

28. (c) is decreasing **29.** (c) is one-to-one
30. Suppose $x_1 > x_2$; then $3x_1 > 3x_2$ and $3x_1 + 5 > 3x_2 + 5$. Therefore, $f(x_1) > f(x_2)$. **31.** Suppose $x_1 > x_2$; then $-3x_1 < -3x_2$ and $-3x_1 + 5 < -3x_2 + 5$. Therefore, $g(x_1) < g(x_2)$.

27.

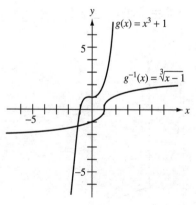

33.

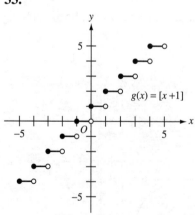

34.

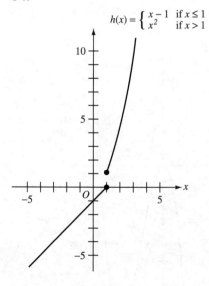

35. $D_g = (-\infty, \infty)$, $R_g =$ the set of all integers **36.** $D_h = (-\infty, \infty)$, $R_h(-\infty, 0] \cup (1, \infty)$ **37.** $y = kuv/w$
38. L varies jointly as w and the square of d and inversely as r. **39.** 80 **40.** about 1.1×10^{-12}

Chapter 5

Section 5.1

1. positive **2.** $(-\infty, \infty)$, $(0, \infty)$ **3.** > 1, < 1 (and > 0) **4.** $1/a$ **5.** one-to-one **6.** true **7.** true **8.** false
9. true **10.** true

11.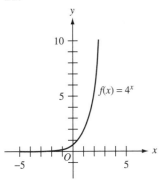

13.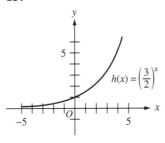

15.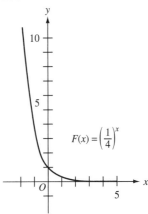

17.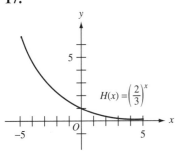

19.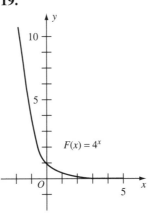

21.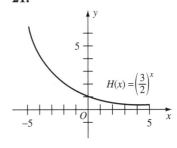

25.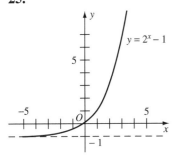

27.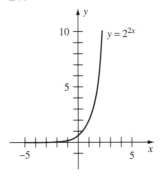

23. The graphs in Exercise 11 and 15, taken together, are symmetric with respect to the y − axis. The same is true for Exercise 12 and 16, 13 and 17, and 14 and 18.

29.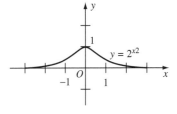

31.

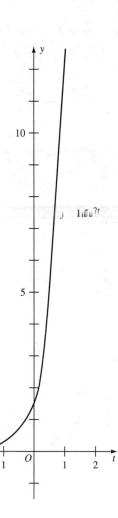

33.

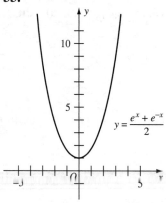

$y = \dfrac{e^x + e^{-x}}{2}$

35. $x =$ any real number
37. $x = 1/6$ **39.** $a^x = 4$
41. $a^x = 1/\sqrt{2}$ **43.** $2^{2\sqrt{2}} + 2^{2\sqrt{2}}$
45. 2^x **47.** a^{-x} **49.** $a^x/(a^x + 1)$

51. With an 8-place calculator, all three numbers agree. With a 10-place calculator, rounded off to 8 places, $(1 + 1/10^6)^{10^6} = .36787963$ and $(1 - 1/10^6)^{10^6} = .36787926$.
53. 1.1051709   **55.** .36666667   **57.** .99004983
59. 3.7371928, 3.9482220, 3.9874538, 3.9913983

Section 5.2
1. a^y **2.** a^x **3.** $(0, \infty), (-\infty, \infty)$ **4.** $> 1, > 0$ and < 1 **5.** $10, e$ **6.** false   **7.** true   **8.** false   **9.** true
10. true   **11.** $\log_2 16 = 4$ **13.** $\log_{10} \sqrt[3]{100} = 2/3$ **15.** $\log_{16} 64 = 3/2$ **17.** $10^2 = 100$ **19.** $3^{-3} = 1/27$
21. $a^0 = 1$ **23.** 4   **25.** 2   **27.** $-19/9, -17/9$ **29.** 100   **31.** -1000 **33.** 5   **35.** 0   **37.** -4

39.

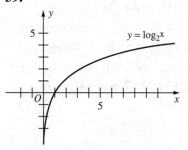

41.

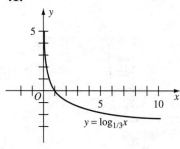

43.

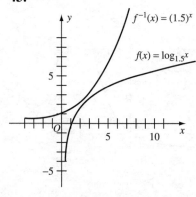

45.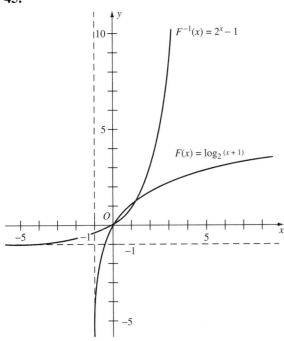

49. $\log_2 100 = (\log_{10} 100)/\log_{10} 2 \approx 6.6439$
51. $\log_6 .25 = (\log_{10} .25)/\log_{10} 6 \approx -.7737$
53. $\log_{17} 4981 = (\log_{10} 4981)/(\log_{10} 17) \approx 3.0049$
61. .795   **63.** 1.47, 7.86

Section 5.3

1. x **2.** x^3 **3.** 2^x **4.** $x^{3/2}$ **5.** $1, a^2$ **6.** false   **8.** false   **9.** true   **10.** true   **11.** 1.5441   **13.** -3.4559
15. -6.6990 **17.** $\log_a[(x-1)^{1/2}/x^7]$ **19.** $\log[(x+2)/(x-2)]$ **21.** $2 \log b + \log c - 4 \log d$
23. $(1/3) \log b + (1/6) \log c - \log d$ **25.** 7   **27.** 100   **29.** 2   **31.** 2   **35.** 2.3802   **37.** 6.1673   **39.** 5.4807
41. 14.2211   **43.** 6.12   **45.** .612
47. $Y = \log_a y = \log_a(ca^{kx}) = \log_a c + \log_a(a^{kx}) = \log_a c + kx = kx + \log_a c = mx + b$, where $m = k$, and $b = \log_a c$ **49.** .470   **51.** 1.26, 14.8

Section 5.4

1. (a) $141.76   (b) $141.90   (c) $141.91   **3.** $1000 at 7.95% compounded daily yields $23.13 more.   **5.** yes
7. $318.41   **9.** 40,960 bacteria   **11.** 640,00 bacteria   **13.** about 287 million people   **15.** about 45 days
17. 15,601 years   **21.** (a) $c \approx 1144$ deer   (b) about 276 deer   **23.** (a) 20 decibels   (b) 100 decibels   **25.** about 60.4 decibels   **27.** 10^{2k+16} watts   **29.** (a) approximately 50,119 joules   (b) approximately 1.585×10^{12} joules   **31.** $10^{2k-4.7}$ joules   **35.** about 106°F   **37.** (a) 6.4   (b) 2.2
39.

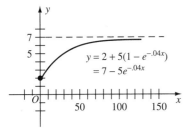

Chapter 5 Review Exercise

1.
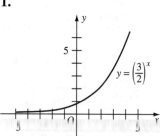

2.
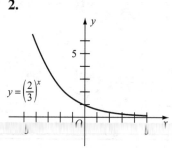

3. 4 **4.** 1 **5.** $a^x - 4$
6. $(1.1)^{10} = 2.5937$,
$(1.01)^{100} = 2.7048$,
$(1.001)^{1000} = 2.7169$,
$(1.0001)^{10,000} = 2.7181$

7.
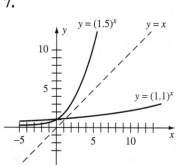

8. $\log_8(1/4) = -2/3$ **9.** $2^{5/3} = 2\sqrt[3]{4}$ **10.** $11, \sqrt{2}$ **11.** $5^{10}/2^{16}$ **12.** $\sqrt{2}$
13. $(1/2)\log x + (1/8)\log y - (3/4)\log z$ **14.** $\log_a[x^2(x+y)/y^3]$
15. Let $y = \log_3 x$ and $z = \log_9(x^2)$. Then $x = 3^y$ and $x^2 = 9^z$. Therefore, $x^2 = (3^y)^2 = 3^{2y} = (3^2)^y = 9^y$. Hence, $9^y = 9^z$ which means $y = z$.
16. $\log_6 28 = (\log_{10} 28)(\log_{10} 6) \approx 1.8597$ **17.** $(3/\ln 10 + 4/\ln 5)\ln x$

18.
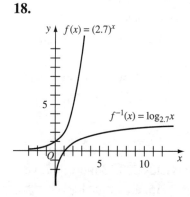

19.
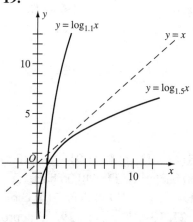

20. $x = e^{-2}$ **21.** $\ln x = -2$ **22.** $1/8$ **23.** $(\log 5)/(2\log 3 - \log 2)$ **24.** approximately 550
25. approximately .0550 **26.** approximately 9.0 **27.** approximately 10.3189 **28.** $4\log 5 - \dfrac{1}{2}\log 7 = 2.37333$
29. $\log 1.4 = \log(7/5) = \log 7 - \log 5 = .14613$
30. $\log_a x = [\log_{a^n}(x)]/[\log_{a^n}(a)] = [\log_{a^n}(x)]/[1/n] = n\log_{a^n}(x)$ **31.** $4/(5\log 1.02) \approx 93$
32. approximately 872.04 **33.** $r \approx .104$ **34.** \$3483.94 **35.** \$3499.08 **36.** \$6376.64 **37.** about 14 years
38. about 830 years **39.** $D_2 = 10\log(10^{16}I_2) = 10\log(10^{16} \cdot 10I_1) = 10\log(10^{16}I_2 \cdot 10) = 10[\log(10^{16}I_1) + \log 10] = 10\log(10^{16}I_1) + 10\log 10 = D_1 + 10$ **40.** 1.0×10^{14} joules

Chapter 6

Section 6.1

1. $n+1$ **2.** n **3.** $n!/[(n-r)!r!]$ **4.** a^4b^8 **5.** $12!/(6!6!)$ **6.** false **7.** true **8.** false **9.** false **10.** false
11. $(x^2 - 2x + 1)(x - 1) = x^3 - 3x^2 + 3x - 1$
13. $(x^2 + 4x + 4)(x + 2)(x + 2) = (x^3 + 6x^2 + 12x + 8)(x + 2) = x^4 + 8x^3 + 24x^2 + 32x + 16$
15. $(x^2 - 2x + 1)(x - 1)(x - 1)(x - 1) = (x^3 - 3x^2 + 3x - 1)(x - 1)(x - 1) =$
$(x^4 - 4x^3 + 6x^2 - 4x + 1)(x - 1) = x^5 - 5x^4 + 10x^3 - 10x^3 - 10x^2 + 5x - 1$ **17.** $x^5, x^4, x^3, x^2, x, 1$
19. $a^4x^4, a^3x^3y, a^2x^2y^2, axy^3, y^4$ **21.** $x^4 + 8x^3 + 24x^2 + 32x + 16$ **23.** $16x^4 + 96x^3 + 216x^2 + 216x + 81$
25. $1 + 6x + 15x^2 + 20x^3 + 15x^4 + 6x^5 + x^6$ **27.** 1120 **29.** (a) 64 (b) 2^n **31.** (a) 24 (b) 120 (c) 5040
33. (a) $(n + 2)(n + 1)$ (b) $n(n + 1)$ (c) $(2n + 2)(2n + 1)$ **35.** (a) $_5C_3 = 10$, $_4C_3 + {}_4C_2 = 4 + 6 = 10$
(b) $_6C_1 = 6$, $_5C_1 + {}_5C_0 = 5 + 1 = 6$ **37.** $x^4 + 4x^3y + 6x^2y^2 + 4xy^3 + y^4$
39. $16x^4 - 96x^3y + 216x^2y^2 - 216xy^3 + 81y^4$ **41.** $x^6 + 12x^5 + 60x^4 + 160x^3 + 240x^2 + 192x + 64$
43. $32x^5 + 60x^4 + 45x^3 + (135/8)x^2 + (405/128)x + (243/1024)$ **45.** 28 **47.** 56 **49.** 60
51. $(1.1)^5 = 1 + 5(.1) + 10(.1)^2 + 10(.1)^3 + 5(.1)^4 + (.1)^5 = 1.61051$
53. $(1.01)^4 = 1 + 4(.01) + 6(.01)^2 + 4(.01)^3 + (.01)^4 = 1.04060401$
55. $(9.8)^3 = (10 - .2)^3 = 10^3 + 3 \cdot 10^2(-.2)^2 + 3 \cdot 10(-.2)^2 + (-.2)^3 = 941.192$
57. $2^n = (1 + 1)^n = {}_nC_0 + {}_nC_1 + {}_nC_2 + \cdots + {}_nC_n$
59. $(x + y)^3 + 3(x + y)^2 z + 3(x + y)z^2 + z^3 = x^3 + y^3 + z^3 + 3x^2y + 3xy^2 + 3x^2z + 3xz^2 + 3y^2z + 3yz^2 + 6xyz$
61. $(1 + .0)^4 \approx \$1.04$ **63.** $100(1.08)^3 \approx \$125.97$ **65.** $_{30}C_3 = 4060$ ways **67.** $_{10}C_1 = 45$ choices
69. $_{100}C_2 = 4950$ ways

Section 6.2

1.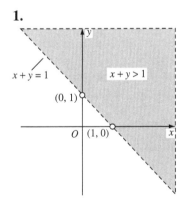

3.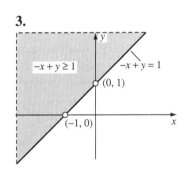

5.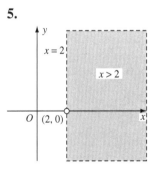

7.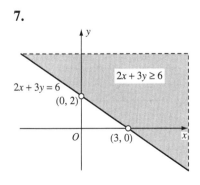

9.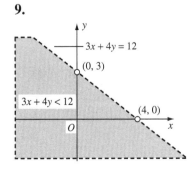

11.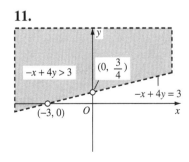

13.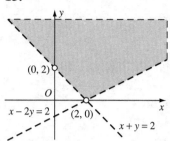

15.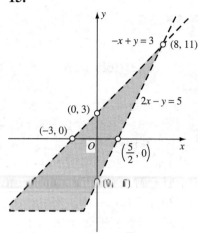

17.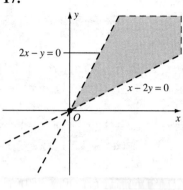

19.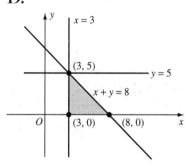

21.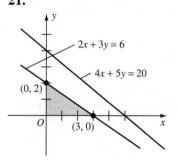

23.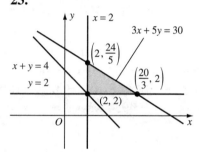

25.
$$x \geq 0$$
$$y \geq 0$$
$$x + 2y \leq 8$$
$$2x + y \leq 10$$

27.
$$x \geq 0$$
$$y \geq 0$$
$$2x + 3y \leq 18$$
$$2x + y \leq 10$$

29.
$$x + y < 12$$
$$x > 4$$
$$y > 3$$

31.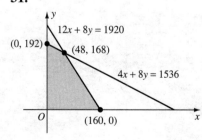

33.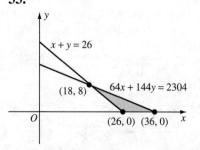

35.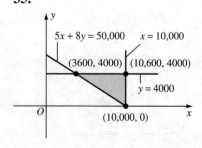

37.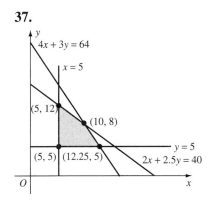

39.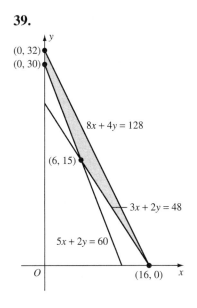

41. (a) $204 (b) $160 (c) $174
43. (a) $70 (b) $52 (c) $52
45. 7600 tickets
47. (a) $252 (b) $230 (c) $246
49. (a) $90 (b) $96 (c) $102

Section 6.3

1. Proper **2.** degree **3.** greater than, equal to, division **4.** real factored form **5.** $n, 2m$ **6.** false **7.** false
8. true **9.** true **10.** true **11.** $\dfrac{3}{x+2} - \dfrac{2}{x-3}$ **13.** $\dfrac{1}{x} + \dfrac{4}{x-4}$ **15.** $\dfrac{3}{x} - \dfrac{1}{x-1} + \dfrac{5}{(x-1)^2}$
17. $\dfrac{1}{x} - \dfrac{1}{x-2} + \dfrac{3}{(x-2)^2} - \dfrac{4}{(x-1)^3}$ **19.** $\dfrac{1}{(x-1)^3}$ **21.** $\dfrac{1}{x-1} + \dfrac{2}{(x-1)^2} + \dfrac{1}{(x-1)^3}$
23. $x + 5 - \dfrac{3}{x-1} + \dfrac{14}{(x-2)}$ **25.** $\dfrac{x}{x^2+1}$ **27.** $\dfrac{1}{x} - \dfrac{x+1}{x^2+x+1}$ **29.** $\dfrac{1}{8(x-2)} - \dfrac{1}{8(x+2)} - \dfrac{1}{2(x^2+4)}$
31. $\dfrac{1}{(x^2+1)^2}$ **33.** $\dfrac{1}{4(x-1)} + \dfrac{1}{4(x-1)^2} - \dfrac{1}{4(x+1)} + \dfrac{1}{4(x+1)^2}$

Section 6.4

1. 2, 2 **2.** triangular **3.** variables, constant term **4.** coefficient, constant **5.** interchanging two rows, multiplying each entry of a row by a nonzero constant, adding a constant times each entry of one (fixed) row to the corresponding entry of another (fixed) row **6.** true **7.** true **8.** false **9.** true **10.** true **11.** $x = -1$, $y = 10, z = 5$ **13.** $x = 2z, y = z + 1$, z = any real number **19.** $x = 2z + 1/2, y = z + 1/2$, z = any real number **21.** $\begin{array}{l} 2x + y = -3 \\ 4x - 5y = 6 \end{array}$ **23.** $\begin{array}{l} x + 2y - 3z + 4t = 5 \\ 2x - y + 7t = 6 \end{array}$ **25.** $x = 2, y = -3, z = 4$ **27.** $\begin{bmatrix} 1 & -1 & | & 6 \\ 2 & 1 & | & 3 \end{bmatrix}$
29. $\begin{bmatrix} 2 & -1 & 3 & | & 8 \\ 7 & 1 & -1 & | & 4 \\ -1 & 2 & 1 & | & 3 \end{bmatrix}$ **31.** $x = 2, y = 3$ **33.** $x = 3, y = -2, z = 1$ **35.** $x = 3 - z, y = 1 + 2z$, z = any real number **37.** $x = 1, y = -2, z = 3$ **39.** no solution **41.** $x = 1, y = -2$ **43.** $x = 2, y = 1, z = 3$
45. $x = -7z - 10, y = -5z - 6$, z = any real number **47.** $x = 3y + 5$, y = any real number **49.** $x = -3, y = 1, z = 2$

Section 6.5

1. rows, columns **2.** same size **3.** cA **4.** columns, rows **5.** I, invertible **6.** false **7.** true **8.** false
9. false **10.** true **11. (a)** 3×1 **(b)** 1×3 **(c)** 2×3 **(d)** 3×3 **(e)** 3×5 **13. (a)** $x = 2, y = 4$
(b) $x = -3, y = 1$ **15.** $\begin{bmatrix} 2 & 1 & 4 \\ 6 & 0 & 4 \\ 10 & 10 & 10 \end{bmatrix}$ **17.** $\begin{bmatrix} 0 \\ 0 \\ 0 \end{bmatrix}$ **19.** A and B cannot be added because they are not the same size.

21. $\begin{bmatrix} -3 & 0 & 3 \\ -6 & 6 & -3 \\ -9 & -12 & 0 \end{bmatrix}$ **23.** $\begin{bmatrix} 1 & -2 & 3 \\ 3/2 & 5/2 & 7/2 \end{bmatrix}$ **25. (a)** $2(3A) = 2\begin{bmatrix} 3 & 9 & -6 \\ 0 & -12 & 9 \end{bmatrix} = \begin{bmatrix} 6 & 18 & -12 \\ 0 & -24 & 18 \end{bmatrix} = 6A$

(b) $2B + 3B = \begin{bmatrix} 4 & -2 & 2 \\ 6 & 0 & 10 \end{bmatrix} + \begin{bmatrix} 6 & -3 & 3 \\ 9 & 0 & 15 \end{bmatrix} = \begin{bmatrix} 10 & -5 & 5 \\ 15 & 0 & 25 \end{bmatrix} = 5\begin{bmatrix} 2 & -1 & 1 \\ 3 & 0 & 5 \end{bmatrix} = 5B = (2+3)B$

(c) $-2(A + C) = -2\begin{bmatrix} 1 & 0 & -1 \\ 2 & -1 & 2 \end{bmatrix} = \begin{bmatrix} -2 & 0 & 2 \\ -4 & 2 & -4 \end{bmatrix}$; $-2A - 2C = \begin{bmatrix} -2 & -6 & 4 \\ 0 & 8 & -6 \end{bmatrix} - \begin{bmatrix} 0 & -6 & 2 \\ 4 & 6 & -2 \end{bmatrix} = \begin{bmatrix} -2 & 0 & 2 \\ -4 & 2 & -4 \end{bmatrix}$

27. $[11]$ **29.** $\begin{bmatrix} -13 & -15 \\ 19 & 5 \end{bmatrix}$ **31.** $\begin{bmatrix} 4 & -3 & 43 \\ 0 & -5 & 9 \\ 10 & 24 & 4 \end{bmatrix}$ **33.** $\begin{bmatrix} 0 & 3 & 6 \\ 4 & 2 & 0 \\ 1 & 8 & 15 \\ -14 & 2 & 18 \end{bmatrix}$

35. (a) $(AB)C = \begin{bmatrix} 1 & 5 \\ 6 & 2 \\ 11 & -1 \end{bmatrix}\begin{bmatrix} -5 & -2 \\ 1 & 6 \end{bmatrix} = \begin{bmatrix} 0 & 28 \\ -28 & 0 \\ -56 & -28 \end{bmatrix}$; $A(BC) = \begin{bmatrix} 1 & -1 \\ 2 & 0 \\ 3 & 1 \end{bmatrix}\begin{bmatrix} -14 & 0 \\ -14 & -28 \end{bmatrix} = \begin{bmatrix} 0 & 28 \\ -28 & 0 \\ -56 & -28 \end{bmatrix}$

(b) $A(B + C) = \begin{bmatrix} 1 & -1 \\ 2 & 0 \\ 3 & 1 \end{bmatrix}\begin{bmatrix} -2 & -1 \\ 3 & 2 \end{bmatrix} = \begin{bmatrix} -5 & -3 \\ -4 & -2 \\ -3 & -1 \end{bmatrix}$; $AB + AC = \begin{bmatrix} 1 & 5 \\ 6 & 2 \\ 11 & -1 \end{bmatrix} + \begin{bmatrix} -6 & -8 \\ -10 & -4 \\ -14 & 0 \end{bmatrix} = \begin{bmatrix} -5 & -3 \\ -4 & -2 \\ -3 & -1 \end{bmatrix}$

39. $\begin{bmatrix} 3/5 & -2/5 \\ 1/5 & 1/5 \end{bmatrix}$ **41.** A is not invertible because $|A| = 0$. **43.** $A^{-1} = \begin{bmatrix} 4/15 & -1/6 & 1/10 \\ 7/15 & 2/6 & -2/10 \\ -1/15 & 1/6 & 1/10 \end{bmatrix}$

45. $\begin{bmatrix} 2 & -1 \\ 8 & 1 \end{bmatrix}\begin{bmatrix} x \\ y \end{bmatrix} = \begin{bmatrix} 4 \\ 1 \end{bmatrix}$; $\begin{bmatrix} x \\ y \end{bmatrix} = \begin{bmatrix} 1/10 & 1/10 \\ -4/5 & 1/5 \end{bmatrix}\begin{bmatrix} 4 \\ 1 \end{bmatrix} = \begin{bmatrix} 1/2 \\ -3 \end{bmatrix}$ **47.** $\begin{bmatrix} 3 & 6 & 1 \\ 3 & -3 & 2 \\ 6 & 9 & 2 \end{bmatrix}\begin{bmatrix} x \\ y \\ z \end{bmatrix} = \begin{bmatrix} 0 \\ 2 \\ 1 \end{bmatrix}$; $\begin{bmatrix} x \\ y \\ z \end{bmatrix} =$

$\begin{bmatrix} -8/3 & -1/3 & 5/3 \\ 2/3 & 0 & -1/3 \\ 5 & 1 & -3 \end{bmatrix}\begin{bmatrix} 0 \\ 2 \\ 1 \end{bmatrix} = \begin{bmatrix} 1 \\ -1/3 \\ -1 \end{bmatrix}$ **49. (a)** 60 type I clocks and 5 type II clocks **(b)** 48 type I clocks
and 20 type II clocks **(c)** 28 type I clocks and 45 type II clocks

Section 6.6

1. $x - 5$ **2.** -3 **3.** $-56, 14, -1$ **4.** $[x - (1 + i)][x - (1 - i)]$ **5.** $6x^4 - 5x^3 + x^2 + 3, x - 4$ **6.** false
7. true **8.** false **9.** false **10.** true **11. (a)** 10 **(b)** -29 **13. (a)** 0 **(b)** 42 **15. (a)** 0 **(b)** 0
17.
$$\begin{array}{r} x + (2x + a) \\ x - a \overline{\smash{\big)}\, x^2 + 2x - 3} \\ \underline{x^2 - ax } \\ (2 + a)x - 3 \\ \underline{(2 + a)x - a(2 + a)} \\ a^2 + 2a - 3 = p(a) \end{array}$$

19.
$$\begin{array}{r} 2x^2 + (2a + 1)x + (2a^2 + a - 4) \\ x - a \overline{\smash{\big)}\, 2x^3 + x^2 - 4x + 5} \\ \underline{2x^3 + 2ax^2 } \\ (2a + 1)x^2 - 4x \\ \underline{(2a + 1)x^2 - a(2a + 1)x } \\ (2a^2 + a - 4)x + 5 \\ \underline{(2a^2 + a - 4)x - a(2a^2 + a - 4)} \\ 2a^3 + a^2 - 4a + 5 = P(a) \end{array}$$

21. quotient $= x^2 - x$; remainder $= -5$ **23.** quotient $= 2x^4 - 2x^3 + 7x^2 - 7x + 4$; remainder $= 0$
25. $P(5) = 874$; $P(-5) = 384$ **27.** $P(5) = 87,978$; $P(-5) = 99,478$ **29.** $P(3/2) = 8.125$; $P(-2.5) = 9.125$
31. $P(-1) = 0$ **33.** $P(2) = 0$ **35.** $P(1/2) = 0$ **37.** $x - 1, x - 2$, and $x + 3$, all of order 1   **39.** $x + 2$ of order 2 and $x - 3$ of order 1   **41.** The complex and real factored form of $P(x)$ is $(x - 2)(x + 2)(x - 1)$.   **43.** The complex and real factored form of $P(x)$ is $(x + 1)(x - 1)(x + 2)(x - 2)$.   **45.** The complex factored form of $P(x)$ is $(x + 2)[x - (1 + i)][x - (1 - i)]$, and real factored form is $(x + 2)(x^2 - 2x + 2)$.   **47.** The complex and real factored form of $P(x)$ is $(x - 1)(x + 1/2)(x - 1/3)(x + 1/4)$.   **49.** The complex factored form of $P(x)$ is $(x - 1/3)^2[x - (3 - 2i)][x - (3 + 2i)]$, and real factored form is $(x - 1/3)^2(x^2 - 6x + 13)$.   **53.** Complex zeros appear is conjugate *pairs*.   **55.** $P(x) = [(2x - 4)x + 5]x + 7$; $P(6) = [(2 \cdot 6 - 4)6 + 5]6 + 7 = 325$
57. $P(x) = \{[(x + 1)x + 12]x - 4\}x - 15$; $P(7) = \{[(7 + 1)7 + 12]7 - 4\}7 - 15 = 3289$ **59.** Exercise 55:

$$6\,\overline{)\,2\ \ -4\ \ \ \ 5\ \ \ \ \ 7}$$
$$12\ \ 48\ \ 318$$
$$\overline{\,2\ \ \ \ 8\ \ \ 53\ \ \ 325 = P(6)}$$

Exercise 57:

$$7\,\overline{)\,1\ \ \ 1\ \ \ 12\ \ -4\ \ -15}$$
$$7\ \ \ 56\ \ \ 476\ \ \ 3304$$
$$\overline{\,1\ \ \ 8\ \ \ 68\ \ \ 472\ \ \ 3289 = P(7)}$$

Section 6.7

1. $\pm 1, \pm 1/2, \pm 3/2$ **2.** $5n$, where n is a nonzero integer   **3.** $3n$, where n is a nonzero integer   **4.** two   **5.** one
6. false   **7.** false   **8.** false   **9.** true   **10.** true   **11.** $1, -1, 2$ **13.** $1, 2, -2, 4$ **15.** $-2, 1/4$ **17.** (a) $1, \pm 2$
(b) $P(x) = (x - 1)(x - 2)(x + 2)$ (c) none   **19.** (a) $1/2$ (b) $P(x) = 2(x - 1/2)(x^2 + 1)$ (c) $\pm i$ **21.** (a) $1, -2$ (b) $\pm\sqrt{3}$, $P(x) = (x - 1)(x + 2)(x - \sqrt{3})(x + \sqrt{3})$ (c) none   **23.** (a) ± 1
(b) $P(x) = (x - 1)(x + 1)(x^2 + 2x + 2)$ (c) $-1 + i, -1 - i$ **25.** $a = -1, b = 2$ **27.** $a = -7, b = 1$
29. three   **31.** three   **33.** two (including 0)   **35.** (a) one positive, two negative   (b) $a = -3, b = 1$ **37.** (a) two positive, one negative   (b) $a = -5, b = 7$ **39.** (a) one positive, three negative   (b) $a = -8, b = 1$

Chapter 6 Review Exercises

1.

$$
\begin{array}{ccccccc}
 & & & 1 & & & \\
 & & 1 & & 1 & & \\
 & & 1 & 2 & 1 & & \\
 & 1 & 3 & & 3 & 1 & \\
 & 1 & 4 & 6 & 4 & 1 & \\
1 & 5 & 10 & & 10 & 5 & 1
\end{array}
$$

$(a + b)^2 = a^2 + 2ab + b^2$
$(a + b)^3 = a^3 + 3a^2b + 3ab^2 + b^3$
$(a + b)^4 = a^4 + 4a^3b + 6a^2b^2 + 4ab^3 + b^4$
$(a + b)^5 = a^5 + 5a^4b + 10a^3b^2 + 10a^2b^3 + 5ab^4 + b^5$

2. (a) 21   (b) 1   (c) 120   **3.** $_8C_4 = 70$; $_7C_4 + _7C_3 = 35 + 35 = 70$ **4.** (a) $x^5 + 10x^4 + 40x^3 + 80x^2 + 80x + 32$ (b) $x^4 - 12x^3 + 54x^2 - 108x + 81$
5. $[(x - y) + 1]^4 = (x - y)^4 + 4(x - y)^3 + 6(x - y)^2 + 4(x - y) + 1 =$
$x^4 - 4x^3y + 6x^2y^2 - 4xy^3 + y^4 + 4x^3 - 12x^2y + 12xy^2 - 4y^3 + 6x^2 - 12xy + 6y^2 + 4x - 4y + 1$

6.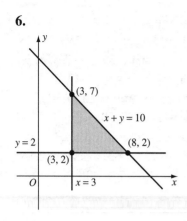

7.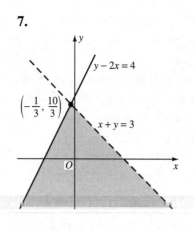

8. 16,000 at \$25 and 4,000 at \$40
9. \$9,400
10. 35 mph and 45 mph
11. 40 of A and 20 of B
12. $\dfrac{1}{x} + \dfrac{2}{x-2} + \dfrac{3}{x+3}$
13. $\dfrac{-1}{x^2} + \dfrac{1}{2(x-1)} - \dfrac{1}{2(x+1)}$
14. $2x + 1 + \dfrac{4}{x+1} + \dfrac{3}{x-2}$
15. $x^2 - 1 + \dfrac{1}{x} + \dfrac{2x+3}{x^2+1}$

16. unique solution: $x = -3$, $y = 2$, $z = 4$ **17.** infinite number of solutions: $x = 1 + z$, $y = 2z$, $z =$ any value
18. no solution **19.** infinite number of solutions: $x = -z/5$, $y = 3z/5$, $z =$ any value **20.** $x = -1$, $y = -8$, $z = 7$, $t = 6$ **21.** Matrices of different sizes cannot be added. **22.** The number of columns of the first matrix is not equal to the number of rows of the second. **23.** The inverse of the first matrix does not exist. **24.** An inverse is defined only for a square matrix. **25.** $\begin{bmatrix} 2 & 12 \\ 13 & 13 \end{bmatrix}$ **26.** $\begin{bmatrix} -12 \\ 16 \end{bmatrix}$ **27.** $\begin{bmatrix} -3 & -4 \\ 6 & 13 \\ 11 & 3 \end{bmatrix}$ **28.** $\begin{bmatrix} 8 & -21 \\ -7 & 17 \end{bmatrix}$

29. $\begin{bmatrix} 0 & 0 \\ 0 & 0 \end{bmatrix}$ **30.** $\begin{bmatrix} 1 & 0 \\ 0 & 1 \end{bmatrix}$ **31, 32.** $\begin{bmatrix} 3/2 & 1/4 \\ -1 & -1/2 \end{bmatrix}$ **33.** $\begin{bmatrix} 1 & 1/2 & 0 \\ -1/2 & -1/2 & 1/2 \\ -1/2 & -1 & 1/2 \end{bmatrix}$ **34.** $x - y = 2$ **35.** $(5, 1, -2)$

36. $\begin{bmatrix} 2 & -1 \\ 1 & 3 \end{bmatrix}\begin{bmatrix} x \\ y \end{bmatrix} = \begin{bmatrix} 7 \\ 0 \end{bmatrix}$; $A^{-1} = \begin{bmatrix} 3/7 & 1/7 \\ -1/7 & 2/7 \end{bmatrix}$; $x = 3$, $y = -1$

37. $\begin{bmatrix} 1 & 1 & 3 \\ 0 & -1 & 1 \\ 1 & 0 & 1 \end{bmatrix}\begin{bmatrix} x_1 \\ x_2 \\ x_3 \end{bmatrix} = \begin{bmatrix} 2 \\ -4 \\ 1 \end{bmatrix}$; $A^{-1} = \begin{bmatrix} -1/3 & -1/3 & 4/3 \\ 1/3 & -2/3 & -1/3 \\ 1/3 & 1/3 & -1/3 \end{bmatrix}$; $x_1 = 2$, $x_2 = 3$, $x_3 = -1$ **38.** $B^t A^t = \begin{bmatrix} a'a + c'b & a'c + c'd \\ b'a + d'b & b'c + d'd \end{bmatrix}$; $(AB)^t = \begin{bmatrix} aa' + bc' & ab' + bd' \\ ca' + dc' & cb' + dd' \end{bmatrix}' = \begin{bmatrix} aa' + bc' & ca' + dc' \\ ab' + bd' & cb' + dd' \end{bmatrix}$ **39.** 20 of type A, 10 of type B, 5 of type C **40.** Every polynomial of degree n has n roots in the complex number system. **41.** The general polynomial equation of degree ≥ 5 cannot be solved by algebraic methods. **42.** remainder $= P(1) = -4$ **43.** remainder $= P(-2) = 8$ **44.** $P(5) = 276$ **45.** $P(-5) = -151$ **46.** $P(\sqrt{2}) = 0$
47. $P(i) = 0$ **48.** $x, x - 1, x + 2$ **49.** $P(x) = 2x^3 - x^2 + 2x - 1$
50. $P(x) = (x+2)(x-3)^2(x-\sqrt{2})[x-(1+2i)][x-(1-2i)]$
51. $P(x) = (x+2)(x-3)^2(x-\sqrt{2})(x^2-2x+5)$ **52.** if $3i$ is a root, the complex conjugate $-3i$ is also a root
53. $1, -1/2$ **54.** The only possible rational roots are ± 1 and ± 2, but $P(1) = 6$, $P(-1) = -16$, $P(2) = 8$, and $P(-1) = -180$. **55.** 2 is an upper bound; -2 is a lower bound. **56.** 5 is an upper bound; -1 is a lower bound.
57. 1 is a root of order 2. **58.** $(x-1)^2(x^2+1)$ **59.** $3n$, where n is any nonzero integer **60.** 0 and ± 1 **61.** 3
62. 1 **63.** 4 **64.** 3 **65.** By Descartes' rule of signs, there are no positive roots and no negative roots. Also, 0 is not a root. **66.** none **67.** $1, 2, \sqrt{2}, -\sqrt{2}$

Index

---------- **A** ----------

Abel, Niels Henrik,, 403
Abscissa, 129
Absolute value, 4, 23, 102
Addition
  of fractions, 29
  of polynomials, 51
  of signed quantities, 17
Additive identity, 7
Additive inverse, 8
Algebraic expression, 4, 41
Algebraic operations, 41
Algebraic process, 41
Algebraic properties of equality, 9, 97
Analytic geometry, 137
Apollonius, 245
Associative property
  for addition, 6, 391
  extended, 13
  for multiplication, 7, 394
Asymptotes
  horizontal, 273, 275, 282
  oblique, 274
  vertical, 271-72, 282
Avogadro's number, 43

---------- **B** ----------

Base
  for an exponential function, 311
  for a logarithmic function, 318-19
  for a power of x, 35
Binomial, 52
Binomial coefficients, 350-52
Binomial theorem, 352
Bounds for roots of polynomial equations
  lower, 418
  upper, 417

---------- **C** ----------

Cartesian (rectangular) coordinates, 129
Circle, standard equation of, 215
Closed system, 4
Coefficients
  binomial, 350-52
  of a polynomial, 50
  of a system of equations, 164
Coincident lines, 151

Combinations, 355
Common denominator, 71
Common factor property, 27
Commutative property
  for addition, 6
  extended, 13
  for multiplication, 6
Completing the square, 178
Completion axiom, 22
Complex factored form, 410
Complex fractions, 31
Complex number system, 59, 185, 187
Complex numbers
  addition and subtraction of, 188
  complex conjugate, 190, 409
  construction of $(a + bi)(c + di)$, 195
  definition of, 187
  division of, 189
  equality of, 188
  geometric representation of, 192-93
  multiplication of, 188
  multiplication by i, 195
  parallelogram rule for addition, 193
Composition of functions, 258
Compound interest, 44, 332, 355
Conditional equation, 32
Conic sections, 214
Conjugates, 86
Constant, 41
Constant terms, 164
Cooling formula, 342
Coordinate(s)
  abscissa (x-coordinate), 129
  ordered pair, 129
  ordinate (y-coordinate), 129
  origin, 129
  rectangular (Cartesian), 129
Coordinate axes, 128-29
Coordinate geometry, 137
Cramer's rule, 158
Cube root, 40

---------- **D** ----------

Dantzig, Tobias, 3
Decibel level, 339
Decimal, 2
Dependent variable, 253

Descartes, Rene, 24, 129, 137, 414, 419
Descartes' rule of signs, 419
Determinant of order 2, 157
Difference of cubes, 67
Difference of squares, 53, 63
Dirichlet, Peter, 286
Discriminant, 182
Distance formula
  along a number line, 24, 131
  in the plane, 132
  from a point to a line, 159
Distributive property
  extended, 13
  left, 7, 394
  right, 7, 394
Dividend, 56
Division
  of absolute values, 23
  definition of, 19
  of fractions, 30
  of nth roots, 41
  of polynomials, 55-56
  of signed quantities, 18-19
  synthetic, 405-07
Division algorithm
  for integers, 55
  for polynomials, 55
Divisor, 56
Domain of a function, 252
Double inverse property
  for addition, 8
  for multiplication, 10

---------- **E** ----------

Ellipse
  definition of, 235
  directrix, 247
  eccentricity, 247
  focal radius, 235
  focus, 235
  major axis, 236
  minor axis, 236
  standard equation of, 235
  with center (x_0, y_0), 238
Empty set, 110

Equality
  algebraic properties of, 9, 97
  of fractions, 28
  logical properties of, 10
  of matrices, 389
Equation
  with absolute values, 102-03
  definition of, 96
  exponential, 331
  with fractional expressions, 99, 197
  of a line, 140, 142, 144-46
  of a plane, 164
  linear, 98, 148
  logarithmic, 331
  quadratic, 176
  with radicals, 100-01, 198
  reducible to quadratic, 196
Equivalent equations, 97
Euler, Leonhard, 286, 315
Exponent
  negative, 38
  positive, 35
  zero, 37
Exponential equation, 331
Exponential function
  definition of, 311
  graph of, 312
  natural base, 315
  properties of, 312, 314-15
Exponential growth and decay, 336
Extraneous roots, 99

---------- F ----------
Factor theorem, 407
Factorial, 351
Factoring polynomials
  common factor, 63
  difference of two cubes, 67
  difference of two squares, 62-63
  grouping, 63
  miscellaneous, 66-67
  perfect squares, 62
  quadratics, 64-65
  by roots, 183
  sum of two cubes, 67
Fermat, Pierre, 137
Field axioms, 14
Fractions
  addition and subtraction of, 29
  complex, 31
  division of, 30
  equality of, 28
  multiplication of, 30
Function(s)
  algebraic, 280
  composition of, 258
  constant, 262
  decreasing, 291
  definition of, 59, 222, 252
  dependent variable, 253
  difference, 256
  domain of, 252
  exponential, 310-11
  even, 265
  graph of, 254
  greatest integer, 285
  increasing, 290
  independent variable, 253
  inverse, 291-92
  linear, 262
  logarithmic, 318-19
  odd, 265
  one-to-one, 288
  piecewise algebraic, 285
  polynomial, 262
  product, 256
  quadratic, 222, 262
  quotient, 256
  range of, 252
  rational, 270
  step, 285
  sum, 256
Fundamental theorem of algebra, 403

---------- G ----------
Galois, Evariste, 403
Gauss, Carl Friedrich, 380, 403, 484
Gaussian elimination, 376-80, 386
Gauss-Jordan elimination, 385-6
General linear eqution, 146
Graphs
  of algebraic functions, 281
  of equations, 140
  of exponential functions, 312
  of functions, 254
  of inverse functions, 293-94
  of logarithmic functions, 320
  of one-to-one functions, 289
  of polynomial functions, 265
  of rational functions, 276
Greater than, 3, 21
Grouping symbols, 12

---------- H ----------
Half-life, 336
Half-plane rule, 357
Half-planes
  left and right, 356
  lower and upper, 358
Horizontal line, equation of, 144
Horizontal line test, 289
Horner's method, 407
Hyperbola
  asymptotes, 243
  definition of, 239
  directrix, 247
  eccentricity of, 247
  focal radii, 239
  foci, 239
  fundamental rectangle, 244
  standard equation of, 240-41
  with center (x_0, y_0), 241-42

---------- I ----------
Imaginary part of a complex
    number, 187
Improper rational expression, 373
Independent variable, 253
Index, 40
Inequality
  with absolute values, 113-14
  definition of, 107
  with fractional expressions, 111
  strict, 107
Insolvability theorem, 403
Integers, 2, 3
Intensity of sound, 339
Intercept equation of a line, 149
Intercepts, 144, 149, 263
Interest
  compound, 44, 332, 355
  continuous, 333
Intersection
  of lines, 155
  of lines, circles, and parabolas, 230
  of rational functions, 276
Interval
  finite, 110-11
  infinite, 109-10
Inverse
  of a matrix, 395
  of a product, 19
  of a quotient, 20
  of a sum, 18
Irrational numbers, 2
Irreducible quadratic, 184, 410

---------- J ----------
Jordan, Wilhelm, 385

---------- K ----------
Kepler, Johannes, 245

---------- L ----------
Learning curve, 343
Least common denominator, 73
Least common multiple, 71-72
Leibniz, Gottfried, 286
Less than, 3
Libby, Willard, 342

Line(s)
  equations of, 140, 142, 144-146
  horizontal, 144
  intersecting, 155
  parallel, 150-151
  Perpendicular, 152-53
  slope of, 140-41
  vertical, 144
Linear equation, 98, 148
Linear expression, 96
Linear form, 98
Linear inequality, 110, 356
Linear programming, 361
Linear system of equations
  geometric meaning of, 166
  solution by determinants (Cramer's rule), 157-58
  solution by elimination method, 157, 376, 380, 386
  solution by Gauss-Jordan method, elimination, 385-86
  solution by Gaussian elimination, elimination, 376, 380, 386
  solution by reduction (substitution) method, 164, 376
  solution by substitution, 157, 164
  in three variables, 163
  in two variables, 150
Logarithm(s)
  change of base, 321-22
  common, 319, 327, 432
  definition of, 318
  natural, 319, 327
  properties of, 324, 326
Logarithmic equation, 331
Logarithmic function
  definition of, 318
  graph of, 320
  properties of, 321
Logistic law, 337
Lower bound for roots of a polynomial equation, 418

---------- M ----------
Malthus, Thomas, 337
Matrix
  addition and subtraction, 389
  augmented, 378
  coefficient, 378
  column, 391
  elementary row operations, 379
  entry (element), 389
  equality, 389
  identity for multiplication, 394
  inverse, 394-95
  invertible, 394
  main diagonal, 389
  multiplication, 391-92
  multiplications by a scalar, 390
  of order n, 389
  row, 391
  size, 388
  square, 389
Median, 5
Menaechmus, 245
Midpoint formula
  along a number line, 134
  in the plane, 135
Minus sign, 17
Mixtures, 121
Motion
  based on average speed, 119, 206
  due to gravity, 207-08
Multiplication
  of fractions, 30
  of matrices, 391-92
  of polynomials, 52
  of signed quantities, 18
Multiplicative identity, 7
Multiplicative inverse, 8

---------- N ----------
Napier, John, 326, 328
Negative multiplier property, 11
Negative numbers, 3
Newton, Sir Isaac, 191, 245, 353, 410
nth power, 35
nth root, 40
Numbers, 2, 3

---------- O ----------
Order
  algebraic properties of, 20-21
  of a term, 349
  transitive property, 22
Ordered set, 3
Ordinate, 129

---------- P ----------
Parabola
  axis of symmetry, 218-19
  definition of, 217
  directrix, 217
  focus, 217
  standard equation of, 217-18
Parallel lines, 150-51
Parametric equations of a line in the plane, 149
Partial Fractions, 368
Pascal, Blaise, 349
Pascal's triangle, 349
Percent, 117, 205
Perfect square, 62-63, 178
Perpendicular lines, 152-53
pH, 343
Planck's constant, 44
Point-slope equation of a line, 142
Polynomial(s)
  addition and subtraction of, 51
  coefficients of, 50
  complex factored form of, 410
  definition of, 50
  degree of, 50
  division of, 55-56
  multiplication of, 52
  real factored form of, 410
  terms of, 50
  zeros of, 59
Polynomial equation, roots of,
  negative, 420
  number of, 403
  positive, 419
  rational, 415
  upper and lower bounds, 417-18
Polynomial function, 58-59
Positive integers, 2
Positive numbers, 3
Power-of-a-power rule, 35, 81, 315
Partial fractions, 368
Preservation property
  for addition, 21, 107
  for multiplication, 21, 107
Prime factor, 72
Priority convention, 12
Pythagorean theorem, 131
Product of polynomials, 52
Proper rational expression, 373
Pure imaginary number, 187
Pythagorean theorem, 131

---------- Q ----------
Quadrants, 130
Quadratic equations
  applications of, 204-10
  complex roots of, 192
  double root (order 2), 177
  simple root (order 1), 177
  solution by completing the square, 178
  solution by factoring, 176
  standard form of, 176
  solution by quadratic formula, 179
Quadratic expression, 176
Quadratic formula, 179
Quadratic function, 222

Quadratic inequalities
  applications of, 208
  complex roots of, 202
  definition of, 200
  distinct real roots of, 200
  equal real roots of, 201
Quotient, 56

---------- R ----------

Radical sign, 40
Radicals
  definition of, 40
  operations with, 83-84
Radicand, 40
Range of a function, 252
Ratio formula in the plane, 139
Rational exponents
  definition of $x^{1/n}$, 78
  definition of $x^{m/n}$ and $x^{-m/n}$, 74
  power-of-a-power rule, 81
  same-base rule, 80
  same-exponent rule, 80
Rational expressions
  addition of, 73
  definition of, 70
  multiplication and division of, 74
  proper, 369
Rational numbers, 2
Rational root theorem, 415
Rationalizing denominators, 84-85
Real factored form, 410
Real number line, 3
Real number system, 4
Real numbers, 3
Real part of a complex number, 187
Rectangular (Cartesian) coordinates, 129
Reflexive property of equality, 10
Region of intersection, 231-32
Remainder, 56
Remainder theorem, 404
Reversal property for multiplication, 21, 109
Richter scale, 340
Roots of an equation, 59
  types of, 40

---------- S ----------

Same-base rule
  for division, 37, 80, 315
  for multiplication, 35, 80, 315
Same-exponent rule
  for division, 36, 80, 315
  for multiplication, 36, 80, 315
Scientific notation, 43, 326

Second-degree equation, 214
Sign rules
  for division, 19
  for multiplication, 18
Significant digits, 429-32
Slope of a line, 140
Slope-intercept equation of a line, 144-45
Solving equations, 96
Solving inequalities, 107-15, 200-03
Square
  of a difference, 53
  of a sum, 53
Square root, 40
Substitution in an equation, 10
Subtraction
  definition of, 18
  of fractions, 29
  of signed quantities, 18
Sum of cubes, 67
Symmetric property of equality, 10
Symmetry, 263-65, 294
Synthetic division, 405-07
System of linear equations *See* Linear
  . system of equations
System of linear inequalities, 356

---------- T ----------

Transitive property
  of equality, 10
  of *greater than*, 22
Translation of axes, 226, 229
Triangle inequality, 24

---------- U ----------

Unknown quantity, 96, 164
Upper bound for roots of a
  polynomial equation, 417

---------- V ----------

Variable
  definition of, 41
  dependent, 253
  independent, 253
Variation
  combined, 300
  direct, 298
  inverse, 299
  joint, 300
Vertical line, equation of, 144
Vertical line test, 255

---------- W ----------

Work, applied problems, 123, 209

---------- Z ----------

Zero of a polynomial, 59
Zero-multiplier property, 8
Zero-product property, 10
Zero-quotient property, 20